Springer Series in
SOLID-STATE SCIENCES 136

Springer
Berlin
Heidelberg
New York
Hong Kong
London
Milan
Paris
Tokyo

Physics and Astronomy ONLINE LIBRARY

http://www.springer.de/phys/

Springer Series in

SOLID-STATE SCIENCES

The Springer Series in Solid-State Sciences consists of fundamental scientific books prepared by leading researchers in the field. They strive to communicate, in a systematic and comprehensive way, the basic principles as well as new developments in theoretical and experimental solid-state physics.

126 **Physical Properties of Quasicrystals**
Editor: Z.M. Stadnik

127 **Positron Annihilation in Semiconductors**
Defect Studies
By R. Krause-Rehberg and H.S. Leipner

128 **Magneto-Optics**
Editors: S. Sugano and N. Kojima

129 **Computational Materials Science**
From Ab Initio to Monte Carlo Methods
By K. Ohno, K. Esfarjani, and Y. Kawazoe

130 **Contact, Adhesion and Rupture of Elastic Solids**
By D. Maugis

131 **Field Theories for Low-Dimensional Condensed Matter Systems**
Spin Systems and Strongly Correlated Electrons
By G. Morandi, P. Sodano, A. Tagliacozzo, and V. Tognetti

132 **Vortices in Unconventional Superconductors and Superfluids**
Editors: R.P. Huebener, N. Schopohl, and G.E. Volovik

133 **The Quantum Hall Effect**
By D. Yoshioka

134 **Magnetism in the Solid State**
By P. Mohn

135 **Electrodynamics of Magneto-active Media**
By I. Vagner, B.I. Lembrikov, and P. Wyder

136 **Nanoscale Phase Separation and Colossal Magnetoresistance**
The Physics of Manganites and Related Compounds
By E. Dagotto

137 **Quantum Transport in Submicron Devices**
A Theoretical Introduction
By W. Magnus and W. Schoenmaker

138 **Phase Separation in Soft Matter Physics**
Micellar Solutions, Microemulsions, Critical Phenomena
By P.K. Khabibullaev and A. Saidov

Series homepage – http://www.springer.de/phys/books/sss/

Volumes 1–125 are listed at the end of the book.

Elbio Dagotto

Nanoscale Phase Separation and Colossal Magnetoresistance

The Physics of Manganites and Related Compounds

With 210 Figures

With Contributions by G. Alvarez, S.L. Cooper, A.L. Cornelius, A. Feiguin, J. Fernandez-Baca, D. Gibbs, J.P. Hill, T. Hotta, S. Ishihara, D. Khomskii, S. Maekawa, A. Moreo, C.S. Nelson, J. Neumeier, T.W. Noh, G. Papavassiliou, H. Rho, P. Schiffer, J.R. Schrieffer, C.S. Snow, Y. Tokura

Springer

Professor Dr. Elbio Dagotto
Physics Department and National High Magnetic Field Lab
Florida State University
Tallahassee, FL 32306, USA

Series Editors:

Professor Dr., Dres. h. c. Manuel Cardona
Professor Dr., Dres. h. c. Peter Fulde*
Professor Dr., Dres. h. c. Klaus von Klitzing
Professor Dr., Dres. h. c. Hans-Joachim Queisser
Max-Planck-Institut für Festkörperforschung, Heisenbergstrasse 1, D-70569 Stuttgart, Germany
* Max-Planck-Institut für Physik komplexer Systeme, Nöthnitzer Strasse 38
D-01187 Dresden, Germany

Professor Dr. Roberto Merlin
Department of Physics, 5000 East University, University of Michigan
Ann Arbor, MI 48109-1120, USA

Professor Dr. Horst Störmer
Dept. Phys. and Dept. Appl. Physics, Columbia University, New York, NY 10023 and
Bell Labs., Lucent Technologies, Murray Hill, NJ 07974, USA

ISSN 0171-1873

ISBN 3-540-43245-0 Springer-Verlag Berlin Heidelberg New York

Library of Congress Cataloging-in-Publication Data.

Dagotto, Elbio, 1959–
Nanoscale phase separation and colossal magnetoresistance: the physics of manganites and related compounds/ Elbio Dagotto; with contributions by G. Alvarez ... [et al.]. p.cm.–(Springer series in solid-state sciences; 136) Includes bibliographical references and index. ISBN 3540432450 (alk. paper) 1. Magnetoresistance. 2. Manganese oxides. 3. Nanostructures. I. Title. II. Series. QC610.7 .D34 2003 630.4'12–dc21
2002030887

Springer-Verlag Berlin Heidelberg New York
a member of BertelsmannSpringer Science+Business Media GmbH

http://www.springer.de

Typesetting: camera-ready copies by the authors
Cover concept: eStudio Calamar Steinen
Cover production: *design & production* GmbH, Heidelberg

Printed on acid-free paper SPIN: 10867797 57/3141/mf 5 4 3 2 1 0

To Carina, Gabi, and Adriana

Preface

Condensed matter is one of the most active areas of research in physics. Every year new materials are discovered with properties that are as challenging as or more than those of the year before. The effort is led mainly by experimentalists, who systematically manage to prepare compounds with exotic new properties, including complex ground states involving nontrivial spin, charge, lattice, and orbital arrangements. This work is typically carried out by a relatively small number of researchers, compared with other disciplines, and the interaction between experimentalists and theorists is quite strong. It is a real pleasure for a theorist, like the author, to have experimental data with which to test proposed ideas in short time scales. This allows for a theory-experiment cross-fertilization that keeps the field very active. These are quite interesting times in condensed matter for sure.

This book is devoted to the study of a family of materials known as manganites. As for other exotic compounds, the fast development of experiments has induced a rapid evolution of the main theoretical ideas. A considerable effort, both on theory and experiments, has led to the currently much-accepted notion that nanoscale phase separation is at the heart of the behavior of electrons in these compounds. This point is described in detail in this book, with a plethora of experimental data, computer simulation results, and analytic calculations supporting that description. However, the main ideas are more universal and we hope that experts in other areas of research of condensed matter will also benefit from the understanding of the manganites. In particular, the study of high-temperature superconductors appears to be heading toward a similarly inhomogeneous picture, at least in the underdoped regime.

While there is now a consensus as to the relevance of inhomogeneities, their origin and consequences are still actively discussed. The final chapter of this book discusses a variety of open questions that could keep researchers busy for a long time. The study of manganites is still work in progress, and further surprises may appear at any time. Since the field is very active and rapidly evolving, even the dominance of phase-separation ideas may be challenged in the near future.

The book is aimed at both theorists and experimentalists. No heavy formalism is used, and the emphasis is on physical results. However, theorists can find enough information to reproduce even the most complex of the simulations. The level is kept simple, centering on qualitative aspects. Notes and comments will hopefully help graduate students or postdocs with the few complicated aspects of the presentation. The book can, in fact, be used as a reference for an advanced graduate course. Some basic knowledge of quantum mechanics and elementary condensed matter physics is assumed. Hundreds of references are included, covering the most relevant literature on the subject of manganites and related compounds. However, it is likely that important papers have been accidentally omitted. I apologize in advance for these oversights, and encourage colleagues whose work is not mentioned to contact me at `dagotto@magnet.fsu.edu`. In future editions new references can be added.

This book also contains invited contributions from many colleagues. The author warmly thanks all of them for kindly providing their time and expertise to help with the very difficult enterprise of summarizing the work of hundreds of researchers in the active area of investigations that is strongly correlated electrons. Most of the contributors also provided very useful comments on the manuscript at large. Those comments were very important for the preparation of the final version of the book.

The author is specially thankful to G. Alvarez, S.L. Cooper, G. Martins, and R. Merlin for a careful reading of most of the book and technical support. Comments from many colleagues are also very much appreciated. The list includes J. Burgy, D. Dessau, A. Feiguin, J.A. Fernandez-Baca, M. Greven, J. Hill, T. Hotta, S. Ishihara, D. Khomskii, S. Kivelson, P. Levy, J. Lynn, A. Millis, V. Martin-Mayor, A. Moreo, C. Nelson, J. Neumeier, T. Noh, A. Oleś, G. Papavassiliou, P. Schiffer, and Y. Tokura. The author is also very grateful to G. Taggart and D. Hess for their support through a grant of the Division of Materials Research of the National Science Foundation (grant DMR-0122523).

Tallahassee, FL, USA, April, 2002 *Elbio Dagotto*

Contents

1. Why Manganites Are Interesting 1
- 1.1 A Dominant Theme of Current Times: Strongly Correlated Electrons 1
- 1.2 The Growing Interest in Manganite Research 2
- 1.3 Nanoscale Phase Separation 4
- 1.4 Complexity in Materials 5
- 1.5 Computational Techniques in Complex Systems 5
- 1.6 Other Books and Reviews 7
- 1.7 The Evolution of Condensed Matter Physics *by J.R. Schrieffer* 7

2. The Discovery of Manganites and the Colossal Magnetoresistance Effect 9
- 2.1 The Early Days of Manganites 9
- 2.2 The Colossal MR Effect 13
- 2.3 Early Theoretical Studies 15
- 2.4 Tolerance Factor 19
- 2.5 High Magnetic Fields 20

3. Phase Diagrams and Basic Properties of Manganites 23
- 3.1 The Intermediate Bandwidth Compound $La_{1-x}Ca_xMnO_3$ 23
- 3.2 Basic Properties of $Pr_{1-x}Ca_xMnO_3$ 30
- 3.3 Basic Properties of $La_{1-x}Sr_xMnO_3$ 35
- 3.4 Half-Metallic Ferromagnets 38
- 3.5 Basic Properties of $Nd_{1-x}Sr_xMnO_3$ and $Pr_{1-x}Sr_xMnO_3$ 41
- 3.6 Summary of Results for Perovskites 42
- 3.7 Basic Properties of Bilayer Manganites 42
- 3.8 Basic Properties of Single-Layer Manganites 48

4. Preliminary Theoretical Considerations: Coulombic and Jahn Teller Effects 51
- 4.1 Basic Multi Orbital Model and On Site Coulomb Interactions 51
- 4.2 Crystal Field Splitting and Jahn Teller Effect 57
  - 4.2.1 Perturbing Potential due to the Crystal 57
  - 4.2.2 The Jahn Teller Effect 63

11. Inhomogeneities in Manganites: The Case of $La_{1-x}Ca_xMnO_3$ 213
11.1 Inhomogeneities in $La_{1-x}Ca_xMnO_3$ 214
11.1.1 Brief Introduction to Neutron Scattering *by J. A. Fernandez-Baca* 214
11.1.2 Neutron Scattering in $La_{1-x}Ca_xMnO_3$ 222
11.1.3 Anomalous Softening of Zone-Boundary Magnons 225
11.1.4 Volume Thermal Expansion 228
11.1.5 Electron Microscopy 230
11.1.6 Brief Introduction to STM 232
11.1.7 Scanning Tunneling Spectroscopy in $La_{1-x}Ca_xMnO_3$ 233
11.1.8 PDF, X-Ray, and Small-Angle Neutron Scattering Experiments 235
11.1.9 Brief Introduction to Nuclear Magnetic Resonance 237
11.1.10 Nuclear Magnetic Resonance in $La_{1-x}Ca_xMnO_3$ 239
11.1.11 The NMR "Wipeout" Effect *by G. Papavassiliou* 241
11.1.12 Influence of Cr Doping on Charge-Ordered Manganites 243
11.1.13 Giant Noise in Manganites 245
11.1.14 Other Results Related with Phase Separation 247
11.2 Inhomogeneities in Electron-Doped Manganites 248
11.3 Probing of Phase-segregated Systems with Low-Temperature Heat-Capacity Measurements *by J.J. Neumeier and A.L. Cornelius* 251

12. Optical Conductivity 255
12.1 $\sigma(\omega)$: Experimental Results in Manganites 255
12.2 $\sigma(\omega)$: Theoretical Results 262
12.3 Brief Introduction to Experimental Aspects of Optical Spectroscopy *by T.W. Noh* 267

13. Glassy Behavior and Time-Dependent Phenomena 273
13.1 Brief Introduction to Spin-Glasses 273
13.2 Glassy Behavior in Manganites 277
13.2.1 Long Relaxation Times 277
13.2.2 Phase Separation at $x = 0.5$ LCMO 278
13.2.3 Manganites: Canonical Glasses or a New Glassy State? 280
13.3 Phase Separation, Time-Dependent Effects, and Glassy Behavior in the CMR Manganites *by P. Schiffer* 281

14. Inhomogeneities in $La_{1-x}Sr_xMnO_3$ and $Pr_{1-x}Ca_xMnO_3$ 287
14.1 Results for $La_{1-x}Sr_xMnO_3$ 287

List of Contributors

Gonzalo Alvarez
National High Magnetic Field Laboratory
and Department of Physics,
Florida State University,
Tallahassee, FL 32306, USA

S.L. Cooper
Dept. of Physics
and Frederick Seitz Materials Research Laboratory,
University of Illinois,
Urbana-Champaign, USA

A.L. Cornelius
Department of Physics,
University of Nevada,
Las Vegas, NV 89154-4002, USA

Adrian Feiguin
National High Magnetic Field Laboratory
and Department of Physics,
Florida State University,
Tallahassee, FL 32306, USA

J.A. Fernandez-Baca
Solid State Division,
Oak Ridge National Laboratory,
P.O. Box 2008,
Oak Ridge, Tennessee 37831-6393, USA

Doon Gibbs
Department of Physics,
Brookhaven National Laboratory,
Upton, NY, USA

J.P. Hill
Department of Physics,
Brookhaven National Laboratory,
Upton, NY, USA

Takashi Hotta
Advanced Science Research Center,
Japan Atomic Energy Research Institute,
Tokai, Ibaraki 319-1195, Japan

Sumio Ishihara
Department of Applied Physics,
University of Tokyo,
Tokyo 113-8656, Japan

D.I. Khomskii
Laboratory of Solid State Physics,
Groningen University,
Nijenborgh 4, 9747 AG Groningen,
The Netherlands

Sadamichi Maekawa
Institute for Materials Research,
Tohoku University,
Sendai 980-8577, Japan

Adriana Moreo
Department of Physics, National High Magnetic Field Lab,
and MARTECH,
Florida State University,
Tallahassee, FL 32306, USA

C.S. Nelson
Department of Physics,
Brookhaven National Laboratory,
Upton, NY, USA

J.J. Neumeier
Department of Physics,
Florida Atlantic University,
Boca Raton, FL 33431, USA

T.W. Noh
School of Physics and Research
Center for Oxide Electronics,
Seoul National University,
Seoul 151-747, Korea

G. Papavassiliou
Institute of Materials Science,
NCSR Demokritos,
153 10 Aghia Paraskevi, Athens,
Greece

H. Rho
Dept. of Physics
and Frederick Seitz Materials
Research Laboratory,
University of Illinois,
Urbana-Champaign, USA

P. Schiffer
Dept. of Physics and Materials
Research Institute,
Penn State University,
University Park, PA 16802, USA

J.R. Schrieffer
National High Magnetic Field Lab
and Physics Department,
Florida State University,
Tallahassee, FL 32306, USA

C.S. Snow
Dept. of Physics
and Frederick Seitz Materials
Research Laboratory,
University of Illinois,
Urbana-Champaign, USA

Y. Tokura
Department of Applied Physics,
University of Tokyo,
Tokyo, Japan

1. Why Manganites Are Interesting

1.1 A Dominant Theme of Current Times: Strongly Correlated Electrons

This book is devoted to the experimental and theoretical description of a family of materials known as manganites. The book also includes information about related compounds with similar physical properties. Manganites are currently attracting much interest, due to their unusual behavior upon the application of relatively small magnetic fields. The possibility of technological applications is often mentioned in this context, but the main driving force for the effort on manganites is triggered by mere scientific curiosity. These materials are without doubt a challenge to our present understanding of electrons in crystals, as remarked by J. R. Schrieffer in the historical perspective at the end of this chapter.

Manganites are members of a broader set of compounds where the effect of correlations among electrons plays a crucial role. This broader framework is widely known as the area of *strongly correlated electrons*. In this context, the simple and well-known perturbative methods typically fail to properly describe the many-body ground state of the system. Perturbation theory is built upon the notion that the kinetic portion of the many-electron Schrödinger equation is the dominant part of the Hamiltonian, and weak correlations are typically added in the form of a finite set of Feynman diagrams. This approach does not capture even the very basic properties of strongly correlated systems, such as the generation of robust energy gaps in the spectra, of an origin different from the well-known bandgaps that lead to insulating behavior when the chemical potential lies in those gaps.

Phenomena such as superconductivity must also be studied beyond the framework of perturbation theory. Fortunately, many materials described by the Bardeen Cooper Schrieffer (BCS) theory of superconductivity [1.1] admit a relatively simple mean-field description due to the large size of the Cooper pairs, which is much larger than the mean distance between electrons. In 1986, the cuprate family of superconductors was discovered by Bednorz and Müller [1.2]. It was rapidly realized that the pair size in those compounds was a few lattice spacings, comparable to intercarrier distances. While the essence of the BCS theory is probably still applicable to these new "high-T_c"

superconductors, quantitative calculations are quite difficult since mean-field theory is not applicable. In fact, even more than 15 years after the discovery, researchers are still arguing about the pairing mechanism at work in these compounds (is it phononic or purely electronic?). This issue is further complicated by the existence of many competing tendencies in the cuprates. The superconducting phase is a relatively small "island" in parameter space, surrounded by an antiferromagnetic insulator, a standard metal known as a Fermi liquid, and even more prominently, by tendencies toward the spontaneous formation of inhomogeneous spin and charge patterns. In the high-T_c arena, these inhomogeneities receive the name of "stripes" [1.3]. In this framework, the holes doped into the system by chemical doping, spontaneously arrange themselves into quasi-one-dimensional lines that run parallel to one another, mainly along the crystal axes. This instability toward stripes complicates enormously the study of the superconducting phase of cuprates. While there is little doubt that a rigid arrangement of stripes destroys superconductivity, it may occur that their partially melted state induces superconductivity, as some have argued. Or it may happen that the critical temperature for superconductivity cannot reach an even larger value due to the competition with dynamical stripes that are still present. This uncertain state of affairs in the cuprates can be summarized by two key concepts that are also of much relevance to the manganites: (i) Competition between drastically different phases is very important. This is a common ingredient of strongly correlated electron systems. Phase competition is difficult to avoid. (ii) In many materials, this phase competition induces stable states where the carriers are *not* homogeneously distributed, but form a variety of inhomogenous patterns. Some of the results on cuprates will be briefly discussed in the last chapter.

1.2 The Growing Interest in Manganite Research

At present, there is enormous interest in the study of manganites, and of strongly correlated systems in general. In the opinion of this author, there are at least three reasons that make manganites an important area for experimentalists and theorists aiming to improve our understanding of the behavior of electrons in crystals.

(i) The first obvious reason is the *unexpectedly large magnetotransport properties* of these compounds, which will be reviewed in Chap. 2. Upon the application of small magnetic fields, the resistivity changes by many orders of magnitude, an effect that carries the name of "colossal magnetoresistance" (CMR). How can the transport properties of a system be so susceptible to external "perturbations"? Ordinary magnetoresistance (MR) already occurs in non magnetic materials. In a magnetic field, the carriers no longer follow straight trajectories, but helical ones with a radius inversely proportional to the magnetic field strength. This increases the resistance since the length of the trajectory is now larger, leading to more collisions. However, this effect is usually small in most metals. Other MR effects, besides the "CMR" of manganites of course, are the "giant" MR and "anisotropic" MR, to be reviewed in Chap. 21. But none of these previously known examples of MR can explain the extremely large magnitude of the MR in manganites.

(ii) A second motivation to study manganites is contained in their *rich phase diagrams*, exhibiting a variety of phases, with unusual spin, charge, lattice, and orbital order. The latter is a characteristic of systems where at least two orbitals per ion are active [1.4, 1.5], and it does not occur in the layered cuprate superconductors. These four active degrees of freedom lead to a rich phenomenology and a plethora of phases that have been identified experimentally using a variety of techniques (see Chap. 3). Phase competition at the boundaries between those phases produces interesting phenomena.

(iii) The third motivation is the most relevant to the main subject of this book. It makes a connection with important physics also unveiled in the cuprates: evidence suggests that even in the best crystals available, manganites are *intrinsically inhomogeneous*. In other words, the states formed in these compounds are dominated by coexisting clusters of competing phases. These phases are typically ferromagnetic and antiferromagnetic. In this book and previous literature, it will be argued that these inhomogeneities are the cause of the unusually large MR found in manganites, such that issues (i) and (iii) are actually closely related. There is a growing body of evidence that the large resistivity of the paramagnetic insulating phase in manganites (see Chap. 3) is associated with nanoscale inhomogeneities. Actually, (ii) is also linked to the other points, since it is due to the proliferation of phases that the phase competition needed for the inhomogeneous states occurs. Note that the study of inhomogeneities in manganites will also likely lead to important information about similar phenomena found in other materials such as cuprates. Lessons learned focusing on Mn oxides can probably be extended to other compounds as well, increasing their importance.

It is clear that some materials receive more attention from the condensed matter community than others. Without doubt, during the 1990s the high-temperature superconductors dominated the study of strongly correlated electrons, and they are still attracting considerable attention in the 2000s. However, manganites and other compounds are rapidly catching up. In fact, Fig. 1.1 shows some crude statistics gathered by the author using the Physical Review Online Archive (PROLA) service. Shown are the number of papers that, in recent years, contain in the abstract the words "cuprate" or "manganite". The number in 2001 suggest a quite similar level of interest between the two families of compounds, and from the slopes perhaps in the near future the curves will cross. Clearly researchers are rapidly realizing that manganites contain physics interesting enough to deserve careful experimental and theoretical studies. The analysis of orbitally degenerate systems may provide us with many surprises.

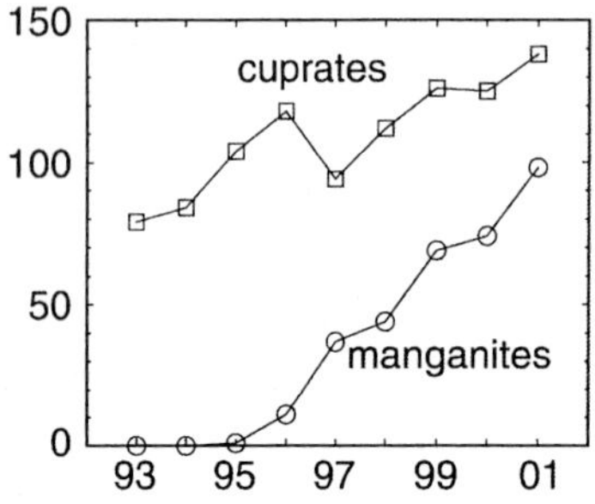

Fig. 1.1. Number of papers (*vertical*) in recent years (*horizontal*) that appeared in Physical Review B, Physical Review Letters, and Review of Modern Physics with the words "cuprate OR cuprates" and "manganite OR manganites" in the abstract. The information was obtained online using the Physical Review Online Archive (PROLA)

However, the popularity of manganites is not our guide for research. Such a success should be a consequence of rational investigations intrinsically showing Mn oxides as materials of much relevance in condensed matter physics. Hopefully, this book will convince the reader of the importance of investigations in the context of CMR materials.

1.3 Nanoscale Phase Separation

As explained in the previous section, an enormous amount of literature on manganites clearly signals the existence of strong analogies with some cuprates [1.3] regarding phase competition. In fact, the effect seems far stronger in manganites. These compounds appear to form "nanoscale phase-separated" states, namely states where two competing phases reach a compromise by forming inhomogeneous patterns. The "nanoscale" term refers to the 10^{-9} m characteristic length scale of those patterns, although some experiments reveal complex behavior at the micrometer scale. The expression "phase separation" is also much used in the cuprate context (see especially papers by Kivelson and collaborators in [1.3]), and it refers to the two competing phases that create the inhomogeneities. These phases may or may not have a different electronic density, but they usually have different symmetry breaking patterns. In manganites they tend to be ferromagnetic (all spins parallel), and charge-ordered and spin antiferromagnetic (a staggered up and down spin pattern). Some researchers are confused by the use of the word "phase" in this context which, strictly speaking, refers to a thermodynamic concept where a large number of electrons are needed. This can be clarified by experimentally observing or theoretically calculating the spin, charge, and orbital patterns formed in the inhomogeneous states. This information shows that *locally*, at the lattice spacing level, the states are quite similar to those observed in other portions of the phase diagram where only one of the competing phases dominates. In other words, the "bubbles", "clusters" or "stripes" found in the inhomogeneous states are like small nanoscale pieces of bulk homogeneous phases found elsewhere in the phase diagram at low temperatures. For this reason, "phase separation" is the terminology often used in reference to states with nanocluster coexistence, even if the language is not fully rigorous.

1.4 Complexity in Materials

The understanding of the complex "self-organized" behavior of transition metal oxides, such as manganites and cuprates, and related compounds is an important topic that has developed into one of the dominant new scientific themes of current times. The previously described coexistence and competition between different kinds of order involving charge, orbital, lattice, and spin degrees of freedom leads to physical "complexity" as a characteristic of their behavior. The tunability of properties can be achieved by changing ionic sizes, chemical compositions, etc. The balance between competing phases is often subtle, and small changes in composition can lead to large changes in the material properties. Elucidating the nature of the colossal magnetoresistance and other exotic phenomena will likely change our perspective of highly correlated electronic materials.

The complexity of many interesting physical systems has recently been addressed by Laughlin and Pines [1.6], who remark that reductionist approaches will not work in such complicated cases. Manganites likely provide another example of complex matter. In fact, Vicsek [1.7] discusses a definition of complex systems in general that applies very well to manganites: they are systems that "can change dramatically and stochastically in time and/or space as a result of small changes in conditions". A first-principles theoretical understanding of the CMR phenomenon seems unlikely. A "ladder" of theories at different energy and length scales will be more illuminating for Mn oxides and other compounds. This is the philosophy followed in the present book and related literature. In fact, basic models will show us the dominant ground-state tendencies, but the study of phase coexistence at long length scales will be carried out with simpler phenomenological models that hopefully capture qualitatively the essence of the problem.

1.5 Computational Techniques in Complex Systems

As already explained, the many-body states emerging from phase competition in real materials are "inhomogeneous". This is clearly a complication for theoretical studies. In standard metals, transforming from real to momentum space simplifies the formalism enormously. However, in a system that is not translationally invariant, it is not useful to carry out such a transformation. The problem has to be addressed directly in real space. Uniform states cannot be proposed as candidates for the ground state, and alternative procedures must be developed. In fact, very exotic uniform states proposed in the cuprate context may simply reflect on the effective form, in which the inhomogeneities manifest themselves when averages over long length scales are made. Proposals insisting on uniformity to describe the regime of phase competition must be revisited.

How do we address theoretically phase competition? The most successful techniques appear to be those based on computational methods. These methods appear crucial to understanding materials such as the manganites where intrinsic inhomogeneities appear unavoidable, and where strong couplings describe the interactions between the constituents. Diagrammatic approaches have limited applicability in such a context. Mean-field approaches are useful, but they need to be contrasted against computational results, and they may not work in inhomogeneous cases. At least on finite size clusters, the computer is able to handle the complexity of the problem and the competing tendencies. Of course, the basic "model Hamiltonian" is an input and must be deduced from a real-space "local" reasoning, analyzing isolated ions and pairs of ions at short distances. Even simple Hamiltonians can lead to complex self-organized patterns, as shown in other chapters of this book.

The tendency toward computational work is a trend in most of physics. Theorists of the new generation are likely to dominate the fine details of computer programming, Fortran or C++, and algorithms for solving numerical problems, while more "classical" theorists would have expertise in integration and differential equation manipulations. The trend toward an increasing amount of computationally related work appears unavoidable, particularly in the area of strongly correlated electrons where even the most abstract of theorists must frequently solve complicated equations with a computer. The relevance of computational physics was recently remarked upon by J. Langer [1.8] in a Reference Frame column for Physics Today. He wrote: "... scientific computation has reached the point where it is on a par with laboratory experiment and mathematical theory as a tool for research in science and engineering." This author totally agrees with this statement.

In addition, it is clear that computational power has been growing very rapidly in recent times, and the trend is expected to continue for at least a few more years. It is remarkable that Moore's law still appears to be applicable even beyond the year 2000. This empirical law states that the number of transistors that could be built on a chip increases exponentially with time (for a discussion see J. Birnbaum and R. Stanley Williams, Physics Today, Jan. 2000, page 38; D. Normile, Science 293, 787 (2001)). However, the cost of manufacturing chips is increasing faster than the market is expanding, and at some point a saturation effect will slow down the exponential growth. In addition, it is estimated that by 2010 individual transistors in circuits will be turned on and off by adding or removing just a few electrons, and the statistics of small numbers will become significant. By 2020, the extrapolation says that less than one electron will be needed, showing that Moore's law will break down sooner or later. These extrapolations illustrate the need for new technologies to continue increasing the power of computers, triggering the current wide interest in nanofabrication, molecular electronics, and quantum computing. However, these comments should not stop our pursuit of computational techniques to study correlated electrons. The computer power will keep on increasing rapidly for at least a few more years, new technologies may soon emerge, and in addition, experience shows that new algorithms are frequently discovered that substantially speed up most codes, surmounting bottlenecks. For this reason, it is reasonable to expect even better results to emerge from computational work in the near future, since in several cases raw CPU time and memory are currently limiting some calculations. The future appears bright for computational physics. Students will greatly benefit from following this line of research, which will not only provide them with unmatched help for the science under discussion, but will also increase their abilities and marketability in other areas of research and technology.

1.6 Other Books and Reviews

The reviews and books on manganites available at present usually bring the perspective of experimentalists to the subject. This is the first book on manganites by a theorist, to the best of the author's knowledge. The presentation of the material in this book follows the organization of a previous review by the author, in collaboration with Hotta and Moreo [1.9]. The notion of phase separation for manganites was already discussed in a brief review by Moreo *et al.* [1.10]. Other previous reviews on manganese oxides include work by Ramirez [1.11], Coey *et al.* [1.12], Tokura and Tomioka [1.13], Ibarra and De Teresa [1.14], Rao et al. [1.15], Moreo [1.16], Ziese [1.17], Salamon and Jaime [1.18], and others. In the latter [1.18], emphasis is also given to the presence of inhomogeneities in manganites, providing a complementary viewpoint. Information about manganites is also contained in the review by Imada *et al.* [1.19], while computational work and the physics of cuprates has been addressed by Dagotto [1.20]. Previous books on manganites, include those edited by Rao and Raveau [1.21], Kaplan and Mahanti [1.22], and Tokura [1.23]. A book on electronic conduction in oxides was presented by Tsuda et al. [1.24]. It is possible that other reviews and books may have escaped the author's attention, but the long list of references at the end of this book hopefully compensates for any unintentional omission.

1.7 The Evolution of Condensed Matter Physics *by J.R. Schrieffer*

The history of condensed matter physics is an excellent example of the interplay of theory and experiment. Early experimental studies of simple metals, semiconductors, and insulators showed that electrons and holes roughly behave as independent particles, which are scattered by lattice vibrations and defects. However, for cooperative effects, such as ferro and antiferromagnetism and superconductivity, while described by phenomenological mean-field approaches, there was no microscopic theory of these phenomena. Gradually, the quantum theory of solids began to develop an understanding of transport and electromagnetic properties of simple solids through self-consistent band theories, although electron electron correlation effects were neglected. Band theory predicted that NiO was a conductor, while it was in fact an insulator. Wigner and later Mott and Hubbard showed how strong Coulomb repulsion can localize electrons to form an electronic insulator. d-band materials presented particular challenges due to the degeneracy of the orbitals and the strong interactions between the electrons. Gradually, it became clear how itinerant carriers can have characteristics of quasi-localized moments with exchange leading to ferromagnetism in Fe, while Fermi surface nesting leads to antiferromagnetism in Cr. Superexchange leads to antiferromagnetic coupling in insulating oxides.

The importance of electron-phonon interactions in the many-body problem became evident when the theory of superconductivity showed that the exchange of phonons between electrons caused an attraction, which bound pairs of electrons into a superfluid condensate.

Manganites pose one of the most interesting and complex systems studied thus far in condensed matter physics. Here, one has strong interactions between electrons that form quasi-localized moments, ferro and antiferromagnetism, metal insulator transitions, orbital ordering effects, strong electron-phonon coupling, charge ordering and spatial separation of these phases. This book strongly makes the case that the colossal magnetoresistance found in these materials arises not from local spin ordering, but rather from the existence of large domains having nearly equal free energy. These domains are affected by the magnetic field, which causes percolation of conducting paths. The physics of these systems clearly spans many length scales with competing interactions, a far more complex and interesting system than the simple solids that were considered at the outset of the field.

The theoretical methods used to study these complex systems increasingly depend on large-scale computation. However, simple analytical models often help gain insight into the numerical results. One can foresee that in the future, other complex materials will be understood by close collaboration between theory and experiment, with simple models continuing to play a key role.

2. The Discovery of Manganites and the Colossal Magnetoresistance Effect

2.1 The Early Days of Manganites

It is widely recognized that the first paper reporting results for manganites was presented in 1950 by Jonker and Van Santen [2.1]. In that first publication, La was replaced either by Ca, Sr, or Ba, and results for polycrystalline samples of (La,Ca)MnO_3, (La,Sr)MnO_3, and (La,Ba)MnO_3 were reported. The main result was the appearance of ferromagnetism in these compounds, as described below. Jonker and Van Santen clarified in their first publication that the term "manganites" is not rigorous, and only compounds containing tetravalent Mn should actually be called manganites. However, they write "For the sake of simplicity the compounds containing trivalent, as well as those containing tetravalent, manganese will be designated as manganites", a convention that it is still followed today.

Jonker and Van Santen found that the manganites they studied crystallize in a "perovskite" structure, as is shown in Fig. 2.1. The A-site of the perovskite contains a large ion, such as Ca^{2+}, Sr^{2+}, La^{3+}, etc., while the B-sites have a small ion, such as Mn^{3+}, Mn^{4+}, Cr^{3+}, or others. The oxygens surround the B-sites forming an *octahedral cage*. Typical ionic sizes, directly quoted from [2.1], are: $Ca^{2+} = 1.06$ Å, $Sr^{2+} = 1.27$ Å, $Ba^{2+} = 1.43$ Å, $La^{3+} = 1.22$ Å for the A-site ion; $Mn^{3+} = 0.70$ Å and $Mn^{4+} = 0.52$ Å for the B-site ion; and $O^{2-} = 1.32$ Å for oxygen. Manganese is the smallest of the ions involved in the manganites.

The Curie temperature was obtained by Jonker and Van Santen [2.1] using magnetization measurements, and some of the results are reproduced in Fig. 2.2a. The early samples of manganites contained an excess of oxygen, and the results they obtained are not accurate compared with the modern versions of the (La,Ca)MnO_3 phase diagram, but the essence of the existence of ferromagnetism can be clearly inferred from Fig. 2.2a. In those early studies, the ferromagnetic phase was attributed to a positive indirect-exchange interaction. This view was soon replaced by the currently more widely accepted *double-exchange* picture, discussed later in this chapter. In the early studies of manganese oxides, it was also noticed that (La,Sr)MnO_3 can only admit up to 70% Sr. Otherwise, a two-phase compound is obtained for geometrical reasons related to the so-called *tolerance factor* (see discussion below). This

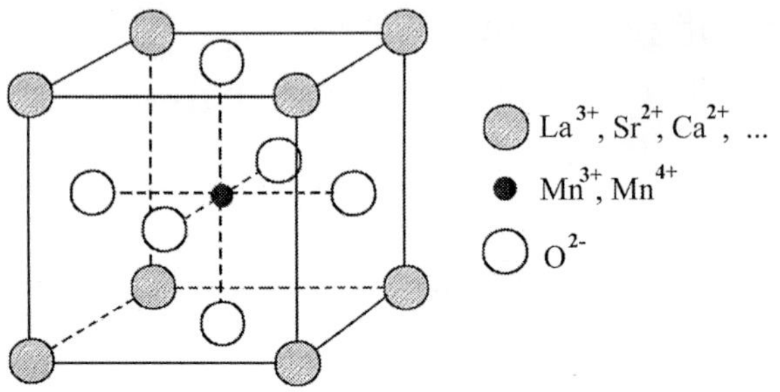

Fig. 2.1. Unit cell of a perovskite. Shown in *gray* are the A-sites (where, for instance, the La ions are located), in *black* the B-sites corresponding to Mn ions, and in *white* the oxygens

factor plays an important role in Mn oxides, and its role will be discussed in several other chapters of this book.

In a subsequent paper by Van Santen and Jonker [2.2] a few months after the original publication, the conductivity of manganites was reported. Anomalies in the conductivity were found at the Curie temperatures. The study of the lattice parameters as a function of hole doping was reported in those early days by Jonker [2.3], observing that near the composition of 100% La the crystal is distorted, while at higher Sr or Ca densities, it is not. These distortions are associated with the Jahn Teller effect, which is very important in manganites, as discussed in the rest of the book.

A few years after the original work of Jonker and Van Santen, the magnetoresistance data on manganites was reported by Volger [2.4]. This author wrote: "Manganites, when in the ferromagnetic state, show a notable decrease of resistivity in magnetic fields." In Fig. 2.2b, the changes in resistance of $La_{0.8}Sr_{0.2}MnO_3$ in magnetic fields at various temperatures are reproduced from [2.4]. It was already noticed in those early studies that standard explanations for the effect did not work, and that the effect was likely related to the favoring of the ferromagnetic state by a magnetic field. However, one should note that the truly enormous magnetoresistance, the now famous "colossal" effect, was discovered much later, in the 1990s.

Among the key efforts widely cited from the early investigations of manganites is the paper by Wollan and Koehler [2.5] using neutron diffraction techniques to characterize the magnetic structures of $La_{1-x}Ca_xMnO_3$ in the entire range of compositions (for an introduction to neutron scattering, see Chap. 11). Wollan and Koehler were among the first to use the technique of neutron scattering to study magnetism in materials. Those authors found that, in addition to the ferromagnetic phase reported earlier by Jonker and van Santen, many other interesting *antiferromagnetic* phases were present in manganites. In some cases, they also reported evidence of charge ordering, concomitant with the antiferromagnetic phases.

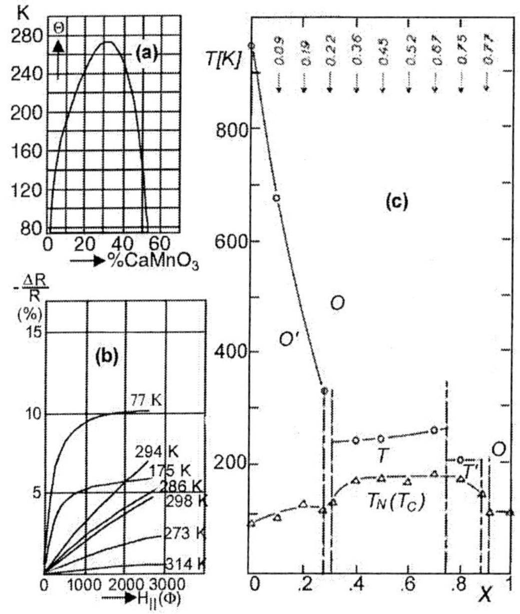

Fig. 2.2. (**a**) Curie temperature θ of mixed crystals (La,Ca)MnO_3, reproduced from the first paper on manganites by Jonker and Van Santen [2.1]. (**b**) Change of resistance of $La_{0.8}Sr_{0.2}MnO_3$ in magnetic fields at several temperatures, from [2.4]. (**c**) Structural and magnetic transition temperatures in $Pr_{1-x}Ca_xMnO_3$, from [2.8]. The notation is standard, but more details can be found in [2.8]. This figure should be contrasted with more modern phase diagrams of $Pr_{1-x}Ca_xMnO_3$, such as Fig. 3.6 of Chap. 3

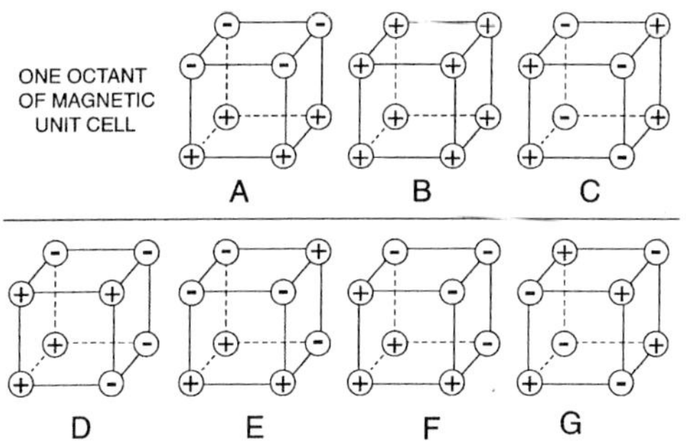

Fig. 2.3. Magnetic structures and their labels, from [2.5]. The circles are the Mn ions, and the signs the orientation of the z-axis spin projection. For example, A-type has planes that are ferromagnetic, with antiferromagnetic coupling between them. The B-type structure is the standard ferromagnetic arrangement, while G-type corresponds to an antiferromagnet in all three directions. The C-type has antiferromagnetism in two directions, and ferromagnetism along the third

A sketch of the magnetic arrangements found in those early neutron studies is presented in Fig. 2.3, which is adapted from one of the original figures of [2.5]. Pattern A (known as A-type antiferromagnetism (AF)) was found at small hole density, including $x = 0$, and corresponds to planes with ferromagnetic alignment of spins, but with antiferromagnetic coupling between those planes. In the other extreme of large x, where most Mn ions are in the 4+ valence state, the dominant pattern was found to be of the G-type, with antiferromagnetism in all three directions. At x around 0.75 or 0.80, the C-type pattern was observed, in which the spins are ferromagnetic along the chains, but antiferromagnetic in the planes. In the region that is ferromagnetic in the three directions, the pattern was called B, and it is also shown for completeness in Fig. 2.3.

Perhaps the most exotic spin arrangement identified by Wollan and Koehler is the one named CE-type state, which following their classification in Fig. 2.3 is a mixture of the C-phase with the E-phase. The arrangement of spins, orbitals, and charge of this highly nontrivial phase will be discussed in Chap. 3 (see, e.g., Fig. 3.4). However, a few details should be given here: (1) In [2.5], from the intensity of the peaks associated with the CE-state, it was already deduced that charge-ordering was present in the state. (2) In agreement with the current observations of mixed-phase tendencies in many manganites, to be discussed extensively in the rest of the book, in [2.5] it was already reported that mixtures of A-type and FM phases, as well as CE-type and FM phases, were observed in the neutron diffraction patterns. (3) Néel temperatures corresponding to antiferromagnetism were inferred from the neutron data, by studying the intensity of the peaks corresponding to the particular A-, CE-, C- or G-type AF spin states, and the temperature at which their intensities departed from zero.

The analysis of another of the very popular manganites, $Pr_{1-x}Ca_xMnO_3$, was initiated by Jirák et al. [2.6] and Pollert et al. [2.7]. These authors used X-ray and neutron diffraction techniques to study the structure of the PCMO manganite, and its magnetic properties. These materials were found to have phases with charge ordering, totally different from the ferromagnetic metallic phases of other manganites. In parallel with this observation, a variety of deviations from the cubic structure were reported, as shown in Fig. 2.2c (from [2.8]). At small hole density x, an orthorhombic structure was found, concomitant with "A-type" AF order (ferro in plane, antiferro between planes, see a more detailed discussion later in Chap. 3). Tetragonal phases dominate near $x = 0.5$, while at $x \sim 1$, the material is nearly cubic. Clearly these compounds are structurally distorted, and the highly nontrivial associated distribution of charge, spins, and orbitals will be discussed in subsequent chapters. The distortions found in PCMO and other compounds were correctly attributed to the Jahn Teller effect in the early studies, and also to buckling effects caused by small central cations in the perovskite structure (tolerance factor less than one, see discussion below).

Several other important results were reported in the first days of investigations in manganites. They include the report of properties of the interesting compound (La,Pb)MnO_3, which has a Curie temperature well above room temperature [2.9], and the more detailed experimental and theoretical

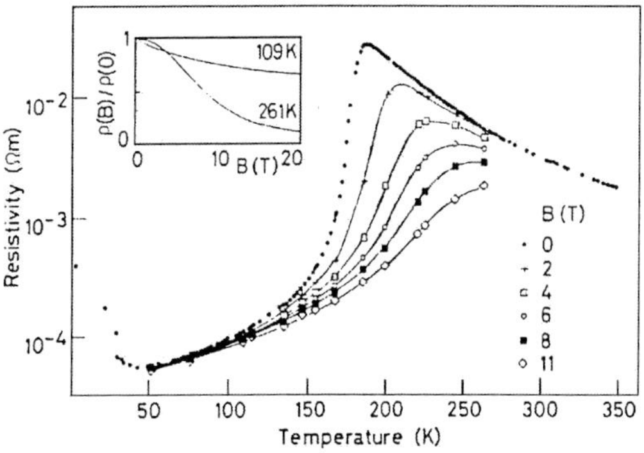

Fig. 2.4. (**a**) Resistivity of $Nd_{0.5}Pb_{0.5}MnO_3$, as a function of temperature and magnetic field, from Kusters et al. [2.11]. *Inset* illustrates the behavior of the magnetoresistance at two temperatures, above and below the ordering temperatures

study of $LaMnO_3$, with its exotic A-type AF state [2.10]. This spin-order arrangement is surprising, since few other materials have such a ground state. Possible theoretical reasons for this behavior are explained in Chap. 9.

2.2 The Colossal MR Effect

Experimental work on manganites dramatically accelerated during the 1990s. This was caused by the observation of large magnetoresistance (MR) effects in these compounds. This property was noticed in the early studies (see Volger [2.4]), as already discussed, but its true magnitude was not appreciated, nor its potential technological use. The early relevant studies in this context include work on $Nd_{0.5}Pb_{0.5}MnO_3$ by Kusters et al. [2.11] (see Fig. 2.4). In that publication, a plot of resistivity vs. temperature at several magnetic fields already reveals the large MR effect of these compounds. Work by von Helmolt et al. [2.12] on $La_{2/3}Ba_{1/3}MnO_x$ also revealed a large MR effect, this time at room temperature, using thin films. The value of the MR was found to be larger than in "giant" MR devices, which are artificially created arrangements of magnetic materials with large magnetoresistance (see Chap. 21). Similar conclusions were reached by Chahara et al. [2.13] using thin films of $La_{1-x}Ca_xMnO_3$ at $x = 0.25$. Ju et al. [2.14] also observed a large MR in films of $La_{1-x}Sr_xMnO_3$ near room temperature.

A defining moment for the field of manganites was the publication by Jin et al. [2.15] of results with truly colossal MR ratios. Those authors studied films of $La_{0.67}Ca_{0.33}MnO_x$, and some results are reproduced in Fig. 2.5. Defining the MR ratio as $\Delta R/R = (R_H - R_0)/R_H$, where R_0 is the resistance without a magnetic field, and R_H is the resistance in a magnetic field of 6 T, Jin et al. reported MR ratio values much higher than previously observed. In Fig. 2.5, a ratio close to 1 500% at 200 K is shown, while at 77 K the reported

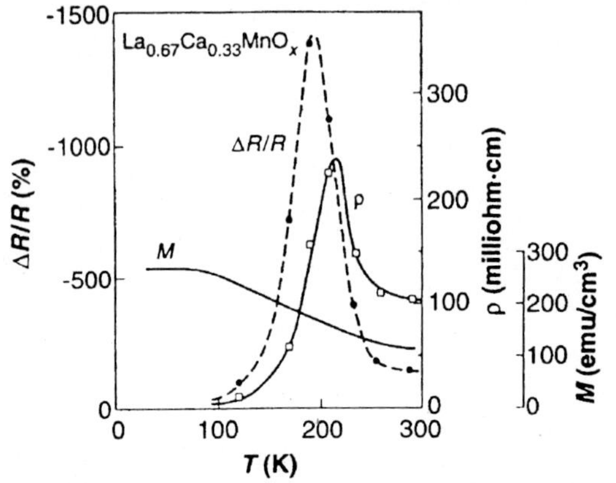

Fig. 2.5. Temperature dependence of $\Delta R/R$, the resistivity ρ, and the magnetization M, as reported by Jin et al. [2.15], for thin films of $La_{1-x}Ca_xMnO_3$ at $x = 0.33$

ratio reaches a value over 100 000% (not shown), after a variety of optimizations of the films were carried out. This enormous factor corresponds to a thousand-fold change in resistivity with and without the field. That record was broken later by Xiong et al. [2.16] when a MR ratio of over 1 000 000% was reported using thin films of $Nd_{0.7}Sr_{0.3}MnO_\delta$ near 60 K, and in the presence of a magnetic field of 8 T. It is difficult to keep track of the current MR records of manganite thin films, but from the discussions in [2.15, 2.16], it seems likely that those early reported records could be broken by even more elaborate treatments of the thin films. The obvious conclusion from the effort of Jin et al. and Xiong et al. was that manganites suddenly were a potential alternative to giant MR systems. From those important results, it was clear that more work in manganites was needed, involving both experiment and theory, and as a consequence an enormous effort in this context started.

After the early studies of manganites described above, Tokura et al. [2.17] proposed that the charge-ordering states observed by Jirák et al. [2.8] were very important for the explanation of the CMR effect. They presented results indicating an abrupt collapse of the charge-order state into a ferromagnetic state under the influence of a magnetic field. These and other crucial contributions to the field of manganites by Tokura and collaborators will be extensively discussed in other chapters of the book. The competition between CO and FM is indeed a key component of the current theories of manganites aiming to explain the CMR phenomenon. It is clear from the experiments and the theory that the CO FM transition should be first-order unless disordering effects smear it into a rapid but continuous transition. The huge CMR effect in some compounds at very low temperatures, such as shown in Fig. 3.7 of Chap. 3, appears to be caused by the CO FM first-order transition induced by magnetic fields. This physics is not contained *at all* in the early

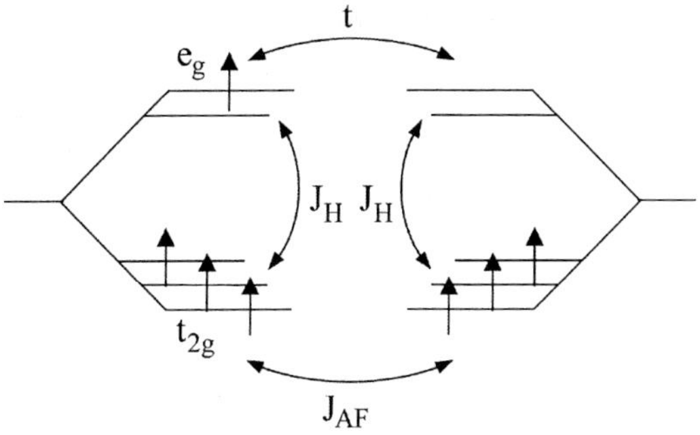

Fig. 2.6. Schematic representation of the ideas of Zener to explain ferromagnetism. Zener envisioned a system with both localized and mobile electrons, which in the manganite language are the t_{2g} and e_g electrons, as indicated. The couplings of relevance are the Hund coupling, J_{H}, the hopping t, and the antiferromagnetic direct exchange J_{AF}. The number of levels shown is realistic for manganese ions. More details are given in Chaps. 4 and 5

theoretical studies of manganites in the 1950s and 1960s, which were based on the so-called double-exchange effects and one-orbital models. Only in the late 1990s and early 2000s has the CO FM competition been identified as the key ingredient of the CMR phenomenon.

2.3 Early Theoretical Studies

The theoretical studies of manganites in the 1950s focused on the origin of the ferromagnetic phase. It was only decades after that initial work that the "colossal" magnetoresistance effects were unveiled, and the excitement around Mn oxides mainly shifted toward understanding the origin of such curious magnetotransport properties, with potential technological applications. Returning to ferromagnetism, the basis for understanding the spin polarization was presented by Zener in 1951, in a couple of seminal papers [2.18,2.19], where mainly qualitative statements and analysis of experiments were presented. In spite of this qualitative character, the essence of the reasoning is certainly correct and Zener's work is widely regarded as providing a proper zeroth-order explanation of manganite ferromagnetism.

Zener interpreted ferromagnetism as arising from an *indirect* coupling between "incomplete d-shells", via "conducting electrons". In the first paper of Zener [2.18], the manganites were not addressed explicitly, but in Mn-oxide language one should think of the incomplete d-shells as the so-called t_{2g} localized spins of manganites, and the conduction electrons are the e_g electrons. A much more formal study of these states will be addressed in Chaps. 4 and 5. For the purposes of this introduction it is sufficient to consider a qualitative description of the splitting of the five d-levels in the presence of a crystal environment, as shown in Fig. 2.6. The Hund's rule in each individual ion or atom is enforced by a ferromagnetic Hund coupling J_H, as in Fig. 2.6. This effect was argued by Zener to play a key role in his mechanism, enforcing the configuration where the unpaired spins are aligned to lower the energy (the nodes generated by the Pauli principle in the many-electron wave function of a polarized state helps in avoiding Coulombic repulsion among the electrons involved). Since the conduction electrons do not change their spin as they move from ion to ion (in a more formal language, the electron hopping maintains the z-projection of the spin), Zener reasoned that those electrons are able to move in the crystal in the optimal manner when the net spin of the incomplete d-shells are all parallel – otherwise, an up electron can land on a down spin ion, and pay an energy proportional to the Hund coupling. In other words, the conduction electrons lower their kinetic energy if the background of d-shell spins, or the t_{2g} spins of manganites, is fully polarized. The kinetic energy is regulated by a hopping amplitude usually denoted by t. The d-shell spins are then indirectly coupled via an interaction mediated by the conduction electrons. Zener remarked clearly in his papers that a *direct* coupling between d-shells (not mediated by conduction electrons, but by the direct virtual hopping of d-electrons) is of opposite sign leading to antiferromagnetism, rather than ferromagnetism. The coupling involved in this direct exchange process is called J_{AF}, and it will be shown to be important in the physics of manganites (see, for instance, Chap. 10).

The ideas of Zener are known today as "double exchange". It is interesting to explain the historical origin of this term. In his second paper in 1951, Zener [2.19] for the first time describes manganites (in his first paper there is no reference to the work of Jonker and van Santen [2.1]), and he concludes that the arguments of his first paper apply to this case, but also writes that it would be interesting to know the precise mechanism by which conduction electrons transfer between manganese ions. He remarks that the transfer must be through the oxygen ion between the manganeses ions, and argues that the true transfer occurs through an actual "double exchange" of electrons. This mechanism is sketched in Fig. 2.7a. It simply amounts to a simultaneous transfer of an electron from the oxygen to the right Mn, and from the left Mn to the oxygen, such that the net transfer is of an electron from left Mn to right Mn. It is also interesting to remark, as Zener did, that the mechanism leading to ferromagnetism that he found should not be confused with the "superexchange" ideas learned in elementary books of condensed matter, which also use an oxygen as a bridge between ions. Zener remarked correctly that the superexchange interaction leads to an *antiferromagnetic* alignment of spins, as discussed also in the previous paragraph.

The original double-exchange idea in [2.19] was discussed not as a procedure to explain ferromagnetism but as a mechanism for electron transfer, that was built in the context of the original idea of ferromagnetism mediated by conduction electrons. To combine both ideas into one, it has become com-

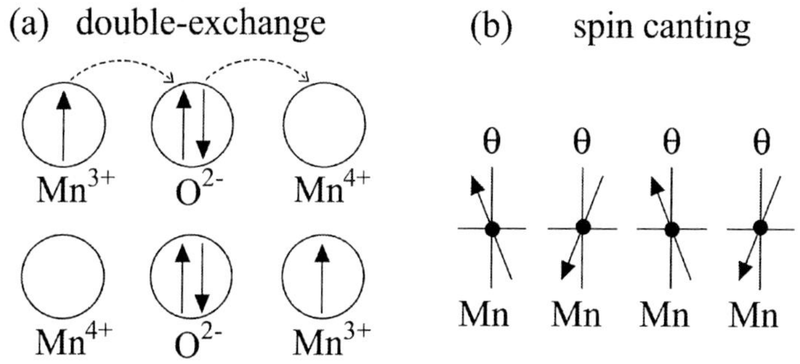

Fig. 2.7. (**a**) Schematic representation of the double-exchange process proposed by Zener. (**b**) Sketch of a spin-canted state, as discussed by de Gennes

mon practice to refer to the whole mechanism for ferromagnetism envisioned by Zener as "double exchange", referring with that name to the ideas of both papers [2.18] and [2.19]. On some occasions, a more proper language of "Zener ferromagnetism" refers only to the results of [2.18], since it is unclear that a simultaneous movement of two electrons between manganese ions, with the oxygen as bridge, truly occurs in real systems. Nevertheless, often in this book ferromagnetic states found in a variety of contexts will be simply called double-exchange-induced for simplicity. The reader should keep the present discussion in mind for a more proper description of the physics behind the words.

The seminal work of Zener was continued by Anderson and Hasegawa [2.20], who studied the proposed mechanism in greater detail. They explained that a better way to describe the motion of electrons from Mn to Mn is to transfer electrons one by one, still using the oxygen as a bridge between ions, rather than simultaneously as believed by Zener. Any perturbative approach in the hopping amplitude t will naturally lead to a one-by-one transfer of electrons. Perhaps the most often-quoted portion of the work of Anderson and Hasegawa [2.20] is the effective hopping t_{eff} of an electron jumping between two nearest-neighbor Mn ions. In fact, a very simple calculation shows that $t_{\mathrm{eff}} = t \cos\frac{\theta}{2}$, where θ is the angle between t_{2g} spins located at the two sites involved in the electron transfer, assuming classical localized spins (the details of the calculation are presented in Chap. 6). Quantum versions of the effective hopping were also studied by Anderson and Hasegawa [2.20], Kubo and Ohata [2.21], Cieplak [2.22], and others. These simple effective hopping ideas guided research in manganites in the early days, but currently it is well known that far more elaborate theories are needed to explain the rich phase diagrams of these compounds, as well as the CMR effect.

Continuing this introduction to theoretical ideas proposed in the early days of manganites, it is interesting to address also the spin-canted state proposed by de Gennes [2.23], as the possible stable state obtained by doping an antiferromagnetic state with holes or electrons (see Fig. 2.7b). This spin-canted idea has been extensively used by experimentalists until very recently, every time they observed some coexistence of ferromagnetic and antiferromagnetic features (the spin-canted state has a net moment, coexisting with a staggered distribution of spins perpendicular to that net moment). However, more modern work has shown that a strong alternative to canting is provided by the tendency to phase separate, as discussed extensively in this book. Currently, it is unclear whether truly canted states can be stabilized in some manganites at zero magnetic field, and this issue will be discussed often in the rest of the book. Bilayer manganites may provide an example.

Another of the theoretical papers still much cited from the early studies of manganites is the analysis by Goodenough [2.24] of the charge, orbital, and spin arrangements in the non ferromagnetic portions of the phase diagram of $La_{1-x}Ca_xMnO_3$, regimes that were discovered experimentally by Wollan and Koehler. Goodenough's paper was based on the notions of "semicovalent bond" and elastic-energy considerations (a simple description of semicovalent bonds can be found in the review [1.9] and, of course, in the original reference of Goodenough). The oxygens play a key role, as well as the Coulomb interactions. In Goodenough's paper there is little use of the couplings introduced in the double-exchange mechanism, referred to as "t", "$J_{\rm AF}$", and "$J_{\rm H}$" in Fig. 2.6, but the arguments were mainly qualitative and based on the above-mentioned elastic and covalent-bond considerations. In [2.24], the phases that compete with the antiferromagnetic ones were not analyzed, namely the states were studied without a discussion of their global stability. In more modern calculations as presented in Chap. 10, these issues of stability are more properly analyzed. However, it is clear that Goodenough was the first to identify the essence of the currently widely accepted "CE-state" at hole density 0.5, and its associated charge and orbital order. In fact, the concept of orbital order already contained in the work of Goodenough will play an important role in theories of manganites and other materials (a review on the subject applied to transition-metal compounds was prepared by Kugel and Khomskii [1.5]).

A reproduction of the original figure of Goodenough of the CE-state is shown in Fig. 2.8. In spite of the success of this prediction, not much work these days is based on covalent bonds; instead, current calculations (or theories) are typically based on one- or two-orbital models with Jahn–Teller and/or Coulombic repulsion. A reason may be that by the use of these models, recently it has been shown that the CE-state is stable in some region of parameter space, and competes with the FM state. This occurs even without oxygen degrees of freedom active in the models, and without long-range Coulomb interactions. A calculation following Goodenough's ideas, where the manganese ions are in extreme states (3)+ and (4)+, rather than, say, $(3.5+\delta)+$ and $(3.5-\delta)+$, appears to correspond to the limit of large Coulombic or electron-phonon couplings. It is difficult to imagine how a metallic ferromagnetic state can be close in energy to the ground state in that regime, as

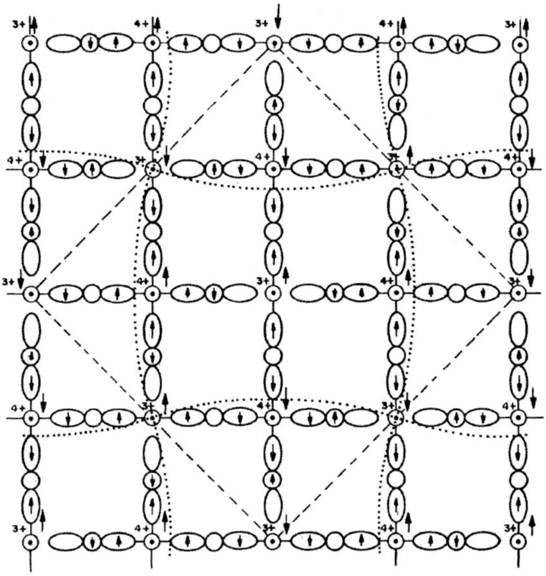

Fig. 2.8. Schematic representation of the CE-state proposed by Goodenough, directly from [2.24], where more details about the notation can be found. This figure is reproduced mainly for historical reasons, to show that spin, charge, and orbital order states were known even in the first studies of Mn oxides

experiments for materials such as LCMO with $x = 0.5$ suggest. The competition between CO and FM is actually a key ingredient of current theories of manganites, and it is important to develop formalisms including both states. For these reasons, no further mention of the covalent-bond ideas will be made in this book. However, it is certainly worth recalling their existence since they may be important for reaching a full understanding of manganites. They were also useful for predicting a highly nontrivial state such as the CE-state in the early studies. In addition, the elastic effects considered by Goodenough are likely important in manganites, and their influence should not be neglected.

2.4 Tolerance Factor

The "tolerance factor" plays an important role in manganites and related materials. This is a geometrical factor defined as $\Gamma = d_{\mathrm{A-O}}/(\sqrt{2}d_{\mathrm{Mn-O}})$. Here, $d_{\mathrm{A-O}}$ is the distance between the A-site, where the trivalent or divalent non-manganese ions are located, to the nearest oxygen (remember that the A ion is at the center of a cube with Mn in the vertices and O in between the Mn ions). $d_{\mathrm{Mn-O}}$ is the Mn O shortest distance. Since, for an undistorted cube with a straight Mn O Mn link, $d_{\mathrm{A-O}} = \sqrt{2}$ and $d_{\mathrm{Mn-O}}$=1, in units of the Mn O distance, then $\Gamma = 1$ in this perfect system. However, sometimes the

A ions are too small to fill the space in the cube centers and for this reason the oxygens tend to move toward that center, reducing $d_{\mathrm{A-O}}$ (in general, $d_{\mathrm{Mn-O}}$ also changes at the same time). For these reasons, the tolerance factor becomes smaller than one, $\Gamma < 1$, as the A radius is reduced, and the Mn O Mn angle θ gets smaller than 180°. The hopping amplitude for carriers to move from Mn to Mn naturally decreases as θ is reduced from 180° (for a 90° bond the hopping involving a p-orbital at the oxygen simply cancels, by orthogonality). As a consequence, as the tolerance factor decreases, the tendencies to charge localization increase due to the reduction in the carrier mobility. This has been observed experimentally in Mn oxides. Since in the general chemical composition for perovskite manganites $\mathrm{A}_{1-x}\mathrm{A'}_x\mathrm{MnO}_3$ there are two possible ions at the "A" site, then the tolerance factor for a given compound can be defined as a density-weighted average of the individual tolerance factors.

Note that the distance Mn Mn is reduced in the situation described so far, while the tolerance factor (monotonically related with the hopping amplitude) is also reduced ($\Gamma < 1$). This is somewhat counter-intuitive since it would be expected that having closer Mn ions would increase the electron hopping between them. However, the hopping amplitude is not only proportional to $1/(d_{\mathrm{Mn-O}})^{\alpha}$, where $\alpha > 1$ [2.25], but also to $\cos\theta$ due to the fact that it is the p-orbital of oxygen that is involved in the process. If this orbital points toward one of the manganese ions, it can not point toward the other one simultaneously for $\theta \neq 180°$.

2.5 High Magnetic Fields

Magnetic fields play a key role in manganites. In this context, it is natural to ask what fields should be considered "large" or "small". In other words, what is really surprising about manganites is that the resistivity changes by many orders of magnitude upon the application of fields of a few Tesla, which in electronic units are actually very small.

To explain this in more detail, first let us recall how to "translate" from energy units natural for electronic processes, such as the electron-volt, eV, to an analogous temperature scale T. The usual relation is through Energy=$k_{\mathrm{B}}T$, where $k_{\mathrm{B}} = 8.617 \times 10^{-5}\,\mathrm{eV/K}$, is the Boltzmann constant. Then, the temperature needed to match 1 eV is $T = 1\,\mathrm{eV}/k_{\mathrm{B}} = 11\,604$ K, which is certainly very high. Let us address now magnetic fields. In the presence of a field of strength H, the Zeeman splitting for free electrons is $2\mu_{\mathrm{B}}H$, where a g-factor equal to 2 has been used. The Bohr magneton can be found (in books) to be $\mu_{\mathrm{B}} = 5.7884 \times 10^{-5}\,\mathrm{eV/T}$. Then, a field of 1 T produces a Zeeman splitting of 1.1577×10^{-4} eV which, based on the conversion from energy to temperature, amounts to 1.34 K. In the table below these numbers are reproduced for easy access. Following this reasoning, it is clear that a magnetic field of the order of 10 T, as usually applied in manganites, produces relatively small changes in the energy expressed in eV units. In fact, 10 T is equivalent to heating up the system by about 13.4 K, which is not much indeed. As a consequence, the fields needed for the CMR effect are considered "small" from this perspective, and it is certainly surprising that a magnetic field of a few T affect so dramatically the properties of electrons in a manganite crystal. On the other hand, from the perspective of applications such as in personal computers, a field of one T is actually quite "large", since other materials with large MR (such as the giant MR devices, see Chap. 21) work at only 0.01 T. What is large or small certainly depends on the context in which the question is formulated.

Table 2.1. Useful numerical relations

1 eV	11 604 K
1 T	1.34 K
Static field	up to 45 T
Pulsed field	60–80 T
Explosive field	1 000 T
Neutron star	10^8 T
Earth's field	0.5 G
1 T	10 000 G

To round up the discussion, let us address now what "large or small magnetic field" means from the laboratory point of view, leaving aside the energy scales of the electrons in materials such as manganites. At present, the world record in magnetic fields that are static, namely, time independent, is about 45 T at the National High Magnetic Field Lab in Florida State University, Tallahassee, Florida. If the field is allowed to be time dependent at the millisecond to 0.1 s scale, then higher fields of about 60 T can be created in several laboratories around the world (these are the "pulsed" magnetic fields). For even shorter periods of time, fields of about 1 000 T can be created for approximately a microsecond using explosive devices (some measurements can be carried out before the shock wave destroys the sample). To locate these numbers in a broader perspective, the field at the surface of a neutron star is 10^8 T, while the Earth's field at the surface of our planet is about 0.5 G which is 1/20 000 T, since 1 T= 10 000 G [2.26].

3. Phase Diagrams and Basic Properties of Manganites

3.1 The Intermediate Bandwidth Compound $La_{1-x}Ca_xMnO_3$

$La_{1-x}Ca_xMnO_3$ is among the most-studied compounds of the manganite family. Among the reasons for its popularity are its robust magnetoresistance effect, larger than in other much-studied materials such as $La_{1-x}Sr_xMnO_3$, and the possibility of analyzing the phase diagram at all values of the Ca concentration (most manganites can only be studied in a more restrictive hole-density range). $La_{1-x}Ca_xMnO_3$ is usually labeled as a "low" bandwidth manganite, but it may be better to consider it as a representative of "intermediate" bandwidth compounds. The reason is that this material still has a ferromagnetic metallic phase in a robust density range, unlike truly low-bandwidth compounds such as $Pr_{1-x}Ca_xMnO_3$ (discussed later) where the metallic phase is only induced by external magnetic fields.

$La_{1-x}Ca_xMnO_3$ was first studied in the full range of densities $0 < x < 1$ by Schiffer et al. [3.1]. These authors measured the magnetization and resistivity, and typical results are shown in Fig. 3.1. The upper panel of this figure shows the magnetization in the range between very low and room temperature, in a density regime where ferromagnetism sets in below ~ 240 K. Note that the magnetization is zero in the absence of magnetic fields, but it rapidly increases as relatively small fields of less than 1 T are switched on. This is typical behavior of ferromagnets attributed to domain formation (different regions of the sample have magnetizations pointing in different directions, separated by domain walls that must be removed to obtain a net magnetization). Note also the substantial effect that a field of 4 T can cause around 300 K, high above the nominal Curie temperature: the material seems easily spin polarizable. In the middle panel of Fig. 3.1, the D.C. resistivity vs. temperature at several magnetic fields is shown. The data at zero field is characteristic of many low- or intermediate-bandwidth manganites: (i) An insulating behavior is found above T_C in the paramagnetic regime, which is exotic since upon further decreasing the temperature a metallic state is eventually obtained. This insulating behavior cannot be explained by the double-exchange mechanism, in which the e_g electrons simply scatter from the disordered localized spins, as this effect is not sufficiently strong to induce

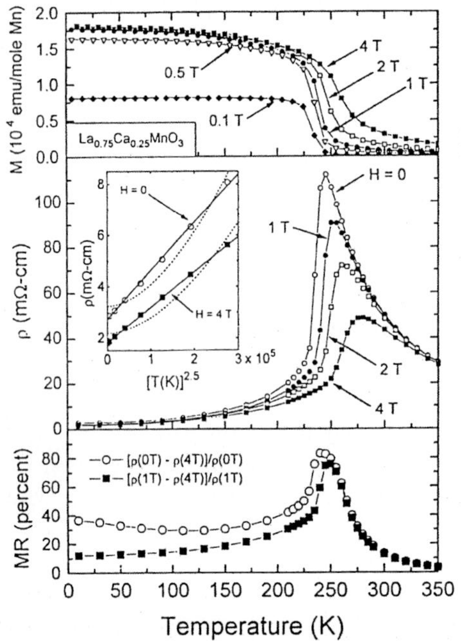

Fig. 3.1. Magnetization, resistivity, and magnetoresistance of $La_{0.75}Ca_{0.25}MnO_3$, as a function of temperature for various magnetic fields (from Schiffer et al. [3.1]). The *inset* in the *middle panel* shows resistivity at low temperature, with a fitting procedure discussed in [3.1]

insulating behavior. (ii) Near T_C a fairly abrupt transition in the resistivity occurs, leading to metallic behavior in the ferromagnetic regime. Then, the Curie temperature is also the metal insulator transition temperature. (iii) As $T \to 0$, the resistivity converges to a fairly large value of the order of $1\,\mathrm{m}\Omega$ cm, far larger than in standard metals. For reference, the resistivity of copper at room temperature is $1.7\,\mu\Omega$ cm. Materials with resistivities comparable to those of the manganites are for instance carbon with $1500\,\mu\Omega$ cm, and high-temperature superconductors, such as Bi2201, with a low-temperature in-plane resistivity of about $100\,\mu\Omega$ cm. This last result suggest analogies between cuprates and manganites, which are discussed at the end of this book in the final chapter.

The most spectacular result of the middle panel of Fig. 3.1 is the substantial effect of magnetic fields on the resistivity. Relatively modest fields of a few Teslas are sufficient to change the resistivity by large factors. In fact, in the lower panel of Fig. 3.1 the MR ratio is shown. This quantity has a large peak at the Curie temperature. The large value of the MR ratio at T_C appears to be caused mainly by the large resistivity of the insulating phase, rather

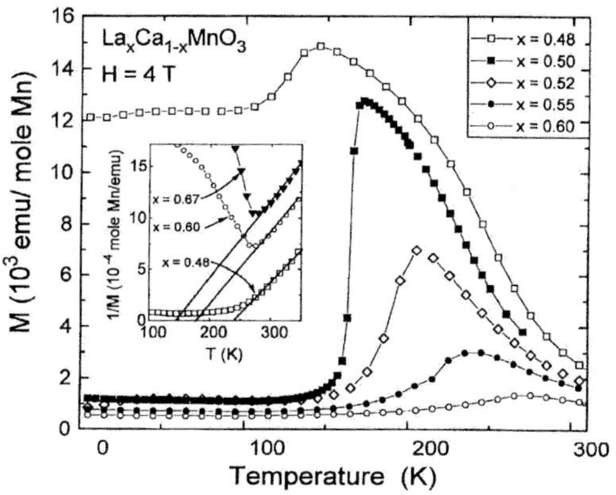

Fig. 3.2. Magnetization of $La_{1-x}Ca_xMnO_3$ at 4 T in the vicinity of $x = 0.5$. The *inset* shows the inverse magnetization vs. temperature, and the *straight lines* indicate ferromagnetic fluctuations at high temperatures at each doping (from [3.1])

than the resistivity of the metallic phase (which is far larger than in standard metals, as already explained). This is a recurrent conclusion of recent literature: to understand the CMR effect the most crucial aspect appears to be the nature of the insulating paramagnetic regime above T_C, and its highly polarizable properties. As time goes by, the emphasis in CMR studies has shifted remarkably from the metal to the insulator. However, as emphasized by the authors of [3.1], note also that the MR ratio is very robust (as large as 35%) even at very low temperatures deep into the ferromagnetic phase. This shows that the ferromagnetic phase is far from a "perfect" ferromagnet, as naively suggested by double-exchange calculations based on just one orbital.

The ground state of $La_{1-x}Ca_xMnO_3$ becomes an insulating antiferromagnet at $x \sim 0.5$, as observed in the early manganite experiments in the 1950s (see Chap. 2). Typical results in this regime are reproduced in Fig. 3.2, where the magnetization is shown. At $x = 0.5$, the region above $\sim$ 160 K is ferromagnetic. Tendencies toward spin polarization are present even above room temperature, as suggested by the inset extrapolation of the inverse magnetization. Remember that the susceptibility of a ferromagnet behaves as $\chi \sim 1/(T-\theta)$, with θ a constant, and that $\chi \sim M/H$, with M the magnetization and H the external field. Then, $1/M$ should behave as $T-\theta$, and $1/M$ should cancel at $T = \theta$, which if properly positive indicates ferromagnetism. Alternatively, if $\theta < 0$, then antiferromagnetic correlations are actually developing in the system. This is a standard method of deciding the FM vs. AF character of a system based on the high-temperature magnetic susceptibility.

The result for LCMO at $x = 0.5$ discussed above is interesting. The system is ferromagnetic at least at short distances near room temperature (this result is confirmed by neutron scattering experiments, to be described later), and there are no precursors of the sharp transition to an AF insulator observed at lower temperatures, at this or other hole density. Many of the transitions in manganites indeed appear to have first-order character. As the temperature is lowered at $x = 0.5$, the AF regime is abruptly reached in what appears to be a first-order transition. At this transition the lattice parameters change, as shown for instance in [3.2], keeping the overall octahedron volume constant (the Jahn–Teller modes Q_2 and Q_3 preserve the volume of the octahedron MnO_6, see Chap. 5). Then, the FM AF transition is *simultaneously* magnetic (the spin order changes), electronic (the resistivity changes), and structural (the lattice is modified at the transition). At hole densities larger than 0.5, the ferromagnetism at low temperatures has vanished and it is replaced by an AF charge-ordered state (the AF character emerges from neutron scattering experiments, while the charge-ordered character can be inferred from inflections in the resistivity plots).

Note: Magnetization. A brief reminder of the units usually employed when measuring magnetizations: From elementary considerations in textbooks it is known that the dipole magnetic moment μ of an ion, which enters into the interaction $-\mu \cdot \mathbf{H}$ with a magnetic field $\mathbf{H}$, is given by $\mu = -g\mu_B \mathbf{S}$, where g is the g-factor of the ion, μ_B is the Bohr magneton $(e\hbar/2mc)$, $\mathbf{S}$ is the spin in dimensionless units, and it was assumed that the orbital angular momentum is quenched ($L = 0$). For more details the reader can consult Ashcroft and Mermin [3.3], Chap. 31. With these conventions, at electronic density 1 in manganites, where only ions Mn^{3+} with total spin 2 are present, the magnetization M (= mean value of the total magnetic moment/number of Mn sites) for a fully polarized state should be $M = \langle \mu \rangle = 4\mu_B$ if a g-factor 2 is chosen. For all Mn^{4+} with spin 3/2, the result should be $M = 3\mu_B$. Experimentally, the results at low temperatures are very close to these values, or simple linear combinations at intermediate densities between 1 and 0.

The complete phase diagram of $La_{1-x}Ca_xMnO_3$, obtained using magnetization and resistivity data, is shown in Fig. 3.3, reproduced from Cheong and Hwang [3.4]. This phase diagram is obtained by working with a relatively small set of samples with varying Ca concentrations. However, this number is sufficient to gather quite reliable results (note that interesting experimental methods to obtain analogous information using thin films and a continuous change of compositions have been proposed [3.5]). Figure 3.3 clearly illustrates the many phases that are in competition in Mn oxides. Very prominent is the ferromagnetic low-temperature metallic phase, which is widely believed to be caused by the double-exchange mechanism, where mobile electrons polarize the localized spins to improve their kinetic energy, as in the Nagaoka phase of cuprates [3.6] (for other references see [1.20]). However, equally prominent is the charge-ordered (CO) regime at hole densities larger than 50%. Recent experiments, to be described for example in Chap. 9, have shown that this state is also orbitally ordered (OO). The FM vs. AF/CO/OO competition is believed to be at the heart of the curious magnetotransport properties of many manganites [1.9], as will be explained elsewhere in this book. Note that the presence of the AF phase clearly shows that the double-exchange model and related ideas are not sufficient to ex-

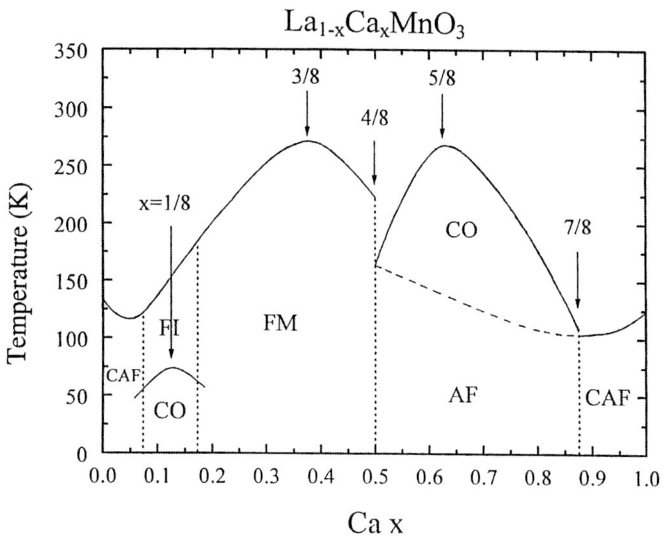

Fig. 3.3. Phase diagram of $La_{1-x}Ca_xMnO_3$, from Cheong and Hwang [3.4]. The notation is standard. The marked hole density fractions appear to have more importance than others. According to modern theories, the canted regimes (CAF) may correspond to mixtures of AF and FM regions

plain the properties of low- or intermediate-bandwidth manganites, since in the double-exchange (DE) context there is no room for such AF/CO/OO phases at low temperature. At low hole density, a complicated regime with FI (ferromagnetic insulator), CO (charge ordering) and CAF (canted antiferromagnetic phases) is reached. The fine details of this complicated regime are still under discussion. Note that the "canted" character of the very small x regime is in doubt in view of the overwhelming theoretical evidence against uniform canted states. This regime may be caused by a mixture of FM and AF phases. The same comment holds for the large hole density limit of x close to 1, as discussed later in this book.

Some comments about Fig. 3.3: (i) In the regime of doping where CMR effects can be found, there is a low-temperature FM metallic phase. The CMR occurs particularly at the boundaries of this metallic phase. This shows clearly that it is the competition between metal and insulator phases that causes CMR, and is likely not an intrinsic property of any of the phases in isolation. In fact, results in the regime of "electron doping" (i.e., large hole density) support this view, for a large MR has been observed at compositions around 85% when the FM and AF phases are in competition [3.7]. It is not necessary to work at the FM metallic phase of low hole density to find a large MR. (ii) Note in Fig. 3.3 the special Ca concentrations (x) at multiples of 1/8, which were highlighted by Cheong and Hwang [3.4] as important for the physics of manganites. This is clearly of relevance at $x = 3/8$ and $5/8$, where the FM and CO maximum critical temperatures are obtained. (iii) The phase diagram also shows that double-exchange ideas, which at large Hund coupling lead to a phase diagram mirror symmetric between $x < 0.5$ and $x > 0.5$, are not suitable to describe these materials. In fact, the phase diagram, Fig. 3.3, has a clear electron-hole asymmetry. (iv) In the two limits of $x = 0.0$ and 1.0, the ground state is antiferromagnetic. At $x = 0.0$ it is A-type (FM in two directions, AF in the other), while at $x = 1.0$ it is G-type (AF in the three directions).

The AF/CO/OO phase has a very complicated arrangement of spin, charge, and orbital order. Figure 3.4 shows the states believed to be stable by Cheong and Hwang [3.4], based on available experimental data. The results at $x = 0.0$ and 0.5 are widely accepted and were discussed by many authors starting in the 1950s. At $x = 0$, the A-type spin AF state has a concomitant orbital order with orbitals $3x^2 - r^2$ and $3y^2 - r^2$ arranged in a *staggered* pattern. This result has been recently confirmed using resonant X-ray experiments (see Chap. 9).

The $x = 0.5$ arrangement is the so-called CE-phase, which has also recently been confirmed experimentally. The charge in the CE-state is *stacked* along the z-axis, according to experiments. This is an unusual feature that is difficult to explain based on theoretical considerations. The origin of the CE-state will be discussed in detail later in this book. At the density $x = 2/3$, the ground state is still controversial. Cheong and Hwang believe that a "bi-stripe" arrangement is stabilized (see Fig. 3.4, see also [3.8]). This pattern has two stripes of charge close to each other, separated by some distance from another pair. States of this nature have been found theoretically (see Chap. 10, for example), and some of its rationalization involves movement of charge perpendicular to the stripes, rather than across them. At densities between $x = 1/2$ and $2/3$, mixtures of those states are believed to be stable.

However, note that other authors (Radaelli et al. [3.9]) have proposed a so-called Wigner crystal (WC) arrangement for $x = 2/3$, where charge is spread as far as possible, without the presence of stripes. The use of the term Wigner crystal is just by analogy with the state of that name where the $1/r$-Coulomb repulsion dominates, spreading charges in a periodic pattern maximizing their distance. This occurs typically in models at very low electronic density, and as a consequence it is somewhat confusing to use such terminology as $x = 2/3$. Nevertheless, a WC is widely accepted as denoting a pattern where charge is at the farthest distance possible, and here it is used in that sense. Recent theoretical results (Chap. 10) have shown that the Wigner crystal and bi-stripe states are indeed very close in energy. Hopefully, further experiments and theories will soon solve the puzzle of the nature of the $x = 2/3$ state.

To complete the present brief description of the main properties of $La_{1-x}Ca_xMnO_3$, two other issues are worth discussing:

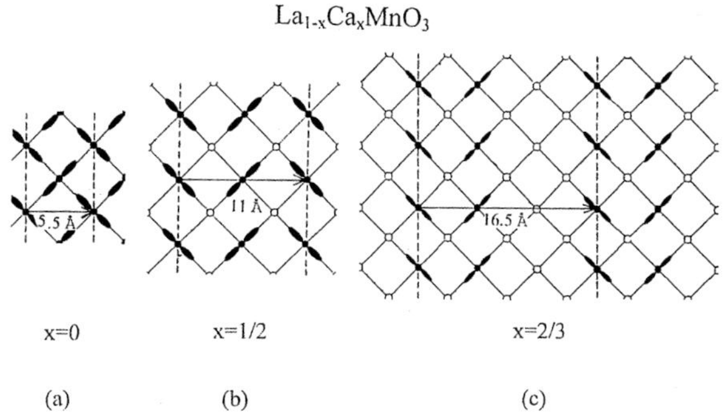

Fig. 3.4. Charge and orbital order in $La_{1-x}Ca_xMnO_3$ at the hole densities indicated, from [3.4]. *Open circles* are Mn^{4+} (no hole). The *lobes* indicate the populated orbital of a Mn^{3+} ion. Note the presence of bi-stripes at $x = 2/3$

(i) *Effect of pressure:* Experimental work on $La_{1-x}Ca_xMnO_3$ has shown that *pressure* has a substantial effect on the electrical conduction process of manganites and on T_C itself [3.10, 3.11]. In these experiments, pressure was applied hydrostatically on the sample inside a pressure cell. In this setup, the effect is uniform in the three directions (contrary to uniaxial pressure devices, where the effect of pressure acts on only one direction). A typical result for the resistivity is shown in Fig. 3.5, where data for Ca $x = 0.21$ are shown. "Zero" pressure is ambient pressure and the critical temperature in this case is 203.3 K, with a resistivity curve shape that is typical of low- and intermediate-bandwidth manganites. When pressure is applied, its effect is quite similar to the changes produced by a magnetic field already discussed, namely, T_C increases with pressure, and the resistivity is drastically reduced. This interesting pressure effect is still not understood, but it is widely believed that pressure affects the electronic hopping amplitude t between Mn ions (reducing their distance and/or modifying the Mn O Mn bond angle), and that this change in t should be responsible for the large changes in the conductivity. The discussion in [3.11] shows that external pressure is similar to "internal" pressure achieved by chemical substitution: when ions with the same valence but different size are substituted for each other, strain is created in the atomic structure. The detailed mechanism of resistivity reduction by pressure is still under discussion.

In [3.10], the rate of increase of T_C with pressure is compared with other compounds. While there are some ferrimagnets that show the same rapid change of T_C as the pressure increases, most of the others show smaller effects and in particular the Curie temperature of Fe and Co are not affected noticeably by pressure. Clearly the manganites are not standard ferromagnets.

(ii) *Effect of isotope substitution:* It has been observed by Zhao et al. [3.12] that upon replacing ^{16}O with ^{18}O, large changes of about 20 K in the Curie

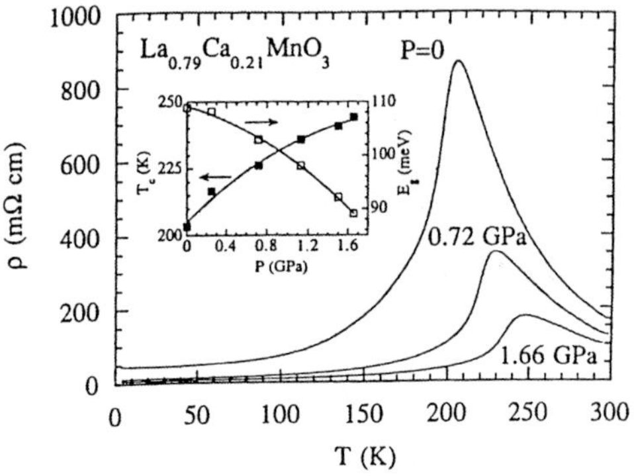

Fig. 3.5. Electrical resistivity vs. temperature at three hydrostatic pressures, for the compound indicated. The *inset* shows the Curie temperature and the activation energy E_g (for details see [3.10]). Note on units: 1 GPa= 10 000 atmos. = 10 kbar

temperature of 20% doped $La_{1-x}Ca_xMnO_3$ were found. This change was referred to as "giant" in [3.12], because it is substantially larger than that observed in other oxides. Similar isotope-effect studies and conclusions have been presented by Isaac and Franck [3.13]. Large changes of properties of materials upon isotope substitution strongly suggest that phonons are important in those materials, since upon oxygen-isotope substitution only the mass of oxygen changes, while Coulombic properties or electronic hopping integrals are believed to remain unaffected. The giant effect found in $La_{1-x}Ca_xMnO_3$ indicates that phonons are important in manganites. The isotope substitution may also alter the value of the electron-phonon coupling λ, described for example in Chap. 5. Similar isotope-substitution experiments were certainly crucial many years ago also in the understanding of superconductors (for details consult any textbook on superconductivity). Since the critical temperature was affected by this procedure, it was clear that phononic mechanisms were needed, and that experimental information eventually led to the BCS theory. In the area of high-temperature superconductors, the situation remains confusing since large effects were observed in the underdoped regime, but not at optimal doping where the critical temperature is maximized. In manganites, there is no doubt that phonons play a key role in describing many properties of these compounds.

3.2 Basic Properties of $Pr_{1-x}Ca_xMnO_3$

Another very important manganite compound is $Pr_{1-x}Ca_xMnO_3$, a material that is widely studied as a representative of low-bandwidth Mn oxides. In this compound there is no stable ferromagnetic metallic phase at low temperature

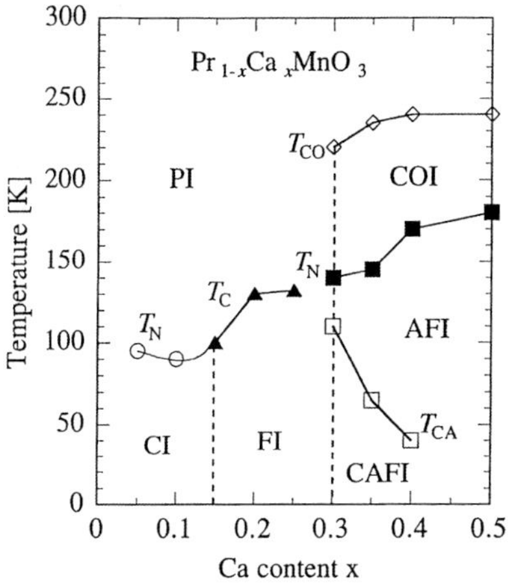

Fig. 3.6. Phase diagram of $Pr_{1-x}Ca_xMnO_3$, from Tomioka and Tokura [3.14] where more details can be found. PI, FI, AFI, COI, and CAFI, denote paramagnetic insulator, ferromagnetic insulator, antiferromagnetic insulator, charge-ordered insulator, and canted-AF insulator states, respectively

at any of the Ca concentrations available. This is usually rationalized as being caused by a small tolerance factor, which roughly translates into a small hopping amplitude. The effect reduces the electronic kinetic energy gained by making the system FM metallic, and increases the tendency to charge localization into a charge-ordered state. The actual phase diagram of this compound is shown in Fig. 3.6. Note that ferromagnetism is observed, but simultaneously with insulating behavior in the resistivity, creating a FM insulator (FI) phase. This phase, and FI regimes found in other manganites, have not been carefully explored for possible charge and/or orbital ordering. This author believes it is an important challenge for the future to clarify the properties of FI phases of Mn oxides. For $x \geq 0.3$, a charge-ordered AF state is stabilized. According to neutron diffraction studies, it appears that the entire CO/AF phase has similarities with the CE-phase discussed before, which is known to be particularly stable at $x = 0.5$. At Ca densities other than 0.5, if a checkerboard charge distribution still remains valid, the amount of charge at each site must be slightly different from the $x = 0.5$ values 3+ and 4+ to accommodate for a global $x \neq 0.5$. It could also occur that the excess charge appears as defects or in a modified (non checkerboard) charge-ordered state away from $x = 0.5$ (for recent information see [3.15]). Note also that at low temperatures between $x = 0.3$ and 0.4, a canted spin state is claimed to be stable. This regime is somewhat complicated, and it may also consist of a mixture of FM and AF clusters according to recent theoretical developments. It will be important to explore microscopically in more detail the characteristics of that proposed canted regime.

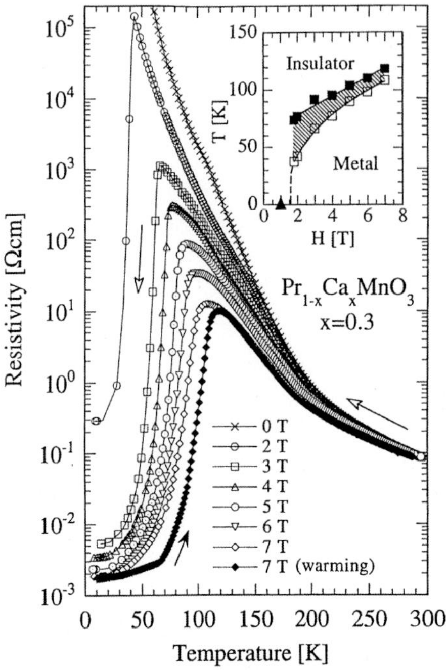

Fig. 3.7. Temperature dependence of the resistivity of $Pr_{1-x}Ca_xMnO_3$ at $x = 0.3$ and various magnetic fields. The *inset* is the phase diagram in the temperature magnetic field plane, with the *hatched region* denoting hysteresis. From [3.14]

Magnetic fields drastically affect the properties of $Pr_{1-x}Ca_xMnO_3$, and generate the FM metallic phase that is absent in the zero-field phase diagram. An impressive result in this context is shown in Fig. 3.7, where the effect of fields on the resistivity of $x = 0.3$ PCMO is shown. As usually found in manganites, relatively small fields of a few Teslas induce important changes in the resistivity, but in this case the changes are truly enormous. For instance, working at 20 K the resistivity changes by about 10 orders of magnitude (extrapolating smoothly to that temperature the zero-field result), an effect far larger than that found in any other compound. The resulting state after the application of the magnetic field is ferromagnetic and metallic. In addition, the abruptness of the result suggests that the transition has first-order characteristics, a recurrent conclusion in several compounds and also in theoretical studies of manganites (see for example Chap. 10). It appears that a first-order transition mixed with percolative characteristics may account for the curious magneto transport properties of manganites, as will be explained later in Chap. 17. It is also remarkable that similar effects as caused by magnetic fields are also produced by external pressure, as shown in Fig. 3.8. Here,

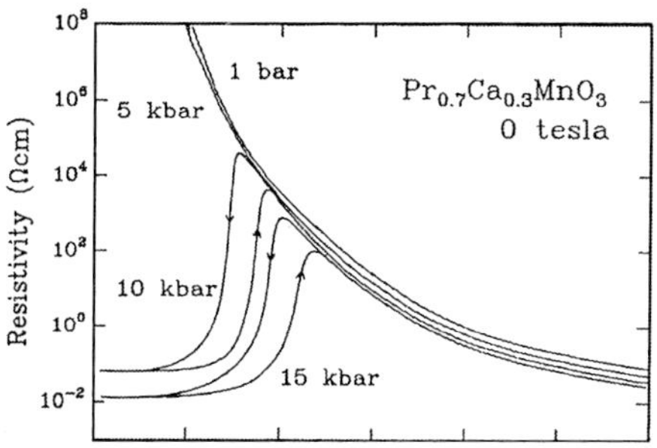

Fig. 3.8. Effect of pressure on $x = 0.3$ PCMO, from [3.16]. Similar results were reported in [3.17]

the abrupt change in the resistivity upon changing the pressure from 5 kbar to 10 kbar is shown, together with hysteresis loops upon warming and cooling (also present in Fig. 3.7).

The behavior observed at $x = 0.3$ also occurs at other densities, as shown in Fig. 3.9. Note that at $x = 0.5$ the CE-state is fairly stable and a field of 12 T is not sufficient to destabilize it (fields as large as 20 or 30 T are needed, see Fig. 3.10 below). At other densities indicated, just 6 T is sufficient to generate a metallic state. Figure 3.10 shows the phase diagram T–H (temperature–external field) for $Pr_{1-x}Ca_xMnO_3$ at various densities. This illustrates the need for using stronger magnetic fields to destabilize the CO-state as the Ca concentration increases toward 0.5. Also note that the shape of the phase diagram changes between $x = 0.4$ and 0.45.

It is also interesting to observe that without varying the density of holes x in the sample, and without introducing a magnetic field, changes from insulator to metal can be obtained by other procedures. This includes chemical substitution of elements with the same valence, such as Ca^{2+} and Sr^{2+}, increasing the applied pressure, and changing the dimensionality from 3D to 2D. In Fig. 3.11a, the resistivity of $Pr_{0.65}(Ca_{1-y}Sr_y)_{0.35}MnO_3$ is shown at several Sr concentrations. A behavior quite similar to that induced by magnetic fields and pressure is found. Similarly, fixing $y = 0.20$ and introducing an external field, once again similar results are obtained (Fig. 3.11b). Clearly there are several ways in which the insulating state can be transformed into the metallic one.

Finally, two features to note as curiosities in these plots: (i) It is surprising that, as a function of external field at fixed chemical composition, the resistivities are nearly identical at room temperature. The temperature at which this effect occurs may be related to the temperature T^* at which clusters of competing phases start forming, as described at length in Chap. 17. (ii) Even at fields so large that the CO-state has entirely melted, the zero-temperature resistivity is still very large compared with that of standard good metals. This may be indicative of residual inhomogeneity in the system. Without a doubt, $Pr_{1-x}Ca_xMnO_3$ is an important member of the manganite family where huge MR ratios have been observed, and where a clear FM AF competition exists.

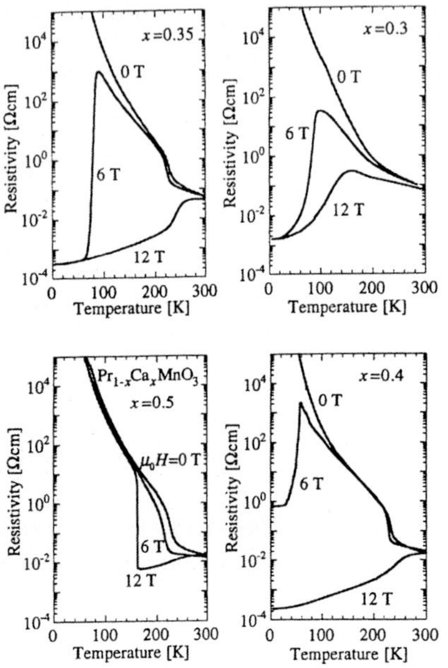

Fig. 3.9. Temperature dependence of the resistivity of $Pr_{1-x}Ca_xMnO_3$, at the Ca concentrations and fields indicated. The results correspond to field cooling (FC) only, thus the hysteresis loops are not present. From [3.18]

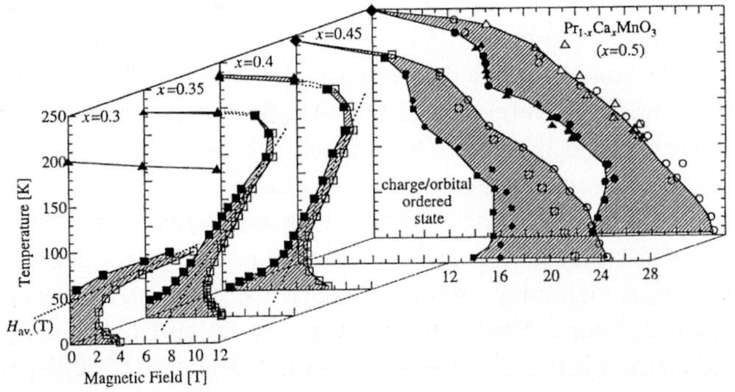

Fig. 3.10. Temperature–magnetic field phase diagram for $Pr_{1-x}Ca_xMnO_3$ at various Ca concentrations. The *hatched area* indicates the hysteresis regions. Note the increase in the magnetic field needed to destabilize the CO-state at low temperature, as the $x = 0.5$ concentration is approached. From [3.14]

It should not be surprising that this material is one of the most intensively studied, in spite of its relatively low Curie temperature.

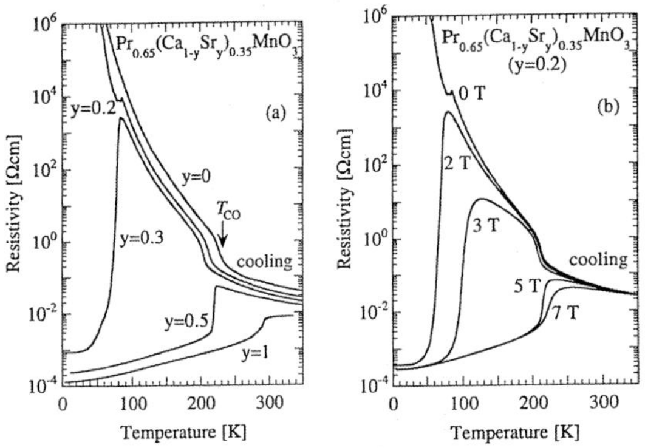

Fig. 3.11. (**a**) Temperature dependence, with varying y, of the resistivity of $Pr_{0.65}(Ca_{1-y}Sr_y)_{0.35}MnO_3$, (**b**) corresponds to $y = 0.2$, changing magnetic fields. From [3.14]

3.3 Basic Properties of $La_{1-x}Sr_xMnO_3$

The manganite $La_{1-x}Sr_xMnO_3$ is one of the Mn oxide compounds that has received the most attention. This material has a large Curie temperature, as high as 370 K at intermediate hole doping, and it is widely considered as a representative of the "large-bandwidth" family. In this material, the electronic kinetic energy is not so reduced by a small hopping amplitude that the ground-state charge orders, as occurs in low- or intermediate-bandwidth compounds. The experimentally determined phase diagram is shown in the lower panel of Fig. 3.12. Here, the description is focused on the results of the comprehensive analysis reported in [3.19] using single crystals. For previous literature, the reader should consult the references listed there. The boundaries of the phase diagram are based on measurements of the resistivity (Fig. 3.12, upper panel) and the total moment of the system (Fig. 3.13). Note that the phase diagram discussed here is somewhat schematic, particularly at low hole density where there is still much discussion about the true character of the ground state. Nevertheless, Fig. 3.12 is sufficient to illustrate the main properties of this compound. In the range from $x \sim 0.17$ to close to $x \sim 0.5$, the system at low temperature is ferromagnetic and metallic, i.e., $d\rho/dT$ is positive, which is usually taken as a definition of a metal in the experimental literature, while the total magnetization is nonzero. The existence of this phase and its explanation in terms of the double-exchange idea motivated most of the work in the early days of Mn oxides. However, note that at temperatures above the Curie temperature, where from resistivity measurements it has been determined that the system is metallic for x larger than 0.26, it turns insulating for smaller hole density (see Fig. 3.12, upper panel). This last property is somewhat puzzling since one would have

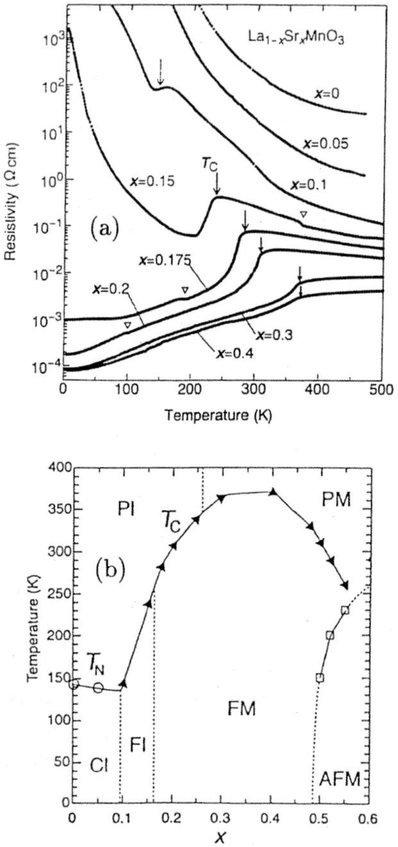

Fig. 3.12. a-b *Upper panel*: Resistivity vs. temperature of $La_{1-x}Sr_xMnO_3$ crystals. The *arrows* indicate the critical temperatures according to measurements of the magnetization. Reproduced from [3.19]. *Lower panel:* Phase diagram of $La_{1-x}Sr_xMnO_3$, courtesy of Y. Tokura and Y. Tomioka, prepared with data from [3.19, 3.20] (see also [3.21]). PM, PI, FM, FI, and CI, denote paramagnetic metal, paramagnetic insulator, FM metal, FM insulator, and spin-canted insulator states, respectively. The AFM phase at large x is what is usually referred to as an A-type AF metal, and it has uniform orbital order. The interpretation of phases such as CI is under much discussion and other explanations, such as mixed-phase states, could be applicable. Both figures are taken from Urushibara et al. [3.19]

expected metallic behavior at high temperature as a precursor of the FM metallic phase to be stabilized at low temperature. This curious behavior is at the heart of the puzzling magnetotransport properties of manganites, and it will be discussed extensively in this book. Also, note the interesting shape of the resistivity curve at $x = 0.15$: a peak exists at the Curie temperature. This peak is characteristic of many lower-bandwidth manganites, and these results show that Sr-based Mn oxides at small hole density are not too different from the Ca-based compounds.

Continuing with the phase diagram, at large hole density an antiferromagnetic state is stabilized. However, note that this state is actually ferromagnetic within planes, and AF only between planes. It is believed that within each plane, the orbital $x^2 - y^2$ is lower in energy, producing a state with a *uniform* orbital order. Such a state is exotic, and illustrates a recursive feature of manganites already addressed in previous sections: not only the spin and charge are active but also the orbital degree of freedom plays an impor-

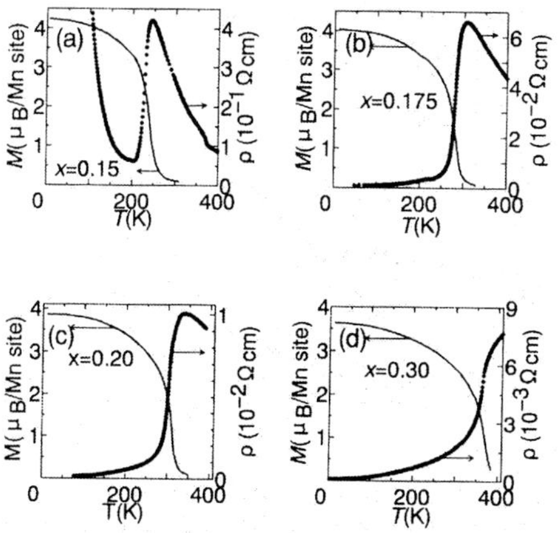

Fig. 3.13. a-d Temperature dependence of the resistivity and magnetic moment of $La_{1-x}Sr_xMnO_3$ at several densities. The magnetization data was taken in a field of 0.5 T. Reproduced from [3.19]. Note the similarity between the many magnetization curves, contrary to the quite different low-temperature behavior of the resistivity, which can be either metallic or insulating

tant role. This is also shown in Fig. 3.13 where spin behavior (magnetization) appears the same at all densities, but the transport is quite different. In the other limit of small hole density, a FM insulating regime was identified using simple transport and magnetic measurements. This state is somewhat puzzling, and analogous FM insulators have been identified in other Mn oxides as well. This state is among the least understood in manganites and more work should be devoted to them. The CI state at very low hole density is a canted insulator, where the word "canted" refers to a state that at the same time has antiferromagnetic and ferromagnetic properties (see Chap. 2). It will be shown in Chap. 6 and others, that such coexistence of FM and AF features may arise from mixed-phase tendencies in the manganite compound, not from a canted state (note that other possible sources of canting, such as the Dzyaloshinksy Moriya [3.22] interaction, are widely considered as negligible in manganites, and will not be discussed further in this book). Moreover, some authors have proposed far more complex phase diagrams in the small hole density region of $La_{1-x}Sr_xMnO_3$, thus the microscopic understanding of that regime is in its early stages at present.

The properties of $La_{1-x}Sr_xMnO_3$ in the presence of a magnetic field are quite interesting, and they illustrate the typical response of manganites to external fields. In Fig. 3.14, results at two typical densities are shown in the presence of fields of a few Teslas. These fields are small in electronic units, but for actual applications in computer devices, unfortunately, they should be considered as high. Both results at $x = 0.15$ and 0.175 show that near the Curie temperature the resistivity peak rapidly changes with the application of a field, which in essence is the large MR effect that has triggered so much work in manganites. A similar effect was found in $La_{1-x}Ca_xMnO_3$, as explained at the beginning of this chapter. For a more quantitative measurement of the effect, the MR value or ratio defined as $-\Delta\rho/\rho(0) = (\rho(0) - \rho(H))/\rho(0)$ is also shown. It is obvious from the figure that the resistivity changes by a large amount upon the application of fields. This occurs *even* in cases such as $x = 0.15$ where apparently there are no metallic ferromagnetic phases, although most of the large MR results reported in the literature involve such a FM metal. In the case of large bandwidth compounds the effect in Fig. 3.14 probably is not as "spectacular" as in lower-bandwidth materials (note that for many other manganites a logarithmic scale is needed for the resistivity), but leaving aside these magnitude considerations the effect is qualitatively the same in $La_{1-x}Sr_xMnO_3$ as in other CMR materials. This suggests a common origin to the effect in all known manganites. Note also that the largest MR ratio is found at the smallest Curie temperature, which itself occurs when the system transforms to another phase upon the modification of some parameter in the problem (in the case at hand, the hole density). Then, a recursive conclusion in the most recent theories to explain CMR is that in order to understand the effect it is *crucial* to incorporate in the considerations the phases that *compete* with the metallic ferromagnet. Information about this crucial point will be discussed extensively in Chap. 17, and in other chapters as well.

3.4 Half-Metallic Ferromagnets

Manganites are compounds that are intrinsically interesting due to their complex behavior, but also for their potentially useful applications. Such applications can be divided into two groups: (i) The CMR phenomenon can be exploited in devices for computers, using Mn oxides as a very sensitive material for reading the small fields of tiny "bits" in a recording medium (for more information, see Chap. 21). Note the disadvantage that the CMR is largest at low temperatures, which is not suitable for large-scale computer applications. (ii) Manganites can be made part of artificially prepared structures, such as spin-valves (Chap. 21), where just the ferromagnetic properties of the compound are of interest. In this last-envisioned application, the notion of "half-metallic ferromagnets" is often invoked.

Half-metallic ferromagnets are systems where spins with a particular direction, let us say "up", have a *partially* occupied band, while the opposite direction, "down", has filled bands separated from the unoccupied ones by a gap. Figure 3.15 shows a simple sketch of the density of states in such a system. For readers that are already familiar with the theoretical ideas discussed in Chap. 2, the concept of a half-metal system is actually quite natural: if the localized spins in the one-orbital model for manganites are up, then conduction electrons with up spins can move freely, while conduction electrons with

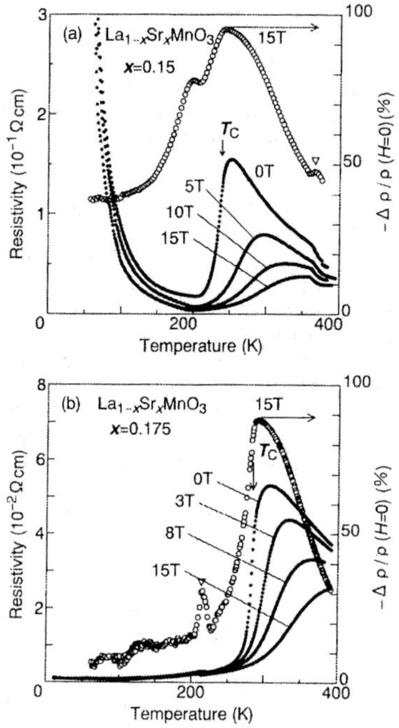

Fig. 3.14. Temperature dependence of resistivity for $La_{1-x}Sr_xMnO_3$ under various magnetic fields. Results are shown for (**a**) $x = 0.15$ and (**b**) $x = 0.175$ [3.19]. The *open circles* are the MR ratio, defined in the text, at a field of 15 T. The Curie temperatures are indicated

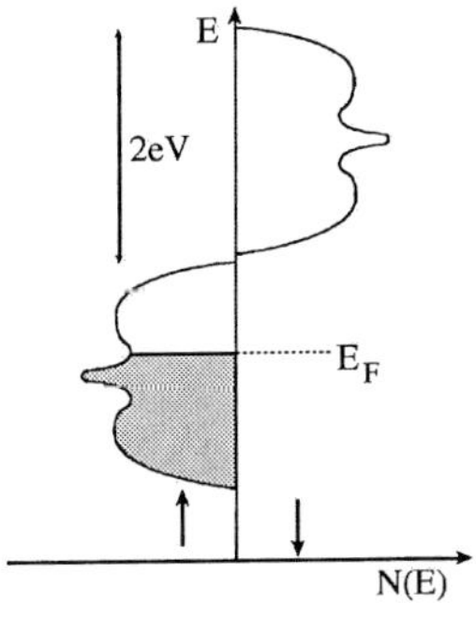

Fig. 3.15. Idealized density of states $N(E)$ for a half-metallic ferromagnet. The chemical potential (Fermi energy) is in the middle of a spin-up band, but in between bands for the spin-down species

down spin cannot readily hop due to the large value of the Hund coupling that prevents their movement. One species is metallic, the other insulating, thus *only half* the available electrons contribute to the conduction. In other words, the conduction electrons at the Fermi level are 100% polarized at low temperature, although in practice only a few materials are close to this ideal total polarization.

Half-metallic systems may find applications in spintronics (Chap. 21), and their properties and history have been reviewed by Pickett and Moodera [3.23]. The basic notion started from the computational work of de Groot

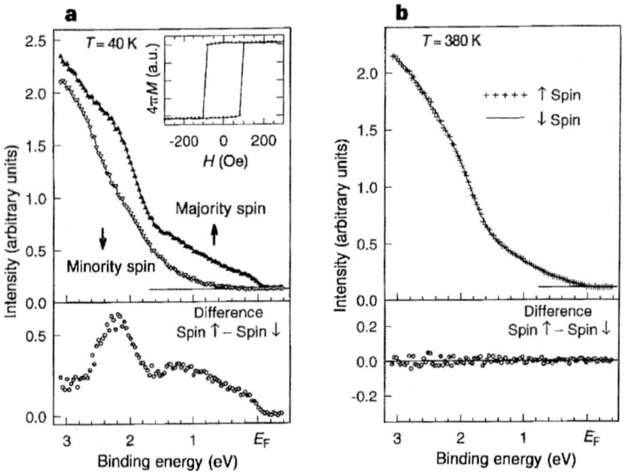

Fig. 3.16. a-b Spin-polarized photoemission spectra of a thin film of $La_{1-x}Sr_xMnO_3$ at $x = 0.3$, near the Fermi energy. Results at $T = 40\,K$ and 380 K, above and below the Curie temperature, are shown. The difference in the low-temperature intensity between up and down spins is clear. From [3.26]

et al. [3.24] applied to Heusler alloys, such as Co_2MnSi. Another widely discussed half-metal ferromagnet is CrO_2, and more recently the double perovskite Sr_2FeMoO_6 [3.25] (with a general chemical formula ABB'O_6, with A an alkaline rare-earth Sr, Ca, or Ba, and B, B' two different transition-metal ions) have also been discussed in this context. Several other half-metal candidates are described in the review article [3.23].

Among the manganites, $La_{1-x}Ca_xMnO_3$ and $La_{1-x}Sr_xMnO_3$ at low temperatures may have half-metallic character, although how close to 100% is the polarization of the conduction electrons is still under debate. Using photoemission techniques (reviewed in Chap. 18), a 90% polarization has been reported in thin films of $La_{1-x}Sr_xMnO_3$ at $x = 0.3$ [3.26]. Figure 3.16 shows the intensity of photoelectrons at low and high temperatures. At low temperature, a gap exists in the spin-down spectrum, while the spin-up band has metallic behavior. It is expected that increasing further the hole density in this compound would lead to an even better half-metal. Currently, several groups are actively working in the use of manganites in multilayer devices, with applications for spintronics as a goal.

However, care must be taken regarding the half-metallic character of the ferromagnetic phase in Sr-based manganites. Optical conductivity measurements in the $x = 0.175$ $La_{1-x}Sr_xMnO_3$ compound [3.27] have shown that the Drude weight is far from optimal, and the effective number of carriers is still changing with temperature even below T_C. There must be scattering mechanisms other than spin acting in the FM phase of this compound. However, its influence could be strongly dependent on x, namely at 0.175 it could be important due to the influence of the nearby insulator. In general,

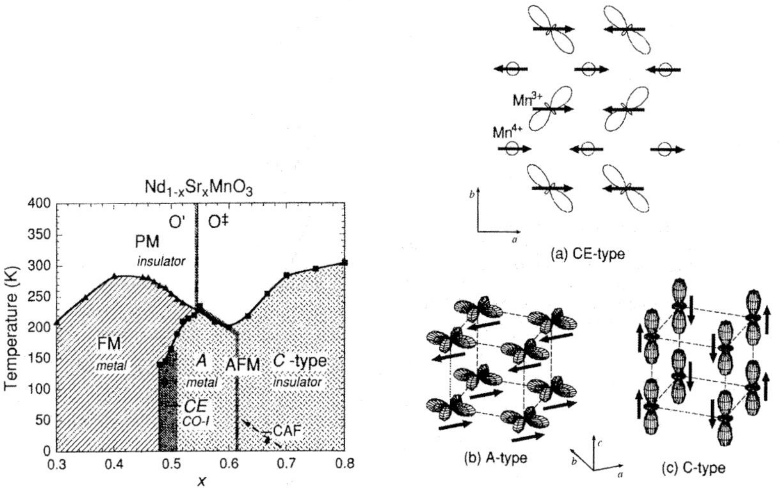

Fig. 3.17. *(Left)* Phase diagram of $Nd_{1-x}Sr_xMnO_3$, from [3.28]. The notation is standard, and more details can be found in [3.28], and references therein. *(Right)* Arrangement of spin, charge, and orbital degrees of freedom in the CE-, A- and C-type states, that appear in the phase diagram of $Nd_{1-x}Sr_xMnO_3$ (from [3.28])

the reader must be alert that the FM phases in manganites are "dirty" in many respects, a recurrent conclusion of the experimental literature reviewed in future chapters.

3.5 Basic Properties of $Nd_{1-x}Sr_xMnO_3$ and $Pr_{1-x}Sr_xMnO_3$

There are manganites "in between" $La_{1-x}Sr_xMnO_3$ and $La_{1-x}Ca_xMnO_3$ regarding their properties near $x = 0.5$. An example is $Nd_{1-x}Sr_xMnO_3$ (NSMO), with the phase diagram shown in Fig. 3.17, from Kajimoto et al. [3.28]. At hole densities smaller than 0.5, the FM metallic phase is stable as in other compounds. In a density window around 0.55, the A-type AF phase is stable, as in $La_{1-x}Sr_xMnO_3$. However, two other phases appear in the phase diagram. One is the CE-state in a very narrow region near $x = 0.5$ (as in LCMO), and the other a C-type insulating phase at large hole density (the shape of these orbital-ordered phases is also depicted in Fig. 3.17).

The actual experimental characterization of the many phases can be carried out using a variety of procedures, but several are just indirect such as searching for lattice distortions (different values of the a, b, and c lattice spacings) compatible with a given state. A more direct probe uses neutron scattering techniques (for a brief review see Chap. 11). If there is long-range order, it is enough to use elastic neutron scattering, and monitor the in-

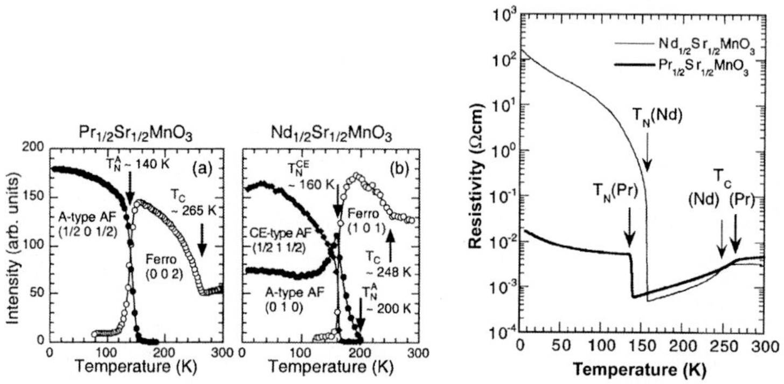

Fig. 3.18. *Left two panels:* Temperature dependence of the FM and AFM Bragg reflections in PSMO and NSMO at $x = 0.5$ [3.29]. *Right panel:* Resistivity vs. temperature of NSMO and PSMO at x=0.5. The transition temperatures are indicated. From [3.30]

tensity of the many peaks related to the possible spin orders, as a function of temperature or densities. Typical results in Fig. 3.18 show data for both $Nd_{1-x}Sr_xMnO_3$ and $Pr_{1-x}Sr_xMnO_3$ (PSMO) at $x = 0.5$ [3.29]. The intensity of the peaks and associated phases are shown. It appears that PSMO has a phase diagram similar to that of NSMO, with the exception of the tiny island of stability of the CE-phase in the latter. This difference shows in the resistivity vs. temperature (Fig. 3.18, right) It is also curious to observe that for NSMO, the CE- and A-type phases appear to coexist.

3.6 Summary of Results for Perovskites

In recent work [3.31], the information corresponding to the several manganites with a perovskite structure described thus far has been "condensed" in a single low-temperature phase diagram, reproduced in Fig. 3.19. This phase diagram nicely illustrates the changes in the ground-state properties as the bandwidth varies. In future chapters, theoretical predictions will be compared against this phase diagram, and a reasonable agreement will be found. It is encouraging that theory and experiments agree in several respects, considering the huge complexity of the Mn oxide compounds. Note that in Fig. 3.19 there is no explicit information about mixed-phase tendencies, which are very important in these materials, at least at intermediate temperatures in the CMR regime.

3.7 Basic Properties of Bilayer Manganites

Manganite compounds with a *layered* structure were first reported by Moritomo et al. [3.33]. This structure can be achieved by inserting rock-salt-type

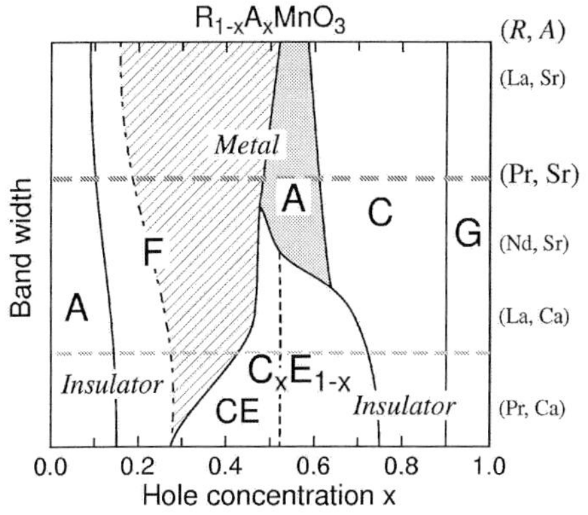

Fig. 3.19. Schematic low-temperature phase diagram of $R_{1-x}A_xMnO_3$ [3.31]. F is the FM state. A, CE, C, and G denote A-type, CE-type, C-type, and G-type AFM states, respectively. The notation C_xE_{1-x}, introduced in [3.32], represents incommensurate charge/orbital ordered states. Note that the F state can be insulating or metallic, depending on the hole density. The mixed-phase tendencies of manganites are not included in this phase diagram

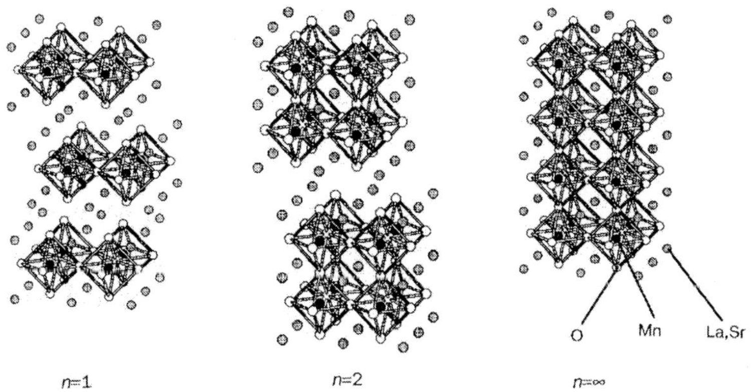

Fig. 3.20. Schematic structures of the perovskites with n equal to 1, 2 and ∞. The *dark circles* are manganese ions. The oxygen octahedra are also shown

block layers ($(La,Sr)_2O_2$) every n MnO_2 sheets. The resulting chemical composition is $(La,Sr)_{n+1}Mn_nO_{3n+1}$. Examples corresponding to $n = 1$ (single layer), $n = 2$ (bilayer) and $n = \infty$ (cubic perovskite) are shown in Fig. 3.20. The hole concentration is controlled by changing the relative amount of La and Sr, leading to $La_{1-x}Sr_{1+x}MnO_4$ for $n = 1$, $La_{2-2x}Sr_{1+2x}Mn_2O_7$ for $n = 2$, and $La_{1-x}Sr_xMnO_3$ for $n = \infty$.

In the pioneering study of Moritomo et al. [3.33] it was found that the bilayer compound at $x = 0.4$ has a Curie temperature of 126 K, separating an

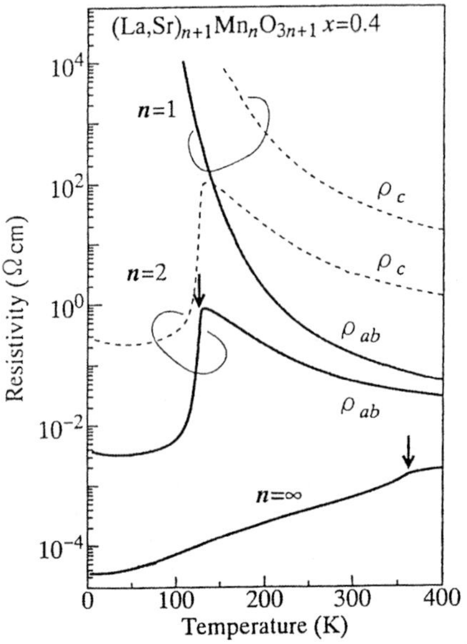

Fig. 3.21. Temperature dependence of the resistivity for layered manganites at nominal hole density $x = 0.4$, with current parallel and perpendicular to the layers. Shown are results at the same composition, but for the cubic perovskite. From [3.33]

insulating state at high temperature, from a ferromagnetic metallic state at lower temperatures. The resistivity of the latter is very high, and it should be considered a "poor" metal. Figure 3.21 shows the resistivity vs. temperature, for n equal to 1, 2, and ∞, for comparison. In the first two cases, there is a natural anisotropy between the (a,b) and c-axis, and their associated resistivities are also different. Note that the tendency toward an insulator state increases as the dimension decreases.

Note. To study the "poor" vs. "standard" character of a metal regarding its resistivity, the Ioffe Regel criterion is often invoked. It establishes the limits of the quasi-classical transport picture, where the ordinary Drude expression for the conductivity $\sigma_0 = (ne^2\tau)/m$ holds in the case of weak scattering. Here $n = k_F^3/3\pi^2$ is the carrier density, k_F is the Fermi wavevector, τ the transport time, and m the effective mass of the carriers. This expression is valid when the electron wavelength $\lambda_F = 2\pi/k_F$ is much smaller than the mean-free path $l = v_F\tau$, with $v_F = \hbar k_F/m$ being the Fermi velocity. This condition implies $k_F l \gg 1$. Using the formula above, the inequality can be transformed into $\sigma_0\lambda_F \gg 5 \times 10^{-5}/\Omega$, where $\hbar/e^2 = 4.1\,\mathrm{k}\Omega$ was used. Defining the resistivity $\rho_0 = 1/\sigma_0$, and using a typical wavelength $\lambda_F = 5\,\text{Å}$, this yields $\rho_0 \ll 10^{-3}\,\Omega$ cm. If the resistivity is larger than this value, then the weak-scattering picture of conductivity is not valid and other effects (such as Anderson localization or percolation, see Chap. 16) may dominate the conductivity. For details and references, see for instance [3.34]. A discussion about the related Mott's maximum metallic resistivity can be found in [3.35].

A very large negative MR effect was observed for the $n = 2$ compound above the Curie temperature. At 129 K, the MR ratio reported reached 200%

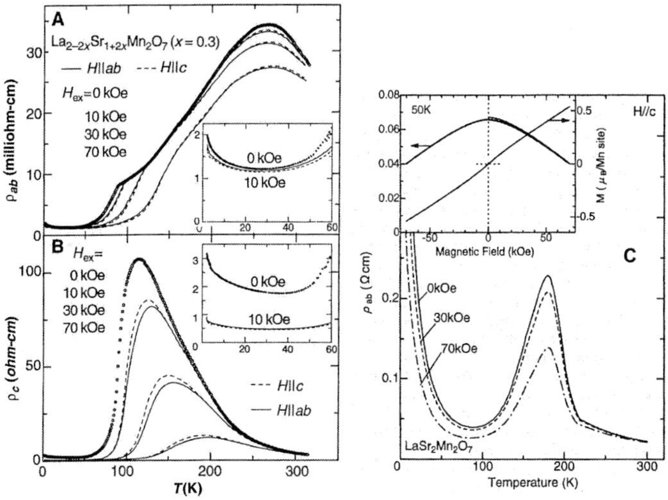

Fig. 3.22. (**a-b**) Temperature dependence of the in-plane ρ_{ab} and out-of-plane ρ_c resistivities, in several fields, for a $x = 0.3$ single-crystal bilayered manganite. The *inset* is an expanded view showing that the effect is more pronounced in ρ_c than ρ_{ab} at 10 kOe [3.36]. (**c**) Temperature dependence of ρ_{ab} in several magnetic fields applied along the z-axis. The crystal is a $x = 0.5$ bilayered manganite [3.38]. The *inset* shows isothermal magnetoresistance and magnetization at 50 K

at 0.3 T, significantly higher than in the equivalent 3D Sr-based compound. The region of substantial changes of resistivity with fields of a few Teslas is very broad, ranging from 100 K to close to room temperature. Later, it will be argued that this range of temperatures is very important in the analysis of the CMR phenomenon, establishing a temperature window where a complex mixed state responsible for the effect is stabilized. In related pioneering work, Kimura et al. [3.36] reported a large MR effect for the $x = 0.3$ bilayer compound, showing that its magneto transport behavior is very different from the $x = 0.4$ compound (see Fig. 3.22a,b). In fact, at $x = 0.3$ the resistivity in-plane and out-of-plane behave qualitatively differently. More information about $x = 0.3$ can be found in [3.37]. Information at $x = 0.5$ in the bilayered manganites is also available. This material undergoes a charge-ordering transition at $\sim$ 210 K. The charge-ordered state is of the $(3x^2 - r^2)/(3y^2 - r^2)$ type, as in the CE-state of cubic perovskites. This state manifests itself as a rapid increase in the resistivity upon cooling, as shown in Fig. 3.22c. However, this state appears to collapse upon further cooling since the resistivity lowers its value, reaching a minimum at $\sim$ 100 K. The $x = 0.5$ compound also exhibits a substantial MR effect, although fields as high as 7 T are not sufficient to entirely destroy the insulating behavior of the charge-ordered state.

A comprehensive study of the phase diagram of bilayer manganites has been presented recently by Ling et al. [3.39]. In this work, results in a wide

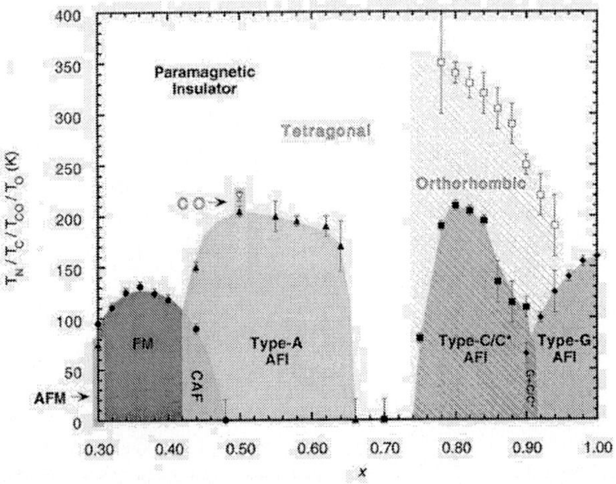

Fig. 3.23. Magnetic and crystallographic phase diagram of $La_{2-2x}Sr_{1+2x}Mn_2O_7$, from Ling et al. [3.39]. *Solid points* represent magnetic transitions, with a notation (partially) explained in the text. *Open points* are crystallographic transitions. For more details the reader should consult [3.39]

range of densities between $x = 0.3$ and 1.0 are reported. The phase diagram was studied using temperature-dependent neutron powder diffraction. The very rich magnetic and crystallographic phase diagram is shown in Fig. 3.23. The many phases observed in this figure are also present in the 3D perovskite manganites, but not in the same location in parameter space as in bilayers. These phases include: (i) an AFM phase only at $x = 0.3$, with an arrangement sketched in Fig. 3.24a. Note once again that AFM spin arrangements do not necessarily mean a staggered pattern in all directions. In this case each bilayer is individually FM, but an overall AFM state is obtained by coupling successive bilayers along the z-axis in an antiferromagnetic form. Thus, locally the system is mainly ferromagnetic. (ii) A three-dimensional FM state appears next in the phase diagram, followed by (iii) an A-type antiferromagnetic state, similar to that reported in $La_{1-x}Sr_xMnO_3$ at a hole density above 0.5, with a dominance of the $x^2 - y^2$ orbitals. Here, the two planes of the bilayer are FM, but their coupling is AF (see Fig. 3.24c).

In between the FM and A-type AF phases, a region (iv) with spin canting was observed (see also [3.40] and [3.41]). Note that the canting of spins occurs between those in different planes of each bilayer, thus, it does not seem directly related to the spin-canted state of deGennes (Chap. 2), but more simply with a two-site effect along the z-axis to increase the electron mobility. In addition, it is possible to imagine canting along the z-axis, coexisting with FM clusters in the basal planes that may produce the poor-metal behavior of the FM phase at smaller densities. In fact, this spin-canted regime can also correspond to a region of coexistence of FM and A-type AF phases in agree-

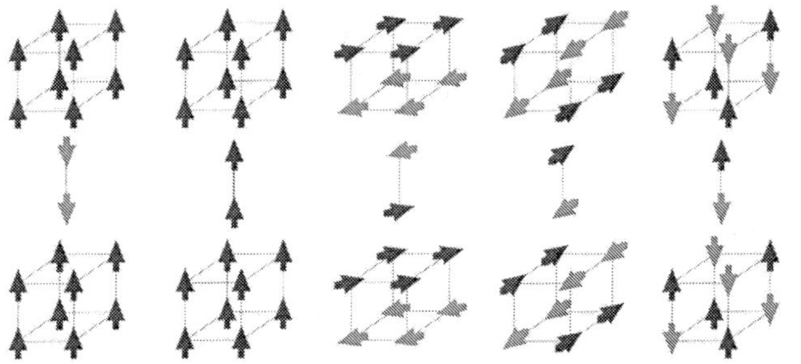

(a) AFM (z) (b) FM (z) (c) Type-A ($x\bar{y}$) (d) Type-C (y) (e) Type-G (z)

Fig. 3.24. a-b Sketch of the spin arrangements in the many phases reported in Fig. 3.23. The reader should consult [3.39] for details

ment with theory described elsewhere in this book (for experimental results along this direction, see [3.42]). More recently, Chun et al. [3.43] reported a detailed study of AC/DC magnetization and longitudinal/transverse transport properties of the $x = 0.4$ bilayer, finding that below 40 K a spin-glass phase is formed. They interpreted their results as supporting a state given by a mixture of FM and AF clusters, in agreement with theoretical expectations. It is clear that more work is needed to clarify the uniform vs. inhomogeneous nature of the "spin-canted" region of bilayer manganites, located between the FM and A-type AF phases. Of particular importance would be the use of experimental techniques that can provide microscopic information, such as NMR (more information is provided in Chaps. 11 and 15). In a narrow region of temperatures and density $x = 0.5$, a charge-ordered CE-type state (v) was identified (Fig. 3.23). The presence of this state may be very important for understanding CMR, as discussed elsewhere in this book, since it appears in the regime where CMR occurs in bilayers.

Continuing with the description of the rich phase diagram of manganites, in a finite density window centered at 0.7 hole doping, it was surprising to observe no indications of long-range spin order, according to Ling et al. [3.39]. This regime may correspond to a complex mixture of two states that are located on the right and left in the phase diagram, namely the A- and C-type AF states. The latter (vi) consists of parallel one-dimensional chains oriented along the b-axis (in plane), with the orbital at Mn^{3+} sites also oriented along that axis ($3y^2 - r^2$). The notation C and C* for the phases refers to the periodicity along the z-axis. Finally, at very large hole density the G-type AF state (vii) is stabilized (with a mixed-phase C–G regime in between). This G-state is believed to exist in a finite window of densities, not only at $x =$ 1.0, with individual e_g electrons forming randomly distributed ferromagnetic clusters, or even stripe structures (see Fig. 13 of [3.39]), with each stripe being a 1D rod that contributes to the C-phase. This interesting picture certainly deserves further work and confirmation. Overall, the complexity of the bilayer phase diagram clearly illustrates one of the main properties of manganites, and many other transition-metal oxides as well, namely the existence of a clear competition among a variety of ordering tendencies. This author believes that this is the main reason for the unusual transport CMR phenomena, as extensively explained in this book.

3.8 Basic Properties of Single-Layer Manganites

The study of single-layer manganites with the composition $La_{1-x}Sr_{1+x}MnO_4$ is important for a variety of reasons. For example, it can be used to compare information against results obtained for other single-layer transition-metal oxides, such as the nickelates and cuprates. In $La_{2-x}Sr_xNiO_4$, it is known that diagonal charge stripes are formed, with a π-shift in the spin AF order parameter across the stripe, while in high-T_c cuprates, such as $La_{2-x}Sr_xCuO_4$, the existence of dynamical stripes is currently under much discussion ([1.3] and references therein). For this reason it is important to clarify the behavior of layered manganese oxide compounds, and compare that against other similar materials. It is also important to contrast results between the single-layer case and the 3D cubic perovskite, to analyze the influence of dimensionality. In fact, starting with a far simpler picture, namely, entirely neglecting electron electron interactions, a two-dimensional gas of electrons on a lattice has a bandwidth ($W = 8t$ on a square lattice) which is smaller than the corresponding bandwidth in 3D (which is $12t$ for a cubic lattice). This suggests that the tendency toward metallicity may be weaker in a layered compound than in a perovskite. This naive expectation is indeed confirmed in the case of manganese oxides since $La_{1-x}Sr_{1+x}MnO_4$ is actually not a metal at any hole density, as discussed below, while $La_{1-x}Sr_xMnO_3$ has a broad metallic regime.

Among the first experimental results in this context, Moritomo et al. [3.44] reported single-crystal data corresponding to the single-layer compound $La_{1-x}Sr_{1+x}MnO_4$ in the broad range $0.0 \leq x \leq 0.7$. Based on susceptibility and magnetization measurements, these authors presented the phase diagram shown in Fig. 3.25a, which contains three regimes: (i) an AF state at small x, (ii) a charge-ordered state (later shown to be of the CE-type) around half-doping, and (iii) a spin-glass intermediate regime. In Fig. 3.25b, the resistivity is shown. At all densities, an insulating behavior is observed. In the regime where a ferromagnetic state could have been expected, based on the behavior of 3D (La,Sr)-based compounds, the resistivity is indeed the lowest. For example, at 150 K the resistivity at $x = 0.4$ is orders of magnitude smaller than at $x = 0$ or 0.5. However, this tendency is not sufficient to induce metallic behavior. The same can be said about the magnetic susceptibility, shown in Fig. 3.25c. At hole densities 0.2, 0.3, and 0.4, there is a tendency to become ferromagnetic since the susceptibility is large. However, the system does not

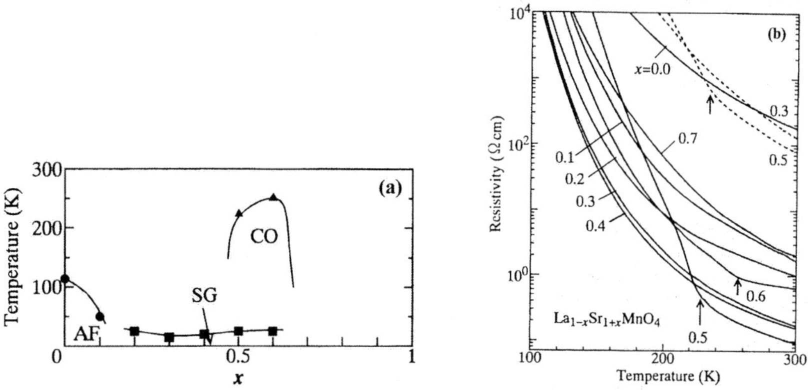

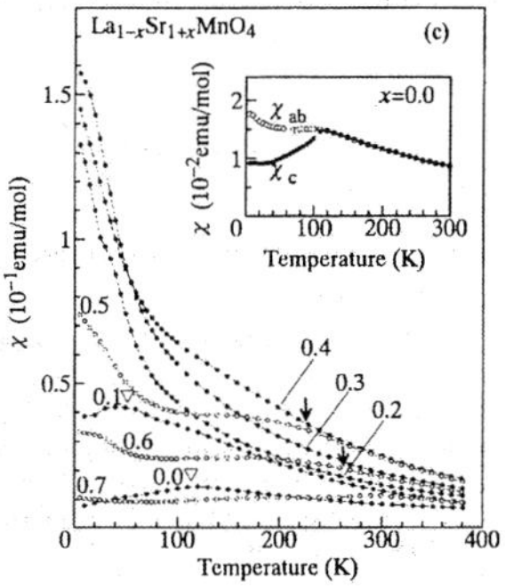

Fig. 3.25. (**a**) Phase diagram, from Moritomo et al. [3.44], of $La_{1-x}Sr_{1+x}MnO_4$. AF, SG, and CO stand for antiferromagnetic, spin-glass, and charge-ordering phases, respectively. (**b**) Temperature dependence of resistivity for single crystals of $La_{1-x}Sr_{1+x}MnO_4$ at various hole concentrations (from Moritomo et al. [3.44]). *Solid* and *broken* curves are for in-plane and out-of-plane components, respectively. *Arrows* indicate steep increases signaling charge ordering temperatures. (**c**) Temperature dependence of the magnetic susceptibility, with field parallel to the c-axis. Further details can be found in Moritomo et al. [3.44]. The *inset* is the result at zero hole density, showing the anisotropy in the susceptibility

develop a true ferromagnetically ordered ground state, at least for the samples analyzed thus far. In the other limit of large hole density, results by Bao et al. [3.45] suggest the existence of mixed-phase tendencies close to $x = 1$. This interesting region of the phase diagram should be further investigated in view of the current wide interest in mixed-phase ideas to explain the behavior of manganese oxides.

Using elastic neutron scattering techniques applied to the half-doped $x = 0.5$ single-layer manganite, Sternlieb et al. [3.46] reported a charge ordering temperature $T_{CO} \sim 217\,K$, where the charges form a checkerboard alternating Mn^{3+}/Mn^{4+} pattern, and a Néel temperature considerably lower at $T_N \sim 110\,K$, leading to the CE-type spin arrangement known to exist in cubic perovskites. The experimental confirmation of orbital ordering occurring together with charge ordering at T_{CO} was provided by Murakami et al. [3.47] using synchrotron X-ray diffraction techniques, by analyzing the anisotropy of the susceptibility tensor. The pattern of orbital order is compatible with the CE-state expected to be stable at $x = 0.5$ from previous considerations.

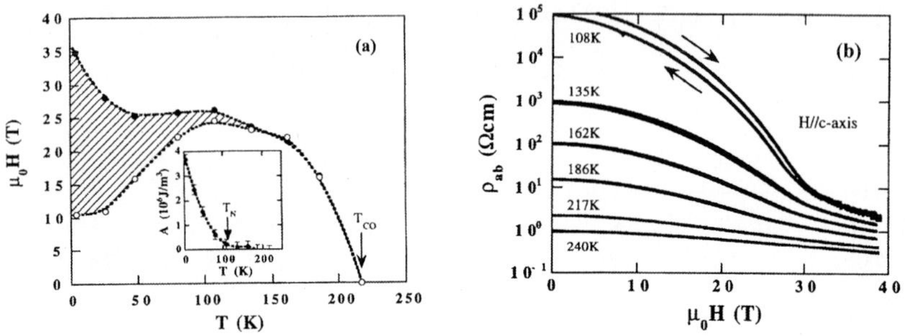

Fig. 3.26. (**a**) Phase diagram of $La_{1-x}Sr_{1+x}MnO_4$ at x=0.5 in the magnetic field H and temperature T plane. The *closed* and *open points* are determined through the inflection point of the magnetization H curves in field-increasing and -decreasing processes, respectively. The *hatched area* is the hysteresis region. The *inset* shows the area of hysteresis in a M–H curve versus temperature. For more details, see Tokunaga et al. [3.48]. (**b**) In-plane resistivity vs. magnetic field along the c-axis

Although there is no ferromagnetic phase in Fig. 3.25a, such a phase can be induced upon the application of magnetic fields. Tokunaga et al. [3.48] studied the half-doped single-layer manganite and observed that at low temperatures a field of 38 T is sufficient to create a fully polarized state. The transition between the charge-ordered CE-state at small fields and the polarized high-field state is of first order. This is illustrated in the phase diagram of Fig. 3.26a, that separates the low-field CE-state regime from the polarized state at high fields. The hatched area is a hysteretic region, compatible with a first-order transition. The in-plane resistivity vs. magnetic field is shown in Fig. 3.26b. Although the system remains an insulator even at high fields, a reduction in the resistivity by many orders of magnitude can be observed. For example, at 108 K the reduction in ρ_{ab} between 0 T and 30 T is more than a factor of 10^4. Huge MR factors can be obtained even without a ferromagnetic metallic metal (at least, an "obvious" FM metallic metal) in the system.

4. Preliminary Theoretical Considerations: Coulombic and Jahn Teller Effects

In this chapter, we start the theoretical study of manganites and related compounds. Instead of directly defining semi phenomenological models, such as the double-exchange model, as frequently done in the manganite literature, we will describe the foundations of those models in detail. This includes the analysis of Coulombic interactions among the five d-levels of relevance for manganites, as well as the Jahn Teller distortions that also are very important in Mn oxides and other materials. The main ideas and calculations presented in this chapter can be applied without many modifications to several compounds, and thus this chapter fulfils one of the goals of the book: using the manganites as a focal point, a variety of general concepts and ideas will be presented that are useful far beyond the Mn oxides.

4.1 Basic Multi Orbital Model and On Site Coulomb Interactions

In this section, realistic models for manganites and related compounds will be constructed, following a standard procedure. The Coulombic interactions will be considered only within a given manganese ion, including in the formulas the five $l = 2$ orbitals of relevance to problems of manganites. Coulombic interactions of a longer range can be easily added to the resulting models if needed. Electrons move from ion to ion through tunneling (or "hopping") between Wannier states localized at nearest-neighbor ions. Although the formalism allows us to write down the resulting couplings as well-defined integrals of Wannier functions, it is common practice to let those parameters be unknown constants to be determined by experiments. The procedure starts with the well-known Hamiltonian in second quantization that corresponds to a general problem of interacting electrons. Following standard texbooks, such as the one of Fetter and Walecka [4.1], it can be shown that the Hamiltonian for the general problem is

$$\hat{H} = \hat{T} + \hat{V} = \int \mathrm{d}^3\mathbf{r}\ \hat{\psi}^\dagger(\mathbf{r}) H_0(\mathbf{r}) \hat{\psi}(\mathbf{r}) + \frac{1}{2} \int\int \mathrm{d}^3\mathbf{r}\ \mathrm{d}^3\mathbf{r}'\ \hat{\psi}^\dagger(\mathbf{r})\hat{\psi}^\dagger(\mathbf{r}') V(\mathbf{r},\mathbf{r}') \hat{\psi}(\mathbf{r}')\hat{\psi}(\mathbf{r})\,, \tag{4.1}$$

where $H_0(\mathbf{r}) = -(\hbar^2/2m)\nabla^2$ (an on site potential $V(\mathbf{r})$ with the symmetry of the lattice could also have been part of H_0), and $V(\mathbf{r},\mathbf{r}')$ is the interaction between two electrons located at coordinates $\mathbf{r}$ and $\mathbf{r}'$. The next step in this general procedure is to write the field operator $\hat{\psi}(\mathbf{r})$ in a complete basis. For the lattice problem considered here, the expansion is

$$\hat{\psi}(\mathbf{r}) = \sum_{\mathbf{i},\gamma,\sigma} w_{\mathbf{i}\gamma}(\mathbf{r}) c_{\mathbf{i}\gamma\sigma} \,, \tag{4.2}$$

where $\mathbf{i}$ denotes lattice sites (where each ion is located), γ runs over the active orbitals at each site, σ denotes spin up and down, $w_{\mathbf{i}\gamma}(\mathbf{r})$ is a Wannier function centered at site $\mathbf{i}$, and $c_{\mathbf{i}\gamma\sigma}$ is the destruction operator for an electron at site $\mathbf{i}$ with the quantum numbers γ and σ (for simplicity, the hat is dropped from these operators from now on). A good introduction to Wannier functions can be found in the book of Ashcroft and Mermin [3.3]: it is sufficient here to recall that these functions at different sites are orthogonal, and that they are fairly localized at the site in which they are defined. The atomic orbitals $\phi_\gamma(\mathbf{r})$ used in the definition of the Wannier functions (4.2) are those of relevance for the case of the transition-metal ions studied here (for details see [3.3]). They are the product of the radial function $R_{32}(r)$ (in the usual hydrogenic atom notation, and with $r = |\mathbf{r}|$) multiplied by suitable (and real) combinations of spherical-harmonics that transform as the functions we are interested in, namely $3z^2 - r^2$ (denoted by $\gamma = 0$ for simplicity), $x^2 - y^2$ ($\gamma = 1$), xy ($\gamma = 2$), yz ($\gamma = 3$), and zx ($\gamma = 4$). These spherical harmonic combinations are, respectively, Y_2^0, $(Y_2^2 + Y_2^{-2})/\sqrt{2}$, $(Y_2^2 - Y_2^{-2})/\sqrt{2}$, $(Y_2^1 + Y_2^{-1})/\sqrt{2}$, and $(Y_2^1 - Y_2^{-1})/\sqrt{2}$.

Replacing (4.2) into the kinetic energy portion of (4.1),

$$\hat{T} = \sum_{\mathbf{j},\mathbf{j}',\gamma,\gamma',\sigma} t_{\mathbf{j}\mathbf{j}'}^{\gamma\gamma'} c_{\mathbf{j}\gamma\sigma}^{\dagger} c_{\mathbf{j}'\gamma'\sigma} \,, \tag{4.3}$$

where the hopping matrix element (or hopping amplitude) is defined as

$$t_{\mathbf{j}\mathbf{j}'}^{\gamma\gamma'} = \int \mathrm{d}\mathbf{r}^3 w_{\mathbf{j}\gamma}^{*}(\mathbf{r}) w_{\mathbf{j}'\gamma'}(\mathbf{r}) \,. \tag{4.4}$$

The Wannier function at $\mathbf{j}$ is fairly localized at that site, and for this reason it is common practice to restrict the hopping amplitude to nearest-neighbor ions (although in some cases longer hopping processes are needed). In addition, in the absence of disorder or site-dependent external potentials that break translational invariance, the hopping depends only on the distance between sites $\mathbf{j}$ and $\mathbf{j}'$. Then, the hopping matrix elements $t_{\mathbf{j}\mathbf{j}'}^{\gamma\gamma'}$ can be replaced by $t_{\mathbf{a}}^{\gamma\gamma'}$, where $\mathbf{a}=\pm\hat{x}, \pm\hat{y}, \pm\hat{z}$ is a unit vector along one of the axes of the lattice.

Regarding the potential term, in principle it should involve a sum over four lattice indices once each wave function operator is expanded in a complete basis, as in (4.2). However, it is widely assumed that the strength of the Coulomb potential between electrons located in the vicinity of different ions is strongly reduced by screening and polarization effects. This assumption is not necessarily correct in all cases, but it is quite common in the literature. For this reason only one sum over lattice sites survives, namely just the Coulombic effects within a given ion are considered (a special case sometimes known as the Hubbard approximation). However, the four sums over orbitals at each ion must certainly be kept in the interaction. Under these approximations, and replacing the Wannier functions by the atomic orbitals at a given ion (due to their localized character), it can be shown that

$$\hat{V} = \frac{1}{2}\sum_{\mathbf{i}}\sum_{\gamma\beta\gamma'\beta'}\sum_{\sigma\sigma'} U(\gamma\beta;\gamma'\beta')c^{\dagger}_{\mathbf{i}\gamma\sigma}c^{\dagger}_{\mathbf{i}\beta\sigma'}c_{\mathbf{i}\beta'\sigma'}c_{\mathbf{i}\gamma'\sigma}\,, \tag{4.5}$$

where

$$U(\gamma\beta;\gamma'\beta') = \int\int \mathrm{d}^3\mathbf{r}\mathrm{d}^3\mathbf{r}'\phi_{\gamma}(\mathbf{r})\phi_{\beta}(\mathbf{r}')V(\mathbf{r},\mathbf{r}')\phi_{\gamma'}(\mathbf{r})\phi_{\beta'}(\mathbf{r}')\,, \tag{4.6}$$

and it has been explicitly assumed that the orbitals of relevance in the problem of interest here are real, and that the potential is spin independent. The number of matrix elements $U(\gamma\beta;\gamma'\beta')$ is, in principle, $5^4 = 625$ in manganites, but the number of independent ones is much smaller. For instance, it is clear that $U(\gamma\beta;\gamma'\beta') = U(\gamma'\beta;\gamma\beta') = U(\gamma\beta';\gamma'\beta) = U(\gamma'\beta';\gamma\beta)$. In addition, it can be shown that matrix elements containing an odd number of orbitals of a certain symmetry type vanish. Then, the actual independent matrix elements are

$$\begin{aligned} U &= U(\gamma\gamma;\gamma\gamma)\,, \\ U' &= U(\gamma\beta;\gamma\beta) \quad (\gamma \neq \beta)\,, \\ J &= U(\gamma\beta;\beta\gamma) \quad (\gamma \neq \beta)\,, \end{aligned} \tag{4.7}$$

which are known as the "Kanamori parameters" [4.2]. Note that the matrix element $J' = U(\gamma\gamma;\beta\beta)$ $(\gamma \neq \beta)$ is associated to a pair-hopping amplitude in the actual potential term (4.5), which is physically different from the role played by other matrix elements in the same potential. This motivates the frequent consideration in the literature of J' as a possible extra parameter. However, from its definition the reader can easily check that $J = J'$, and that the number of independent matrix elements remains three. In fact, a relation to be derived below will reduce this number to just two.

In order to calculate the matrix elements, it greatly reduces the complexity of the problem to use a description that neglects crystal-field effects, namely one that considers the ion as if it were in isolation (note that crystal-field effects are very important in manganites, but temporarily it can be neglected to estimate the on site Coulomb interactions, which are typically the largest energy scale in the problem). In such a case the potential to be used in the calculation of matrix elements is rotationally invariant (Coulombic or screened Coulombic, for instance). Such a function can be expanded in spherical harmonics (see, for instance, Sect. 58 of [4.1], or any other elementary text on quantum mechanics),

$$V(\mathbf{r},\mathbf{r}') = V(|\mathbf{r}-\mathbf{r}'|) = \sum_{l,m} V_l(r,r') Y_l^{m*}(\hat{\mathbf{r}'}) Y_l^m(\hat{\mathbf{r}}) \,, \tag{4.8}$$

where Y_l^m are the spherical harmonics, with $m = -l, -l+1, ..., l-1, l$, $r = |\mathbf{r}|$, and $r' = |\mathbf{r}'|$. The radial functions $V_l(r,r')$ depend on the particular potential used. The unit vectors $\hat{\mathbf{r}}$ and $\hat{\mathbf{r}'}$ in the spherical harmonics indicate the pair of angles that are being considered in each one.

To explicitly obtain the value of the first Kanamori parameter U, we can choose any of the $l = 2$ orbitals we are interested in, since all orbitals lead to the same result (this is not obvious at first sight). For our purposes, the orbital leading to the simplest calculation is $\gamma = 0$, namely when the orbital of the $3z^2 - r^2$ type is used. In this case, U becomes,

$$\begin{aligned} U &= U(0,0;0,0) \\ &= \sum_l U_l \sum_m \left[\int \mathrm{d}\Omega' Y_2^{0*} Y_l^{m*} Y_2^0 \right] \left[\int \mathrm{d}\Omega Y_2^{0*} Y_l^m Y_2^0 \right] \,, \end{aligned} \tag{4.9}$$

where the integral over Ω' (Ω) involves integration over the angles related to the coordinate $\mathbf{r}'$ ($\mathbf{r}$), and we have introduced the definition

$$U_l = \int_0^\infty \mathrm{d}r r^2 \int_0^\infty \mathrm{d}r' r'^2 V_l(r,r') R_{32}^2(r) R_{32}^2(r') \,. \tag{4.10}$$

Expressing the product of two spherical harmonics in terms of the Clebsch Gordan (CG) series (4.24), as well as a table of CG coefficients, both integrals over angles can be carried out. It can be shown that the allowed values for l are only 0, 2, and 4. The explicit result is

$$U = \frac{1}{4\pi} \left(\frac{2}{7} \right)^2 [U_0 + 5U_2 + 9U_4] \,. \tag{4.11}$$

It is left as an exercise to the reader to show that the same result for U is obtained using other d-shell orbitals, namely using $\gamma = 1$ to 4. Equation (4.11) is usually expressed in terms of Racah parameters (see [4.3–4.5]), but that extra step is not needed for the present discussion. Regarding U', its value can also be obtained from a special case. For instance, if we are interested in the subspace involving $\gamma = 0$ and 1, which forms the "e_g sector" after the introduction of the crystal-field splitting, then

$$\begin{aligned} U' = U(0,1;0,1) &= \sum_l \frac{U_l}{2} \sum_m \left[\int \mathrm{d}\Omega' Y_2^{0*} Y_l^{m*} Y_2^0 \right] \\ &\times \left[\int \mathrm{d}\Omega (Y_2^2 + Y_2^{-2})^* Y_l^m (Y_2^2 + Y_2^{-2}) \right] \,. \end{aligned} \tag{4.12}$$

It is clear that m must be zero to avoid the cancellation of the first integral. Carrying out the Clebsch Gordan coefficient manipulations, the result is

$$U' = \frac{1}{4\pi} \left(\frac{2}{7} \right)^2 \left[U_0 - 5U_2 + \frac{3}{2} U_4 \right] \,. \tag{4.13}$$

Note that unlike U, the integral U' is different if either γ or β are different from 0 or 1. More information about the explicit values of U' beyond the e_g-sector under consideration here, can be found in [4.5].

The procedure followed thus far can be repeated for the last integral of relevance, namely J, using

$$J = U(0,1;1,0) = \sum_l \frac{U_l}{2} \sum_m \left[\int \mathrm{d}\Omega' Y_2^{0*} Y_l^{m*} (Y_2^2 + Y_2^{-2}) \right] \times \left[\int \mathrm{d}\Omega (Y_2^2 + Y_2^{-2})^* Y_l^m Y_2^0 \right] . \tag{4.14}$$

The result is

$$J = \frac{1}{4\pi} \left(\frac{2}{7} \right)^2 \left[5U_2 + \frac{15}{4} U_4 \right] . \tag{4.15}$$

Once again, (4.15) is only valid for orbitals of the e_g-sector, i.e., if other orbitals of the $l = 2$ set are used, the results for U' and J will be different (but those for U remain the same for the entire $l = 2$ manifold). The values of U' and J for orbitals of the "t_{2g} sector", or when mixing e_g and t_{2g} subspaces, are provided in [4.5], with results written in terms of Racah parameters. For our purposes below, it is sufficient to know that J is always positive for all combinations of the orbitals involved. This favors the alignment of spins in non filled $l = 2$ shells, effectively leading to the well-known Hund's rule of atomic physics.

A very important observation arising from the explicit forms of U, U', and J in terms of the integrals U_0, U_2, and U_4, is that the former parameters are *not* independent. In fact, using (4.11, 4.13, and 4.15) it can be shown that

$$U = U' + 2J , \tag{4.16}$$

which is an important widely quoted, but rarely derived, relation that reduces by one the number of parameters in calculations. Remember also that calculations of U, U', and J usually have large error bars and, as a consequence, it is common practice to simply consider them as free parameters to be determined from experiments, although still using (4.16).

Considering the information gathered in this section, the potential term $\hat{V}$ in (4.5) specialized to the case of manganites, and including the electron electron repulsion within the same ion (but still without considering lattice effects), can be explicitly written. If the orbitals are restricted to those of the e_g sector formed by $3z^2 - r^2$ and $x^2 - y^2$, namely $\gamma = 0$ and 1 in our convention, then we can use just one value of U' and J as in (4.12, 4.14) and the Hamiltonian is given below in (4.17). On the other hand, if all the five orbitals must be included, then U' and J (but not U) should carry a pair of orbital indices. The Hamiltonian in the e_g sector, or for the full $l = 2$ subspace if supplemented by the crude but qualitatively correct extra approximation of considering U' and J to be orbital independent, becomes

$$\hat{V} = \frac{U}{2}\sum_{\mathbf{i}}\sum_{\gamma,\sigma\neq\sigma'} n_{\mathbf{i}\gamma\sigma}n_{\mathbf{i}\gamma\sigma'} + \frac{U'}{2}\sum_{\mathbf{i}}\sum_{\sigma,\sigma',\gamma\neq\gamma'} n_{\mathbf{i}\gamma\sigma}n_{\mathbf{i}\gamma'\sigma'}$$
$$+\frac{J}{2}\sum_{\mathbf{i}}\sum_{\sigma,\sigma',\gamma\neq\gamma'} c^{\dagger}_{\mathbf{i}\gamma\sigma}c^{\dagger}_{\mathbf{i}\gamma'\sigma'}c_{\mathbf{i}\gamma\sigma'}c_{\mathbf{i}\gamma'\sigma}$$
$$+\frac{J'}{2}\sum_{\mathbf{i}}\sum_{\sigma\neq\sigma',\gamma\neq\gamma'} c^{\dagger}_{\mathbf{i}\gamma\sigma}c^{\dagger}_{\mathbf{i}\gamma\sigma'}c_{\mathbf{i}\gamma'\sigma'}c_{\mathbf{i}\gamma'\sigma}\,, \quad (4.17)$$

where $n_{\mathbf{i}\gamma\sigma} = c^{\dagger}_{\mathbf{i}\gamma\sigma}c_{\mathbf{i}\gamma\sigma}$ (remember that $J = J'$ and $U = U' + 2J$). The first term in (4.17) corresponds to a standard Hubbard-like repulsion for electrons in the same site and orbital. The second term of (4.17) is also a repulsive interaction between electrons in the same ion, but when they are in different orbitals. The third term plays the role of a *ferromagnetic* coupling between electrons in different orbitals, which enforces in the system the standard first Hund's rule. The last term appears exotic, and it is sometimes neglected entirely although mathematically $J = J'$. It corresponds to hopping of pairs of electrons between two orbitals of the same ionic site. Since U is typically the largest scale in the problem, double occupancy of a given orbital is suppressed and the pair-hopping term is heavily penalized. This pair hopping rarely plays an important role.

To understand better the Hund coupling term, at an arbitrary site $\mathbf{i}$ and for a fixed pair of different orbitals γ and γ', let us carry out the explicit calculation of the effect of $(J/2)\sum_{\sigma,\sigma'} c^{\dagger}_{\mathbf{i}\gamma\sigma}c^{\dagger}_{\mathbf{i}\gamma'\sigma'}c_{\mathbf{i}\gamma\sigma'}c_{\mathbf{i}\gamma'\sigma}$, on the states $|\uparrow\uparrow\rangle$, $|\uparrow\downarrow\rangle$, $|\downarrow\uparrow\rangle$, $|\downarrow\downarrow\rangle$, where the first spin refers to γ and the second to γ'. The calculation results in a 4×4 Hamiltonian, with a spin-triplet eigenstate of energy $-J/2$ and a singlet of energy $J/2$. Then, clearly this portion of the full Hamiltonian aligns the spins. As usually explained in rationalizing the Hund's rule, a spin-triplet forces the space portion of the wave function to be antisymmetric with nodes spreading the charge apart, reducing the effect of the Coulomb repulsion.

Regarding the kinetic-energy operator ($\hat{T}$ of (4.1)), its form in second quantization language is

$$\hat{T} = \sum_{\mathbf{j},\mathbf{a},\gamma,\gamma',\sigma} t^{\mathbf{a}}_{\gamma\gamma'} c^{\dagger}_{\mathbf{j}\gamma\sigma}c_{\mathbf{j}'\gamma'\sigma}\,, \quad (4.18)$$

with $\mathbf{j}' = \mathbf{j}+\mathbf{a}$, within the approximation of considering hopping only between nearest-neighbors ions, and assuming that the system has translational invariance in multiples of the lattice spacing. The actual values of the hopping amplitudes certainly depend on the orbitals involved in the problem, and for the case of manganites they will be explicitly evaluated in Chap. 5.

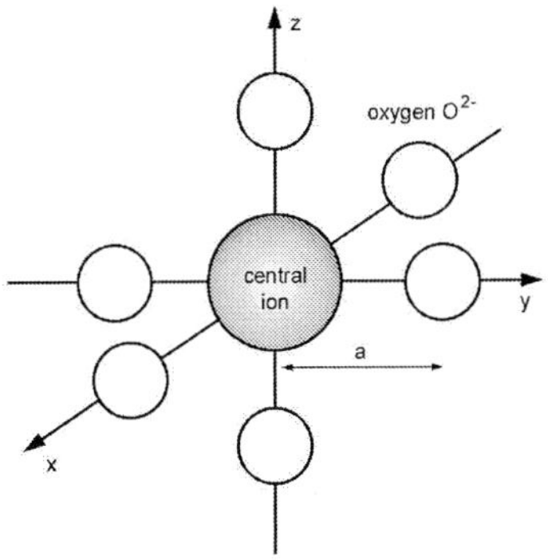

Fig. 4.1. Schematic representation of the crystal environment in which transition-metal ions, such as Mn, are immersed

4.2 Crystal Field Splitting and Jahn Teller Effect

After considering purely Coulombic interactions in the previous section, with ions in a fixed regular position, let us now address the Jahn Teller effect, which is caused by the motion of ions to reduce the total energy.

4.2.1 Perturbing Potential due to the Crystal

Considered in isolation, the ions of Mn, Cu, or other elements in the transition-metal row of the Periodic Table have an active d-shell with five degenerate levels. The degeneracy is present due to rotational invariance within the angular momentum $l = 2$ subspace, as we learn in elementary quantum mechanics. In empty space, any direction or axis is the same as any other. The situation drastically changes once the ions are part of a crystal structure, since now the directions of the crystal axes are certainly special compared with other directions. Full rotational invariance is lost, but a subgroup remains. This effect results in a particular splitting of the five $l = 2$ levels (called the *crystal-field splitting*), which is very important for the physics of transition-metal oxides. The calculation showing how the splitting occurs is conceptually simple for anyone with a background in quantum mechanics and group theory, and the results are frequently quoted in the literature. However, it is rare to find in textbooks the calculation carried out in sufficient detail to be easily followed by readers, unless specialized books devoted to the subject (probably with too many details) are consulted. In addition, the notations are somewhat confusing among different books. For these reasons, here the crystal-splitting calculation for the case of Mn is presented starting from scratch, and ending in the famous splitting "$10Dq$" that most experts frequently quote, but rarely derive [4.4].

Intuitively, the result of the effect of the crystal environment on the d-levels is clear. By symmetry, the d_{xy}, d_{yz}, and d_{zx} orbitals discussed in the introduction must all be affected similarly, if the three axes (x, y, z) are equivalent (namely in a cubic environment). Thus, they must form a triplet at the end of the calculation. Regarding the other two orbitals, at first sight they appear different enough that it may be tempting to conclude that they will suffer separate splitting. However, $(3z^2 - r^2)$ can be rewritten as $(z^2 - x^2) + (z^2 - y^2)$ and in this respect it is natural that it will be degenerate, forming a doublet with $(x^2 - y^2)$. In addition, the doublet must have a higher energy than the triplet since the orbitals of the doublet point along the directions where the negative oxygen ions are located, thus increasing the energy due to Coulombic repulsion. But the explicit value of the crystal-field-induced doublet triplet gap can only be obtained after an explicit quantum-mechanical study. The actual calculation is usually carried out under the assumption that the electric field caused by the charges that surround the ion under consideration are sufficiently small, that they can be considered as a perturbation. This is a crude approximation, but it can be used as a starting point to illustrate the principles involved in the calculation. At the lowest order in degenerate perturbation theory the problem reduces to finding the matrix elements of the perturbing Hamiltonian in the basis of degenerate states, which then can be easily diagonalized. Let us start the calculation considering simply one electron in the d-shell, which is valid for d^1 elements. The case of manganese with a configuration d^4, i.e., four active electrons in the d-shell of the Mn^{3+} ion, can be easily obtained from the d^1 case, as discussed later. For just one electron, the unperturbed basis to be used is $R(r)_{(n=3,l=2)} Y_2^m(\theta, \phi)$, following the standard quantum-mechanical notation for the hydrogen atom. The quantum number m takes the values -2, -1, 0, 1, and 2, and the corresponding five spherical harmonics Y_2^m are the important ingredients of the basis, since they carry the angular dependence. Note that the electrons of Mn with principal quantum number $n = 1$ and 2 (filled shells) are neglected in the problem, since it is assumed that the perturbation acting on closed shells is not important. Note also that it is implicitly assumed that the wave functions of the "active" electrons (those of the partially filled shells) are not dramatically distorted by the crystal environment. Those electrons in the compounds of interest here remain in a fairly localized state centered at their ions, even in a crystal. In addition, issues related to the relative strength of the crystal-field effects vs. the spin-orbit coupling must be addressed, as discussed later in this section.

The perturbing potential is produced by the six oxygen ions of charge Zq (typically $-2q$, if q is defined positive), located at a distance a from the center of the Mn ion (see Fig. 4.1). The distance between an electron at (x, y, z) and a perturbing charge at, e.g., $(a, 0, 0)$ is $\sqrt{(x-a)^2 + y^2 + z^2}$, which can be rewritten as $\sqrt{r^2 + a^2 - 2ax}$, with $r^2 = x^2 + y^2 + z^2$. Then, the potential due to the six charges becomes $V = V_x + V_y + V_z$, with

$$\frac{V_x(x,y,z)}{Zq} = \frac{1}{\sqrt{r^2 + a^2 - 2ax}} + \frac{1}{\sqrt{r^2 + a^2 + 2ax}} . \tag{4.19}$$

The other two terms in the potential are obtained by replacing x by y and z, in the obvious way. Next, it is assumed that the electron can only be located close to the origin, allowing for an expansion in powers of x/a in (4.19). Following Hutchings [4.6], V_x can be rewritten as

$$\frac{V_x(x,y,z)}{Zq} = \frac{1}{\sqrt{A}}\left[\frac{1}{\sqrt{1-X}} + \frac{1}{\sqrt{1+X}}\right], \tag{4.20}$$

where $X = 2ax/A$, and $A = r^2 + a^2$. Using the Taylor expansion $1/\sqrt{1+X} \approx 1-(1/2)X+(3/8)X^2-(5/16)X^3+(35/128)X^4$, it can be easily shown that

$$\frac{1}{\sqrt{1+X}} + \frac{1}{\sqrt{1-X}} \approx 2 + \frac{3}{4}X^2 + \frac{35}{64}X^4. \tag{4.21}$$

Using again the expansion of $1/\sqrt{1+X}$ quoted above but applied to $1/\sqrt{A}$, and expanding in powers of r^2/a^2, we arrive to $(1/\sqrt{A}) \approx 1/a[1-(r^2/2a^2)+(3r^4/8a^4)]$. Collecting results for V_x, V_y and V_z,

$$\frac{V(x,y,z)}{Zq} \approx \left[\frac{6}{a} + \frac{35}{4a^5}\left(x^4+y^4+z^4-\frac{3}{5}r^4\right)\right]. \tag{4.22}$$

Note that for our purposes it was sufficient to carry out an expansion up to fourth order in the coordinates. Other more complicated problems need the potential up to order six (see Hutchings [4.6]). Note also that the constant contribution in (4.22) does not lead to level splittings, and it will not be included in the rest of the discussion.

The next step in our calculation is to write the perturbing potential in terms of spherical harmonics, which will simplify the task of evaluating the matrix elements in the basis $\{Y_2^m(\theta,\phi)\}$. For this purpose, the spherical coordinate representation of the cartesian coordinates $x = r\sin\theta\cos\phi$, $y = r\sin\theta\sin\phi$, and $z = r\cos\theta$ is used, as well as the normalized spherical harmonics that can be found in any elementary book of quantum mechanics. Here, it is sufficient to remember that $Y_4^{\pm 4} = (3/16)\sqrt{35/(2\pi)}\sin^4\theta e^{\pm i4\phi}$, and $Y_4^0 = (3/(16\sqrt{\pi}))(35\cos^4\theta - 30\cos^2\theta + 3)$ (see Hutchings [4.6], p. 239). After some algebra, it can be shown that

$$(x^4+y^4+z^4-\frac{3}{5}r^4) = \frac{16}{3}\sqrt{\frac{2\pi}{35}}\frac{r^4}{8}(Y_4^4+Y_4^{-4}) + \frac{4}{15}\sqrt{\pi}r^4 Y_4^0, \tag{4.23}$$

which can be easily combined with the rest of (4.22) to construct the perturbing potential.

With the perturbation expressed in terms of spherical harmonics, the matrix elements in the $\{Y_2^m(\theta,\phi)\}$ basis can be evaluated. For this purpose, the Clebsch Gordan (CG) series is used (see Merzbacher [4.7], p. 432) applied to the product of two spherical harmonics, which in the general case of arbitrary angular momenta l_1 and l_2, and z-projections m_1 and m_2 is

$$Y_{l_1}^{m_1}(\theta,\phi)Y_{l_2}^{m_2}(\theta,\phi) = \sum_l \sqrt{\frac{(2l_1+1)(2l_2+1)}{4\pi(2l+1)}}$$
$$\times\langle l_1 l_2 00|l_1 l_2 l0\rangle\langle l_1 l_2 m_1 m_2|l_1 l_2 l(m_1+m_2)\rangle Y_l^{m_1+m_2}(\theta,\phi), \tag{4.24}$$

where the matrix elements denote CG coefficients. For the particular case of interest, this formula is useful to calculate the product of the perturbing potential times the basis functions. For example, it can be easily shown that

$$(Y_4^4 + Y_4^{-4})Y_2^{\mp 2} = \sqrt{\frac{9}{4\pi}}\langle 4200|4220\rangle\langle 42,\pm 4,\mp 2|422,\pm 2\rangle Y_2^{\pm 2}\,, \tag{4.25}$$

if only $l = 2$ is of relevance, as in our case. With the same restriction, $(Y_4^4 + Y_4^{-4})Y_2^{m'}$ cancels for $m' = \pm 1$ and 0. The term proportional to Y_4^0 in the potential (4.23) produces a diagonal contribution, through the formula

$$Y_4^0 Y_2^m = \sqrt{\frac{9}{4\pi}}\langle 4200|4220\rangle\langle 420m|422m\rangle Y_2^m\,. \tag{4.26}$$

The explicit value of the Clebsh–Gordan coefficients can be found in tables, or can be searched for on the web (for instance at `http://venusv.kek.jp:8000/info/cgcal.html`). Using this procedure, it can be shown that $\langle 4200|4220\rangle = \sqrt{2/7}$ and $\langle 42,\pm 4,\mp 2|422,\pm 2\rangle = \sqrt{5/9}$.

Finding the CG coefficients of the diagonal contribution, and recalling the multiplicative constants that appear in the potential (4.23) in front of the spherical harmonics, the matrix representation of the perturbation $(x^4 + y^4 + z^4 - \frac{3}{5}r^4)$ in the $\{Y_2^m(\theta,\phi)\}$ basis becomes

$$\frac{V}{Zq} = \frac{2\langle r^4\rangle}{105}\begin{pmatrix} 1 & 0 & 0 & 0 & 5 \\ 0 & -4 & 0 & 0 & 0 \\ 0 & 0 & 6 & 0 & 0 \\ 0 & 0 & 0 & -4 & 0 \\ 5 & 0 & 0 & 0 & 1 \end{pmatrix}. \tag{4.27}$$

The mean value in front of the 5×5 matrix is an integral of the form

$$\langle r^4\rangle = \int_0^\infty dr r^6 R_{32}^2(r)\,. \tag{4.28}$$

The 5×5 matrix (4.27) can be easily diagonalized. The eigenvalues are (i) $(12/105)\langle r^4\rangle$ with degeneracy 2 and associated normalized-to-one eigenvectors Y_2^0 and $(Y_2^2 + Y_2^{-2})/\sqrt{2}$ (these two are the orbitals $d_{3z^2-r^2}$ and $d_{x^2-y^2}$, respectively), and (ii) $-(8/105)\langle r^4\rangle$ with degeneracy 3 and normalized eigenvectors $(Y_2^2 - Y_2^{-2})/\sqrt{2}$, $(Y_2^1 + Y_2^{-1})/\sqrt{2}$, and $(Y_2^1 - Y_2^{-1})/\sqrt{2}$ (which are the d_{xy}, d_{yz}, and d_{zx} orbitals, respectively). The difference between the two eigenvalues is $(20/105)\langle r^4\rangle$. Recalling that to obtain the non constant portion of the actual potential (4.22) we have to multiply these results by $(35Zq/4a^5)$, the actual splitting of energies becomes

$$\Delta = \frac{5Zq}{3a^5}\langle r^4\rangle\,. \tag{4.29}$$

This result is usually expressed in textbooks as $\Delta = 10Dq$ employing an old notation originally due to van Vleck. From the explicit form of Δ, in principle one could estimate the expected splitting, but since such a calculation is

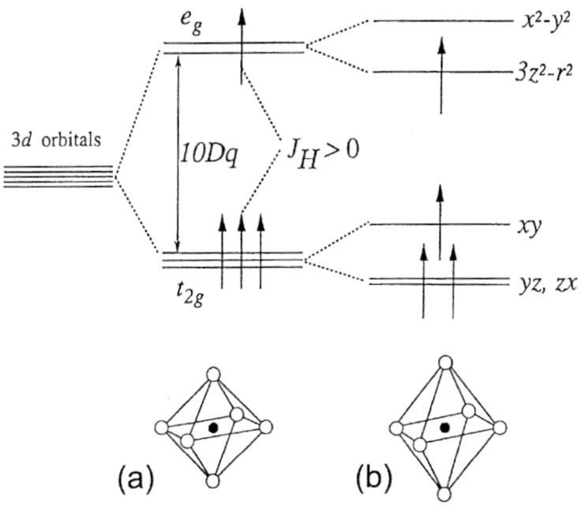

Fig. 4.2. Schematic representation of the crystal-field splitting of the d-levels of a transition-metal ion (for details see text). (**a**) corresponds to a cubic, while (**b**) corresponds to a tetragonal environment with lattice spacing c larger than a and b. From [3.27]

not very accurate it is more standard to simply determine Δ experimentally. In fact, while symmetry considerations are correct in the analysis discussed here, the approximations used are somewhat crude. In practice, typical results for the gap Δ are between 1 and 2 eV. In Fig. 4.2, the splitting we have found is schematically shown. Remember that in the calculation a cubic environment was assumed, which leads to the doublet and triplet structure of the split levels. If the crystal were *tetragonal*, the lattice spacing along one direction (say, the z-axis) is different from the other two ($a = b \neq c$), and as a consequence the splittings would be different. For example, in such a case the orbitals d_{yz} and d_{zx} would remain degenerate by symmetry, but d_{xy} will have a different energy. The same will occur with $d_{x^2-y^2}$ (which has most of its relevant weight on the $x-y$ plane) and $d_{3z^2-r^2}$, which elongates along the z-axis. Their energies will now be different. In Fig. 4.2, a tetragonal splitting is shown for the case where the lattice spacing c is longer than $a(=b)$. The splitting would be the opposite if c is smaller than $a(=b)$.

Although we have arrived at the well-known result for the crystal-field splitting in a cubic environment, the calculation presented in this section has been carried out for the d^1 configuration, i.e., just one electron in the d-shell. However, the case of interest here, Mn^{3+}, has four electrons. In principle, one should repeat the calculation using the same perturbation but a more complicated basis, with four particles. Fortunately, such an extra calculation is not needed since the electrons in Mn^{3+} are in a state of maximum spin $S = 2$ due to the strong Hund coupling. The importance of this coupling will be discussed in more detail in Chap. 5, and the order of magnitude of the crystal-field effect and the Hund coupling will be compared, showing that the Hund effect in general dominates. Then, the Mn^{3+} system is equivalent to having just one "hole" in a polarized band. To understand this, consider Fig. 4.2 and start filling up orbitals from the bottom up, with the same z component of the spin. Only one orbital of the e_g manifold will not have

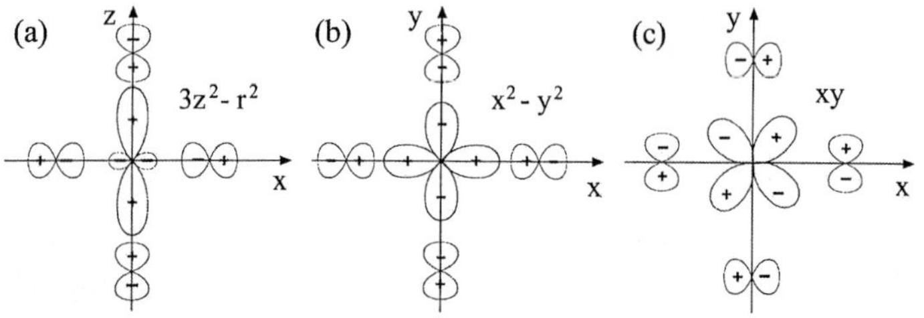

Fig. 4.3. Schematic representation of the central ion and ligand ion relevant orbitals corresponding to (**a–b**) e_g symmetry and (**c**) t_{2g} symmetry

an electron. In the hole language for this case, where just one particle with positive charge is considered, the splitting would be the same as for one electron, but upside down. To return to the electron language the spectra must be inverted once again, and we end up with exactly the same result for the cases d^1 and fully polarized d^4. Then, for manganese Mn^{3+} the splitting is also $\Delta = 10Dq$.

Another detail: For the 3*d* ions under consideration here, the crystal-field splitting is usually *larger* than the spin-orbit (SL) interaction, compatible with the discussion above. Interestingly, for 4*f* rare-earth ions, the situation is the inverse and the calculation differs in detail from the one discussed here. For instance, the angular-momentum operators that are used in the calculations [4.6] involve the sum of the orbital and spin components (J), rather than just the total orbital angular momentum L. Thus, in general, given an ion to analyze it is important to decide first whether crystal-field effects or spin-orbit coupling dominates, to set up the perturbative formalism [4.8]. Further information about these topics can be found in the reviews and books by Hutchings [4.6], Yosida [4.9], and Abragam and Bleaney [4.10].

To complete the discussion, note that the orbitals whose splitting we are discussing involve symmetry-related combinations of the *d*-levels of the central ion of the octahedron and the *p*-level orbitals of the neighboring ligands. The particular ligand orbitals that are mixed to a central *d*-orbital, depend on the symmetry of those *d*-orbitals. Typical examples are shown in Fig. 4.3 for the cases of central $d_{x^2-y^2}$, $d_{3z^2-r^2}$, and d_{xy} orbitals. The first two form σ-bonding orbitals and they belong to the symmetry class e_g, while the last one forms π-bonding orbitals and belongs to the symmetry class t_{2g}. In principle, one should carry out the calculations indeed involving combinations of both the central ion and ligand orbitals, as described here, but those calculations are complicated and beyond the scope of this book. In addition, symmetry considerations are sufficient to obtain the correct splittings of relevance. The price to pay for this simplication is that a small number of parameters, which could be calculated from first principles, must instead be determined from experiments. This is not an important restriction in our studies. Thus, from now on we will frequently use either the more proper language of e_g and t_{2g}

orbitals to refer to particular levels of relevance in the problem of manganites, or the less careful language of their bare d-shell orbitals.

In general, the estimation of energy splittings due to a crystal environment (cubic or noncubic) for ions with arbitrary angular momentum quantum numbers coupled to ligand ions is a complicated task. The calculations presented here provides just a glimpse of such complexity. The reader interested in learning more details about these manipulations should consult the literature quoted in this section. For our purposes, it is more important to understand how other perturbations can further split the e_g sector. This is described next.

4.2.2 The Jahn Teller Effect

In the cubic symmetry considered in the previous section, the splitting due to crystal-field effects leads to an e_g doublet and a t_{2g} triplet. The remaining degeneracy is usually broken by the lattice motion. The ligand ions surrounding the transition-metal ion under consideration (say, the oxygens around manganese) can slightly readjust their locations, creating an asymmetry between the different directions that effectively removes the degeneracy. The lifting of degeneracy due to the orbital-lattice interaction is called the Jahn Teller cooperative effect. This effect tends to occur spontaneously because the energy penalization of the lattice distortion grows as the square of that distortion, while the energy splitting of the otherwise degenerate orbitals is linear, as we will see explicitly in our calculations below. For this reason, it is energetically favorable to spontaneously distort the lattice, thus removing the degeneracy. A review of these concepts when applied to transition-metal compounds can be found in [1.5]. The Jahn Teller lattice distortion can be "static", as occurs in manganites at small hole density, or "dynamic", namely a given ion is not frozen in one distorted configuration but evolves among several configurations as a function of time. This gives a null time-averaged effect, but instantaneous snapshots of the ion locations show a distorted lattice.

In the case of manganites, the Jahn Teller effect is very important. Let us allow the six oxygens around the Mn ion (Fig. 4.1) to move by a small amount. Assume that each oxygen has an associated coordinate X_1, X_4, Y_2, Y_5, Z_3, and Z_6, considered for simplicity just along the axes corresponding to the corresponding Mn O Mn bonds (Fig. 4.4).

Working only at first order in the shifts of oxygens with respect to their equilibrium positions, the resulting change in the potential caused by the ligands on the electrons orbiting around the central ion is

$$\Delta V = -\frac{1}{2}\left(\frac{9}{a^4}\right)\left[x^2(X_1 - X_4) + y^2(Y_2 - Y_5) + z^2(Z_3 - Z_6)\right] . \qquad (4.30)$$

In deriving (4.30), a constant term has been omitted since it produces an overall shift of the energy levels without inducing splitting. For the same reason, a contribution proportional to the electronic coordinate r^2 is also not

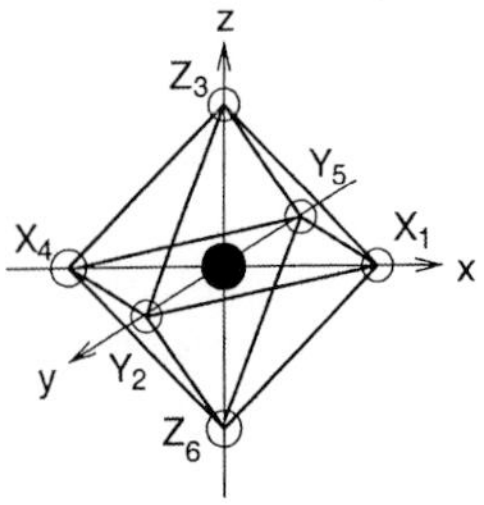

Fig. 4.4. Schematic representation of the coordinates used to label the shifts of oxygen ions from their equilibrium positions

important for our purposes and it is not considered. Finally, terms of order, e.g., $X_1 x$ do not contribute in perturbation theory since the corresponding matrix elements vanish, as shown in detail below. Hence, the nontrivial portion of the modified potential (4.30) is of first order in the oxygen shifts and *second* order in the electronic coordinates. However, even with this simplified form dealing with the full perturbation (4.30) would be too complicated. Fortunately, group-theory arguments can simplify the calculation substantially. Let us consider the qualitative aspects of the group-theory discussion. Suppose that we allow temporarily for the individual oxygens to move in any direction, not only along the Mn O Mn bond directions. Including the movement of the Mn ion itself, this amounts to having 21 degrees of freedom (7 ions $\times 3$ degrees of freedom each). However, 6 must be subtracted since they are due to pure rotations and translations. Following Orbach and Stapleton [4.11], the remaining 15 can be grouped into normal modes of vibration following standard procedures. Of those 15, 9 are "odd" modes with respect to inversion through the origin. These modes are not important in a perturbative calculation since, for instance, the doublet and triplet of Fig. 4.2 have the same symmetry under inversion within each multiplet, and the "odd" modes would lead to vanishing matrix elements of the perturbation. The remaining 6 modes are shown in Fig. 4.5. For our purposes, the Q_4, Q_5, and Q_6 modes are not important since they correspond to displacements of the oxygens *perpendicular* to the direction of the Mn O Mn bond related to that oxygen. These displacements do not contribute to the splitting of the $d_{x^2-y^2}$ and $d_{3z^2-r^2}$ orbitals in which we are interested, since they change sign under reflection with respect to a plane, while those orbitals are invariant under the same operation. Then, imposing $X_4 = -X_1$, $Y_5 = -Y_2$, and $Z_6 = -Z_3$, and defining $Q_1 = \frac{1}{\sqrt{3}}(X_1 + Y_2 + Z_3)$, $Q_2 = \frac{1}{\sqrt{2}}(X_1 - Y_2)$ and $Q_3 = \frac{1}{\sqrt{6}}(2Z_3 - X_1 - Y_2)$ (see Fig. 4.5), it can be shown that the even sector of the potential (4.30) becomes

$$\Delta V|_{\text{even}} = -\gamma[x^2 X_1 + y^2 Y_2 + z^2 Z_3] = -\gamma\left[\frac{1}{\sqrt{3}}(x^2+y^2+z^2)Q_1 + \frac{1}{\sqrt{2}}(x^2-y^2)Q_2 + \frac{1}{\sqrt{6}}(2z^2-x^2-y^2)Q_3\right], \tag{4.31}$$

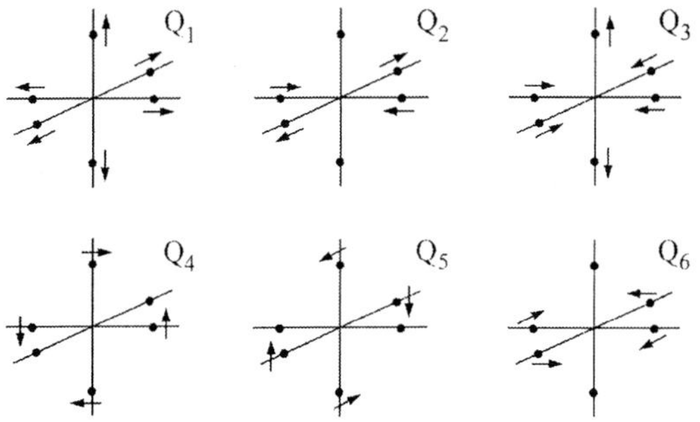

Fig. 4.5. Normal modes of the oxygens surrounding the transition-metal ion. Shown are only those modes, that are even under inversions. The $\{Q\}$-coordinates of relevance for the Jahn Teller splitting of the e_g sector are Q_2 and Q_3

where $\gamma = 9/a^4$. In addition, the Q_1 mode being uniform in all directions also does not contribute to the splitting in the e_g sector. For this reason, only Q_2 and Q_3 are needed to understand how the e_g degeneracy is removed. Then, the important portion of the perturbing potential is

$$\Delta V|_{\mathrm{JT}} = -\sqrt{\frac{2}{3}}\gamma \left[\sqrt{\frac{3}{2}}(x^2 - y^2)Q_2 + \frac{1}{2}(2z^2 - x^2 - y^2)Q_3 \right] , \tag{4.32}$$

where, remember once again that the displacements of the oxygens are along the corresponding Mn O Mn axes, otherwise the normal modes Q_4, Q_5, and Q_6 should also be included (see Yoshida [4.9]), as already explained.

To calculate the effect of this perturbation, we focus later on the e_g manifold, and as a consequence the final perturbing Hamiltonian will simply be a 2×2 matrix. The problem will be presented here in a slightly more general form for completeness, involving in principle the five d-levels. The procedure, similar to that followed above to analyze the crystal splitting, starts using the representation of cartesian coordinates using spherical coordinates. With this transformation, and a table of normalized spherical harmonics, it can be easily shown that

$$\begin{aligned} 2z^2 - x^2 - y^2 &= 4r^2\sqrt{\frac{\pi}{5}}Y_2^0(\theta,\phi) , \\ x^2 - y^2 &= 2r^2\sqrt{\frac{2\pi}{15}}(Y_2^2(\theta,\phi) + Y_2^{-2}(\theta,\phi)) , \\ xy &= -\mathrm{i}r^2\sqrt{\frac{2\pi}{15}}(Y_2^2(\theta,\phi) - Y_2^{-2}(\theta,\phi)) , \end{aligned} \tag{4.33}$$

with the last relation quoted only for completeness, and the corresponding ones for yz and zx left as exercises to the readers. Working in the standard $\{Y_2^m(\theta,\phi)\}$ basis of spherical harmonics, the next step involves the use of the CG series (4.24) and a table of CG coefficients, as in the case of the previously addressed crystal-field-splitting calculation. Since the operators that contribute to the potential can be written in terms of just a few spherical harmonics, we will be interested in decomposing $Y_2^{m'}(\theta,\phi)Y_2^m(\theta,\phi)$, with m' restricted to ± 2 and 0 (quantum numbers that appear in (4.33)), and m taking its five possible values. By a straightforward calculation it can be shown that

$$Y_2^0(\theta,\phi)Y_2^m(\theta,\phi) = \frac{n_m}{7}\sqrt{\frac{5}{4\pi}}Y_2^m(\theta,\phi) + \dots\,, \tag{4.34}$$

where $n_2 = n_{-2} = -n_0 = -2$, $n_1 = n_{-1} = 1$, and the dots indicate spherical harmonics not belonging to the $l = 2$ subspace.

At this point one could proceed directly into the construction of the 2×2 perturbing Hamiltonian, following a procedure similar to that outlined in the case of the crystal splitting. Instead, it is instructive to proceed by rewriting the perturbing potential in terms of "equivalent" angular-momentum operators (see Hutchings [4.6]). This is a very useful and commonly used trick to carry out calculations when they are much more complex than those addressed in this section. The manipulations below involving these "equivalent operators" can be set in a more formal language through the use of the Wigner–Eckart Theorem [4.12], which in its simplest form states that the matrix elements of any vector operator are proportional to those of the angular momentum operator [4.13]. The formalism of equivalent operators can be seen as just a rephrasing of the Wigner Eckart Theorem, and its widespread use has historical reasons (dating to pioneer work on magnetic resonance). Unfortunately, some inconsistencies of notation, such as the use of unnormalized spherical harmonics, have caused confusion in the literature. The interested reader is referred to the works of Buckmaster [4.14] and Smith and Thornley [4.15], where the widely used so-called Stevens' operators [4.16] are redefined in a way to be consistent with the Wigner Eckart Theorem.

For the case of the $(2z^2 - x^2 - y^2)$ portion of the perturbation, the proof that its effect is equivalent to that of an operator based on angular momenta is simple, since it is well known from elementary quantum mechanics that

$$(3\hat{L}_z^2 - 6)Y_2^m = (3m^2 - 6)Y_2^m\,, \tag{4.35}$$

where $\hat{L}_z$ is the z-component of the angular-momentum operator ($\hbar$ is considered to be 1 in the manipulations described here). Then, comparing (4.33) with (4.35), and using (4.34), it is clear that the multiplicative application of $2z^2 - x^2 - y^2$ over the $l = 2$ basis or the use of the $(3\hat{L}_z^2 - 6)$ operator produce equally diagonal results, and in addition working out the coefficients on the right-hand side of (4.33, 4.35) it can be shown that both operators are actually proportional to each other. Collecting all the factors, the relevant relationship is

$$2z^2 - x^2 - y^2 = -\frac{2}{21}r^2(3\hat{L}_z^2 - 6)\,, \tag{4.36}$$

if restricted to the l=2 subspace.

The analogous relation for the case of $x^2 - y^2$ is more tedious to elaborate. As in the previous cases, the use of the CG series allow us to arrive to

$$
\begin{aligned}
(Y_2^2 + Y_2^{-2})Y_2^m &= -\frac{2}{7}\sqrt{\frac{5}{4\pi}}Y_2^0 + \dots \quad (m = 2, -2)\,, \\
(Y_2^2 + Y_2^{-2})Y_2^m &= -\sqrt{\frac{6}{7}}\sqrt{\frac{5}{4\pi}}Y_2^{-m} + \dots \quad (m = 1, -1) \\
(Y_2^2 + Y_2^{-2})Y_2^0 &= -\frac{2}{7}\sqrt{\frac{5}{4\pi}}(Y_2^2 + Y_2^{-2}) + \dots\,.
\end{aligned}
\tag{4.37}
$$

On the other hand, the well-known properties of raising and lowering operators $\hat{L}_\pm Y_l^m = \sqrt{l(l+1) - m(m \pm 1)}Y_l^{m\pm 1}$ can be used to show that

$$
\begin{aligned}
(\hat{L}_x^2 - \hat{L}_y^2)Y_2^m &= \sqrt{\frac{24}{2}}Y_2^0 \quad (m = 2, -2)\,, \\
(\hat{L}_x^2 - \hat{L}_y^2)Y_2^m &= \sqrt{\frac{36}{2}}Y_2^{-m} \quad (m = 1, -1)\,, \\
(\hat{L}_x^2 - \hat{L}_y^2)Y_2^0 &= \sqrt{\frac{24}{2}}(Y_2^2 + Y_2^{-2})\,.
\end{aligned}
\tag{4.38}
$$

Comparing (4.37) with (4.38), once again a proportionality can be shown to exist. With all factors in the proper place, we arrive at the relation

$$
x^2 - y^2 = -\frac{2}{21}r^2(\hat{L}_x^2 - \hat{L}_y^2)\,. \tag{4.39}
$$

As an exercise, the reader should derive other operatorial equivalences, such as

$$
xy = -\frac{2}{21}r^2\frac{(\hat{L}_x\hat{L}_y + \hat{L}_y\hat{L}_x)}{2}\,, \tag{4.40}
$$

valid within the $l = 2$ subspace. Terms containing xy appear in the perturbing Hamiltonian if oxygen movements perpendicular to the main axes are considered. However, even if such an effect is taken into account, the resulting 2×2 Hamiltonian would not be affected since the matrix elements of xy, yz, and zx in the e_g states can be shown to cancel (this is left as an exercise for the reader).

After writing the perturbation in the form of equivalent angular momenta operators, we are now ready to construct the perturbing Hamiltonian in the e_g sector. For such a purpose, consider the states $|\alpha\rangle$ and $|\beta\rangle$, given by $(1/\sqrt{2})(Y_2^2 + Y_2^{-2})$ and Y_2^0, respectively. Recalling the relations (4.33), these two states are proportional, respectively, to $(x^2 - y^2)$ and $(2z^2 - x^2 - y^2)$, which correspond to the orbitals we are perturbing with the potential $\Delta V|_{\mathrm{JT}}$ (4.32). It is easy to show that $\langle\alpha|(\hat{L}_x^2 - \hat{L}_y^2)|\alpha\rangle = \langle\beta|(\hat{L}_x^2 - \hat{L}_y^2)|\beta\rangle = 0$ and $\langle\alpha|(\hat{L}_x^2 - \hat{L}_y^2)|\beta\rangle = \sqrt{12}$, while $\langle\alpha|(3\hat{L}_z^2 - 6)|\alpha\rangle = -\langle\beta|(3\hat{L}_z^2 - 6)|\beta\rangle = 6$, and $\langle\alpha|(3\hat{L}_z^2 - 6)|\beta\rangle = 0$.

Then, collecting the contributions to the Hamiltonian of the Q_2 and Q_3 portions of the perturbation, (4.32), we arrive at the important relation

$$\Delta V|_{\mathrm{JT}} = \frac{2\sqrt{6}}{21}\gamma\langle r^2\rangle \left[Q_2\begin{pmatrix} 0 & 1 \\ 1 & 0 \end{pmatrix} + Q_3\begin{pmatrix} 1 & 0 \\ 0 & -1 \end{pmatrix}\right], \tag{4.41}$$

valid in the e_g manifold. The mean-value affecting r^2 should be understood as in (4.28). This perturbation (4.41) induced by the Jahn Teller distortions plays a very important role in the analysis of manganites.

Note. To complete the calculation neatly, we would have to justify that the contribution to the potential ΔV that is linear in the coordinates does not contribute to the problem. The proof is similar to that found in books of quantum mechanics when sum rules for transitions between atomic levels with the emission or absorption of one photon are discussed. The main mathematical reason is that $Y_1^{m'} Y_2^m$, with $m' = 0, \pm 1$ does not have a component in the $l = 2$ sector. This can be shown using the CG series described before, since $\langle 1200|1220\rangle = 0$ (while naively it would appear that mixing $l = 1$ and 2 angular momenta, a resulting $l = 2$ state would be possible). In general, $\langle n2m'|\mathbf{r}|n2m\rangle = 0$, with $\mathbf{r} = (x, y, z)$.

The total energy corresponding to the distortion we are studying should also contain the energy penalization for the distortion itself, due to its deviation from the equilibrium position. If M is the ligand ion mass, and ω the frequency associated to the quadratic potential around equilibrium for the normal modes under consideration, the full 2×2 Hamiltonian is

$$H = -g[Q_2\sigma_x + Q_3\sigma_z] + \frac{1}{2}M\omega^2(Q_2^2 + Q_3^2), \tag{4.42}$$

where the Pauli matrices σ_x and σ_z are used (note that σ_y, which is imaginary, does not appear in the Hamiltonian), and $g = -(2\sqrt{6}/21)\gamma\langle r^2\rangle$. This Hamiltonian can be diagonalized easily, at least for the case of a single octahedra (e.g., MnO_6). Changing variables into polar coordinates using

$$Q_3 = Q\cos\theta\,, \quad Q_2 = Q\sin\theta\,, \quad Q = Q_3\cos\theta + Q_2\sin\theta\,, \tag{4.43}$$

the eigenvalues of H are

$$E = \mp gQ + \frac{1}{2}M\omega^2 Q^2\,. \tag{4.44}$$

The static distortion arising from (4.44) is obtained by minimizing the energy of the octahedron with respect to Q. The result is

$$Q_{\mathrm{optimal}} = \frac{|g|}{M\omega^2}, \tag{4.45}$$

which has an associated energy $E_{\mathrm{optimal}} = -(g^2/2M\omega^2)$.

Note that (4.44) is *independent* of the angle θ, namely any distortion gives the same energy, which is somewhat surprising. This no longer occurs if anharmonic terms are included, such as those of third order in the oxygen

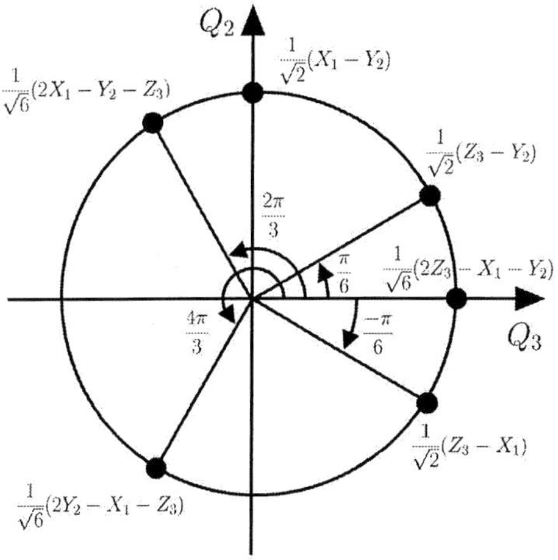

Fig. 4.6. Important special cases of distortions of an isolated oxygen octahedron obtained for the angles θ indicated

displacements (we neglect those terms for simplicity). Quantum effects in the oscillations around the equilibrium positions are also neglected.

Following Kanamori [4.17], there are special cases of Jahn Teller distortions that are interesting to consider. For instance, if $\theta = 0$, the distortion is purely of the Q_3 form and it corresponds to an elongation of the octahedra along the z-axis (contraction along the other axes). If $\theta = \pi$, it is also a pure Q_3 case, with the elongation and contractions reversed. Distortions that are purely Q_2 occur for $\theta = \pm\pi/2$. The case $\theta = 2\pi/3$ is interesting. In this situation, $Q = Q_3 \cos\theta + Q_2 \sin\theta = \frac{1}{\sqrt{6}}(2X_1 - Y_2 - Z_3)$, which is like a Q_3 elongation but along the x-axis. If $\theta = 4\pi/3$, then the Q_3-like elongation is along the y-axis, namely $Q = \frac{1}{\sqrt{6}}(2Y_2 - X_1 - Z_3)$. If $\theta = \pi/6$, then $Q = \frac{1}{\sqrt{2}}(Z_3 - Y_2)$, while $\theta = 5\pi/6$ gives $Q = \frac{1}{\sqrt{2}}(Z_3 - X_1)$, namely distortions that are as those found for a pure Q_2 case but with the axes rotated. This information is summarized in Fig. 4.6.

In future chapters, the case where more than one octahedron must be considered, such as in a lattice arrangement, will be discussed in detail. In such a context, *cooperative* effects between oxygens are important: each ligand ion belongs to two nearest-neighbor octahedra and if one of them elongates along the direction where that ion is located, then the neighboring octahedra must contract in that direction (for a review see Kugel and Khomskii [1.5]). This will be shown to be of particular importance for materials such as $LaMnO_3$. Certainly, once a lattice is considered, not all angles θ will have the same energy. On a lattice, very specific distortions will be associated to ground-state properties.

5. Models for Manganites

In the previous chapter, an electronic model was derived for the case of several nearly degenerate orbitals in a given ion. The resulting model, which included on site Coulombic interactions (4.17, 4.18), was found to be very general, and its applicability goes beyond the manganites into ruthenates, nickelates, and other multiorbital compounds. Also, the discussion in the last chapter on the Jahn Teller effects was general as well, and suitable for a variety of materials. In this chapter, the model-writing effort will be entirely focused on manganites. This effort will not simply lead to trivial special cases of the previous results, but further work will be needed since the Hamiltonians of Chap. 4 are, unfortunately, too complicated for an accurate many-body treatment. Simpler models must be derived (or guessed!) under reasonable assumptions, to make progress on the theoretical aspects of manganese oxides. The setup of these simpler models is the main goal of the present chapter.

5.1 Coulombic Terms

The complicated interactions contained in the Hamiltonians presented in the previous chapter are virtually impossible to handle accurately, even with modern many-body techniques. Finding a simplified form of these Hamiltonians is crucial. Fortunately, the discussion of Chap. 4 already shows that in the presence of the crystal field the five d-orbitals are split into two groups: the e_g and t_{2g} states. The former multiplet is higher in energy than the latter, and electrons are removed from the e_g sector upon hole doping. The t_{2g} electrons at each site are not affected by doping. Their population remains constant, and the well-known Hund rule of atomic physics (which is contained in the multiorbital model derived in Chap. 4) enforces alignment of the three t_{2g} spins into a $S = 3/2$ state. Then, the t_{2g} sector of the problem can be replaced by a *localized spin* at each manganese ion, enormously reducing the complexity of the original five-orbital model. Moreover, the value 3/2 for the net spin is high enough that the use of *classical* spins could be regarded as a good approximation. In fact, the spin the most "quantum" is 1/2, with only two allowed projections along the z-axis. As the spin magnitude grows, the number of projections increases and the limit of a classical spin (with no restrictions on the z-axis projections) is approached. The numerical results discussed later in Chap. 6 confirm the expectation that $S = 3/2$ quantum and classical spins give similar results.

How does the resulting classical localized t_{2g} spin interact with the e_g electrons? The U'-term between e_g and t_{2g} electrons is of no relevance, since the occupancy of the latter is fixed at each site, after the localized spin assumption is made. The only important interaction between the two sectors can involve at most the spin, not the charge, and it is contained in the Hund interaction of the multiorbital Hamiltonian that favors a ferromagnetic alignment. To account for this effect, the phenomenological term

$$H_{\mathrm{Hund}} = -J_{\mathrm{H}} \sum_{\mathbf{i}} \mathbf{s_i} \cdot \mathbf{S_i}\,, \tag{5.1}$$

is introduced, where $\mathbf{s_i} = \sum_{\gamma\alpha\beta} c^{\dagger}_{\mathbf{i}\gamma\alpha} \sigma_{\alpha\beta} c_{\mathbf{i}\gamma\beta}$ is the spin of the e_g electrons, γ is restricted to the e_g sector, $J_{\mathrm{H}}(> 0)$ is the Hund coupling between localized $t_{2\mathrm{g}}$ spins and mobile e_{g} electrons, and $\sigma = (\sigma_x, \sigma_y, \sigma_z)$ are the Pauli matrices. The magnitude of J_{H} is of the order of the Hund coupling in the multiorbital model. The classical t_{2g} spin $\mathbf{S_i}$ is usually normalized as $|\mathbf{S_i}| = 1$ (although some authors use conventions such as 3/2, and the reader must be alert for this or other possible normalizations). The direction of the classical $t_{2\mathrm{g}}$ spin at site $\mathbf{i}$ is defined through

$$\mathbf{S_i} = (\sin\theta_{\mathbf{i}} \cos\phi_{\mathbf{i}}, \sin\theta_{\mathbf{i}} \sin\phi_{\mathbf{i}}, \cos\theta_{\mathbf{i}})\,, \tag{5.2}$$

by using the polar angle $\theta_{\mathbf{i}}$ and the azimuthal angle $\phi_{\mathbf{i}}$. Note that the introduction of the Hund term is not the entire Coulombic contribution to the problem, since electrons within the e_g sector can still interact through the U, U' and J couplings. The actual form of this interaction, denoted $H_{\mathrm{el-el}}$, is the same as in Chap. 4, simply constraining the number of orbitals to two.

5.2 Heisenberg Term

In the limit when all the e_g electrons are removed from the system ($x = 1$), the model becomes trivial since the interaction between localized spins is only mediated by the e_g electrons. Without mobile electrons, the localized spins simply become "free" or noninteracting. But, it is well known that manganites have a G-type AF state at $x = 1$ (as occurs for example in $\mathrm{CaMnO_3}$). Then, the localized spins must have an antiferromagnetic coupling among themselves to account for the G-type state. This is not too surprising, since the t_{2g} electrons can induce an AF Heisenberg interaction through virtual hopping between nearest-neighbors, leading to the standard exchange mechanism. To account for this effect, the term

$$H_{\mathrm{AF}} = J_{\mathrm{AF}} \sum_{\langle \mathbf{i},\mathbf{j} \rangle} \mathbf{S_i} \cdot \mathbf{S_j}\,, \tag{5.3}$$

is introduced, where $J_{\rm AF}$ is the AF coupling between nearest-neighbor $\langle \mathbf{i}, \mathbf{j} \rangle t_{2g}$ spins. The value of $J_{\rm AF}$ should be small compared with hopping amplitudes of e_g electrons, and the Hund coupling as well. However, the discussion in the following chapters shows that $J_{\rm AF}$ plays a key role in manganites, where there are competing states with similar energies (and as a consequence the relevance of "small" terms is amplified).

5.3 The Electron–Phonon Interaction

The basic interaction between the Jahn Teller modes of oscillation of the MnO_6 octahedron, and the e_g electrons is in (4.42). The 2×2 Pauli matrices in that equation reflect the two-dimensionality of the e_g sector. The rewriting of (4.42) in a more formal second-quantization language is straightforward. With the use of the "pseudospin" operators

$$\begin{cases} T_{\mathbf{i}}^x = (1/2) \sum_\sigma (c_{\mathbf{i}a\sigma}^\dagger c_{\mathbf{i}b\sigma} + c_{\mathbf{i}b\sigma}^\dagger c_{\mathbf{i}a\sigma}) \,, \\ T_{\mathbf{i}}^y = -(i/2) \sum_\sigma (c_{\mathbf{i}a\sigma}^\dagger c_{\mathbf{i}b\sigma} - c_{\mathbf{i}b\sigma}^\dagger c_{\mathbf{i}a\sigma}) \,, \\ T_{\mathbf{i}}^z = (1/2) \sum_\sigma (c_{\mathbf{i}a\sigma}^\dagger c_{\mathbf{i}a\sigma} - c_{\mathbf{i}b\sigma}^\dagger c_{\mathbf{i}b\sigma}) \,, \end{cases} \tag{5.4}$$

the local electron JT phonon interaction at a given site $\mathbf{i}$ becomes

$$H_{\mathbf{i}}^{\rm JT} = -2g(Q_{2\mathbf{i}} T_{\mathbf{i}}^x + Q_{3\mathbf{i}} T_{\mathbf{i}}^z) + (k_{\rm JT}/2)(Q_{2\mathbf{i}}^2 + Q_{3\mathbf{i}}^2) \,, \tag{5.5}$$

where the spring constant $k_{\rm JT}{=}M\omega^2$ was introduced. The factor 2 multiplying g arises from the (1/2) contained in the definition of the pseudospin operators. Summing over all sites of the lattice leads to the complete electron phonon coupling $H_{\rm el-ph} = \sum_{\mathbf{i}} H_{\mathbf{i}}^{\rm JT}$. The definition of Q_2 and Q_3 in terms of the displacements of the oxygens can be found above (4.31) in Chap. 4. Below that equation, the Q_1 *breathing* mode is also mentioned. This mode does not remove the e_g degeneracy but it can be important in some studies, and in principle it should be included in realistic Hamiltonians. The breathing mode was discussed in [18.12] and [9.7]. Some of these references address the subtle issue of the different couplings and spring constants for JT and breathing modes.

In the context of theories for manganites, phonons are usually described as *classical.* This is a tremendous simplification that allows for Monte Carlo studies of the model for two orbitals, to be defined in more detail later. This assumption can be justified a posteriori, given the success of the two-orbital model to reproduce the properties of manganites. However, a more refined discussion on the relevance of quantum effects in the phononic sector is not available. Such study would be difficult, since reliable calculations with quantum phonons are quite difficult. For the time being, it is reasonable to proceed with the formalism using classical phonons, but the reader must keep in mind the existence of this extra approximation in the theory.

The spring constant k_{JT} is usually absorbed into the Q^2s of the quadratic purely phononic term in H_{el-ph}, through a mere rescaling. It reappears in the electron phonon term of H_{el-ph}, involving the pseudospin operator and the phonons. The dimensionless combination $g/\sqrt{k_{JT}t}$, with t the hopping scale of the kinetic term of the Hamiltonian, emerges naturally in this context, and it is usually denoted as λ. The constant λ is more frequently used than g to judge the strength of the electron JT phonon coupling.

Another very important detail is the use of *cooperative* or *non cooperative* phonons in the actual implementation of the calculations. Several examples discussed later simply neglect the breathing mode and take Q_2 and Q_3 as fundamental and independent degrees of freedom. In this approximation, the fact that an oxygen is shared by two nearest-neighbor manganese ions is lost. On the other hand, if the coordinates of the oxygens are explicitly used, then the Qs are no longer independent, leading to the cooperative phonon treatment (see [9.7, 18.12] and references therein). In this context, usually the displacements of oxygens from their equilibrium positions along the main axes, denoted by $\{u\}'s$, are considered, but in principle displacements of each oxygen in any direction are possible. Non cooperative phonons, crude as they may seem, often provide reliable results. It is only recently that the effect of cooperative phonons is starting to be properly analyzed in numerical simulations. Plenty of work remains to be done!

5.4 A Simple but Realistic Model for Manganites

Summarizing, the many steps followed in the construction of a simple Hamiltonian for manganites lead to a model that contains five terms: (i) A kinetic energy for the e_g electrons, usually denoted by $H_{\rm kin}$ or $\hat{T}$. This term is shown in (4.18) of Chap. 4 (the actual hopping amplitudes are evaluated later in this chapter). The t_{2g} electrons are assumed to be localized, and only their spin degree of freedom is kept. In addition, these spins are assumed to be classical. (ii) The Hund coupling between the e_g electrons and the t_{2g} spins, $H_{\rm Hund}$, discussed already in Sect. 5.1. (iii) The antiferromagnetic Heisenberg exchange coupling between nearest-neighbor spins, $H_{\rm AF}$. To induce this term it is necessary to allow for the t_{2g} electrons to move between nearest-neighbors (as in any perturbative calculation of exchange interactions that can be found in textbooks). However, once the effective Heisenberg interaction is derived, the t_{2g} kinetic energy can be dropped. (iv) The coupling between electrons and the Jahn–Teller distortions of the local MnO_6 octahedron, H_{el-ph}. This term should include also the breathing mode, as discussed before. The actual implementation can involve cooperative or non cooperative phonons. The phonons are usually assumed to be classical. (v) The Coulomb interaction among the e_g electrons, H_{el-el}. In this term, interactions beyond the on site terms should in principle be included, but often "screening" or "polarization" is invoked to truncate the interaction to just the on site contribution. It is a matter of hot debate whether this is a correct procedure, not only in manganites but in transition-metal oxides in general. Leaving these discussions aside, a fairly realistic model for manganites is then

$$H = H_{\mathrm{kin}} + H_{\mathrm{Hund}} + H_{\mathrm{AF}} + H_{\mathrm{el-ph}} + H_{\mathrm{el-el}} \,. \tag{5.6}$$

Even with all these approximations this model is still quite difficult to study in practice. This is a frustrating, but recurrent problem in many-body physics. Writing reasonable Hamiltonians can be usually accomplished since the basic interactions among the several degrees of freedom are known. However, the subsequent solution of the model is often a formidable task. Further simplifications are still needed to arrive at a useful model that truly illuminates the physics of manganites.

5.5 The One-Orbital Model

A widely used model for manganites, and the one that historically was the first to be proposed and studied, is the "one-orbital" model. This model is actually more often called the "ferromagnetic Kondo" model (the reason will be explained in Chap. 6). It is also often called the "double-exchange" model, since it contains enough degrees of freedom to manifest the phenomenon of double-exchange ferromagnetism, as envisioned in the early days of manganite studies (see Chap. 2). Here, to establish a clear distinction between this model and the more realistic model involving two orbitals, the "one-orbital model" notation will be used the most, but the reader should keep a reasonable flexibility in this respect when consulting the literature, since the other two names are certainly very popular.

The one-orbital model is obtained from (5.6) by (i) neglecting the coupling with phonons, (ii) neglecting the Coulomb interaction among e_g electrons, and (iii) assuming that a static Jahn Teller distortion leads to a spatially uniform splitting of the degenerate e_g levels, allowing us to keep only one active orbital in the Hamiltonian. As explained in the following section, these approximations are very drastic, making the model not suitable to explain most of the phenomenology of manganites. However, the model is sufficient to understand the mechanism of ferromagnetism, and in recent years it has been found that insulating phases can be generated as well, leading to a metal insulator competition. This competition is at the heart of modern theories of manganites. This is why the study of the one-orbital model continues to attract much attention, and why it is the simplest model that most researchers accept as reasonable. The one-orbital Hamiltonian is

$$\hat{H} = -t \sum_{\langle ij \rangle,\sigma} (a^{\dagger}_{i,\sigma} a_{j,\sigma} + H.c.) - J_{\mathrm{H}} \sum_{i,\alpha,\beta} a^{\dagger}_{i,\alpha} \sigma_{\alpha\beta} a_{i,\beta} \cdot \mathbf{S}_i + J_{\mathrm{AF}} \sum_{\langle ij \rangle} \mathbf{S}_i \cdot \mathbf{S}_j \,, \tag{5.7}$$

where $a_{i,\sigma}$ is the destruction operator for an electron with spin z-projection σ located at site i. There are no orbital indices, since only one orbital is involved.

The site label i shoud be considered a vector with as many components as the dimension of the lattice under study. Equation (5.7) is quadratic in the fermionic operator (since the Coulomb interaction is not considered) allowing for a variety of many-body treatments of the problem including the Monte Carlo simulations, discussed later in Chap. 7.

At this point it is important to briefly discuss the consequences of neglecting the Coulomb interaction. This approximation appears to be very drastic. In fact the on site Coulomb repulsion (i.e., the famous Hubbard U) is the key ingredient of models for high-temperature superconductors, which also involve one orbital and a transition-metal oxide. However, in (5.7) the localized spin degree of freedom saves the day. In the case of a large Hund coupling (the widely accepted realistic regime of manganites), if two electrons are located in the same e_g orbital, then one can have a spin parallel to the localized spin optimizing, at least locally, the Hund interaction. Due to the Pauli principle the other spin must point antiparallel to the localized spin, and no net gain in energy is obtained when the two electrons are in the same site. However, if the two electrons spread and occupy one site each, then both can have a spin parallel to the corresponding localized spin direction. For this reason, the large Hund coupling plays a role analogous to a large Hubbard U in the sense that it favors the spreading of charge, and the drastic suppression of double occupancy. This fact has been verified in many numerical studies, and, as a consequence, the neglect of the Hubbard U term appears to be a reasonable approximation.

5.6 The Limit of $J_{\mathrm{H}} = \infty$

Since a large Hund coupling is widely assumed in models for manganites, another much used simplification involves the limit $J_{\mathrm{H}} = \infty$. In this case, the e_g electron necessarily *must* have a spin perfectly parallel to that of the localized (assumed classical) t_{2g} spin at the same site, effectively suppressing entirely the antiparallel configuration. As a consequence, it is possible to mathematically "rotate" the e_g operators at each site, such that the parallelism among the spins e_g-t_{2g} is enforced. After that rotation, the spin index can be dropped. This is an important reduction in the number of degrees of freedom, allowing for simpler numerical and analytical studies of the one-orbital model, without drastically altering its original physics. The approximation is often applied to the two-orbital model as well (to be discussed in the next sections), since it does not depend on the coupling to phonons or whether one or two orbitals are active.

Formally, the transformation from the operators that appear in (5.7), and those with a spin parallel to the localized spin is

$$\tilde{a}_{\mathrm{i}\gamma} = \cos(\theta_{\mathrm{i}}/2) a_{\mathrm{i}\gamma\uparrow} + \sin(\theta_{\mathrm{i}}/2) \mathrm{e}^{-\mathrm{i}\phi_{\mathrm{i}}} a_{\mathrm{i}\gamma\downarrow} , \tag{5.8}$$

where the orbital index γ was kept since the transformation can be applied to both the one- or two-orbital model, as explained before. The angles θ and ϕ at every site correspond to the orientation of the localized classical spin. More details can be found in Chap. 6, where the historically famous "cos$\theta/2$" effective hopping is derived. The one-orbital Hamiltonian is now much simplified since the Hund term becomes an irrelevant constant, and the spin degree of freedom can be dropped. The reader can easily show that in the limit $J_{\mathrm{H}} = \infty$, if the coupling with phonons and Coulomb interactions are neglected, then the one-orbital or two-orbital models become

$$H^{\infty} = -\sum_{\mathbf{i}\mathbf{a}\gamma\gamma'} S_{\mathbf{i},\mathbf{i}+\mathbf{a}} t^{\mathbf{a}}_{\gamma\gamma'} \tilde{a}^{\dagger}_{\mathbf{i}\gamma} \tilde{a}_{\mathbf{i}+\mathbf{a}\gamma'} + J_{\mathrm{AF}} \sum_{\langle \mathbf{ij} \rangle} \mathbf{S}_{\mathbf{i}} \cdot \mathbf{S}_{\mathbf{j}} , \tag{5.9}$$

where the complex number $S_{\mathbf{i},\mathbf{j}}$, sometimes referred to as a Berry phase, is

$$S_{\mathbf{i},\mathbf{j}} = \cos(\theta_{\mathbf{i}}/2)\cos(\theta_{\mathbf{j}}/2) + \sin(\theta_{\mathbf{i}}/2)\sin(\theta_{\mathbf{j}}/2)\mathrm{e}^{-\mathrm{i}(\phi_{\mathbf{i}}-\phi_{\mathbf{j}})} . \tag{5.10}$$

More details about this transformation can be found in [5.1]. As mentioned before, the $J_{\mathrm{H}} = \infty$ is widely used in the literature, since it simplifies substantially the calculations without altering the essence of the physics.

5.7 The Relevance of Jahn–Teller Phonons

With hindsight, it is actually quite surprising that the one-orbital or double-exchange model was considered to be the main model for manganites for so many years, before the revival of interest on these materials in recent times. It is true that the model explained fairly well the ferromagnetic phase, but it provided no help for understanding the other competing phases – at least with the limited many-body techniques used until recently – and certainly no progress was made until the 1990s in trying to understand the large MR effects. Although the one-orbital model can not have orbital ordering by mere construction, recently phenomena such as phase separation and some hints of charge ordering have started to emerge. These new discoveries in the one-orbital context make the model fairly interesting since competition between phases, the "key concept" emerging in manganite and other areas of research, can be studied with less effort than in more realistic models. But it is clear that to be realistic and quantitative, the two e_g orbitals are needed. This issue was clear since the pioneering work of Goodenough and others reporting orbital ordering in Mn oxides. In addition, Millis et al. [5.2] estimated the D.C. resistivity of the one-orbital model, and concluded that it is far from realistic, strongly suggesting that two orbitals are crucial for a proper description of manganites, as remarked also by Kanamori and others (see discussion in Chap. 4). Note that arguments related to the actual value of the Curie temperature may not be sufficient to exclude the one-orbital model since realistic numbers can be found with little effort (see Chap. 6). But the other arguments presented in this paragraph are sufficient to motivate the use of the more complicated two-orbital models.

Millis et al. [5.3] and [5.4] argued that the physics of manganites is much influenced by a strong electron Jahn–Teller phonon coupling. Röder et al. [5.5] also remarked on the potential relevance of phonons in the manganite context. The electron-JT-phonon coupling leads to a static Jahn Teller distortion at small hole density, well known from experiments, but it was argued by Millis et al. that it also produces a dynamical Jahn-Teller effect at other densities (namely the phonons are active, but no long-range order is established for the particular lattice distortion that dominates). These dynamical fluctuations can strongly affect the physics of carriers in such *local* environments. At finite hole density, and using the infinite-dimensional mean-field approximation, Millis et al. argued that very small polarons were created, and their potential localization could lead to large magnetoresistance. These issues are further discussed in Chap. 22 where the theories for CMR are summarized, and the small-polaron approach of Millis is contrasted with quite different theories that exploit phase competition.

5.8 The Two-Orbital Model

Considering the previous discussion on the relevance of the Jahn–Teller phonons and orbital ordering, it is imperative to define a simple model that includes explicitly the two active orbitals of the e_g sector, and a coupling to those JT phonons. The argument that a large Hund coupling is sufficient to account for the on site Hubbard U repulsion still holds even with two orbitals, and for that reason the Coulomb term $H_{\mathrm{el-el}}$ of (5.6) is dropped, an important simplification for both analytical and numerical treatments. The resulting model is

$$\begin{aligned} H &= H_{\mathrm{kin}} + H_{\mathrm{Hund}} + H_{\mathrm{AF}} + H_{\mathrm{el-ph}} \\ &= -\sum_{\mathbf{ia}\gamma\gamma'} t^{\mathbf{a}}_{\gamma\gamma'} a^{\dagger}_{\mathbf{i}\gamma} a_{\mathbf{i+a}\gamma'} - J_{\mathrm{H}} \sum_{\mathbf{i}\gamma} \mathbf{s}_{\mathbf{i}\gamma} \cdot \mathbf{S}_{\mathbf{j}} + J_{\mathrm{AF}} \sum_{\langle \mathbf{ij} \rangle} \mathbf{S}_{\mathbf{i}} \cdot \mathbf{S}_{\mathbf{j}} \\ &\quad -2g \sum_{\mathbf{i}} (Q_{2\mathbf{i}} T^x_{\mathbf{i}} + Q_{3\mathbf{i}} T^z_{\mathbf{i}}) + (k_{\mathrm{JT}}/2) \sum_{\mathbf{i}} (Q^2_{2\mathbf{i}} + Q^2_{3\mathbf{i}}) \,. \end{aligned} \tag{5.11}$$

The kinetic part, H_{kin}, was discussed earlier in this chapter and in the previous one (4.18). The actual values of the hopping amplitudes for the e_g sector are obtained in the next section. The AF coupling between the localized spins, H_{AF}, was presented in Sect. 5.2, and the Hund coupling H_{Hund} was given in Sect. 5.1 (simply including now a sum over the two orbitals). Finally, the electron phonon coupling was presented in Sect. 5.3. This last term contains two "ambiguities" that must be clarified when using or citing the model: (i) either cooperative or non cooperative phonons can be used, and (ii) the breathing-mode term may or may not be incorporated in the analysis.

The reader should be alert to these subtleties. The two-orbital model defined in (5.11) will be extensively discussed in the rest of the book. The electron phonon coupling will be often denoted by λ instead of g, as explained in Sect. 5.3.

5.9 Hopping Amplitudes

In the kinetic energy portion of the Hamiltonian, the electron-hopping processes are regulated by matrix elements of Wannier functions centered at different sites of the lattice. The use of Wannier functions was important in the setup of the problem, since the second-quantization wave function operators must be expanded in an orthogonal basis (plain atomic orbitals at different sites are not orthogonal to each other). The calculation of the matrix elements relevant to electron hopping has been addressed by Slater and Koster in 1954 [5.6], and there are many reasonable simplifications that can be introduced in that calculation. In particular, they argued that both for non orthogonal atomic orbitals, as well as for orthogonal Wannier functions, the integrals that appear in the hopping term of the Hamiltonian decrease rapidly with the distance between the two sites involved in that integral (for the actual definition of the integrals, see Chap. 4). Then, a natural simplification is to restrict the sum to nearest-neighbor (NN), or at most second- or third-nearest-neighbor sites. In this chapter, the NN approximation will be made (although to be quantitative in the comparison between theory and experiments, results in the context of cuprates have shown that the second-NN and third-NN contributions cannot be neglected [5.7], and perhaps a similar extension could be needed for manganites as well).

Slater and Koster carried out the calculation of the hopping integrals using a variety of symmetry arguments to reduce their number. Along the line linking the two sites (i,j) that appear in the integral, the two Wannier orbitals involved in the problem (example, p_x, and d_{xy}) can be decomposed as combinations of rotated orbitals with quantization axis defined along that line, introducing geometrical factors in the results. After such decomposition a small number of basic integrals must still be calculated. These are overlap integrals between the p-orbitals of the oxygens and the d-orbitals of manganese, and two of them are shown in Fig. 5.1. No other details will be provided here about the procedure and discussion related to the estimation of hopping integrals. The reader can find additional information in [5.6].

The results for the matrix elements that are of interest for our problem can be read directly from the original paper of Slater and Koster [5.6]. The integral between p_x- and $d_{x^2-y^2}$-orbitals is given by

$$E_{x,x^2-y^2}(\ell,m,n) = (\sqrt{3}/2)\ell(\ell^2 - m^2)(pd\sigma) + \ell(1 - \ell^2 + m^2)(pd\pi)\,, \tag{5.12}$$

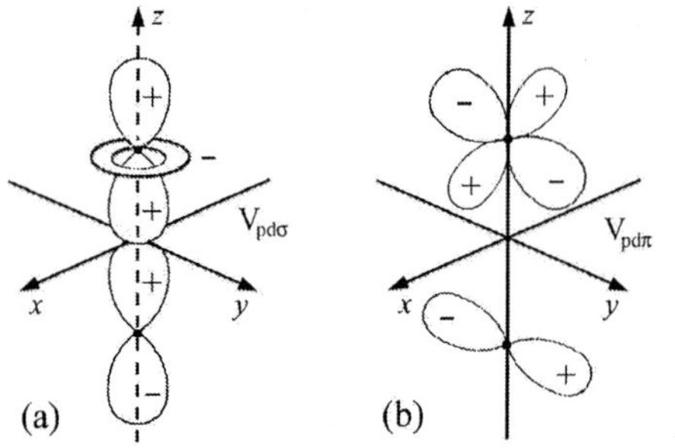

Fig. 5.1. (**a**, **b**) Schematic representation of overlap integrals that appear in the Slater Koster matrix elements for electron hopping. Details can be found in [5.6]

where $(pd\sigma)$ $[(pd\pi)]$ is the overlap integral between the $d\sigma$ and $p\sigma$ orbitals [$d\pi$ and $p\pi$ orbitals] (see Fig. 5.1). Note that these integrals actually depend on the distance between the pairs of sites considered (see [5.6]). The integers (ℓ, m, n) correspond to the components of the *unit* vector along the direction from oxygen to manganese. Similarly, the overlap integral between p_x- and $d_{3z^2-r^2}$-orbitals is expressed as

$$E_{x,3z^2-r^2}(\ell,m,n) = \ell[n^2 - (\ell^2+m^2)/2](pd\sigma) - \sqrt{3}\ell n^2(pd\pi)\,. \qquad (5.13)$$

Using these expressions, the effective hopping amplitude between adjacent (i.e., NN) manganese ions along the x-axis, with hopping via the oxygen $2p_x$-orbitals, can be evaluated. The results are

$$\begin{aligned} -t^{\mathrm{x}}_{\mathrm{aa}} &= E_{x,x^2-y^2}(1,0,0) \times E_{x,x^2-y^2}(-1,0,0) = -\frac{3}{4}(pd\sigma)^2\,, \\ -t^{\mathrm{x}}_{\mathrm{bb}} &= E_{x,3z^2-r^2}(1,0,0) \times E_{x,3z^2-r^2}(-1,0,0) = -\frac{1}{4}(pd\sigma)^2\,, \\ -t^{\mathrm{x}}_{\mathrm{ab}} &= E_{x,x^2-y^2}(1,0,0) \times E_{x,3z^2-r^2}(-1,0,0) = +\frac{\sqrt{3}}{4}(pd\sigma)^2\,, \qquad (5.14) \end{aligned}$$

where a $= x^2-y^2$ and b $= 3z^2-r^2$. Note that the minus sign in front of $t^{\mathrm{x}}_{\gamma\gamma'}$ on the left-hand side of (5.14) is due to the definition of the hopping amplitudes in H_0 (in [5.6], no sign is introduced in the definition of the hopping term). The minus sign in the argument of the second of the integrals for each of the hopping amplitudes in (5.14) arises from the fact that the unit vector (ℓ, m, n) is from the left state to the right state in the integral, by convention. Also note that at the distance of one lattice spacing, the $(pd\pi)$ integral does not contribute, and the same occurs in the other two directions y and z, as shown below.

The same procedure can be followed to construct the integrals along the y- and z-axis. Using the results of Slater and Koster [5.6],

$$\begin{aligned} E_{y,x^2-y^2}(\ell,m,n) &= (\sqrt{3}/2)m(\ell^2-m^2)(pd\sigma) - m(1+\ell^2-m^2)(pd\pi)\,, \\ E_{y,3z^2-r^2}(\ell,m,n) &= m[n^2-(\ell^2+m^2)/2](pd\sigma) - \sqrt{3}mn^2(pd\pi)\,, \\ E_{z,x^2-y^2}(\ell,m,n) &= (\sqrt{3}/2)n(\ell^2-m^2)(pd\sigma) - n(\ell^2-m^2)(pd\pi)\,, \end{aligned}$$

$$E_{z,3z^2-r^2}(\ell,m,n) = n[n^2 - (\ell^2+m^2)/2](pd\sigma) + \sqrt{3}n(\ell^2+m^2)(pd\pi)\,. \tag{5.15}$$

For the NN-site hopping, we obtain

$$\begin{aligned}
-t^{\mathbf{y}}_{\mathrm{aa}} &= E_{y,x^2-y^2}(0,1,0) \times E_{y,x^2-y^2}(0,-1,0) = -\frac{3}{4}(pd\sigma)^2\,,\\
-t^{\mathbf{y}}_{\mathrm{bb}} &= E_{y,3z^2-r^2}(0,1,0) \times E_{y,3z^2-r^2}(0,-1,0) = -\frac{1}{4}(pd\sigma)^2\,,\\
-t^{\mathbf{y}}_{\mathrm{ab}} &= E_{y,x^2-y^2}(0,1,0) \times E_{y,3z^2-r^2}(0,-1,0) = -\frac{\sqrt{3}}{4}(pd\sigma)^2\,,\\
-t^{\mathbf{z}}_{\mathrm{aa}} &= E_{z,x^2-y^2}(0,0,1) \times E_{z,x^2-y^2}(0,0,-1) = 0\,,\\
-t^{\mathbf{z}}_{\mathrm{bb}} &= E_{z,3z^2-r^2}(0,0,1) \times E_{z,3z^2-r^2}(0,0,-1) = -(pd\sigma)^2\,,\\
-t^{\mathbf{z}}_{\mathrm{ab}} &= E_{z,x^2-y^2}(0,0,1) \times E_{z,3z^2-r^2}(0,0,-1) = 0\,.
\end{aligned} \tag{5.16}$$

Overall, the hopping amplitudes along the x-, y- and z-axis satisfy

$$\begin{aligned}
t^{\mathbf{x}}_{\mathrm{aa}} &= -\sqrt{3}t^{\mathbf{x}}_{\mathrm{ab}} = -\sqrt{3}t^{\mathbf{x}}_{\mathrm{ba}} = 3t^{\mathbf{x}}_{\mathrm{bb}} = 3t_0/4\,,\\
t^{\mathbf{y}}_{\mathrm{aa}} &= \ \ \sqrt{3}t^{\mathbf{y}}_{\mathrm{ab}} = \ \ \sqrt{3}t^{\mathbf{y}}_{\mathrm{ba}} = 3t^{\mathbf{y}}_{\mathrm{bb}} = 3t_0/4\,,\\
t^{\mathbf{z}}_{\mathrm{bb}} &= t_0\,, \quad t^{\mathbf{z}}_{\mathrm{aa}} = t^{\mathbf{z}}_{\mathrm{ab}} = t^{\mathbf{z}}_{\mathrm{ba}} = 0\,,
\end{aligned} \tag{5.17}$$

where t_0 is defined by $t_0 = (pd\sigma)^2$. Note that in [5.6] and other references, procedures to evaluate $(pd\sigma)$ are outlined. In particular, Harrison [2.25] has developed formulas where the integrals are shown to be proportional to the distance d between oxygen and manganese raised to the power -3.5. This information is useful if problems with anisotropic lattice parameters are considered, perhaps caused by Jahn Teller effects. However, in our context, a different philosophy is followed. Instead of attempting to calculate the hopping amplitude t_0, a difficult task that usually leads to numbers with substantial errors, we will simply consider it as a free parameter to be fixed only after a comparison between theory and experiments is carried out. Typical values for hoppings between nearest-neighbor transition-metal ions in oxides, such as manganites and cuprates, are approximately a fraction of an eV.

5.10 Noninteracting e_g-electron Model

An interesting limit is obtained by retaining only the kinetic portion of the full Hamiltonian, namely, by considering the case of "free" electrons that do not interact with other degrees of freedom or among themselves. Although this is certainly an oversimplification for describing the complex nature of manganites, it can be a starting model to study the transport properties of these compounds (see [5.8]). This approximation could be used in the ferromagnetic region at low temperature in which experiments show that the static Jahn Teller distortion does not occur, and the effect of Coulomb interactions is expected to be renormalized into quasi-particle properties.

In the free-electron limit, the kinetic term can be easily diagonalized using the standard Fourier transform to momentum-space of the real-space creation and destruction operators, namely, introducing $a_{\mathbf{k}\gamma\sigma} = (1/\sqrt{N}) \sum_{\mathbf{j}} \mathrm{e}^{\mathrm{i}\mathbf{j}\cdot\mathbf{k}} a_{\mathbf{j}\gamma\sigma}$, with N the number of sites. Applying this transformation, the free-electron portion of the Hamiltonian, denoted by H_0, becomes

$$\begin{aligned} H_0 &= - \sum_{\mathbf{j}\mathbf{a}\gamma\gamma'\sigma} t^{\mathbf{a}}_{\gamma\gamma'} a^{\dagger}_{\mathbf{j}\gamma\sigma} a_{\mathbf{j}+\mathbf{a}\gamma'\sigma} \\ &= -\frac{1}{N} \sum_{\mathbf{j}\mathbf{a}\gamma\gamma'\sigma} t^{\mathbf{a}}_{\gamma\gamma'} \sum_{\mathbf{k}} \mathrm{e}^{\mathrm{i}\mathbf{j}\cdot\mathbf{k}} a^{\dagger}_{\mathbf{k}\gamma\sigma} \sum_{\mathbf{p}} \mathrm{e}^{-\mathrm{i}(\mathbf{j}+\mathbf{a})\cdot\mathbf{p}} a_{\mathbf{p}\gamma'\sigma} \\ &= - \sum_{\mathbf{k}\mathbf{p}\mathbf{a}\gamma\gamma'\sigma} t^{\mathbf{a}}_{\gamma\gamma'} a^{\dagger}_{\mathbf{k}\gamma\sigma} a_{\mathbf{p}\gamma'\sigma} \mathrm{e}^{-\mathrm{i}\mathbf{a}\cdot\mathbf{p}} \frac{1}{N} \sum_{\mathbf{j}} \mathrm{e}^{\mathrm{i}\mathbf{j}\cdot(\mathbf{k}-\mathbf{p})} \\ &= + \sum_{\mathbf{k}\gamma\gamma'\sigma} \varepsilon_{\mathbf{k}\gamma\gamma'} a^{\dagger}_{\mathbf{k}\gamma\sigma} a_{\mathbf{k}\gamma'\sigma} , \end{aligned} \tag{5.18}$$

where $\varepsilon_{\mathbf{k}\gamma\gamma'} = - \sum_{\mathbf{a}} t^{\mathbf{a}}_{\gamma\gamma'} \mathrm{e}^{-\mathrm{i}\mathbf{k}\cdot\mathbf{a}}$, and we used $\frac{1}{N} \sum_{\mathbf{j}} \mathrm{e}^{\mathrm{i}\mathbf{j}\cdot(\mathbf{k}-\mathbf{p})} = \delta_{\mathbf{k}\mathbf{p}}$ (note also that the vector $\mathbf{a}$ points in $2D$ directions in a D-dimensional lattice, i.e., it provides a link to sites forward and backward along the crystal axes from the site $\mathbf{j}$ under consideration). Due to its translationally invariant character, H_0 is already diagonal in the momentum degree of freedom (the Fourier trick will not be useful in a disordered system where translational invariance is lost). H_0 is also diagonal in spin, but regarding the orbital degree of freedom, 2×2 blocks remain to be diagonalized. The coefficients $\varepsilon_{\mathbf{k}\gamma\gamma'}$, needed to fully solve the problem, can be obtained explicitly from the hopping amplitudes quoted in the previous section of this chapter. Providing these coefficients explicitly as a 2×2 matrix $\hat{\varepsilon}_{\mathbf{k}}$ in the orbital index, they are

$$\hat{\varepsilon}_{\mathbf{k}} = \frac{t_0}{2} \begin{pmatrix} -3(C_x + C_y) & \sqrt{3}(C_x - C_y) \\ \sqrt{3}(C_x - C_y) & -(C_x + C_y + 4C_z) \end{pmatrix} , \tag{5.19}$$

where $C_\mu = \cos k_\mu$ ($\mu{=}x$, y, and z). After the straightforward diagonalization of $\hat{\varepsilon}_{\mathbf{k}}$, two bands are obtained with energies

$$\begin{aligned} E^{\pm}_{\mathbf{k}} = -t_0 \Big(& C_x + C_y + C_z \\ & \pm \sqrt{C_x^2 + C_y^2 + C_z^2 - C_x C_y - C_y C_z - C_z C_x} \Big) . \end{aligned} \tag{5.20}$$

Note that the cubic symmetry can be observed clearly in $E^{\pm}_{\mathbf{k}}$, namely all three directions are equivalent. This, in spite of having hopping amplitudes that appear to be quite anisotropic, due to the choice of a particular basis for the d-orbitals. In spite of such arbitrary basis selection, the Hamiltonian obeys the rotational invariance in orbital space, as shown in this example of the free theory. Other bases for the d-orbitals certainly lead to the same result. This effect is somewhat similar to the choice of a "gauge" in electromagnetism: the most important results are gauge invariant.

5.11 Estimation of Couplings

To analyze the physics contained in the models discussed in this chapter, it is important to estimate the magnitude of their coupling constants. With this information, the dominant processes for the behavior of electrons in manganites can be clarified. In addition, if some couplings are clearly stronger than others, further simplifications can be made in the models while still keeping their main properties intact. However, the estimations are difficult. One may think that a simple comparison with experiments would be sufficient for such estimations, but the reader must realize that carrying out reliable calculations with the many-body Hamiltonians described in previous sections is truly a formidable task. In several cases only rough estimates of observables can be made. In addition, experiments are sometimes difficult to analyze, since it is not always clear what observables theorists should calculate to compare with the data. For this reason, the values quoted here should only be taken as rough estimates, to guide the intuition to be developed in later chapters about the physics of the multiorbital models coupled with phonons, which must be studied in the manganite context.

The largest energy scale is typically that of the on site Hubbard repulsion U. This is reasonable since the Coulombic repulsion between two electrons in the same orbital (with different spins) is naturally expected to be substantial. In addition, other transition-metal oxides such as the high-T_c cuprates also have a large Hubbard on site repulsion (see [1.20]). One way to estimate the coupling U is by using photoemission techniques, to be described later in this book. Usually the comparison between photoemission data and theory is reliable for very "local" quantities such as U: a small cluster with many active orbitals is solved exactly, and the results compared with the photoemission data. Using this method, U was estimated to be between ~ 5 and $\sim 3.5\,\mathrm{eV}$, in [5.9]. On the theory front, using density-functional calculations it has recently been argued that U should be $\sim 7\,\mathrm{eV}$ for some manganites [5.10]. The order of magnitude estimated from experimental and theoretical considerations is quite similar.

Note. Density-functional theory is a powerful tool to carry out ab initio calculations of many properties of solids. The "local density approximation" (LDA) has been widely applied to transition-metal oxides. In general, lattice constants, atomic positions, elastic and phononic properties are accurately predicted by these techniques. Hopping amplitudes and properties of metallic phases can be studied properly within LDA, while the analysis of phases with insulating characteristics is more complicated. More information about these methods can be found in [5.11].

In [5.9] the "charge-transfer" gap Δ of manganites was estimated to be $\sim 3\,\mathrm{eV}$. This coupling Δ was not described in the previous discussion, but it is important in transition-metal oxides. Its origin can be traced to the obvious fact that the manganites have oxygen ligands between the manganese ions. Their effect is sometimes incorporated into the analysis of cooperative phonons, but even without phonons the oxygens influence on electronic be-

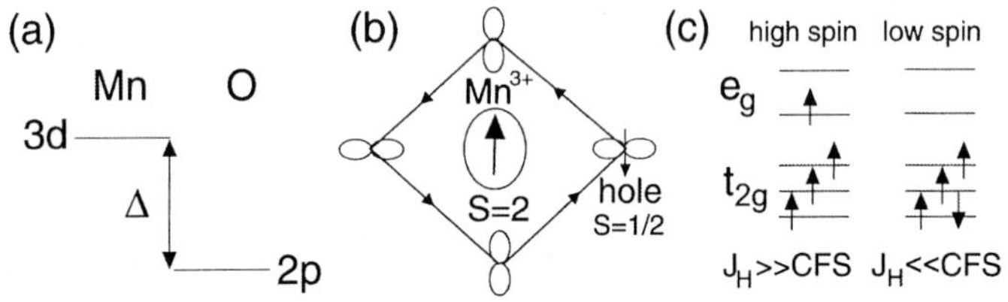

Fig. 5.2. (**a**) The charge-transfer gap Δ between the d-levels of a Mn ion and the p-levels of an O ion. (**b**) Schematic representation of a Zhang Rice quadruplet. At the center is a spin-2 Mn^{3+} ion. (**c**) The competition between the Hund coupling J_{H} and the crystal-field splitting (CFS) leads either to high-spin or low-spin configurations. In manganites, the high-spin state lowers the energy

havior. For instance, as shown in Fig. 5.2a the oxygen p-orbitals (the active ones for oxygen) are higher in energy by an amount Δ with respect to the active e_g orbitals of manganites. The movement of electrons, which have the oxygens as bridges between Mns, is affected by the value of Δ. Moreover, when holes are added by removing electrons, those electrons may come from the p-orbitals of the oxygens. Then, the holes may effectively be located at the oxygens rather than at the manganese sites. All this appears to be sufficiently important that oxygens should under no circumstances be neglected from the theoretical considerations. That may well be true. However, adding the oxygen orbitals to the electronic models complicates enormously the theoretical studies, which are already quite difficult even with only Mn active sites.

The rationale out of the apparently drastic assumption that oxygens can be dropped from the models lies in the concept of "Zhang Rice" singlets. This idea was introduced a long time ago in the area of cuprates [1.20] where a similar neglect of oxygens is widely done in theoretical studies that rely on the t–J or one-band Hubbard models. The main concept, adapted to the case of manganites, is sketched in Fig. 5.2b. Here, is shown the spin-2 Mn^{3+} ion (the spin is the largest due to the large Hund coupling), and a hole of spin-1/2 located at a neighboring oxygen. The idea is that, by symmetry, the hole can be equally located in any of the four neighboring oxygen sites next to the center Mn. In addition, the hole and Mn spins will couple in a state of the minimum spin, which is $S = 3/2$ here (see Fig. 5.2b), Antiparallel coupling is usually the case when superexchange produces antiferromagnetic interactions between nearest-neighbor sites (in this case Mn O). For cuprates, the center spin is 1/2 (involving Cu^{2+}), creating a singlet spin-0 when coupled with the spin 1/2 of the oxygen holes, justifying the "singlet" name of the Zhang Rice state for the high-T_{c} models.

The expectation is that these states, the "Zhang Rice $S = 3/2$ quadruplet" in the manganite case, become mobile due to the nonzero overlaps between these quadruplets when centered at nearest-neighboring Mn sites. This produces a band that can be decoupled from the spin-5/2 sector of maximum Mn O spin. By working in the reduced Hilbert space of spin-3/2 Zhang Rice quadruplets, a model with degrees of freedom only centered at the Mn sites should be recovered. This resulting model will be of the variety described in the previous analysis in this chapter. This approximation closes the logical rationale in favor of dropping the oxygens from the theoretical considerations. However, it is important to remark that the actual formal study of the stability of the Zhang Rice quadruplets has not been carried out to the best of my knowledge, and the issue has not received as much attention as in the case of the cuprates. Until detailed studies are presented, the reader should consider the neglect of oxygens as a guess that still needs a firmer basis, although it has a reasonable rationale behind. Nevertheless, the use of models without active oxygens (with the exception of those cases involving cooperative Jahn Teller phonons) is so widely spread, that the rest of this book is built upon models where oxygens and the charge-transfer gap are not explicitly considered for the electronic motion. However, a recent paper claims that the actual location of holes in oxygens is important to explain ferromagnetism [5.12].

The large value of the on site Hubbard coupling basically forbids the double occupancy of the same orbital. But this constraint is not sufficient to understand the physics of manganites. Another very important coupling is the Hund coupling $J_{\rm H}$, which appears in basic models such as the one- and two-orbital models. Estimations of this coupling gives values between 1 eV and 2 eV. The evidence comes from photoemission [5.13], optical conductivity experiments [5.3, 5.14–5.16] (the technique of optical conductivity is very important in studies of oxides and other materials: its main properties are described in Chap. 12), and band-structure calculations [5.17]. The Hund coupling should be compared with the crystal-field splitting, since their relative value is important for the total spin of the individual ions. In the case of manganites, the separation between the e_g and t_{2g} levels is about 1 eV [3.27,5.18], which is not sufficient to dominate over the Hund coupling tendency toward ferromagnetism. As a consequence, the total spin is maximum. In Fig. 5.2c, a sketchy representation of Hund coupling dominated and crystal-field dominated states is provided. Not all transition-metal oxides are dominated by the Hund term, and there are cases where the minimum spin configuration lowers the energy.

A key parameter in all the models studied is the hopping amplitude "t", which regulates how mobile the electrons are, at least at short distances. It is particularly difficult to obtain this number since it requires a careful comparison between theory and experiments, and many of the calculations are not accurate. Nevertheless, it is widely accepted that t is just a fraction of 1 eV, and a quick average of information in the literature locates its value around 0.5 eV. However, wide fluctuations in the reported numbers, or in the ad hoc values selected by theorists to compare with experiments, exist.

Table 5.1. Approximate values for the couplings that appear in the one- and two-orbital models for manganites. These are rough estimations, average over numbers quoted in the literature. However, the order of magnitude is widely accepted as correct. The numbers are in units of eV, with the exception of λ, which is dimensionless

U	J_{H}	CFS	t	λ	J_{AF}
6	2	1	0.5	1–1.6	0.05

Hoppings as small as 0.2 eV or as large as 1 eV are not rare in papers. Some estimations can be found in [5.19].

As emphasized in the manganite literature, the electron phonon coupling may be large, particularly when considering Jahn Teller phonons (this issue is discussed extensively later in this book). Photoemission experiments [5.13] estimate the value of the dimensionless coupling λ, the JT electron coupling, between 1 and 1.6. This is in agreement with other studies [5.3, 5.20].

The smallest coupling is J_{AF}, which, according to some estimations, is $\sim 0.1\,t$ [5.21]. Using $t = 0.5\,\mathrm{eV}$, this leads to $J_{\mathrm{AF}} = 0.05\,\mathrm{eV}$. However, it will be discussed in future chapters that even with a small value, J_{AF} may play a key role in stabilizing interesting experimentally observed phases of manganites. In particular, neglecting this coupling entirely would fail to produce a G-type AF phase at 100% hole doping, as in $CaMnO_3$.

In Table 5.1, the main results of this section are summarized. The numbers quoted are simply rough estimations, which should be considered together with generous error bars.

6. The One-Orbital Model: Phase Diagram and Dominant Correlations

This chapter describes the main results available in the context of the one-orbital model for manganites. This model appears in the literature also as the ferromagnetic Kondo model or the double-exchange model. We use the name "one-orbital model" to highlight the differences with the more realistic two-orbital model with Jahn Teller phonons, also analyzed extensively in this book. The information provided in this chapter includes phase diagrams and dominant ordering tendencies in the ground state. It also includes the regimen of *electronic phase separation* believed to be of relevance for Mn oxides. A similar phenomenon, although not involving ferromagnetism, has been widely discussed in the Cu oxide literature as well [1.3]. This chapter also contains a brief review by A. Moreo of related spin-fermion models used in the context of high-T_c superconductors. A simple model of interacting localized spins and mobile electrons clearly finds applications in several contexts, including also heavy fermions and diluted magnetic semiconductors.

6.1 Phase Diagram at $J_{AF} = 0.0$

The zero-temperature phase diagram of the one-orbital model in the absence of an explicit coupling J_{AF} between the localized spins is shown in Fig. 6.1a–c, for 1D, 2D, and infinite dimension. The procedure followed to obtain the phase diagram, and the intuitive origin of the many phases in the problem are explained in the following sections. Here we simply remark that *four* different regimes or phases have been systematically observed (again, as long as $J_{AF} = 0.0$). These phases show (i) antiferromagnetism (AF) at a density of mobile electrons $\langle n \rangle = 1$, (ii) ferromagnetism (FM) at intermediate electronic densities, (iii) phase separation (PS) typically between FM and AF phases (although other phases can also participate), and (iv) a regime with spin incommensurability (IC) at not too large Hund coupling. Understanding the origin of these phases leads to a better rationalization of the physics of manganite models. The details leading to Fig. 6.1a–c are explained below.

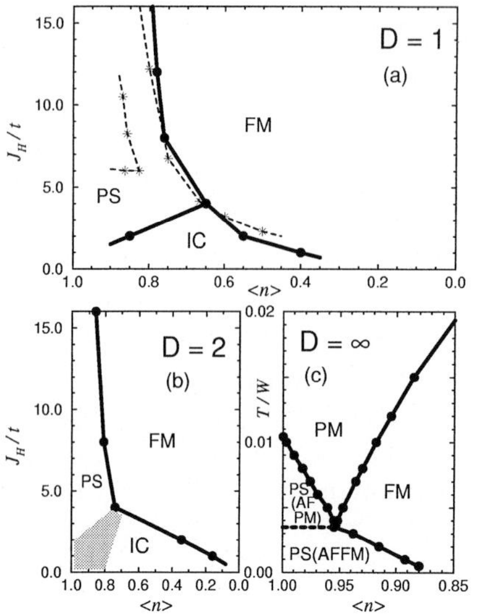

Fig. 6.1. Phase diagram of the one-orbital model for manganites, also known as the FM Kondo model, reproduced from Yunoki et al. [6.1]. AF, FM, PS, and IC denote regimes with antiferromagnetic correlations, ferromagnetic correlations, phase separation, and spin-incommesurate correlations, respectively. For details on cluster sizes used, boundary conditions, temperatures, the reader should consult the original reference [6.1]. (**a**) and (**b**) were obtained using Monte Carlo techniques, at very low temperatures. AF (not shown) occurs at $\langle n \rangle = 1$. The *dashed lines* in (**a**) are results with quantum-localized spins discussed in [6.1]. (**c**) was obtained using the dynamical mean-field approximation (also known as the infinite-dimensional mean-field approximation), with W being the half-width of the density of states used in the calculation

6.2 Ferromagnetism

As discussed in previous chapters, it is well known that ferromagnetism is an important ingredient of the physics of manganites. For this reason, any realistic model for Mn oxides must have some portion of the phase diagram with spin ferromagnetic properties. At least qualitatively, the discussion in this chapter shows that the one-orbital model has properties that makes it fairly realistic in spite of its lack of orbital degeneracy, thus it should have such a spin-ordered phase. To find out if the lowest-energy state of the model has "ferromagnetic" properties, we have to calculate equal-time spin spin correlations in that state, which are denoted by $\langle \boldsymbol{S_i} \cdot \boldsymbol{S_j} \rangle$, involving spins at the lattice sites $\boldsymbol{i}$ and $\boldsymbol{j}$. The mean value denotes both thermal and quantum averages. For the type of models under consideration, it is usually equivalent to consider as "spin" in the mean value defined above, either the classical localized spin or the quantum mobile one. Due to the strong Hund coupling, any of these spins should signal the presence of ferromagnetism. However, as the electronic density is reduced toward zero, we must be careful since the mobile-electrons contribution can be weak.

Returning to the spin spin correlations, if two spins located at nearest-neighbors sites are aligned perfectly parallel to each other, then their spin spin correlation will be positive and of maximum value. However, in general, due either to finite temperature effects and/or quantum fluctuations, this correlation at distance one is smaller than its maximum value, although in a ferromagnetic state it typically remains substantial. Although knowing the strength of this correlation at one lattice spacing is qualitatively a fairly clear indication of FM "tendencies", it is not sufficient to label the state as ferromagnetic. For this, what must occur is that the spin spin correlations at *any* distance must remain positive and finite. In such a case, we say that the system has developed "long-range" order. If the correlations at long distances vanish, while those at short distances are positive and finite, the system has instead "short-range" order. Crudely, we can visualize the latter as consisting of regions or islands of ordered spins each with a net moment, but pointing in different directions, leading to an overall vanishing total magnetic moment. The size of these regions is the correlation length ξ, which is defined through the exponential decay of the spin spin correlations with distance. The state is not frozen, but very dynamical in that the positions and sizes of these regions change with time (unlike the situation in explicitly disordered systems, discussed in Chap. 17, where the clusters can be nearly frozen in time).

As representative of calculations that search for ferromagnetism, consider the spin spin correlation shown in Fig. 6.2a, obtained using the Monte Carlo technique (to be discussed in detail in Chap. 7) at the couplings, temperature, and densities shown, for a one-dimensional chain of 24 sites. In the regime studied, the correlations are all positive and finite, even up to distances of 12 lattice spacings, signaling a state with strong tendencies toward ferromagnetism. Figure 6.2b shows the dependence on temperature. It is clear that at the lower temperatures of the plot, the correlations are robust at least up to 5 lattice spacings. At the larger temperatures, these correlations decay rapidly and nearly vanish at a finite distance from the spin of reference. In this case, the state is not globally ferromagnetic, although it has a tendency to such a state at short distances. This is characteristic of many systems. As the temperature is reduced, correlations start developing at short distance of the type of order that eventually will become globally stable at low temperatures. Usually the decay of the correlations along the chain can be fit with an exponential $e^{-\ell/\xi}$, with ξ the correlation length denoting the typical size of an spin-ordered island. This correlation length grows with decreasing temperature and for continuous transitions (second order) they diverge at the Curie temperature, T_C, where global ferromagnetism appears in the system. Note that for real manganites, the first- vs. second-order nature of the ferromagnetic transition is much debated, but the one-orbital model appears to have a continuous transition at least for $J_{AF} = 0$.

When spin-order patterns are searched for, it is very common to use the spin structure factor $S(q) = (1/L) \sum_{j,m} e^{i(j-m)q} \langle \boldsymbol{S}_j \cdot \boldsymbol{S}_m \rangle$, namely the transform of real-space spin correlations into momentum space (the expression is valid in 1D, for higher dimensions just use vector indices for sites and momenta). By this procedure, if the spin correlations are positive and finite, then at momentum zero $S(q)$ is the largest. Other momenta induce cancellations in the sums. In addition, when ferromagnetic long-range order is

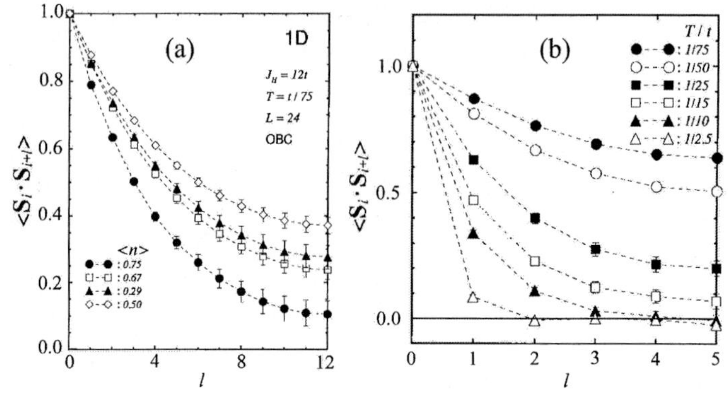

Fig. 6.2. (**a**) Spin spin correlations among the localized spins (classical) obtained with the Monte Carlo technique applied to a chain of 24 sites (figure reproduced from [6.2]). Densities, temperature and couplings are shown (J_{AF} is zero). (**b**) Spin spin correlations among localized spins on a 10-site cluster at e_g density 0.7 (mobile electrons), and with $J_{\mathrm{H}} = 8t$. Temperatures are indicated

established, $S(q = 0)$ must diverge as the number of sites grows, since in its definition there is a double sum over sites of correlations that are positive. A commonly used procedure to search for long-range order is to monitor the size dependence of $S(q = 0)$. In Fig. 6.3a, results for chains of increasing size L are shown. It is clear that at low temperatures $S(q = 0)$ increases rapidly as L grows, indicating a strong tendency toward ferromagnetism. The characteristic temperature, T_{C}^*, at which $S(q = 0)$ starts growing rapidly is around $0.1t$, just a small fraction of the total hopping. The physics behind the ferromagnetic tendencies is the "double-exchange" mechanism of Chap. 2. Here, carriers polarize the background of localized spins to improve their kinetic energy. A similar mechanism is also at work in the case of the Nagaoka phase of high-temperature superconductors [3.6] (references can also be found in [1.20]). For extra references on the early work on ferromagnetism in the Hubbard model see [6.3].

The characteristic temperature T_{C}^* is very similar in one-, two- and three-dimensional systems, as will be shown below. In practice, it is a good approximation to consider it as a critical transition temperature. However, one must be careful with such statements. There are exact theorems that have established the absence of long-range order at finite temperature in 1D and 2D systems for short-range interactions in the Hamiltonian (a more detailed discussion can be found in Chap. 7). On the other hand, very often we have observed that even in these dimensions the spin correlations are very large, larger than the clusters that can be analyzed. In this case the finite size system behaves as if it has long-range order. In addition, in real systems a small coupling among chains or layers is always to be expected, rendering the system truly three-dimensional. In this case, the T_{C}^* is expected to become a

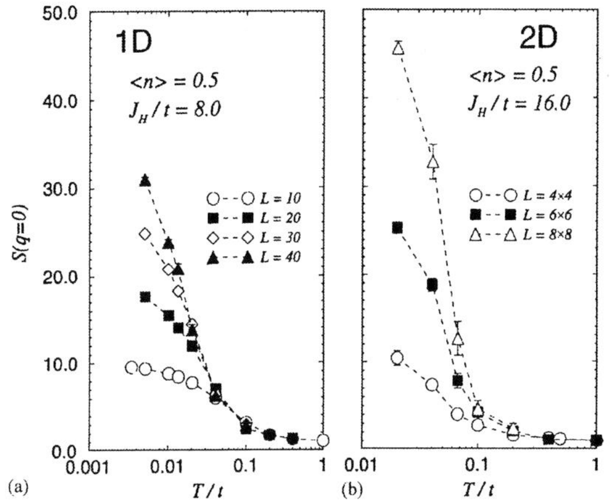

Fig. 6.3. Fourier transform of the spin-spin correlations at zero momentum vs. temperature T (in t units). Results are shown in 1D (**a**) and 2D (**b**) for the lattice sizes indicated. Density and coupling J_{H} are also shown. For more details see [6.2]

true critical temperature transition, although explicit calculations are difficult. Then, in general, we must be cautious about statements in the literature regarding long-range ordering in these systems. On the one hand, it is improper to label T_{C}^* as a true critical temperature. It is better to refer to it as a characteristic temperature at which correlation lengths start growing very rapidly with decreasing temperatures, but not truly diverging until zero temperature is reached. On the other hand, the system behaves virtually as if it were long-range ordered in 1D and 2D, and neglecting that physics would be misleading. Labeling the states in Figs. 6.2 and 6.3 as "paramagnetic" would give the false impression that magnetism is not present at all in the problem. A fine balance must be obtained. Our experience is that in general the behavior in 1D, 2D, and 3D is similar in numerical studies, and at least the intuitive physical trends can be obtained safely in any dimension, regardless of whether we have true critical temperatures or just crossover ones. Labeling T_{C}^* colloquially as a critical temperature is not such a bad approximation after all, and that language will be sometimes used in the rest of the book.

The results shown thus far were obtained using 1D and 2D systems. Similar work can be repeated in three dimensions. However, remember that the overall limitation in CPU time of the Monte Carlo method is the total number of sites $N = L^d$, where L is the linear dimension of the d-dimensional cube that is being considered. As a consequence, the largest linear dimension L that can be handled goes down rapidly with increasing dimension. In spite of this problem, reliable work has been carried out in three-dimensional systems, as shown in the next section.

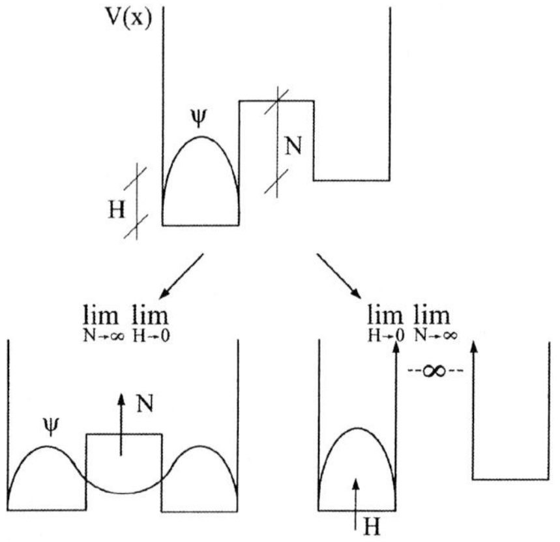

Fig. 6.4. Sketch crudely illustrating the phenomenon of symmetry breaking for a toy model of one electron in a one-dimensional potential. The notation is explained in the text

Note: Spontaneous Symmetry Breaking. Here, a brief reminder is provided for graduate students of the subtle concept of "spontaneous symmetry breaking" that occurs in many-body problems in the bulk limit. The case of ferromagnetism is an example, namely the invariance under rotations of the spin degree of freedom in the Hamiltonian is spontaneously broken in the ground state, which chooses a particular direction for the magnetization. The ground state and the Hamiltonian do not have the same symmetries. In a finite system, such as studied here, this issue becomes subtle since even in a case where the ground state is ferromagnetic the magnetization may rotate very slowly as the Monte Carlo time increases, leading, after a long period of time, to a vanishing time-averaged overall magnetization. This is a general phenomenon: on a finite cluster we say that the system can "move", although with difficulty, between equivalent minima of the free energy. Another common expression is that tunneling can occur between those equivalent minima. This is why the study of spin spin correlations is so important for finite clusters, since these correlations are the same in any of those states (the dot product between spins is rotationally invariant), while the magnetization is not (it can change sign if all spins are reversed, while a spin spin correlation does not change sign).

A quite simple illustration of how spontaneous symmetry breaking works can be found even in a toy model of quantum mechanics. Consider a potential with two asymmetric minima as shown in Fig. 6.4, and suppose the particle in the potential satisfies the Schrödinger equation. There are two parameters to play with, namely, the asymmetry in the depth of the minima H, and the size of the "wall" between them N. A situation that mimics spontaneous symmetry breaking even at the level of one electron can be set up by initially locating the electron in the ground state, which is in the deepest of the two minima, as in the upper part of the figure. The associated wave function is shown. Now let us consider two different processes. If first the constant H that regulates the asymmetry is sent to zero, then tunneling can occur (as in finite systems) and the symmetry is restored, namely the wave function is even. No symmetry breaking occurs, and the ground state is invariant under a reflection with respect to the center of the potential. After the symmetry is recovered, the barrier in between can be sent to ∞ with not much change to the system. However, if the barrier is made to diverge *before* the asymmetry constant H is sent to zero, then the electron will remain trapped on the original minima even if H is sent to zero later, since the infinite wall in between prevents the tunneling. The symmetry has been broken. A similar process occurs in many-body systems with the barrier representing the number of electrons involved in the problem (certainly a huge number) and H a small external field that can be introduced to break the symmetry, that is sent to zero at the end of the calculations. The limits $N \to \infty$ and $H \to 0$ simply do *not* commute. One has to be careful with these subtleties when analyzing spontaneous symmetry-breaking phenomena in finite systems.

6.2.1 Critical Temperature of the 3D One-Orbital Model in the Ferromagnetic Regime

The discussion in previous sections has been mainly semi quantitative. The existence of many qualitatively different regimes (ferro, antiferro, phase separated, etc.) has been established in the context of the one-orbital model, based on clear tendencies of the correlations at short distances. This is not in doubt. However, it is quite difficult to refine quantitatively the analysis to search for accurate predictions for the critical temperatures of each phase. The subtleties already mentioned related to long-range order vs. a large but finite ξ in 1D and 2D complicates these matters substantially. Even in 3D some authors consider that it may not be worth devoting enormous resources to find accurate critical temperatures, given the fact that the models we study are certainly idealized models that do not consider many important effects (a variety of phononic modes, several classes of disorder, Coulomb interactions at short and long distances, influence of oxygen ions, etc.). Then, a comparison with experiments of a necessarily simplistic model even if solved with high precision may not teach us much. On the other hand, it is important to carry out very accurate studies at least in some particular cases of simple models in order to gauge whether rougher but faster approximations have captured the essence of the problem. In addition, experiments for some manganites or other compounds may have reached sufficient precision such that *critical exponents* may be available, and if there is some universality in the results perhaps devoting a considerable effort to a particular model may pay. The best way to calculate critical exponents theoretically, with some confidence, is through extensive computational studies, with increasing lattice sizes that can accommodate increasing correlations lengths near critical points (for cases where the transition is of second order). For these reasons, recent computational studies of critical temperatures in the context of the one-orbital model are certainly welcomed.

Using recently developed Monte Carlo techniques, Alonso et al. [6.4] and Motome and Furukawa [6.5] carried out an extensive analysis of the Curie temperature T_C of the 3D one-orbital model, at $J_\mathrm{AF} = 0$, and for large values of the Hund coupling. The reported critical temperature at density $x = 0.5$ is shown in Fig. 6.5. The critical exponent β found numerically by those authors appears consistent with the expected result for a 3D Heisenberg model, namely, $\beta = 0.365$. Note that $x = 0.5$, $J_\mathrm{H} = \infty$, and $J_\mathrm{AF} = 0.0$ is in the regime of parameter space that is the farthest from competing phases,

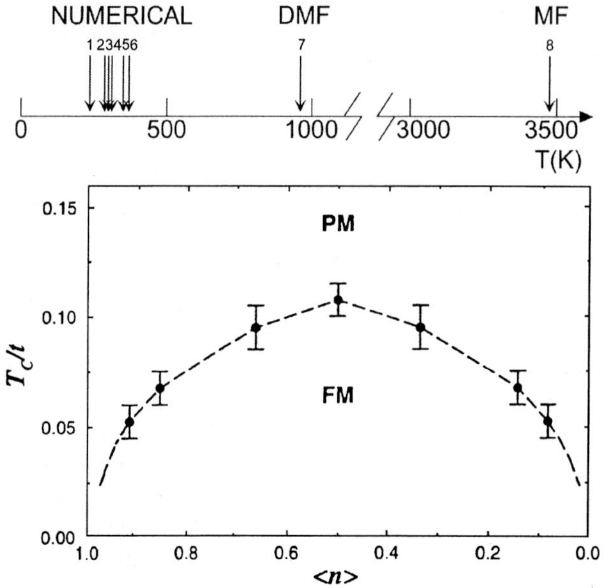

Fig. 6.5. (**a**) Sketch of predictions using the one-orbital model for the 3D Curie temperature at $J_{\mathrm{H}} = \infty$, and $J_{\mathrm{AF}} = 0.0$. The density is $x = 0.5$. "Numerical" denote results obtained with Monte Carlo or high-temperature expansion techniques. "DMF" are results obtained with the dynamical mean-field approximation, while "MF" were found with a standard mean-field approach. The result denoted by (1) corresponds to [6.1], (2) to [6.6], (3) to [6.4], (4) to [6.9], (5) to [6.8], (6) to [6.7], (7) to [6.10], and (8) to [5.2]. (**b**) Rough estimation of critical temperatures (from Yunoki et al. [6.1]) at $J_{\mathrm{H}} = \infty$, and $J_{\mathrm{AF}} = 0.0$, using the one-orbital model. More refined calculations are by now available (see (**a**)), but the essence of this result is still valid, namely the Curie temperature is a small fraction of the hopping, it is maximized at half-doping, and it behaves as a function of density symmetrically with respect to $x = 0.5$, for the one-orbital model

such as antiferromagnetism. Thus, here the critical Curie temperature is the largest and the behavior is expected to be the most standard regarding the formation of ferromagnetism. The analysis may become more complicated at other densities and couplings, with a decreasing Curie temperature.

An important practical issue is the comparison of results for T_{C} among the different available techniques. A sketch summarizing the main predictions is shown in Fig. 6.5a, where all the results using different conventions have been normalized to a bandwidth $W = 1\,\mathrm{eV}$. In the relatively low temperature region, there is an accumulation of results coming from numerical studies (including one point obtained with the high-temperature expansion, which also demands a computational effort). These numerical studies are certainly the most accurate available for the model. Although between the lowest and largest prediction there is a factor of roughly two, all results capture the essence of the problem, namely there is a transition to ferromagnetism at temperatures in the vicinity of room temperature. Some of the results in that group, for instance those of [6.1], were obtained with crude procedures aiming at finding the order of magnitude of the critical temperature (Fig. 6.5b). It is comforting that even with those crude procedures, the result was off not by a large margin. In Fig. 6.5a, results using other tecniques are also shown. They include predictions from the dynamical mean-field approximation (which is exact in infinite dimensions), with results close to 1000 K. The reason that this method incorrectly produces a T_C substantially larger than more accurate techniques is that fluctuations in space of the spin correlations are not incorporated. On the other hand, the approach takes into account dynamical fluctuations, namely the time dependence of the problem is not ignored. Finally, results corresponding to the standard mean-field approximation, where fluctuations in both space and time of the spin correlations are not incorporated in the formalism, lead to results that are too large by a factor 10. These conclusions are general and they are valid for a variety of problems. Effects of fluctuations are needed to accurately predict the onset of long-range order. For this reason our studies using the mean-field approximation (see for example Chap. 8) have been mainly focused on zero-temperature properties, where the fluctuations are the smallest. In this case, the technique is expected to capture at least the essence of the problem. However, note that there are a large variety of mean-field approximations, with many degrees of sophistication. In particular, the mean-field prediction for T_C in [6.50] is very accurate (see also [2.23] and Arovas and Guinea [6.13], for other fairly accurate predictions).

6.3 Antiferromagnetism at Electronic Density $\langle n \rangle = 1.0$

Working within the one-orbital model, the ferromagnetic phase appears simply as a natural way to optimize the movement of e_g electrons when the Hund coupling is large. A t_{2g}-polarized background is the best in this respect since, as explained before, electrons jumping from site to site do not pay Hund coupling energy in such an environment. However, while this picture is certainly valid for intermediate e_g densities between 0 and 1, it breaks in those limits. At e_g density zero, there are no mobile electrons and the only relevant coupling is the small and positive J_{AF}, which favors antiferromagnetism among the localized spins. The other limit of density 1 is slightly more subtle, and also leads to antiferromagnetism. Imagine what would occur at such densities if the system were ferromagnetic, as in Fig. 6.6a. In such a configuration the strong on site Hund coupling is properly accounted for since the e_g and t_{2g} spins are parallel, but due to the Pauli principle the mobile electrons simply cannot move. Thus, the kinetic energy is zero in the absence of holes, and the argument that a ferromagnetic background would favor the electronic

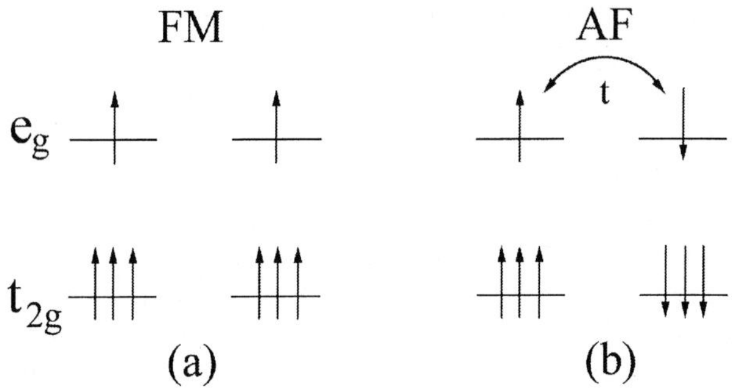

Fig. 6.6. (**a**) Ferromagnetic state at large Hund coupling and e_g density 1. The Pauli principle freezes electrons. (**b**) At density 1, an antiferromagnetic state has a lower energy since the electrons can move in the exchange processes

kinetic energy is no longer valid. On the other hand, consider an antiferromagnetic configuration Fig. 6.6b. The on site Hund coupling is also satisfied by this arrangement, and in this configuration electrons can move at least to nearest-neighbors, as shown in the figure. In such a movement they must pay a large energy of order $J_{\rm H}$, but at least such process reduces the e_g kinetic energy from zero, improving the energy with respect to the ferromagnetic state. A small effective antiferromagnetic Heisenberg coupling is generated, which regulates the behavior of the resulting spin staggered ordered state. The presence of antiferromagnetism at e_g density 1 has been extensively confirmed by a variety of techniques. For instance, in the inset of Fig. 6.7a (next section) the spin spin correlations in real-space calculated using Monte Carlo techniques are shown for two densities, including $\langle n \rangle = 1$. The alternation in the sign of the correlation, indicative of antiferromagnetism, is clear from this figure. This tendency toward antiferromagnetism is strong in a variety of mean-field approximations as well.

6.4 Phase Separation

From the previous discussion, it is clear that the one-orbital model has antiferromagnetic properties at densities of e_g electrons either 1 or 0, while at intermediate densities, such as 0.5, the ferromagnetic ground state is preferred. A natural question arises: how do we interpolate from one to the other as the density is reduced, for example, from 1 to 0.5? The question is non trivial since the states in competition are very different, and it is difficult to imagine a simple way in which short-range correlations of one of the two phases can start building up in the other, in the vicinity of the transition. This suggests that a *first-order* transition may occur between these two states. In fact, first-order transitions smeared by quenched disorder are at the heart of possible explanations of the CMR effect that will be discussed

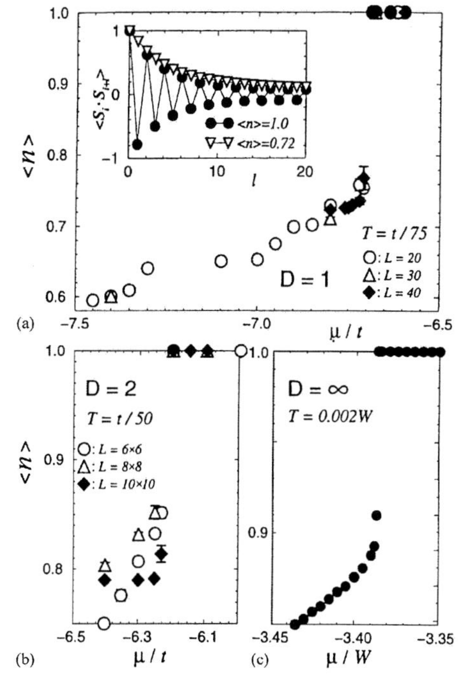

Fig. 6.7. Electronic density $\langle n \rangle$ of e_g electrons vs. chemical potential μ, reproduced from Yunoki et al. [6.1]. The results have been obtained using Monte Carlo techniques at $J_{\mathrm{AF}} = 0$. (**a, b**) and (**c**) correspond to dimensions one, two, and ∞, respectively. Temperatures and cluster sizes are indicated. The coupling is $J_{\mathrm{H}}/t = 8.0$ in (**a**) and (**b**) and $J_{\mathrm{H}}/W = 4.0$ in (**c**) (W is the half-width of the density of states). In (**a**) the *inset* shows the spin spin correlation in real space at densities 1.0 and 0.72, i.e. at the extremes of the jump in μ, showing that phase-separation occurs between FM and AF regions

in Chap. 17. Limiting the discussion for the time being to the case in hand, namely the one-orbital model and the interpolation from FM to AF as the density changes, how does such an abrupt transition occur in practice? The answer is through a first-order transition in the external parameter that regulates the density, namely through an abrupt change in the behavior of the system as the *chemical potential* μ is varied. This phenomenon leads to the notion of electronic *phase separation*, a concept that is much discussed these days in many areas of research, including high-temperature superconductors.

Consider the Monte Carlo results obtained in simulations of the one-orbital model shown in Fig. 6.7a for the case of a one-dimensional chain. Results very similar to those obtained in 1D also appear in studies carried out with 2D and 3D clusters, and also in investigations of the problem in the limit of infinite dimension (Figs. 6.7b and c). From elementary statistical mechanics, we know that in the grand-canonical ensemble the density of electrons is not fixed. Only its mean value is known as μ varies. This relation density μ is analogous to the relation between energy and temperature, the latter both in the canonical and grand-canonical ensembles. As μ is changed in Fig. 6.7 as a continuous external parameter, a *discontinuous* behavior is observed in the output, in particular in the density of e_g electrons. This is characteristic behavior of first-order abrupt transitions. No matter how carefully we try to tune μ in the region of the discontinuity, it is simply not possible to obtain the intermediate densities as an output. Then, for the case of Fig. 6.7a, the range of densities between 1 and approximately 0.78 is not accessible. We say that these densities are *unstable*. By measuring the spin spin correlations at the extremes of the density discontinuity, it has been observed that as μ changes across its critical value the properties of the system change from an AF at density 1, to a FM at density 0.78 or lower. Then, this is indeed like a first-order transition between ferromagnetic and antiferromagnetic states. The results in all dimensions are quite similar, showing that the phenomenon is robust and not sensitive to fine details of the coupling selection or the technique used.

In addition, a variety of mean-field and variational calculations also lead to phase separation. This shows that the evidence of its presence is not restricted to computational methods. In fact, using approximate analytical techniques, Kagan et al. [6.11] and Khomskii [6.12] also observed the existence of phase separation, instead of a canted state, at low hole density, in agreement with the results of the computational studies for both classical and quantum-localized spins. Phase separation was also found by Arovas and Guinea [6.13], using the Schwinger boson formalism, and it was also obtained in studies in the $D = \infty$ limit (see Nagai et al. [6.14]), as well as in several other calculations [6.15–6.20].

More intuitively, and more compatible with the words "phase separation", if one would force the system to have one of the unstable overall densities, then spontaneously the ground state will generate a fraction of its volume occupied by a FM hole-doped state and the rest will be filled by an AF hole-undoped state, as schematically represented in Fig. 6.8. This can be realized by working in the canonical ensemble where the number of particles is exactly fixed, such as in a cluster exact diagonalization procedure. This phase-separation phenomenon is reminiscent of what happens in alloying when we try to mix Cu and Ag: we simply cannot, and regions dominated by one element are formed embedded in regions dominated by the other, without a microscopic true mixture being generated. The system separates into two macroscopic phases.

The phase separation in the context of manganites is usually referred to as *electronic*, since it involves phenomena related to the behavior of electrons and since the phases in competition have different electronic density. Similarly phase separation occurs in models for high-temperature superconductors, such as in the famous t–J model at intermediate J/t, although involving in this case an undoped antiferromagnetic and a hole-doped state, which is not ferromagnetic (perhaps a superconductor). In fact, phase separation and inhomogeneities are widely discussed in the context of high-temperature superconductors (see [1.3, 6.21] for a brief list of references). However, in the work of Emery, Kivelson and Lin cited in [1.3] a phase separation between an antiferromagnetic and a ferromagnetic state at small J/t in the t–J model was also discussed. This phenomenon was also mentioned by Visscher [6.22]

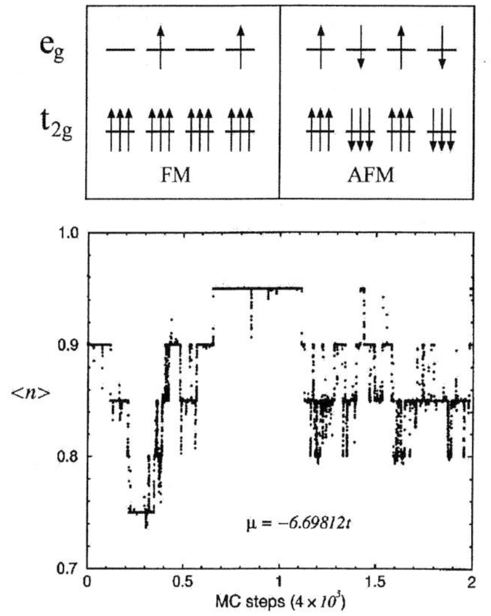

Fig. 6.8. (Upper panel) Sketch of a phase-separated state for the case of an "unstable" density, namely a density that cannot be stabilized by a simulation in the grand-canonical ensemble. (**b**) Monte Carlo time evolution of the density $\langle n \rangle$ at the particular value of μ where the discontinuity takes place, at coupling $J_{\rm H}/t = 8.0$, temperature $T = t/75$, and using a chain of 20 sites. Frequent tunneling events are observed, among the possible densities, including those that are "unstable"

several years ago, and it has clear analogies with the phase separation found in manganite models between FM and AF states. In fact, the phenomena in manganites and cuprates have a large number of common features, particularly at the phenomenological level, although there are differences as well. For a more detailed discussion and comparison between phase separation in cuprates and manganites, the reader is referred to the Appendix of Hotta et al. [6.23].

If μ is selected to be almost exactly at the location of the first-order transition, then the two states are nearly degenerate and a Monte Carlo time evolution of the density will show a complex behavior, with "tunneling" between sectors with different numbers of particles. An example of such behavior is reproduced in Fig. 6.8 (lower panel). These tunneling events are characteristic of any first-order transition, and their rate depends on the lattice size (if the lattice is very large the probability of tunneling will be very small). Note also that at low temperatures, the simulation carried out in the grand-canonical ensemble nevertheless provides information about an integer number of electrons compatible with the lattice size (see the clear "plateaus" in the density of Fig. 6.8 (lower panel), as it evolves in Monte Carlo time). This allows for the study of properties of some of the unstable regions *even* within the grand-canonical approximation.

It is important to remark that historically the possibility of this "electronic" phase separation (see discussion below), was already addressed by Nagaev [6.24] many years before it became a popular subject in the context of widely studied compounds, such as the high-temperature superconductors and more recently in the manganites. Its original application, as envisioned by Nagaev, was to antiferromagnetic semiconductors, where the doping of electrons is expected to create ferromagnetic-phase regions embedded in an AF matrix. Nagaev [6.25] remarked that if the two phases have opposite charge, the Coulombic forces will break the macroscopic clusters into microscopic ones, typically of nanometer-scale size. The calculations of Nagaev [6.26] were carried out for one-orbital models and usually in the limit where the hopping t of the conduction electrons is much larger than the Hund coupling (although Nagaev expected the results to qualitatively hold even in the opposite limit $J_{\rm H} > t$). Also, a low density of carriers was assumed, and many calculations were performed mainly for the one-electron problem (magneto-polaron), and then rapidly generalized to many electrons. The formation of lattice polarons is not included in the approach of Nagaev. These parameters and assumptions are reasonable for AF semiconductors, and Nagaev argued that his results can explain a considerable body of experimental data for EuSe and EuTe (Eu-based semiconductors are discussed in more detail in Chap. 20). However, note that Mauger and Mills [6.27] have shown that self-trapped FM polarons (ferrons) are *not* stable in three dimensions. Instead, they [6.27] proposed that electrons bound to donor sites induce a ferromagnetic moment, and they showed that bound magnetic polarons can account for the FM clusters observed in EuTe. Free carriers appear "frozen" at low temperatures in these materials, and there are no ferron-like solutions of the underlying equations in the parameter range appropriate to Eu chalcogenides. In addition, note that the manganites have a large $J_{\rm H}$ and a large density of electrons. Calculations such as those described above must be carried out using more realistic parameters, if the results can indeed apply to manganites. These calculations are difficult without the aid of computational techniques. In addition, it is clearly important to consider two orbitals to address the orbital ordering of the manganites, the possibility of orbital phase separation, and the influence of Jahn–Teller or Coulombic interactions that lead to charge-order AF states. Disorder also appears to play a key role in manganites. Nevertheless, although not quantitative, it is fair to acknowledge that the ideas of Nagaev have contributed to the study of a variety of compounds, and they helped in the current wide effort in the analysis of phase-separated systems.

6.4.1 Compressibility and Phase Separation

The compressibility K is defined as

$$\frac{1}{K} = -V\left(\frac{\mathrm{d}P}{\mathrm{d}V}\right)_N , \tag{6.1}$$

where P is the pressure, V is the volume, and N is the number of particles. The multiplication by V makes K independent of the size of the system. If the volume is reduced, then the pressure usually increases. Noticing the negative sign in the definition of the compressibility, it is concluded that K is usually positive. Actually, for the stability of the system K *must* be positive, otherwise if reducing the volume would reduce the pressure, then the system would simply collapse. Finding a negative compressibility in a calculation indicates that the system under scrutiny is unstable.

The definition of K can be transformed into an expression involving derivatives of the energy with respect to the density $n = N/V$, which is more useful in some contexts. First note that

$$\frac{1}{K} = -\frac{V}{N}\left(\frac{\mathrm{d}P}{\mathrm{d}(V/N)}\right)_N = -\frac{1}{n}\left(\frac{\mathrm{d}P}{\mathrm{d}(1/n)}\right)_N = n\left(\frac{\mathrm{d}P}{\mathrm{d}n}\right)_N . \tag{6.2}$$

Then, remember that the pressure itself is defined as $P = -(\mathrm{d}E/\mathrm{d}V)_N$, which can be rewritten as $n^2(\mathrm{d}(E/N)/\mathrm{d}n)_N$, following similar steps as in the derivation of the previous equation. Replacing the last formula for the pressure into (6.2), we obtain $1/K = 2n^2\mathrm{d}(E/N)/\mathrm{d}n + n^3\mathrm{d}^2(E/N)/\mathrm{d}n^2$, which can be rewritten in the simpler form

$$\frac{1}{K} = n^2\frac{\mathrm{d}^2(E/V)}{\mathrm{d}n^2} . \tag{6.3}$$

Working at constant volume on a finite system, this simple formula allows for the calculation of compressibilities based on the energy of different number of particles, which can be obtained, for instance, with exact diagonalization techniques [1.20] at zero temperature. The equation used in practice is

$$\frac{1}{K} = \frac{N^2}{V}\frac{\mathrm{d}^2E}{\mathrm{d}N^2} \approx \frac{N^2}{V}\left[\frac{E(N+2)+E(N-2)-2E(N)}{4}\right] . \tag{6.4}$$

The last step is obtained using (formally) an expansion $E(N \pm 2) \approx E(N) \pm 2\mathrm{d}E/\mathrm{d}N + (2^2/2)\mathrm{d}^2E/\mathrm{d}N^2$. Equation (6.4) is the natural expression for calculations of the compressibility when using techniques that directly work in the canonical ensemble, with a fixed number of particles. On the other hand, if work is carried out in the grand-canonical ensemble then it can be shown that

$$\frac{1}{K} = n^2\frac{\mathrm{d}\mu}{\mathrm{d}n} , \tag{6.5}$$

which can be evaluated numerically from the μ vs. density curve obtained, for instance, from a Monte Carlo simulation. The proof of the previous equation starts with the chemical potential μ defined as the difference in energy between having $N+1$ and N particles, i.e., $\mu = E(N+1) - E(N)$. But the energy with one more particle admits again a (formal) expansion $E(N+1) \approx E(N) + (\mathrm{d}E/\mathrm{d}N) = E(N) + (1/V)(\mathrm{d}E/\mathrm{d}n) = E(N) + (\mathrm{d}(E/V)/\mathrm{d}n)$. Then, $\mu \approx \mathrm{d}(E/V)/\mathrm{d}n$, which in combination with (6.3) produces the result (6.5). All these expressions are quite general. They have been applied in particular to manganite models, in the context of the phase-separated regimes.

Note:. If working in the canonical ensemble the compressibility is found to be negative for some finite system, then this system has a tendency to phase separate, as explained before. This can be understood from Fig. 6.9, where the energy of a *homogeneous* hypothetical finite system is shown as a function of the fraction of carriers in the system, x. In this example, most of the region between $x = 0$ and

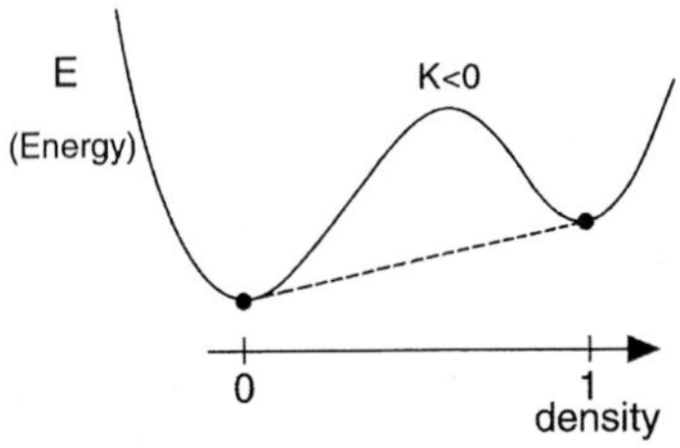

Fig. 6.9. Energy vs. density x of an example of a finite system with phase separation tendencies. See text for the discussion

$x = 1$ has negative curvature in the energy (i.e., the compressibility is negative), and at a density x it is energetically advantageous for the system to separate into a region with $x = 0$ (occupying a fraction 1-x of the system) and a region with $x = 1$ (occupying a fraction x of the system). This corresponds to the dashed line shown in the figure, which is lower than the solid line. In the bulk limit, a calculation in the canonical ensemble will directly lead to that straight-line energy, since the system spontaneously phase separates. The negative curvature can actually be found only working on a finite system. This line of reasoning is usually known as the "Maxwell construction" procedure to detect phase separation, and it is discussed in elementary books of thermodynamics.

In the context of phase separation in general, well-known examples involve *alloys*. A typical case is the combination of Cu and Ag, elements that cannot be homogeneously mixed below $\sim 700°$C. The atoms are not ionized and there is no Coulombic penalization for the separation, contrary to what occurs in electronic systems, as described later in this chapter. In alloys, the separation can be truly macroscopic. In this context, as the system is cooled from the homogeneous high-temperature regime into the phase-separated region, the process of separation occurs either through a rapid *spinodal decomposition*, in the average density between the two minima of the free energy where the curvature is truly negative, or through a more complicated *nucleation* process, when that curvature remains positive in the vicinity of the two minima. In the latter, the system is stable against small fluctuations, and only large fluctuations allow the system to find the true phase-separated state. In view of the effect of Coulombic interactions that reduce the electronic phase separation from macroscopic to microscopic, the differences between spinodal decomposition and nucleation have not been much discussed in the one-orbital model, and in practice it is the spinodal case that one has to consider, at least intuitively, when discussing phase separation in manganites. The reader can find an elementary discussion about these issues in the book of Marder [6.28].

6.4.2 High Compressibility near Phase Separation

As an application of the equations found in the previous subsection, in Fig. 6.10, lines of constant compressibility K are shown, obtained numerically from the density vs. μ curves. As long as K is finite the system is stable, but by mere continuity it is expected that the compressibility will grow as the regime of phase separation (where K diverges) is approached.

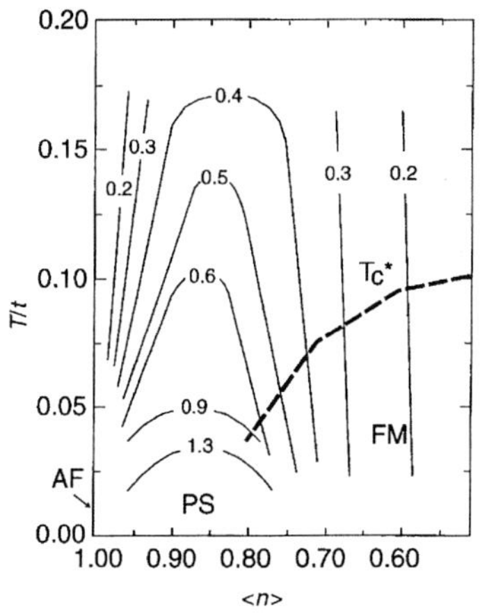

Fig. 6.10. Lines of constant $K\langle n\rangle^2 = \mathrm{d}\langle n\rangle/\mathrm{d}\mu$, for the one-orbital model on a chain with $J_\mathrm{H} = \infty$ and $J_\mathrm{AF}/t = 0.05$, in the plane temperature density. Results were obtained using Monte Carlo simulations. The compressibility grows as the phase-separated regime is approached. T_C^* is a characteristic temperature where ferromagnetic correlations start growing very rapidly with decreasing temperature, and it may become a true critical temperature in a 3D system. From Moreo et al. [1.10]

Intuitively, this means that in the vicinity of phase separation there must be strong fluctuations in the charge, since the two competing phases have different electronic densities. In fact, analyzing "by eye" the Monte Carlo configurations (snapshots) generated at density around 0.90 and $T \sim 0.05t$, it is clear that coexisting *clusters* with ferromagnetic and antiferromagnetic properties are generated. These clusters are *dynamical*, namely their location is not fixed in space (the model is translational invariant), they move, and may even disappear or fuse with other clusters in a complicated way as a function of time. Dynamical cluster coexistence is characteristic of regions with high compressibility, and such effects may have an influence on the physics of manganites where a variety of experiments, reviewed later in this book, have revealed cluster-formation tendencies in manganese oxides.

6.5 Spin Incommensurability

In the phase diagram of Fig. 6.1, shown at the beginning of this chapter, a spin-incommensurate (IC) phase was indicated. This phase appears at values of J_H/t that do not seem sufficiently large to be of relevance for manganites. As a consequence, the spin IC regime has not been studied in much detail, at least until recently when indications of stripes in manganite models, and their concomitant spin IC were reported. Since issues related to stripes will be presented in Chap. 10, here the discussion is limited to the numerical evidence for spin incommensurability, as provided in Fig. 6.11. In this figure, the Fourier transform $S(\boldsymbol{q})$ is shown at not too large values of J_H/t, at several densities. Results in 1D and 2D are provided, the former both with classical and quantum-localized spins. In all cases, the spin IC manifests itself as a peak in $S(\boldsymbol{q})$ located at values of the momentum that deviate from both the AF and FM states, which in 1D have strong peaks at momentum π and

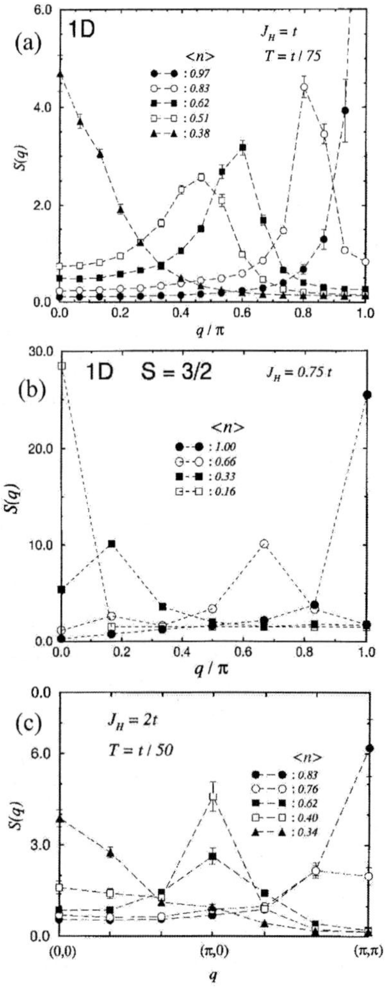

Fig. 6.11. Fourier transform $S(\boldsymbol{q})$ of the spin spin correlations between the localized spins vs. momentum, for the one-orbital model. Densities are indicated. (**a**) Results obtained with the Monte Carlo technique (classical spins) working on 30-site chains, $J_{\rm H} = t$, and at low temperature. (**b**) Zero-temperature results found with the DMRG method using quantum $S = 3/2$ localized spins on a 12-site chain. (**c**) Results in 2D obtained with the Monte Carlo method (classical spins) using 6×6 clusters. The coupling and temperature are indicated. All results are from [6.2]

0, respectively. The actual location of the peak changes with the density smoothly, interpolating between the extreme cases of AF and FM [6.29].

What causes such spin IC effects? There are a variety of possible reasons and insufficient work has been carried out to totally identify the dominating effect in manganite models. However, at least in two-dimensional spin-fermion models for cuprates (which are quite similar to those of manganites, and are reviewed by A. Moreo in this chapter), the origin are stripes. The issue of stripes is important in a variety of transition-metal oxides. Other possible reasons for spin IC can be found in studies of t–J models, where "across the hole" antiferromagnetism was identified [5.7]. More work is needed to fully clarify the origin of spin IC in the manganite context.

6.6 Phase Diagram using Quantum $S = 3/2$ Spins

As explained earlier, it is common practice to replace the quantum-localized t_{2g} spin by a classical spin (i.e., $S = \infty$), instead of the value $S = 3/2$ corresponding to three electrons with aligned spin due to a strong Hund coupling. This approximation appears reasonable since a spin $S = 3/2$ is believed to be sufficiently large that quantum effects are sometimes not crucial. However, it is important to check such an assumption at least in particular examples. The problem here is that a truly quantum-localized spin complicates computational techniques enormously. In particular, the powerful and fairly simple Monte Carlo approximation that has been so useful in determining properties of the one-orbital model with classical t_{2g} spins (to be discussed in Chap. 7), can no longer be used when the spins are quantum. Truly quantum Monte Carlo approaches could be employed, but they are typically plagued by the so-called sign problems. Then, accurate computational studies using quantum $S = 3/2$ localized spins are restricted only to one-dimensional systems, where powerful computational methods are available, such as the exact diagonalization [1.20] and density-matrix renormalization group (DMRG) techniques [6.30]. Results of such an analysis are shown in Fig. 6.12, which is reproduced from [6.2], in the form of a phase diagram. Very prominent AF (not shown, it exists at density 1), FM, PS, and spin IC phases are obtained, in excellent agreement with the phase diagram already discussed for the case of classical localized spins. Perhaps the most notorious difference is the opening of a window between PS and FM at large Hund coupling. In this regime, the computational analysis reveals a ferromagnetic state, but this state does not have a full polarization, only a partial one ("not fully saturated ferromagnet"). In addition, PS for the quantum case does not seem to occur at small and intermediate values of the Hund coupling. Of course, as usual with fully quantum studies restricted to fairly small lattices, it is possible that size effects may be masking the fine details of the phase diagram in the bulk limit. However, overall it is clear that the dominant features in the quantum and classical cases are similar (for an analogous conclusion see [6.31]). This result is important, showing that the classical spin approximation fortunately captures the essence of the problem. Monte Carlo approximations can thus be applied to the localized classical spin limit with reasonable confidence.

Note: Spin-Gaps. There is a subtlety that the reader should be aware of when studying one-dimensional or quasi-one-dimensional quantum spin systems. Some models develop a "spin-gap" in the spectrum of spin excitations. This occurs without the charge playing an important role, namely there are examples where the phenomenon is present in undoped systems. The best-documented cases correspond to the nearest-neighbors (NN) Heisenberg chain with *spin* 1 (or integer), and the "two-leg" (or even number of legs) *ladder* with spin 1/2 degrees of freedom interacting also through a NN Heisenberg Hamiltonian. Both develop a spin-gap in the ground state, making the properties of such systems quite different from spin-gapless models, such as the spin 1/2 Heisenberg chain (for a recent review see [6.32], and references therein). In manganite chains, there are spins 2 and 3/2 degrees of

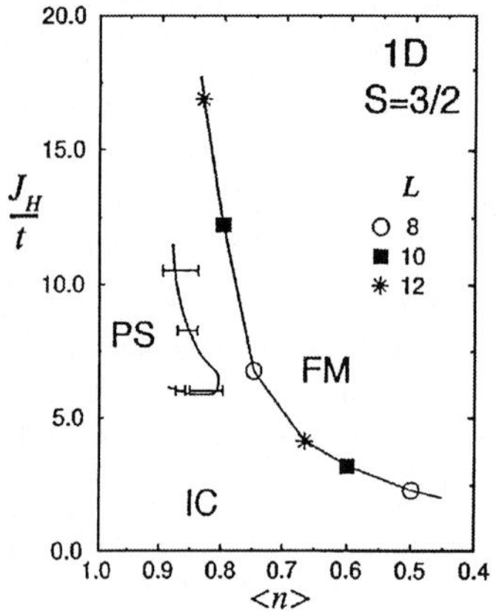

Fig. 6.12. Phase diagram of the one-dimensional zero-temperature FM Kondo model (also known as one-orbital model) using $S = 3/2$ localized t_{2g} spins. The phase diagram was obtained with Lanczos and DMRG techniques (see [6.2] for details). The notation is standard. Lattice sizes are indicated. The phase diagram is similar to that obtained using classical localized spins

freedom, and regimes of couplings and parameters with a spin-gap in the ground state could develop. This is a potentially interesting area of research, that has not as yet received much attention since manganites in one dimension have not been synthesized. In cases with a spin-gap, the classical approximation for the localized spin is not a good approximation and the full quantum problem must be studied.

An analysis including the quantum nature of the localized spin has been carried out by Riera et al. [6.33], for one-dimensional systems that allows for accurate numerical studies. In this analysis, a large Hund coupling was considered. In the limit $J_{\mathrm{H}} = \infty$, the quantum model was rewritten exactly in an elegant way as

$$H^{\text{quantum}}_{J_{\mathrm{H}}=\infty} = -t \sum_{ij} P_{ij} Q_S(y_{ij}) , \tag{6.6}$$

where $y_{ij} = \boldsymbol{S}_i \cdot \boldsymbol{S}_j / S(S-1/2)$. S is in this case the total spin of the parallel combination of the electron and localized spins in the original model. A similar model, the quantum version of the double-exchange model, was studied some time ago by Kubo and Ohata [2.21]. In the case of manganese oxides, S is 2 (parallel alignment of $S = 3/2$ of localized spins and $S = 1/2$ of the mobile electron). The polynomial $Q_S(y)$ is just $(1 + y)/2$ for $S = 1$ (which can mimic nickel), and it becomes $-1 - (5/4)y + (7/4)y^2 + (3/2)y^3$ for manganese oxides with $S = 2$. This model can be supplemented with a Heisenberg AF interaction among the S-spins. Such an interaction does exist in the limit of zero hole density. Let us denote by J the strength of the Heisenberg interaction (which scales as the inverse of the Hund coupling). Numerical techniques allowed for investigations of the phase diagram of this model, and the results for $S = 1$ and $S = 2$ are shown in Fig. 6.13. The phase diagram contains phases quite similar to those found in the investigations described before, namely a very prominent ferromagnetic phase, antiferromagnetism at $x = 0$ (although not explicitly shown in the figure), and a large

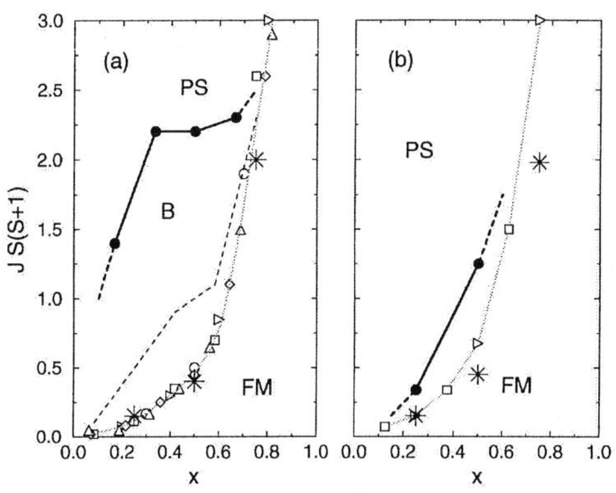

Fig. 6.13. Phase diagram of t–J-like models for manganites in 1D corresponding to (**a**) nickelates and (**b**) manganites. J is the coupling between Heisenberg spins at each site in the large Hund coupling limit, and x is the hole density. PS, B, and FM, denote, phase separation, hole binding, and ferromagnetic phases, respectively. The reader can find additional information, including the meaning of the symbols not explained here directly in [6.33], from where this figure is reproduced

regime of phase separation. This also shows that nickelates and manganites have several features in common.

An extra observation contributes to the excitement induced by the results of Fig. 6.13. Note that there are regions of *hole binding* in the phase diagram, at least for $S = 1$. This effect has not been investigated much beyond cuprates in the context of transition-metal oxides, but it is conceivable that carriers doped into any antiferromagnetic background can create spin polarons in its vicinity, and a couple of holes can further reduce their energy by sharing the spin-polaronic regions, thus effectively binding into pairs. This type of effect has been widely mentioned in the context of cuprates (see [1.20]). Can they occur in models with a spin larger than 1/2? The possibility of pair formation, and its associated potential superconductivity, in nickelates and manganites is an open problem that certainly deserves further theoretical work. Unfortunately, superconductivity has not yet been found experimentally in doped nickelates and manganites! However, certainly not all of the parameter space has been investigated in these compounds. If it is experimentally confirmed that these compounds do not superconduct, then a fundamental difference exists between cuprates and nickelates and manganites. If we use the phase diagram of the 1D system as guidance, the key difference appears to be the presence of ferromagnetism in Ni and Mn oxides, caused by the existence in both cases of two or more active orbitals. This robust ferromagnetic phase is not present in the famous t–J model for cuprates, which has only one active orbital, compatible with the layered Cu oxides structure with a static Q_3 Jahn–Teller distortion, where only the $x^2 - y^2$ orbital is active. However, the t–J model has a small region of ferromagnetism (the so-called Nagaoka ferromagnetism) at very small J/t, and the analogies between all the transition-metal oxides discussed here could be stronger than anticipated.

Note: Polarons. In the previous discussion a "polaron" was mentioned. This concept is very basic in condensed matter and here it is assumed that the reader

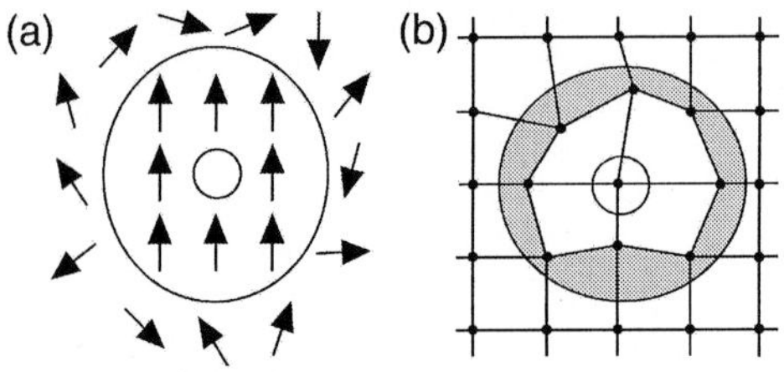

Fig. 6.14. Qualitative representation of a spin polaron (**a**) and a lattice polaron (**b**). The *small circle* in the middle is the carrier, the *larger circle* is the distorted region around that carrier

is familiar with polarons. As a quick reminder, electrons in a solid are in general simply not free to move but instead they interact with other electrons and with other degrees of freedom, a process usually called "dressing". The original electron, or particle under study in general, becomes a "quasi-particle", with properties, such as the mass, which can be quite different from those of the original "bare" particle. The polarons are quasi-particles, and they are visualized as the original carrier (electron or hole) surrounded by a "cloud" or finite region around, where the properties of the system deviate from those in the absence of the carrier. For instance, for "spin polarons", the cloud around the carrier can have magnetic properties different from the undoped system. The cloud may be ferromagnetic and the background paramagnetic, indicative of a ferro-polaron Fig. 6.14a, or the background can be antiferromagnetic and the cloud have either ferro properties or a suppressed antiferromagnetism. The cloud can also contain lattice distortions ("lattice polaron", which historically was the first polaron considered), as shown in Fig. 6.14b. The distortions are sometimes described as virtual clouds of phonons, or a combination of spin and lattice ("spin-lattice polaron", with a concomitant virtual cloud of phonons and magnons), or even the orbital degree of freedom may be involved ("orbital polaron"), or any other variable could contribute to the distortion around the carrier. For example, in a plasma or gas of charged particles, a charge carrier can attract a cloud of oppositely charged particles, leading to the usual screening phenomenon. Polarons are important in manganites and they are frequently mentioned in this context, both in theoretical and experimental discussions.

In spite of decades of effort, some of the properties of just one polaron are only qualitatively understood. It is clear that depending on the strength of the electron–phonon coupling the size of the polaron will be small or large. Perhaps the simplest model to study these effects is the Holstein model (for a recent reference with numerical work, see [6.34]), but even in this model many-body effects are quite nontrivial. The simple cartoonish picture Fig. 6.14b will be invoked when polarons in the context of manganites are later discussed from the perspective of experiments. The conclusion will be that most of the structure referred to as "polaronic-like" found experimentally, actually corresponds to larger and more complex "clusters", perhaps with several carriers inside. As a consequence, the use of the term polaron in that context may be confusing.

6.7 Influence of Long-Ranged Coulomb Repulsion on Phase Separation

The electronic phase-separation phenomenon described here has been obtained using Hamiltonians where the long-range portion of the Coulomb interactions has been neglected (the "$1/r$" tail of the interaction). This approximation is very common in the context of transition-metal oxides, but it is in part caused by a practical technical fact: including these long-ranged interactions is just very difficult! There are more elegant rationalizations for neglecting these correlations. It can be argued that polarization effects, particularly in oxygens, are large and they can reduce the strength of the Coulomb interaction through the generation of a substantial dielectric constant (actually a more refined analysis should consider a full dielectric function $\epsilon(\boldsymbol{q},\omega)$, carrying both momentum and frequency dependency). Also, metallic screening can contribute to the reduction of the Coulomb interaction, as described in elementary textbooks. This type of study in the context of cuprates has been briefly discussed by Gazza et al. [6.35] in calculations of the influence of Coulomb interactions on doped two-leg ladders, and the reader can find references about these (rather subtle) issues in that paper. It is widely accepted that the strength of long-range Coulomb interactions can be substantially reduced in transition-metal oxides, by large factors. However, the overall macroscopic accumulation of charge of an electronically phase-separated regime is difficult to avoid, even with a large dielectric function. As a consequence, it is also widely accepted that the two macroscopic regions that form an electronic phase-separated state, cannot survive the inclusion of the $1/r$ tail of the Coulomb interactions. These macroscopic regions will break up into smaller pieces due simply to energetic considerations. The resulting state is still expected to be highly inhomogeneous, but it now involves coexisting clusters with a typical size on the order of nanometers (this length scale often arises in cluster-size estimations using realistic parameters). A sketch of possible states, reproduced from Moreo et al. [1.10], is shown in Fig. 6.15.

There are few computational calculations addressing the effect of off site Coulomb repulsion. One of them was presented by Malvezzi et al. [6.36], using the one-orbital model in its quantum version (i.e., with quantum-localized spins). To reduce the complexity of the calculation, those quantum spins were considered as having spin $S = 1/2$ instead of $3/2$, and the calculation was carried out on a one-dimensional system. On site Hubbard repulsion U, as well as nearest-neighbor repulsion V, were included. In the absence of V, with and without U, phase separation was found, as expected from the considerations described in this chapter. In the canonical-ensemble context in which the calculations of Malvezzi et al. were carried out, phase separation can be monitored by calculating the compressibility. When this quantity becomes negative the system is unstable, as already explained. Using chains with open-boundary conditions (for technical reasons related to the DMRG technique) Malvezzi et al. found that in the PS regime one of the competing phases

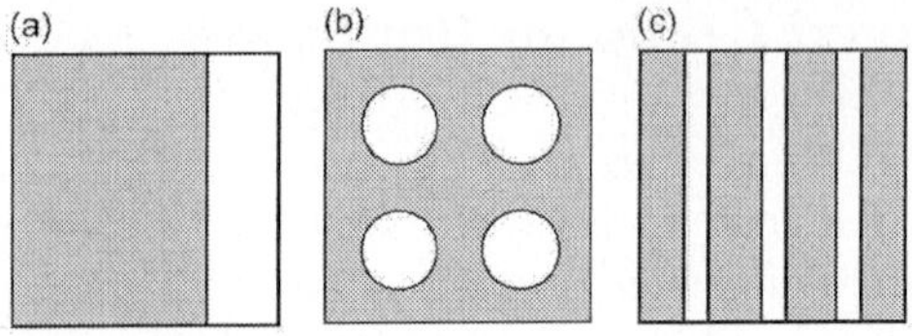

Fig. 6.15. Qualitative representation of (**a**) a macroscopic phase-separated state, *gray* and *white* being the two phases involved. Also shown are possible charge inhomogeneous states that could be stabilized by the competition of Coulomb long-range forces (that tends to spread the charge), and the "attractive" effective forces that produce the phase-separated state. In (**b**) and (**c**), droplets and stripes are shown. The figure is reproduced from Moreo et al. [1.10]. Detailed calculations starting from realistic models are quite difficult, and only rough estimations can be made about the state emerging from the combination of strong short-range attraction and weak long-range repulsion

typically became stable at the center of the chain, and the other at the two extremes in a symmetrical way.

Upon switching on the repulsion V, Malvezzi et al. [6.36] observed that the phase-separated regime was substantially affected. As V was increased, eventually a charge-density-wave (CDW) state was stabilized. Instead of having carriers accumulating on some portion of the chain, now they tend to spread periodically, each one forming a small bubble or polaron with spin ferromagnetic order inside. The resulting phase diagram is very complex (see for instance Fig. 14 of [6.36]), and it will not be reproduced here. Sufficient for our purposes is to mention a few qualitative facts that are robust enough that they are likely to be valid in more general cases: (i) Long-range interactions destabilize the *electronic* phase-separated regime (note the emphasis on electronic, since other types of phase-separation involving equal-density phases to be described later will not be much affected by Coulomb interactions). (ii) Small "islands" of one phase embedded into the other replace the phase-separated regime. Each island has the size of a few lattice spacings. Since the latter is about 4 Å in many transition-metal oxides, then a typically accepted scale for the new stable domains is the *nanometer* scale. However, phases that may be in competition could have very similar densities, and adding that to screening and polarization effects, it could occur that structures larger than nanometer scale could also be stable. (iii) The new arrangement of clusters or islands of one phase embedded in the other tends to be *periodic*, at least in models that are translationally invariant, as those used thus far here. A charge-density-wave arrangement of charge tends to be formed. This state will be contrasted against experiments when other forms of phase separation are discussed later in the book. It will be concluded that the more chaotic and fractal-like shape of coexisting clusters found in some manganites may need to be supplemented by other explanations besides the electronic phase-separation ideas. However, in regimes such as low hole-density it is conceivable that small modifications of the electronic PS scenario would be sufficient to explain the experimental data.

Note: CDW. A charge-density-wave (CDW) is a periodic arrangement of charge that emerges out of interactions from a charge uniform state. A CDW state breaks the invariance under translations of the Hamiltonian. The reader can find information about CDWs in the review article S. Brown and G. Grüner, in Scientific American, April 1994, p. 50, and references therein. In this book, periodic arrangements of charge will often be found. However, they tend to go together with a similarly periodical distribution of spins and orbitals. A plain CDW state is typically associated with charge periodicity only, thus the charge-ordered states of manganites are usually not considered standard CDWs, and more complex descriptions and notations are usually employed.

Other authors have also addressed the typical size of clusters once Coulomb interactions are included. For recent calculations, see for instance Lorenzana et al. [6.37]. Earlier in the development of these ideas, Nagaev [6.38] also presented estimations of the sizes of the hole-doped ferromagnetic bubbles that can form in an antiferromagnetic background. Overall, the estimation of having nanometer-scale coexisting clusters appears to be correct. Note that these considerations, namely the formation of small structure from tendencies to electronic phase separation, have also been extensively discussed in the context of high-temperature superconductors (see [1.3, 6.21] for some references). Interesting issues related to charge ordering due to a large nearest-neighbor Coulomb repulsion in a one-band model, and re-entrant charge ordering at finite temperature have been also addressed recently by several authors [6.39].

6.8 Phase Diagram Including the J_{AF} Coupling

When estimations of the actual values of the coupling constants in the model Hamiltonians were discussed in Chap. 5, it was clear that the smallest was J_{AF}, the exchange coupling between localized t_{2g} spins. However, its relevance in the problem and influence on the phase diagram is much stronger than one can naively expect. As a matter of fact, in models based on Jahn Teller phonons or directly within the more simple double-exchange ideas, this coupling is *crucial* to generate insulating phases that are important for the physics of manganites, such as the CE-phase at half-doping $x = 0.5$, as well as the charge-stacking effect. It appears that in the limit $J_{\mathrm{AF}} = 0$, the models often used are too biased in favor of ferromagnetic tendencies, even at large electron phonon couplings. To mimic the experiments where nonferromagnetic phases have been found, it is certainly very important to consider a nonzero J_{AF} in manganite models.

The computational (Monte Carlo) study of the one-orbital model including J_{AF} started with the study of Yunoki and Moreo [6.40] in one dimension. In this early study it was already noticed that at intermediate values of the coupling "exotic" phases could be generated, which are typically insulating. The important role of phase-separation tendencies in the regime of *low* e_g-electronic density was also observed. For instance, Fig. 6.16a shows the e_g density vs. chemical potential at a relatively small value of $J_{\rm AF}$ and large Hund coupling. The stable region centered at half-doping is ferromagnetic, according to spin spin correlation measurements. The figure shows two regions labeled as PS1 and PS2 where phase separation occurs, manifested as a discontinuity in the density. The full low-temperature phase diagram reported by Yunoki and Moreo [6.40] is in Fig. 6.16b. Prominent phase-separation regions are present, and their importance grows together with the coupling between localized spins. A ferromagnetic phase exists at small $J_{\rm AF}$, while an antiferromagnetic phase is present at density 0.5 and large $J_{\rm AF}$ (reasonable, since as $J_{\rm AF}$ grows it will eventually become the dominant term in the Hamiltonian, generating antiferromagnetism). However, there is also an intermediate phase between the ferromagnetic and antiferromagnetic regimes, which is insulating. This last property is clear from its vanishing Drude weight [see Fig. 6.16c (issues related to optical conductivity and Drude weights will be discussed in more detail in Chap. 12)]. The density of states has a prominent gap at the chemical potential (Fig.6.16d). The spin arrangement corresponding to this intermediate phase can be found from the spin spin correlations or by visual inspection of low-temperature Monte Carlo configurations, described in the next section. The existence of intermediate phases, such as the one discussed here, will be addressed in more detail below, and also when the more realistic model with two orbitals is discussed. These intermediate phases play a very important role in models for manganites, particularly those in the orbital degenerate model.

6.8.1 Exotic Phases Induced by $J_{\rm AF}$

Island Phases. As described in the previous section, the AF coupling $J_{\rm AF}$ among the localized spins favors an *insulating phase* and, thus, it can destabilize the ferromagnetic phase of the one-dimensional (and presumably "any" dimensional) one-orbital large-$J_{\rm H}$ model at $x = 0.5$. These insulators naturally have staggered spin order (antiferromagnetic) at sufficiently large $J_{\rm AF}$, where this coupling dominates. However, at intermediate values of $J_{\rm AF}$, in the regime where ferromagnetism and antiferromagnetism are in close competition, more exotic phases can be stabilized. This effect is actually a precursor in the one-orbital case of the stabilization of the famous CE-phase of half-doped manganites when a more realistic model with two orbitals is considered. This CE-phase will be discussed later in Chap. 10. In such a case, the CE-phase also appears in the regime of competition between ferro and antiferro tendencies, as explained before. Returning to the one-orbital case, the intermediate phase that is stabilized at low temperature has an equal number of ferro and antiferromagnetic links, with the pattern $\uparrow\uparrow\downarrow\downarrow$. This structure was observed

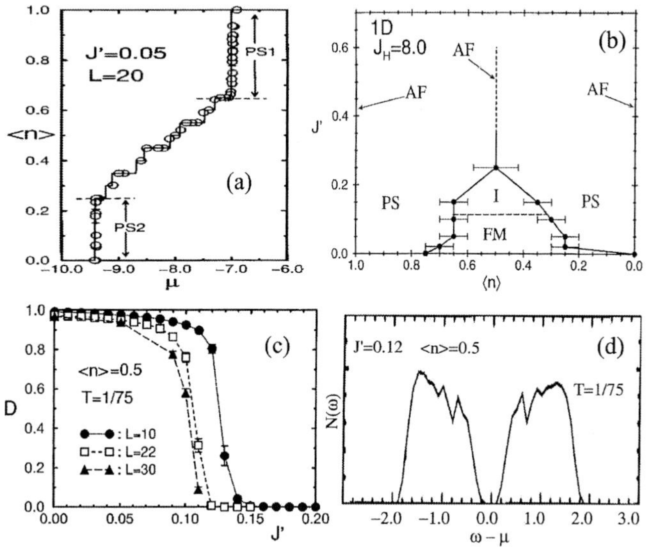

Fig. 6.16. Results obtained by Yunoki and Moreo [6.40] in their Monte Carlo study of the influence of J_{AF} on the one-orbital phase diagram in one dimension, working at $J_{\mathrm{H}} = 8$ $(t = 1)$. (**a**) Density vs. chemical potential on a 20-site chain, at $J_{\mathrm{AF}} = 0.05$ (here and below J_{AF} is denoted by its alternative symbol J'). PS1 and PS2 are two regions of phase separation. The intermediate stable regime is ferromagnetic. (**b**) Phase diagram at low temperature. "I" denotes the intermediate phase discussed in the text, with insulating properties and a ↑↑↓↓ periodic arrangement of spins. The remaining notation is standard. (**c**) Drude weight D obtained from calculations of the optical conductivity at e_g density 0.5, as a function of J_{AF} using different chain lengths. A vanishing D corresponds to an insulator. (**d**) Density of states of the intermediate phase of part (**b**) at e_g density 0.5, $J_{\mathrm{AF}} = 0.12$, and low temperature

by Moreo et al. [6.41] in their studies of the influence of disorder on first-order transitions, and also by Yunoki et al. (see footnote 17 of [6.42]) when searching for the CE-state in the two-orbital model. Independently, Garcia et al. [6.43] studied the intermediate phase between the ferro and antiferro states in much more detail, observing that the spin pattern found at $x = 0.5$ can be generalized to other densities. For instance, at $x = 2/3$ a state with the periodic arrangement ↑↑↑↓↓↓ was identified. These exotic states suggest that charge can be trapped in the tiny ferromagnetic regions, simply following the standard double-exchange argument where the hopping cancels for antiparallel spins along a given link. For the pattern found at $x = 2/3$, Garcia et al. remarked in addition that charge ordering can be produced since in an effective 3-site well, charge would be more likely found at the center site. Then, charge ordering could be obtained simply as a competition between double exchange and superexchange between localized spins. This line of reasoning is compatible with other calculations carried out by Hotta et al. [3.32]. They used the two-orbital model to show that the pattern of spins of the famous CE-phase can exist even in the absence of Hubbard Coulomb or electron phonon coupling, as long as the Hund coupling is large and J_{AF} is nonzero (to be discussed in Chap. 10). Clearly in the one-orbital model there are precursors of results that are found in the more realistic two-orbital one, showing that the former deserves a detailed analysis (for more work on the relevance of J_{AF} see [6.44]).

Phases similar to those described thus far in one-dimensional systems have also been found in two dimensions. The spin pattern of typical 2D phases reported by Aliaga et al. [6.45], where the name "island" phases was introduced, are shown in Fig. 6.17a,b. They amount to generalizations of the results of one dimension (Fig. 6.17a), with antiferromagnetic links between the chains, or truly 2D islands (Fig. 6.17b). The actual phase diagram obtained with Monte Carlo techniques (shown in Fig. 6.17c) is very rich, with a complex structure at intermediate values of J_{AF}. Phase-separation tendencies are quite clear in this and other studies, as remarked extensively in the present chapter. On the other hand, there are some isolated densities that are stable, particularly $n = 1/2$, $1/3$, and $1/4$ in special ranges of J_{AF}. These stable densities are separated by unstable regions. The trend is quite similar to those found in more complex studies of models with two orbitals and cooperative phonons (see later the discussion about stripes in Chap. 10), where similar particularly stable densities were also reported. It is likely that this phenomenon is general, and there are densities in manganite models that are dominant over others. As remarked by Hotta et al. [3.32], and as appears also in Fig. 16 of Aliaga et al. [6.45], *stripes* may be formed in these models at particularly stable densities. This is also reminiscent of experimental results that have established special densities (1/8, 1/4, 3/8, etc.) as being more relevant than others in the manganites (see Chap. 3).

Flux Phase. In the 1D system at $x = 0.5$, increasing J_{AF} induces a transition between a FM metal and a nonconventional AF insulator, as described before. What is the analog of this AF insulator for two dimensions? This issue was addressed by Yamanaka et al. [6.46] working at $J_H = \infty$ and the result is an arrangement of spins called the "flux" phase, schematically shown in Fig. 6.18 together with the zero-temperature phase diagram of [6.46]. As in one dimension, prominent ferromagnetic and phase-separated regimes were observed, together with a flux phase near $x = 0.5$. The latter has a plaquette flux arrangement that changes sign between nearest-neighbor plaquettes. Thus, it is actually a *staggered* flux phase. Agterberg and Yunoki [6.47] studied this flux phase with Monte Carlo techniques, emphasized the fact that spin current circulates around plaquettes, and analyzed the spin-wave spectrum.

Skyrmion Phase. Recent studies of the one-orbital model in three dimensions [6.48], using the hybrid Monte Carlo method [6.4], have reported a rich phase diagram at $J_H = \infty$, due to the nontrivial influence of the coupling J_{AF} among localized spins. The zero-temperature phase diagram is shown

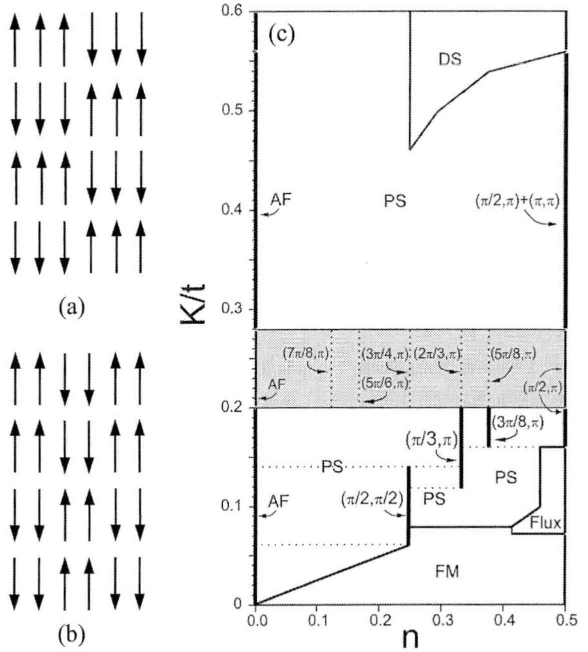

Fig. 6.17. (**a**,**b**). Schematic representation of two island phases that appear in the one-orbital model at intermediate values of J_{AF}, as reported by Aliaga et al. [6.45] using Monte Carlo techniques. (**c**) Phase diagram of the 2D one-orbital model, reproduced from Aliaga et al. [6.45]. Here, K is the coupling more usualy referred to as J_{AF}, and n is the e_g density. Since the work was done with $J_{\mathrm{H}} = \infty$ only densities from 0 to 0.5 are shown (the phase diagram is symmetric with respect to $n =$ 0.5). AF, FM, PS, and Flux, denote the antiferromagnetic, ferromagnetic, phase-separated, and flux phases (the latter will be explained in the next subsection). DS is a spiral-ordered phase. The *thick vertical lines* are island phases. The momenta shown indicate the structure. For instance, the phases in (**a**) and (**b**) are $(\pi/3, \pi)$ and $(\pi/2, \pi/2)$, respectively. A more complete discussion can be found in [6.45]

in Fig. 6.19a. In agreement with the calculations in lower-dimensional systems, prominent ferromagnetic and phase-separated regimes are found. This reinforces the notion that phase diagrams for manganite models are qualitatively similar for the various dimensions of interest. In addition, phases that only appear in 3D systems have also been reported. For instance, in the figure a small island of A-type antiferromagnetism (see Chap. 2 for experiments, and Chap. 9 for theoretical aspects) was identified. As discussed before, this A-type AF is important to explain experiments for $LaMnO_3$. A more detailed description of real manganites is provided by the two-orbital model since, concomitant with spin order, there is a prominent orbital order in the system. In Fig. 6.19a there are also (i) island phases, which appear in lower-dimensional systems, (ii) the expected G-type AF phase in the absence

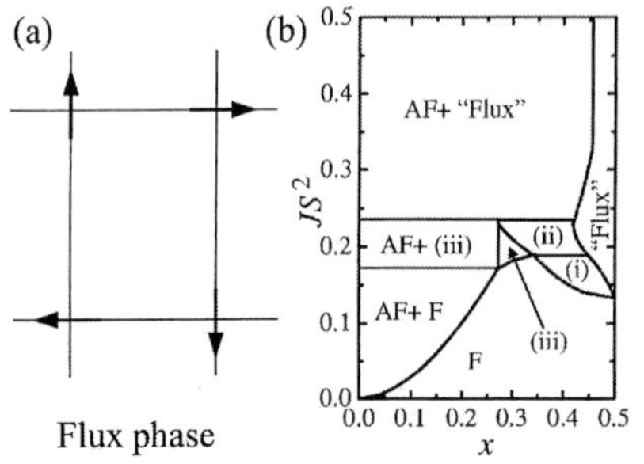

Fig. 6.18. (**a**) Sketch of the flux phase. (**b**) Phase diagram of the $J_{\rm H} = \infty$ 2D one-orbital model, from [6.46]. The $x = 0$ line is antiferromagnetic (AF), F denotes ferromagnetism, and "Flux" the phase described in the text. The rest are regimes of phase separation among two phases, such as AF and F

of mobile electrons, (iii) a twisted phase that interpolates between A- and C-type antiferromagnets, and (iv) a flux phase that is a three-dimensional version of the analogous phase described in the previous subsection. The most exotic phase reported by Alonso et al. [6.48] was a *skyrmion* phase that is stable in a narrow region of parameters, with the hedgehog shape arrangement of spins shown in Fig. 6.19b. This arrangement, which has no analog in lower-dimensional systems, can be considered as caused effectively by a magnetic monopole at the center of a cube, with magnetic charge changing sign between nearest-neighbor cubes. Calculating the electronic dispersion for a fixed background of classical spins in the flux and skyrmion phases, it can be shown that it resembles the dispersion of Dirac massless particles near momenta $(\pm\pi/2, \pm\pi/2, \pm\pi/2)$. It is also interesting to note that the critical temperatures reported [6.48] for the more exotic phases, such as flux and skyrmion, are rather small, on the order of 0.04–$0.02t$, even smaller than those characteristic of ferromagnetism. The low critical temperatures may be caused by the competition among many tendencies observed in these systems. It is conceivable that the skyrmion phase, as well as the other exotic arrangements such as the flux and island phases, could have an analog in experiments. It would be desirable to carry out similar detailed calculations using the two-orbital formulation to search not only for spin, but also for exotic charge and orbital arrangements. It is clear that the phase diagrams of models for manganites are extremely rich. Although much has been learned in recent years of the physics of Mn oxides and other transition-metal oxides compounds, only a small fraction of the many phases has been unveiled. Much work remains to be carried out in the exploration of the rich phase structure of realistic manganite models.

First-order PM–FM Transitions for Nonzero $J_{\rm AF}$. Using a variational technique, Alonso et al. [6.49] have argued that the transition from the high-temperature paramagnetic state of the one-orbital model at large Hund coupling to the low-temperature ferromagnetic can become of first order if the superexchange $J_{\rm AF}$ is nonzero. The effect occurs as the Curie temperature $T_{\rm C}$ decreases, due to the proximity of competing phases as $J_{\rm AF}$ grows from zero. Further work by Alonso et al. [6.50] has shown that the Berry phase

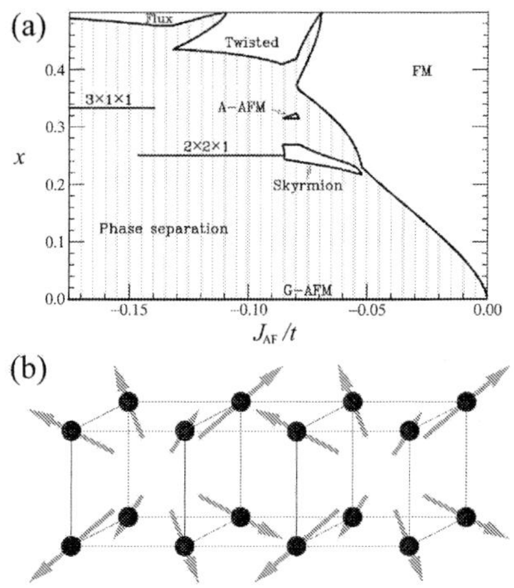

Fig. 6.19. (**a**) Zero-temperature phase diagram of the one-orbital model in 3D, reproduced from Alonso et al. [6.48]. The notation is explained in the text. All transitions are first order. (**b**) Spin arrangement found in the skyrmion phase

(discussed before in Chap. 5, and also in [6.51]) may be crucial to induce a first-order transition, confirming that the complex nature of the effective hopping at $J_H = \infty$ in the double-exchange model (see Appendix I below) cannot be neglected. These important results – the appearance of PM FM first-order transitions – should be confirmed with a full Monte Carlo approach. The subject is important since experimentally there is evidence that the transition PM FM becomes of first order, as T_C is reduced.

Summarizing, it is clear that the one-orbital model is far richer than previously anticipated, and similar surprises are likely awaiting investigations for the case of two orbitals as well. A variety of interesting techniques have been used for the analysis of the model, with very interesting results. However, the theoretical effort in the study of the FM Kondo model or one-orbital model for manganites is simply too large to be described in its entirety in this chapter. The reader should consult additional literature on the subject beyond that provided here. A partial list of additional interesting references can be found in [6.52].

6.9 The Spin–Fermion Model for Cuprates
by A. Moreo

The traditionally used microscopic Hamiltonians to model the high-T_c cuprates, such as the Hubbard and t–J models, are very difficult to study both analytically and numerically. The well-known "sign problem" that arises when four-fermion operators have to be decoupled to carry out simulations, prevents the study of low temperatures ($T < t/6$) at interesting dopings using quantum Monte Carlo techniques. Other methods, such as the Lanczos or exact diagonalization algorithms, only allow for the study of small systems (less than 30 sites in general) [1.20]. As an alternative, here a spin-fermion (SF) model is described. Although this model is phenomenological in nature, it has the advantage that it can be studied numerically without sign problems, such that temperatures of the order of $t/100$ can be reached at all values of dopings, in clusters with more than 100 sites.

The SF model is constructed as an interacting system of electrons and spins, mimicking phenomenologically the coexistence of charge and spin degrees of freedom in the cuprates [6.56]. Its Hamiltonian is given by

$$H = -t \sum_{\langle \boldsymbol{ij} \rangle \alpha} (c^{\dagger}_{\boldsymbol{i}\alpha} c_{\boldsymbol{j}\alpha} + h.c.) + J \sum_{\boldsymbol{i}} \boldsymbol{s_i} \cdot \boldsymbol{S_i} + J' \sum_{\langle \boldsymbol{ij} \rangle} \boldsymbol{S_i} \cdot \boldsymbol{S_j} \,, \tag{6.7}$$

where $c^{\dagger}_{\boldsymbol{i}\alpha}$ creates an electron at site $\boldsymbol{i} = (i_x, i_y)$ with spin projection α, $\boldsymbol{s_i} = \sum_{\alpha\beta} c^{\dagger}_{\boldsymbol{i}\alpha} \sigma_{\alpha\beta} c_{\boldsymbol{i}\beta}$ is the spin of the mobile electron, the Pauli matrices are denoted by σ, $\mathbf{S_i}$ is the localized spin at site $\boldsymbol{i}$, $\langle \boldsymbol{ij} \rangle$ denotes nearest-neighbor (NN) lattice sites, t is the NN-hopping amplitude for the electrons, $J > 0$ is an antiferromagnetic (AF) coupling between the spins of the mobile and localized degrees of freedom (DOF), and $J' > 0$ is a direct AF coupling between the localized spins. The density $\langle n \rangle$=1-x of itinerant electrons is controlled by a chemical potential μ. The unit of energy will be $t = 1$. J' and J are fixed to 0.05 and 2.0, respectively, values shown to be realistic in previous investigations [6.57–6.60]. This model is clearly related to the one-orbital model for manganites, discussed extensively in this chapter. The values and signs of couplings are the difference between the manganite and cuprate contexts. The properties of the ground states in both cases are very different.

To simplify the numerical calculations, avoiding the sign problem, the localized spins are assumed to be classical (with $|\mathrm{S}_{\boldsymbol{i}}| = 1$). This approximation is not drastic, and it was already discussed in detail in [6.57].

Many of the observed properties of this model are in qualitative agreement with experimental data for the cuprates. Magnetic and charge incommensurability occur upon hole doping. The incommensurability is due to the formation of AF domains separated by metallic stripes of holes. A π-shift is observed between the domains, and the stripes are partially filled with an electronic density of about 0.5. A snapshot of the stripes is displayed in Fig. 6.20c. Charge and spin incommensurability appear correlated, and stripe-like configurations are obtained independently of the boundary conditions used, and without the long-range Coulomb repulsion. The effect arises from the strong charge fluctuations between densities 0.5 and 1.0, which in addition produces a clear pseudogap in the DOS, shown also in Fig. 6.20d. Phase separation has not been observed in this study, as demonstrated in parts (a) and (b) of Fig. 6.20.

Indications of robust d-wave pairing correlations upon doping were also found [6.60]. The results indicate that static AF order and d-wave pairing compete. The latter is enhanced when AF is replaced by magnetic incom-

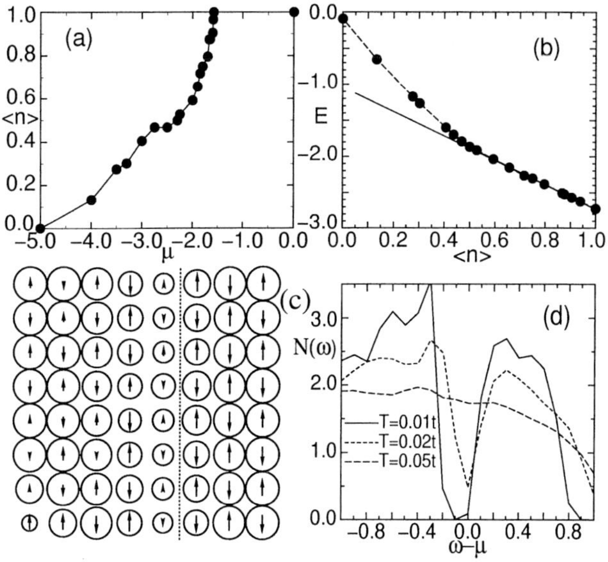

Fig. 6.20. (**a**) Density $\langle n\rangle$ vs. μ for $J = 2$, $J' = 0.05$ and temperature $T = 0.01$, on an 8×8 lattice with PBC. (**b**) Energy versus density for the same parameters as in (**a**). (**c**) Spin and charge distribution for a MC snapshot at $\langle n\rangle \approx 0.85$, and the same parameters as in (**a**) using OBC. The *arrows* are proportional to the z-component of the classical spins, and the radius of the circles to the electronic density. (**d**) Density of states as a function of $\omega - \mu$ at $\langle n\rangle = 0.75$ for different temperatures. The remaining parameters are as in (**a**)

mensurability. Static, stripe-like charge inhomogeneities only decrease the strength of the pairing correlations, as compared with the effect of dynamic charge inhomogeneities. Snapshots showing dynamic stripe configurations are presented in Fig. 6.21a–c, while the d-wave correlations as a function of distance at $\langle n\rangle = 0.75$ are shown in Fig. 6.21d.

In addition to the results presented in this short note, dynamical studies have shown that the band structure is not rigid, contrary to the behavior found in mean-field studies, and it changes with doping. Both mid gap (stripe-induced) and valence band states determine the Fermi surface [6.59]. Doping with non magnetic impurities has also been studied with this model [6.58]. Many features of the optical conductivity, as well as the linear behavior of the resistivity at optimal doping, are also well reproduced by this remarkably simple, yet very rich, spin-fermion model [6.61].

6.10 Appendix I: The Effective Hopping $t \cos \frac{\theta}{2}$

It is interesting to calculate the effective hopping amplitude of conduction electrons in the frozen background of classical spins using the one-orbital

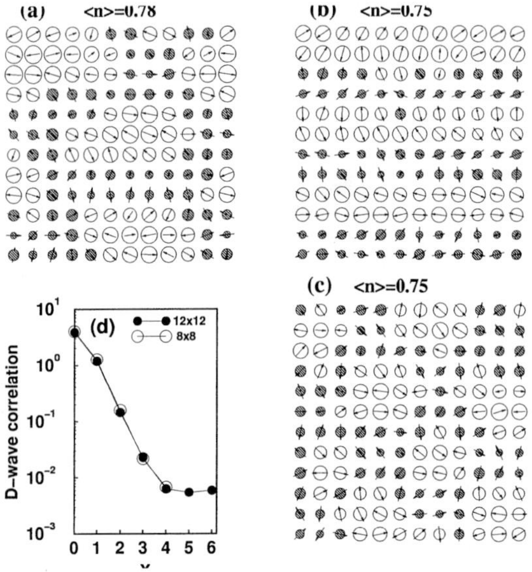

Fig. 6.21. (**a**) Representative snapshot of the spin and charge degrees of freedom on a 12×12 cluster at $\langle n \rangle = 0.78$, in the regime of *dynamic* stripes. The meaning of *circles* and *arrows* is explained in the previous figure. Here, in addition, local densities less than average have *gray* circles. (**b**) Same as (**a**) for $\langle n \rangle = 0.75$. (**c**) Snapshot for the same parameters as in (**b**) obtained at a later time during the simulation. (**d**) *d*-wave pairing correlations in the direction perpendicular to the stripes, using two lattice sizes

model in the large Hund coupling limit. This calculation leads to the famous renormalization of the conduction electron hopping by a "$\cos\frac{\theta}{2}$" factor, roughly explaining the relation between transport of conduction electrons and the order in the spin background (θ is the angle between neighboring localized spins). To carry out the calculation consider just two sites, 1 and 2 (Fig. 6.22), and assume that the classical spin associated with site 1 points in the z-direction for simplicity, while the classical spin of site 2 points in an arbitrary direction in the x–z plane, i.e., $\boldsymbol{S}_1 = S(0,0,1)$ and $\boldsymbol{S}_2 = S(\sin\theta, 0, \cos\theta)$. In the limit of a large Hund coupling only the spin component along the z-axis will matter in site 1, and the spin component along the direction of the field $\boldsymbol{S}_2$ will only matter at site 2. To describe the hopping of electrons at large Hund coupling, it is convenient to rotate the operators describing the conduction electrons, such that at each site their spin points in the direction of the classical spin. After the rotation only the "up" spins will be kept. In site 1 the classical spin is pointing along the z-axis and, as a consequence, the standard basis of creation and destruction operators is already suitable. However, in site 2 a rotation must be carried out as follows.

First consider the spinor defined as

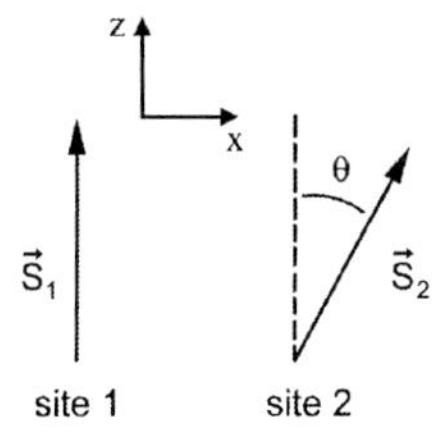

Fig. 6.22. Relative orientation of the classical spins at nearest-neighbor sites 1 and 2

$$\chi_\theta = \begin{pmatrix} \cos(\theta/2) \\ \sin(\theta/2) \end{pmatrix} . \tag{6.8}$$

This spin points in the x–z plane along the direction specified by the angle θ (which, as special cases, is 0 for the positive z-axis direction and $\pi/2$ for the x-axis), since the mean values of the spin operators satisfy $\langle S_z \rangle = (\hbar/2)\cos\theta$, $\langle S_x \rangle = (\hbar/2)\sin\theta$, and $\langle S_y \rangle = 0$. If the spinor χ_θ is rotated using the 2×2 matrix

$$H = \begin{pmatrix} \cos\frac{\theta}{2} & \sin\frac{\theta}{2} \\ -\sin\frac{\theta}{2} & \cos\frac{\theta}{2} \end{pmatrix} , \tag{6.9}$$

we obtain a spin-up spinor along the z-axis. Then, the matrix amounts to a rotation by an angle θ of the system of coordinates. Let us now apply these properties to the rotation of creation and destruction operators that appear in the original Hamiltonian of the one-orbital model at site 2. When this rotation acts over arbitrary operators $c^\dagger_{\sigma,2}$ and $c_{\sigma,2}$ for site 2, new operators $\tilde{c}^\dagger_{\sigma,2}$ and $\tilde{c}_{\sigma,2}$ are found. The relationship between them is $c_{\uparrow,2} = \cos\frac{\theta}{2}\tilde{c}_{\uparrow,2} - \sin\frac{\theta}{2}\tilde{c}_{\downarrow,2}$ and $c_{\downarrow,2} = \cos\frac{\theta}{2}\tilde{c}_{\downarrow,2} + \sin\frac{\theta}{2}\tilde{c}_{\uparrow,2}$.

Using this transformation, it can be shown that the Hund coupling term now contains only the z-component of the spin, both at sites 1 and 2, the latter with the new operators such that the z-axis has rotated to match the direction of the classical spin in that site. In the large Hund coupling limit, only the spinor with positive z-component matters. States in the Hilbert space with negative z-projected spins are discarded. The most interesting result emerging from this simple exercise is the new form of the hopping term in the rotated basis. Replacing the old $c_{\sigma,2}$ and $c^\dagger_{\sigma,2}$ operators by the new ones $\tilde{c}^\dagger_{\sigma,2}$, $\tilde{c}_{\sigma,2}$, and dropping all terms with down spins, the hopping term becomes

$$H_{\text{hopp}} = -t\cos\frac{\theta}{2}(c^\dagger_{\uparrow,1}\tilde{c}_{\uparrow,2} + \text{h.c.}) . \tag{6.10}$$

Hence, an effective hopping $t_{\text{eff}} = t\cos\frac{\theta}{2}$ regulates the movement of carriers. It is intuitively obvious that if the classical spin at site 1 is also allowed to have an arbitrary orientation in the x–z plane, then the angle θ will now be the angle between the classical spins at sites 1 and 2. This effective hopping is maximized for a FM configuration of classical spins, but it cancels for an AF configuration. This simple calculation is the main result contained in the 1951 paper of Anderson and Hasegawa. This work has guided research in manganites until recently when its form was found to be insufficient to explain either the insulating state at high Curie temperature, charge and orbital ordering, or the CMR effect.

With the guidance of the previous example, the case in which the classical spin at site 2 points in an arbitrary direction (i.e., now using two angles θ and ϕ for its determination in the usual spherical coordinates notation), can be easily worked out. Assuming that the angles for spin 1 are θ_1 and ϕ_1, and for spin 2 they are θ_2 and ϕ_2, the effective hopping can be shown to be [5.1]

$$t_{\mathrm{eff}} = t[\cos(\theta_1/2)\cos(\theta_2/2) + \sin(\theta_1/2)\sin(\theta_2/2)\mathrm{e}^{\mathrm{i}(\phi_2-\phi_1)}] , \qquad (6.11)$$

which for $\phi_1 = \phi_2$ reduces to the more traditional result discussed before in this Appendix. Clearly the new hopping is *complex*, and its influence may lead to interesting physics [5.1] that has not yet been fully explored.

6.11 Appendix II: Brief Introduction to the Kondo Effect

The one-orbital model for manganites is also referred to as the "ferromagnetic Kondo" model, as already remarked before. Here, a very brief primer on Kondo physics is presented, with the purpose of allowing newcomers to condensed matter to become acquainted with the physics behind the name of the model that is extensively used in this book.

When the coupling J_{H} in the one-orbital model becomes *negative*, i.e., when the mobile and localized spins tend to be antiparallel, and considering quantum-localized spins, the resulting model is widely known as the "Kondo lattice model" (for a review, see [6.53]). This model is a lattice version of the original Kondo model used to explain the interaction between magnetic impurites and conduction electrons [6.54] (for a recent brief review and results, see [6.55]). The famous Kondo effect arises from the interaction between a single magnetic atom or ion with a nonzero spin, such as cobalt, and the mobile electrons of the metal in which that atom is immersed. As shown in Fig. 6.23a, experimentally it was found that metals become insulators at low temperature when magnetic impurities are added. The temperature at which this effect becomes prominent (the minimum in the resistivity) is called the Kondo temperature. In 1964, Kondo [6.54] showed that electrons can be scattered strongly from a magnetic ion, thus increasing the resistivity. Subsequent work showed that at temperatures below the Kondo temperature the magnetic moment (spin) of the impurity is entirely *screened* by the spins of the electrons in the metal.

This phenomenon is the result of a spin-exchange process. The spin of the impurity can interact with the spins of the mobile electrons. For instance, it is possible for the electrons of the impurity (those responsible for its spin) to

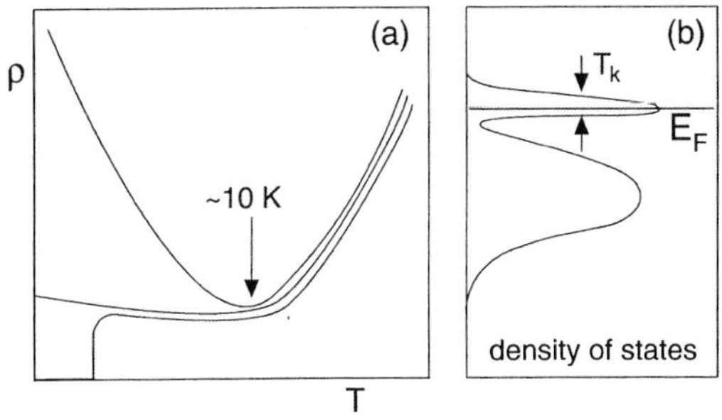

Fig. 6.23. (**a**) Resistivity vs. temperature for a metal with and without magnetic impurities. Shown are metallic (*middle*), superconducting (*bottom*) and insulating resistivities of a metal where the Kondo effect occurs (*top*). The Kondo temperature is shown. (**b**) Density of states showing schematically the Kondo resonance, and its width (T_K is the Kondo temperature). Figures reproduced from Kouwenhoven and Glazman [6.55]

tunnel from the impurity to the Fermi level of the metal (but not to lower energy levels, since they are occupied). Usually this costs a large amount of energy, but it is possible. This electron can return to its original position without altering its spin, or another electron of the Fermi sea can occupy the empty state at the impurity. This last process can occur with an overall spin-flip process, namely the spin of the electron that eventually arrives at the impurity could be the opposite to the one that the original electron had (these type of process are very similar to those used to describe antiferromagnetism in an ionic crystal; they both occur through an exchange process in which actual spins are interchanged). As a consequence, an effective antiferromagnetic interaction between the localized spin and the mobile electrons is generated. Calculations show that this interaction can be sufficiently important that a singlet can be established between the spin of the impurity and electrons at the Fermi level. Such a singlet is equivalent to making the impurity spin "disappear". This state is a new state for the electrons at the Fermi sea. Since these electrons contribute the most to the low-temperature resistivity, the latter is strongly affected by impurities. Electrons spend considerable time "trapped" near the impurities, effectively increasing the resistivity. The new state of electrons at the Fermi level (Kondo resonance) manifests itself in a peak in the density of states (Fig. 6.23b).

Clearly, an antiferromagnetic coupling between localized and mobile electrons is important. The Kondo lattice model is obtained when many impurities are studied. In this model, the coupling is present at each impurity, the location of the impurities is assumed to be regular instead of random, and a term to describe the kinetic energy of electrons in the metal is considered. These considerations show that the so-called one-orbital model for manganites is simply like a Kondo lattice model but with a ferromagnetic local coupling as well as classical spins. Models with separate spin (localized) and charge (mobile) degrees of freedom are very common. They have been used not only for manganites and Kondo systems (in the area of heavy fermions) but for the so-called diluted magnetic semiconductors (DMS) and even the high-T_c cuprates (see the contribution of A. Moreo in this chapter, as well as the discussion on cuprates and DMS in the last chapter).

7. Monte Carlo Simulations and Application to Manganite Models
with G. Alvarez and A. Feiguin

Partition functions and functional integrals (path integrals) reduce many-body problems to complicated multidimensional integrals or sums. Monte Carlo simulations are tools to evaluate in an approximate way these integrals/sums, allowing for fairly precise calculation of observables. The results are exact within statistic and controlled systematic errors. This is a powerful procedure that supplements analytical calculations. A large variety of physical properties can be studied in a wide range of parameters with relatively small effort, allowing the investigation of low, intermediate, and large coupling regimes, where analytical techniques cannot be controlled. The simulations give results that can often be directly compared with experiments, offering a method to determine if a given model contains the essential physics of a system. This allows for a systematic exploration of the relationship between microscopical models and experimental observations, varying the parameters of a model.

The Monte Carlo technique has its origins in the numerical evaluation of integrals. The technique was later generalized to calculate the partition function and the mean value of observables in classical systems. A similar idea can be applied to quantum systems through the Feymann's path-integral representation. The properties of a model can be obtained summing an exponential of the action over classical paths in phase space.

In this chapter, we define the basic concepts, introduce the method of Monte Carlo integration, and the importance-sampling technique. Later, we describe a simple algorithm for the classical Ising model, and give a brief introduction to quantum algorithms applied in condensed matter physics. Although the goal of this chapter is to give a brief but comprehensive outline of the main ideas, the reader is encouraged to consult the numerous reviews and books in the literature, for a more detailed description of the method and its different variations (see, for instance [7.1–7.5]).

7.1 Numerical Integration

Most integrals cannot be solved explicitly, and must be calculated numerically. In some cases this is carried out employing some kind of quadrature procedure. However, for multidimensional integrals, or when the integrand is a very complicated function containing sharp peaks, Monte Carlo methods are a better alternative tool.

7.1.1 Simple Sampling

We start this brief review with the "simple sampling" Monte Carlo method for evaluating integrals. Let us assume that we need to calculate the integral of a function f in the interval $[a,b]$. For this, we first generate a sequence of random numbers x_i $(i = 1, ..., N)$ uniformly distributed in $[a,b]$; the probability density of these points is

$$P = \begin{cases} 1/(b-a) & \text{if} \quad x \in [a,b] \\ 0 & \text{if} \quad x \notin [a,b] \,. \end{cases}$$

Then, we can calculate the expectation value of the function $f(x)$ according to a probability distribution P as:

$$E(f(x)) = \int_a^b f(x)P(x)\mathrm{d}x = \frac{1}{(b-a)}\int_a^b f(x)\mathrm{d}x = \frac{I}{(b-a)},$$

leading to $(b-a)E(f(x)) = I$. From these considerations, we define the "Monte Carlo Integral" I_{MC} as

$$I_{\mathrm{MC}} = (b-a)\sum_{i=1}^{N}\frac{f(x=x_i)}{N} \,.$$

According to the *central limit theorem*, I_{MC} approaches I as N goes to infinity:

$$E[I_{\mathrm{MC}}] = \frac{1}{N}E[(b-a)\sum_{i=1}^{N} f(x_i)] = E[(b-a)f(x)] = I\,,$$

with a variance $s^2[I_{\mathrm{MC}}] = (b-a)(\langle f^2\rangle - \langle f\rangle^2)/N$, where $\langle f\rangle = \sum_{i=1}^{N} f(x_i)/N$. The square root of this equation, s, is frequently referred to as the "Monte Carlo error". For large N, the error varies as $1/\sqrt{N}$, implying that if we wish to increase the precision by a factor 10, the number N should be increased by a factor $10^2 = 100$. If N is already large, improving the precision could be a difficult task.

7.1.2 Importance Sampling

In the previous section we assumed that the election of the points in the integration region was completely random. However, such an election is not practical if the function has large fluctuations within the integration range, for instance if it has a large narrow peak. If we wished to evaluate the integral $\int_0^1 \mathrm{e}^{-10x}\mathrm{d}x$ using randomly distributed points in $[0,1]$, many of these points would be lost because the integrand is very small for $x \geq 0.1$, and we would need a very large number of points N in order to obtain a reasonable good result. In this kind of situation, it is preferable to choose the random numbers in such a way that the majority of them fall in the region that truly contributes to the integral. This process is called "importance sampling". For this purpose, we choose our points distributed according to a specially chosen, positive, normalized, probability density $P(x)$, defined in the interval $[a, b]$. To calculate the integral we now define I_{sel} as

$$I_{\text{sel}} = \frac{1}{N}\sum_{i=1}^{N} \frac{f(x_i)}{P(x_i)},$$

whose expectation value is

$$E[I_{\text{sel}}] = \frac{1}{N}\sum_{i=1}^{N} E\left[\frac{f(x_i)}{P(x_i)}\right] = \frac{1}{N}\sum_{i=1}^{N} \int_a^b \frac{f(x)}{P(x)} P(x)\mathrm{d}x = I\,.$$

The variance of the sampling is $s^2[I_{\text{sel}}] = [\langle (f/P)^2\rangle - \langle f/P\rangle^2]/N$. It is clear that we can exploit the freedom in the election of P in order to find the best possible function for our particular problem. In particular, we would like to choose P in such a way that s is minimized, which effectively occurs if P is proportional to $f(x)$. In this situation, the ratio $f(x)/P(x)$ becomes a constant, resulting in a zero variance. This option has only academic relevance, because the normalization of the probability would require us to know $\int_a^b f(x)\mathrm{d}x$ a priori, which is precisely the quantity to be determined. In practice, we will try to approximate f with a simple function $P(x)$ that imitates its behavior, and is easy to normalize in the range of interest.

7.2 Monte Carlo Simulations

7.2.1 Importance Sampling of Physical Quantities

The most common application of these ideas is to compute the statistical average of observables. By definition, the thermodynamic average $\langle O\rangle_T$ of an observable $O(\overline{x})$, where $\overline{x}$ is a point in the phase space of the problem, is given by

$$\langle O\rangle_T = \frac{1}{Z}\int_V O(\overline{x})\mathrm{e}^{-S(\overline{x})}\mathrm{d}\overline{x}\,,$$

with the partition function Z defined as $Z = \int_V \mathrm{e}^{-S(\overline{x})}\mathrm{d}\overline{x}$. In these equations, V is the phase space volume, and S is the action of the system, which in classical systems corresponds to βH. We could immediately apply the ideas of simple sampling explained before and calculate the arithmetic average summing over configurations of $\overline{x}$ randomly generated, but this is clearly very inefficient. Instead, to evaluate these integrals we choose a probability distribution $P(\overline{x})$, and using the concepts of importance sampling learned in the previous sections we arrive at:

$$\langle O\rangle_T \approx \frac{\sum_{i=1}^{N} O(\overline{x}_i)P^{-1}(\overline{x}_i)\mathrm{e}^{-S(\overline{x}_i)}}{\sum_{i=1}^{N} P^{-1}(\overline{x}_i)\mathrm{e}^{-S(\overline{x}_i)}}\,.$$

If the probability density is proportional to the equilibrium distribution $P(\overline{x}) \sim \mathrm{e}^{-S(\overline{x})} = P_{\mathrm{eq}}(\overline{x})$, then

$$\langle O\rangle_T \approx \frac{1}{N}\sum_{i=1}^{N} O(\overline{x}_i)\,. \tag{7.1}$$

A very important obstacle is that the normalization of the probability would require knowing Z. In quantum problems we will also find that the quantity $S(\overline{x})$ is not easy to evaluate. However, in the following section we describe a method for sampling that does not require calculating explicitly the probabilities involved.

7.2.2 Markov Processes

As we have seen in the previous section, the equilibrium distribution is not known explicitly. For instance, in the situation that concerns us, the Boltzman factor is normalized by the partition function, which is not known a priori. To overcome this obstacle, Metropolis et al. in 1953 [7.6] developed a method based on the concept of a "random walk" through the phase space in the form of a sequence of values $\{\overline{x}_i\}$ such that, independently of the initial distribution, the probability $P(\overline{x}, t)$ of the system being found in a configuration $\overline{x}$ at a time t converges to the equilibrium distribution when $t \to \infty$, i.e., $\lim_{t\to\infty} P(\overline{x}, t) = P_{\mathrm{eq}}(\overline{x})$. This technique is based essentially on the properties of the Markov processes. A Markov process is defined by a transition probability $W(\overline{x} \to \overline{y})$ from one configuration to another. If our election of $W(\overline{x} \to \overline{y})$ satisfies certain conditions, the system will reach the equilibrium after a sufficiently long enough period of time (proof can be found in books devoted to Monte Carlo processes). The conditions are:

$$\begin{array}{ll}
\text{(i)}\ \ W(\overline{x} \to \overline{y}) \geq 0 & \forall \overline{x} \neq \overline{y} \\
\text{(ii)}\ \ \sum_{\overline{y}} W(\overline{x} \to \overline{y}) = 1 & \\
\text{(iii)}\ \ P_{\mathrm{eq}}(\overline{x})W(\overline{x} \to \overline{y}) = P_{\mathrm{eq}}(\overline{y})W(\overline{y} \to \overline{x})\,. &
\end{array} \tag{7.2}$$

The condition (ii) states that from any given starting state, it must be possible to evolve the system to any other configuration, i.e., all configurations are accessible. This is often called the "ergodicity hypothesis". The probability (iii) is the condition of "microscopic reversibility" or "detailed balance".

During the "Markov" process, the evolution of $P(\overline{x}, t)$ is given by the "master equation":

$$P(\overline{x}, t + \Delta t) - P(\overline{x}, t) = \sum_{\overline{x}} W(\overline{x} \to \overline{y}) P(\overline{x}, t) - \sum_{\overline{y}} W(\overline{y} \to \overline{x}) P(\overline{y}, t) ,$$

where the discrete time slices correspond to a step in the Markov process. In the thermodynamic equilibrium, the balance equation assures that we reach the stationary state with $P(\overline{x}, t) = P_{\text{eq}}(\overline{x})$ and, as a consequence, $P(\overline{x}, t + \Delta t) = P(\overline{x}, t) = P_{\text{eq}}(\overline{x})$, or

$$\sum_{\overline{x}} W(\overline{x} \to \overline{y}) P_{\text{eq}}(\overline{x}) = \sum_{\overline{y}} W(\overline{y} \to \overline{x}) P_{\text{eq}}(\overline{y}) ,$$

from which we deduce

$$\frac{W(\overline{x} \to \overline{y})}{W(\overline{y} \to \overline{x})} = \frac{P_{\text{eq}}(\overline{y})}{P_{\text{eq}}(\overline{x})} = \mathrm{e}^{-[S(\overline{y}) - S(\overline{x})]} . \tag{7.3}$$

Note that while the ratio of transition rates is determined by the balance condition, there is freedom to choose W as a convenient positively defined function $W(\overline{x} \to \overline{y}) = F\left(\mathrm{e}^{-[S(\overline{y}) - S(\overline{x})]}\right)$, where F must satisfy $F(R)/F(1/R) = R$, $0 \leq F(R) \leq 1$ for all R. Although the equilibrium is always reached, the convergence time depends on the particular W, and the properties of the system itself.

7.2.3 Sampling Techniques

The sampling algorithm consists of selecting a starting configuration $\overline{x}$, and then generating another possible configuration $\overline{y}$ with probability $P_{\text{eq}}(\overline{x}) \sim \mathrm{e}^{-S(\overline{x})}$. For this purpose we employ von Neumann's "rejecting method", commonly used to generate random numbers with a given distribution P (see for instance, Numerical Recipes [7.7]). In our approach, we choose randomly a number between 0 and 1; if this number is lower than P, then the new configuration is accepted, otherwise it is rejected. This method is repeated for the new configuration.

At this point we notice again that we do not know the normalization factor of P. Fortunately, there are very efficient ways to address this problem. We can generate a Markov chain as follows: starting from a configuration $\overline{x}$, we randomly choose a new state $\overline{y}$ differing slightly from the previous one, say by $\Delta\overline{x}$. Then we can apply one of the following alternatives:

1) The "Metropolis algorithm" [7.6]: We calculate the ratio $R = W(\overline{x} \to \overline{y})/W(\overline{y} \to \overline{x}) = \mathrm{e}^{-[S(\overline{y}) - S(\overline{\overline{x}})]}$, and we apply the rejecting method with a probability $p = F(R) = \min(1, R)$.

2) The "heat bath algorithm": In systems with local action, such as spins on a lattice, we consider a single spin immersed in a thermal bath, with the rest of the lattice acting as an "instantaneous environment". Then, we can sample the action of the spin considering the rest of the degrees of freedom frozen. In this case, we accept or reject the new spin configuration with probability $p = F(R) = R/(1 + R)$. This "heat bath" algorithm is more convenient when the changes are so small that $R \sim 1$.

The closer the quantity p is to the value 1/2, the higher will be the acceptance ratio, and the more efficient the algorithm will be.

It is important to assure that the proposed moves (the strategy to choose a new state in the "neighborhood" of another) are "ergodic", i.e., that all the possible states are accessible, and that they are compatible with the requirement of microscopic reversibility. If this were not the case, the simulation would be biased, and the result would be spurious.

It is clear that the configuration obtained in one iteration strongly depends on what the state was in the previous one. As a consequence, two subsequent configurations are not independent of each other, and their physical properties are highly correlated. This is a downside of the approach that is overcome by performing several iterations between successive measurements. In practice, the first few hundred iterations are discarded since they could be influenced by the starting configuration. Once we have proved that the equilibrium has been reached (for instance observing the variance of a physical quantity), the next iterations are used to measure the observables. These measurements will be finally averaged to get the estimation of the mean values, with their corresponding statistical errors.

Note: The Sign Problem. In quantum Monte Carlo we usually have to deal with fermionic anticommutation rules, or frustrated magnetic systems. In these situations it frequently occurs that the the transition probabilities have a negative sign due to the exchange of two particles. This is not a problem of the algorithm, but a simple manifestation of the statistic of the particles involved or frustration effects. The so-called "sign problem" is often fatal for the simulation, producing large fluctuations in the statistical averages, and it is a pathological problem that affects a wide variety of models. However, in certain cases there is a simple trick that allows us to perform the simulation in spite of this problem. By definition, a probability cannot be negative, and the simulations have to be done with the absolute value of p. Then, the mean value of an observable becomes $\widehat{O} =< \mathrm{sgn}\widehat{O} > / < \mathrm{sgn} >$. In the case of large fluctuations of the sign, i.e., $< \mathrm{sgn} > \sim 0$, the algorithm loses efficiency due to the cancellations of numerator and denominator. However, it can occur that, for certain values of the parameters of the model, these effects are less severe or negligible, and the simulation results can sometimes be satisfactory. A more detailed discussion of the sign problem and references are given in [1.20].

7.3 The Ising Model

Let us consider as a specific example the Ising model, which is an ideal prototype to illustrate the Monte Carlo method (a variation of this model is used as a "toy" model in Chap. 17 to describe the competition between ordered states in manganites). The Hamiltonian is given by

$$H = \sum_{<ij>} J_{ij} s_i s_j \, ,$$

where the sum runs over first neighbors on a square lattice with periodic boundary conditions, and the spins s_i can assume the values +1 (up) and −1 (down). In what follows, we consider the couplings to be uniform and negative $J_{ij} = -J$, but adding "disorder" in the form of a nontranslationally invariant coupling is fairly easy if needed. The ground state of the model corresponds to all the spins pointing in the same direction, whose energy is $E_0 = zNJ/2$, where z is the coordination number of the lattice (4 in the two-dimensional case), and N is the number of sites. The Hamiltonian is diagonal in this representation, and the space of configurations is given by all the possible combinations of spins $\{\overline{s}\} \equiv \{s_0, s_1, ..., s_N\}$ with values +1 and −1. It is obvious that the number of such configurations is equal to 2^N, which grows rapidly with the size of the lattice.

The problem typically consists of calculating physical quantities such as the energy E and magnetization M, defined as:

$$E = \langle H \rangle_T; \quad M = \langle \sum_i s_i \rangle_T / N\,,$$

where the brackets mean thermal average, defined as:

$$\langle O \rangle_T = \frac{1}{Z} \sum_{\overline{s}} O\left(\overline{s}\right) \exp(-\beta E\left(\overline{s}\right))\,, \tag{7.4}$$

with the partition function of the model Z given by $Z = \sum_{\overline{s}} \exp(-\beta E\left(\overline{s}\right))$. Here, $\beta = 1/k_{\mathrm{B}}T$ is the inverse of the temperature, $\{\overline{s}\}$ are all the possible configurations of spins $\{s_0, s_1, ..., s_N\}$, and $E\left(\overline{s}\right)$ are the energies $\langle \overline{s} | H | \overline{s} \rangle = \sum_{<ij>} J_{ij} s_i s_j$. The calculation presented here has analogies with a grand-canonical ensemble study because the total magnetization (equivalent to the number of particles in this analogy) is not fixed. Thus, we have to sample the entire phase space. This is possible in small lattices (an interesting exercise for the reader would be to generate the $2^{16} = 65536$ configurations of spins of a 4×4 lattice, and calculate the energy and magnetization *exactly* as a function of the temperature), but the problem becomes unmanageable when N is large. It is clear that in order to calculate accurate properties of the system and carry out a finite-size scaling analysis, we have to perform a Monte Carlo simulation.

We can perform a simple sampling straightforwardly: we generate random configurations of spins, and calculate the arithmetic averages:

$$\overline{O} = \frac{1}{\overline{Z}} \sum_{\overline{s}} O\left(\overline{s}\right) \exp(-\beta E\left(\overline{s}\right)); \; \overline{Z} = \sum_{\overline{s}} \exp(-\beta E\left(\overline{s}\right))\,.$$

However, this method is highly inefficient, as already discussed, since especially at low temperatures only a few configurations close in energy to the ferromagnetic ground state are expected to contribute most of the weight to the sums. At temperatures below $kT/J = 1$, for instance, the weight of the two configurations with all the spins up and all the spins down accounts for 99% of the total weight!

The next step consists in applying the idea of importance sampling. For this purpose we have to choose the new configuration randomly, but according to a convenient probability distribution P. It is natural to choose $P = \exp(-\beta E(\overline{s}))$, since this is the equilibrium distribution, and the averages reduce to (see (7.1)):

$$\overline{O} = \frac{1}{N} \sum_{\overline{s}} O(\overline{s}) .$$

Notice that the new sums run over configurations chosen with probability density P. The task is then to generate a Markov chain to sample all these configurations stochastically. For that purpose, instead of choosing the spin configurations independently of each other, we define a "random walker", an entity that will sweep the space of configurations $\{\overline{s}\}$. The walker traces a path through the phase space from a configuration $\overline{s}$ to another $\overline{s}'$, according to a transition probability p. Following Metropolis we choose (we could alternatively use the "heat bath" expression):

$$p = \begin{cases} \exp[-\beta(E(\overline{s}') - E(\overline{s}))] , & \text{if } E(\overline{s}') > E(\overline{s}) \\ 1 , & \text{if } E(\overline{s}') < E(\overline{s}) \end{cases} .$$

Notice that if the change in energy is lower that zero, i.e., $E(\overline{s}') < E(\overline{s})$, according to this definition the move will always be accepted. On the other hand, large and positive changes in $\Delta E = E(\overline{s}') - E(\overline{s})$ will lead to values of p that may produce small transition probabilities. If we update several spins simultaneously this would occur quite often, leading to a small acceptance rate, i.e., most of the attempted moves will be rejected, affecting the efficiency of the method. Thus, we have to perform moves where the configurations $\overline{s}'$ and $\overline{s}$ differ only by a few spins. Although clever and efficient multispin algorithms have been devised [7.8, 7.9], these are beyond the scope of this book, and here we will update one spin at a time. A scheme for the single-spin flip algorithm for the Ising model is presented below:

(1) Select a site i of the lattice.
(2) Compute the energy change ΔE associated to the spin flip $s_i \to -s_i$.
(3) Calculate the transition probability p.
(4) Generate a random number r in the range $[0, 1]$.
(5) If $r < p$ flip the spin, otherwise leave it unchanged.

There are several ways to choose a spin in step (1). One could select a spin randomly, but we would have to generate an extra random number. Thus, we simply sweep the lattice sequentially. This can be done by rows, columns, or checkerboard arrangements, since the result should be independent of this election. After selecting any of these procedures, we define a "Monte Carlo step" as a complete sweep through the lattice.

It is clear that the transition probabilities depend on the local change in energy defined by the z spins interacting with S_i, through the nearest-neighbors terms in the Hamiltonian, since the rest of the lattice remains unaltered. The possible "local energies" are given by $E_i^{\text{local}} = J \sum_{\delta=1}^{z} s_i s_{i+\delta}$, where the sum is performed over the first neighbors of the spin at the site i. In the square lattice $z = 4$, and there are $2^5 = 32$ possible values. This is a small number and the transition probabilities can be stored in a lookup table, translating in a considerable gain in efficiency.

After these considerations, the pseudocode for a Monte Carlo sweep is:

```
for all the sites of the lattice:
    retrieve the transition probability p from the lookup table
    compute a random number r
    if (r < p) s_i → −s_i
```

As explained before, typically the first few hundred iterations are discarded since they could be influenced by the starting configuration. This number can be much larger in regimes where many states compete. Once it is believed that equilibrium has been reached, for instance by monitoring the behavior of observables as a function of Monte Carlo steps, we can start the actual measurement of physical quantities. Then, the complete Monte Carlo algorithm can be described as follows:

```
(1) Initialize variables, lookup tables and random number generator.
    Generate initial configuration.
(2) Perform a sufficient number of iterations for thermalization.
(3) Carry out measurements and store the associated numbers.
(4) Compute total averages and statistical errors.
```

In the Appendix of this chapter, the reader will find the Fortran90 code for the Ising model on the square lattice and a single walker. In Fig. 7.1, we show results for the absolute value of the magnetization and the specific heat for different lattice sizes, where it can be seen how the curves converge to the exact result for the thermodynamic limit, with a transition from the ferromagnetic to the paramagnetic phase at $T_C = 2.26$ [7.10], in units of J. The code for the calculation of the specific heat is left as an exercise.

To conclude this section, we would like to remark that Monte Carlo is the paradigm of parallelization since it becomes clear that we can use several random walkers to sweep the phase space independently, and add their contributions to the final averages. This is a very common practice, since it allows us to take full advantage of modern parallel computers, reducing the simulation time by several orders of magnitude.

7.4 A Brief Summary of Quantum MC Techniques

There are many MC techniques that can be applied when the model under study is quantum, rather than classical as in the Ising case. Reviewing these many techniques would be a formidable task, thus here only a brief sketch of a few techniques is provided, mainly for completeness. The reader should

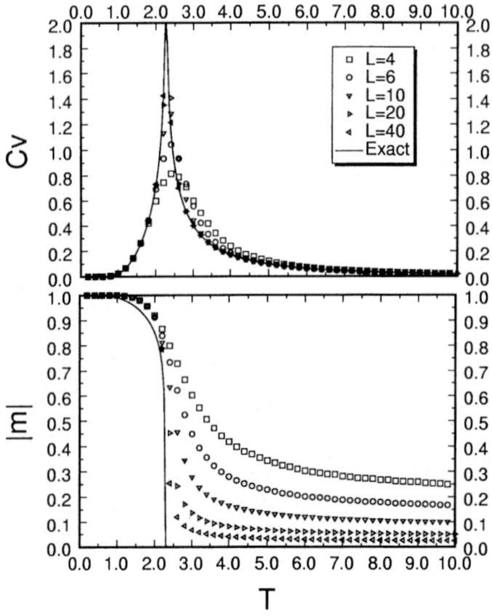

Fig. 7.1. Specific heat (*upper panel*) and absolute value of the magnetization (*lower panel*) of the 2D Ising model calculated for different lattice sizes $L \times L$ using the program listed in this section, with 100 000 measurements taken every 5 MC sweeps. Onsager's exact results [7.10] are also plotted for comparison. The error bars are smaller than the symbol size

consult the literature if details are needed. These quantum MC techniques have still not been applied extensively to models of manganites, since much can be done with classical localized spins and classical phonons, but we expect that in the future they will.

(A) Variational Monte Carlo. We begin our description of the quantum Monte Carlo method with the variational Monte Carlo (VMC) technique [7.5, 7.11, 7.12]). Here, the goal is to find a good representation of the actual ground state of a model, using a convenient wave function derived from our physical insight of the system under study. Thus, in this approach, instead of sampling the action of a system, we sample the trial wave function. This wave function $|\psi_T(\overline{\eta})\rangle$ depends on a set of parameters $\overline{\eta}$, and it is represented in terms of a basis of orthogonal states $|x\rangle$, namely $|\psi_T(\overline{\eta})\rangle = \sum_x \langle x|\psi_T(\overline{\eta})\rangle|x\rangle = \sum_x C_x(\overline{\eta})|x\rangle$, where the coefficients of the parametrization are known functions of $\overline{\eta}$. Finding the best wave function means finding the set of parameters $\overline{\eta}$ that maximize the overlap with the actual ground state. In practice, this is impossible since we do not know the ground state a priori. To carry out the calculation instead we apply the variational principle, which states that the variational energy of the trial state is always greater than or equal to the exact energy of the ground state:

$$\langle E\rangle_T = \frac{\langle\psi_T \mid H \mid \psi_T\rangle}{\langle\psi_T \mid \psi_T\rangle} \geq E_0\,, \tag{7.5}$$

and we use the criterion of minimizing the variational energy. To do that, we require to calculate this quantity for different sets of parameters $\overline{\eta}$, and once we found a proper wave function, we can calculate the physical quantities of interest. The important point for our purposes is that the calculation of the expectation value in the previous equation can be carried out with Monte Carlo techniques, as explained in [7.5, 7.11, 7.12]), and references therein. The variational simulations are simple to perform and very stable. Since the probabilities do not depend on the statistics of the particles involved, they do not suffer from the sign problem. However, the results certainly depend on the quality of the variational wave function, which guides the simulation.

(B) Determinantal Monte Carlo. This method is typically applied to the quantum-mechanical many-body problem of interacting electrons on a lattice, working in the grand-canonical ensemble. The basic idea of this approach was presented a long time ago by Blankenbecler et al. [7.13] and White et al. [7.14]. Quantum Monte Carlo, in general, relies on the fact that d-dimensional quantum problems can be interpreted as classical problems in $(d+1)$ dimensions through the Feymann's path integral representation. The task consists in identifying scalar fields that would allow us to evaluate the partition function and mean values using the same Metropolis algorithm we have used before in the equivalent classical model. Suzuki [7.15] was the first to apply these concepts after generalizing an idea by Trotter [7.16].

Suppose we want to evaluate the expectation value of a physical observable $\hat{O}$, at some finite temperature $T = 1/k_{\mathrm{B}}\beta$. If $\hat{H}$ is the Hamiltonian of the model, the expectation value is defined as

$$\langle \hat{O} \rangle_T = \frac{\mathrm{Tr}(\hat{O}\mathrm{e}^{-\beta \hat{H}})}{\mathrm{Tr}(\mathrm{e}^{-\beta \hat{H}})}\,, \tag{7.6}$$

where the notation is standard. The Hamiltonian of the Hubbard model, a special but important case, with the addition of a chemical potential, can be naturally separated into two terms as

$$\hat{K} = -t \sum_{\langle \mathbf{ij} \rangle,\sigma} (c^{\dagger}_{\mathbf{i}\sigma} c_{\mathbf{j}\sigma} + c^{\dagger}_{\mathbf{j}\sigma} c_{\mathbf{i}\sigma}) - \mu \sum_{\mathbf{i}} (n_{\mathbf{i}\uparrow} + n_{\mathbf{i}\downarrow})\,, \tag{7.7}$$

$$\hat{V} = U \sum_{\mathbf{i}} \left(n_{\mathbf{i}\uparrow} - \frac{1}{2} \right) \left(n_{\mathbf{i}\downarrow} - \frac{1}{2} \right)\,. \tag{7.8}$$

Discretizing the inverse temperature interval as $\beta = \Delta\tau L$, where $\Delta\tau$ is a small number, and L is the total number of time slices, we can apply the well-known Trotter's formula to rewrite the partition function as,

$$Z = \mathrm{Tr}\left(\mathrm{e}^{-\Delta\tau L \hat{H}} \right) \sim \mathrm{Tr}\left(\mathrm{e}^{-\Delta\tau \hat{V}} \mathrm{e}^{-\Delta\tau \hat{K}} \right)^{L}\,, \tag{7.9}$$

where a systematic error of order $(\Delta\tau)^2$ has been introduced, since $[\hat{K}, \hat{V}] \neq 0$. In order to integrate out the fermionic fields the interaction term $\hat{V}$ has to be made *quadratic* in the fermionic creation and annihilation operators by a decoupling Hubbard Stratonovich transformation. This decoupling procedure introduces scalar variables that can be simulated by the Monte Carlo method. Details and additional references can be found in the review [1.20]. This method has allowed for a careful study of the Hubbard model at half-filling, i.e., one electron per site. At other densities away from half-filling, such as those important for high-temperature superconductors, many investigations have been carried out and progress was made [1.20], although the sign problem unfortunately prevents accurate studies at the low temperatures of relevance for superconductivity.

(C) Projector Monte Carlo. A simple modification of the "determinantal" algorithm allows for the calculation of ground-state properties in the canonical ensemble, i.e., with a fixed number of electrons. This approach is called "projector Monte Carlo." Consider the ground state $|\psi_0\rangle$ of a system, and let $|\phi\rangle$ be a trial state with a nonzero overlap with the actual ground state. If we apply the operator $P = \mathrm{e}^{-\tau H}$ a number of times m over the state $|\phi\rangle$, and we insert the identity operator in the basis $|\alpha\rangle$ of eigenstates of H, we obtain:

$$\begin{aligned} |\phi^m\rangle = P^m|\phi\rangle &= \sum_\alpha \mathrm{e}^{-m\tau H}|\alpha\rangle\langle\alpha|\phi\rangle = \sum_a \left(\mathrm{e}^{-\tau E_\alpha}\right)^m |\alpha\rangle\langle\alpha|\phi\rangle \\ &= \mathrm{e}^{-\beta E_0}\sum_a \left(\mathrm{e}^{-\tau(E_\alpha - E_0)}\right)^m |\alpha\rangle\langle\alpha|\phi\rangle\,. \end{aligned} \tag{7.10}$$

In the limit $\beta \to \infty$, the projection will filter out all the excited states, and only the ground state will survive, i.e., $\lim_{m\to\infty} P^m|\phi\rangle = |\,\psi_0\rangle$. Thus, the ground-state energy is given by:

$$E_0 = \frac{\langle\phi|H|\psi_0\rangle}{\langle\phi|\psi_0\rangle} = \lim_{m\to\infty}\frac{\langle\phi|H\mathrm{e}^{-m\tau H}|\phi\rangle}{\langle\phi|\mathrm{e}^{-m\tau H}|\phi\rangle}\,, \tag{7.11}$$

while for any other physical observable $\hat{O}$, the expectation value can be exactly written as:

$$\frac{\langle\psi_0|\hat{O}|\psi_0\rangle}{\langle\psi_0|\psi_0\rangle} = \lim_{m,m'\to\infty}\frac{\langle\phi|\mathrm{e}^{-m'\tau H}\hat{O}\mathrm{e}^{-m\tau H}|\phi\rangle}{\langle\phi|\mathrm{e}^{-(m'+m)\tau H}|\phi\rangle}\,. \tag{7.12}$$

The steps necessary to Monte Carlo simulate (7.12) are similar to those in the previous section. First, the imaginary time axis is discretized in a finite number of slices, then the Trotter approximation, as well as the Hubbard Stratonovich decoupling are used. Fermions are integrated out, and all observables are finally expressed in terms of the spin fields, which are treated using a Metropolis algorithm [7.14]. A variant called Green function Monte Carlo (GFMC) is much used (see [7.4, 7.5, 7.11, 7.12] and references therein).

7.5 Basic Monte Carlo Formalism for Manganite Models

In this section, details related to Monte Carlo simulations in the context of manganites are provided. The work reported here was carried out in collaboration with S. Yunoki, who performed the first Monte Carlo simulations of models for manganites a few years ago. For simplicity, let us focus on one-dimensional systems. Also, the case of the one-orbital model will be mainly discussed. Generalizations to higher dimensions and two orbitals are straightforward, and left as an exercise to the reader. Note that the one-orbital Hamiltonian is simply denoted by $\hat{H}$ in this section, and β indicates the inverse temperature, i.e., $\beta = 1/k_{\mathrm{B}}T$. As discussed in Chap. 5, the Hamiltonian for the 1D one-orbital model is given by

$$\hat{H} = -t \sum_{\langle ij \rangle,\sigma} (a^{\dagger}_{i,\sigma} a_{j,\sigma} + \mathrm{H.c.}) - J_{\mathrm{H}} \sum_{i,\alpha,\beta} \mathbf{s}_i \cdot \mathbf{S}_i + J_{\mathrm{AF}} \sum_{\langle ij \rangle} \mathbf{S}_i \cdot \mathbf{S}_j \, , \qquad (7.13)$$

The spin of the mobile electron is defined as $\mathbf{s}_i = a^{\dagger}_{i,\alpha} \sigma_{\alpha\beta} a_{i,\beta}$, $\sigma = (\sigma_x, \sigma_y, \sigma_z)$ are the standard Pauli matrices, and the localized spins are considered classical and of length one ($S = |\mathbf{S}_i| = 1$). Hence, they admit the parametrization $\mathbf{S}_i = (\sin\theta_i \cos\phi_i, \sin\theta_i \sin\phi_i, \cos\theta_i)$.

It is important to remark that in this problem the Hamiltonian only contains terms that are *quadratic* in the fermionic operators, namely there are no explicit Coulombic interactions in the problem (although the Hund coupling plays a qualitatively similar role, see Chap. 5). This allows for the enormous simplification of having to solve only the *one*-electron problem even if a large density is studied, since such density can be reached by simply filling levels of the one-body problem from the bottom of the energy spectrum up. In other words, although our results bear on problems with an arbitrary number of particles (compatible with the Pauli principle), any fermionic number subspace can be reached exactly from the one-body case by moving up or down the chemical potential. Note that as soon as the Coulomb interaction is switched on, the exact relation among Hilbert spaces of different number of electrons is lost and one encounters the problem of exponentially growing Hilbert spaces for exact treatments, or the infamous "sign" problem in Monte Carlo analysis. In manganites we are fortunate that important information can be obtained using electrons that only interact with localized spins and Jahn–Teller distortions. As already remarked, these couplings are in many respects qualitatively similar to the Coulomb interaction (Chap. 5).

Since in practice the solution of the problem at hand involves only gathering information about the "one-electron" sector, with the total density regulated by the chemical potential μ, then for the case of a chain with L sites the basis to be used can be considered as $a^{\dagger}_{1,\uparrow}|0\rangle, \ldots, a^{\dagger}_{L,\uparrow}|0\rangle$, $a^{\dagger}_{1,\downarrow}|0\rangle, \ldots, a^{\dagger}_{L,\downarrow}|0\rangle$. In this one-electron sector, $\hat{H}$ can be represented by a $2L \times 2L$ matrix for a fixed configuration of the classical spins.

The partition function in the grand canonical ensemble can be written as

$$Z = \prod_i^L \left(\int_0^\pi d\theta_i \sin\theta_i \int_0^{2\pi} d\phi_i \right) Z_g(\{\theta_i, \phi_i\}) \,. \tag{7.14}$$

Here g denotes conducting e_g electrons and $Z_g(\{\theta_i, \phi_i\}) = \mathrm{Tr}_g(\mathrm{e}^{-\beta\hat{K}})$, where $\hat{K} = \hat{H} - \mu\hat{N}$, with $\hat{N}$ the number operator, and the trace is taken with respect to the mobile electrons in the e_g orbital, which are created and destroyed by the fermionic operators $a^\dagger$ and a. It will be shown that Z_g can be calculated in terms of the eigenvalues of $\hat{K}$ in the one-particle sector, denoted by ϵ_λ ($\lambda = 1, \cdots, 2L$). The diagonalization is performed numerically using standard library routines.

Since $\hat{K}$ is a hermitian operator, it can be represented in terms of a hermitian matrix, which itself can be diagonalized by a unitary matrix U such that

$$U^\dagger K U = \begin{pmatrix} \epsilon_1 & 0 & \dots & 0 \\ 0 & \epsilon_2 & \dots & 0 \\ \vdots & \vdots & \ddots & \vdots \\ 0 & 0 & \dots & \epsilon_{2L} \end{pmatrix} . \tag{7.15}$$

The base that diagonalizes K is $u_1^\dagger|0\rangle, \dots, u_{2L}^\dagger|0\rangle$, where the fermionic operators are obtained from the original operators through $u_m = \sum_{j\sigma} U^\dagger_{m,j\sigma} a_{j\sigma}$, with m running from 1 to $2L$.

Defining $u_m^\dagger u_m = \hat{n}_m$ and denoting by n_m the eigenvalues of $\hat{n}_m$, the trace can be written as

$$\begin{aligned} \mathrm{Tr}_g(\mathrm{e}^{-\beta\hat{K}}) &= \sum_{n_1,\dots,n_{2L}} \langle n_1 \dots n_{2L} | \mathrm{e}^{-\beta\hat{K}} | n_1 \dots n_{2L} \rangle \\ &= \sum_{n_1,\dots,n_{2L}} \langle n_1 \dots n_{2L} | \mathrm{e}^{-\beta \sum_{\lambda=1}^{2L} \epsilon_\lambda n_\lambda} | n_1 \dots n_{2L} \rangle \,, \end{aligned} \tag{7.16}$$

since in the $\{u_m^\dagger|0\rangle\}$ basis, the operator $\hat{K}$ is $\sum_\lambda \epsilon_\lambda \hat{n}_\lambda$, and the number operator can be replaced by its eigenvalues. The exponential is now a number, equivalent to a product of exponentials such that

$$Z_g = \sum_{n_1} \langle n_1 | \mathrm{e}^{-\beta\epsilon_1 n_1} | n_1 \rangle \dots \sum_{n_{2L}} \langle n_{2L} | \mathrm{e}^{-\beta\epsilon_{2L} n_{2L}} | n_{2L} \rangle \,, \tag{7.17}$$

which can be written compactly as

$$Z_g = \prod_{\lambda=1}^{2L} \mathrm{Tr}_\lambda(\mathrm{e}^{-\beta\epsilon_\lambda n_\lambda}) \,. \tag{7.18}$$

Note the clear formal similarity of this formula with those that are learned in the study of noninteracting fermions in elementary courses. The difference is that the eigenvalues ϵ_λ cannot be obtained in closed form, due to the presence of the classical spins in the problem, which break the translational invariance.

Since the particles are fermions, the occupation numbers are either 0 or 1, and the sum in (7.17) is restricted to those values,

$$Z_g = \prod_{\lambda=1}^{2L} \sum_{n=0}^{1} \mathrm{e}^{-\beta\epsilon_\lambda n} = \prod_{\lambda=1}^{2L} (1 + \mathrm{e}^{-\beta\epsilon_\lambda}) . \tag{7.19}$$

Thus, combining (7.14) and (7.19), we get

$$Z = \prod_{i}^{L} (\int_0^\pi \mathrm{d}\theta_i \sin\theta_i \int_0^{2\pi} \mathrm{d}\phi_i) \prod_{\lambda=1}^{2L} (1 + \mathrm{e}^{-\beta\epsilon_\lambda}) . \tag{7.20}$$

The integral over the angular variables can be performed using a classical Monte Carlo simulation. The eigenvalues must be obtained for each classical spin configuration using library subroutines. Finding the eigenvalues is the most time-consuming part of the numerical simulation. Note that the integrand here is clearly positive. Thus, "sign problems" in which the integrand of the multiple integral under consideration can be non positive, are fortunately not present in our study. If classical phonons are also considered, then the final result is similar to the formula above, with the addition of integrals from $-\infty$ to $+\infty$ of the phononic modes or oxygen coordinates under consideration, and the concomitant changes in the Hamiltonian.

Before addressing the technical details in the next section, the reader should know that there are already more sophisticated Monte Carlo techniques discussed in the literature. These techniques include those of Furukawa, Motome, and Nakata [6.5]. Their method has the important advantage that it can be optimized for parallel computations. Other techniques include the hybrid Monte Carlo approach by Alonso et al. [6.4]. Results using both these methods were discussed in the previous chapter. However, even if the alternative Monte Carlo methods described in this paragraph are implemented, it is important to contrast results and debug the programs against the simpler MC approach described in this chapter.

7.5.1 Calculation of Static Observables

We now discuss how to calculate physically relevant quantities in the Monte Carlo procedure. The first step is to obtain the mean value of the operator of relevance for the quantity that is investigated. Considering first equal-time or static observables, they have associated operators $\hat{O}(\{a_i, a_i^\dagger\})$, whose mean values are given by

$$\begin{aligned} \langle \hat{O} \rangle &= \frac{1}{Z} \prod_{i}^{L} \left(\int_0^\pi \mathrm{d}\theta_i \sin\theta_i \int_0^{2\pi} \mathrm{d}\phi_i \right) \mathrm{Tr}_g \left(\hat{O} \mathrm{e}^{-\beta\hat{K}} \right) \\ &= \frac{1}{Z} \prod_{i}^{L} \left(\int_0^\pi \mathrm{d}\theta_i \sin\theta_i \int_0^{2\pi} \mathrm{d}\phi_i \right) Z_g \langle \tilde{O} \rangle , \end{aligned} \tag{7.21}$$

where $\langle \tilde{O} \rangle = \mathrm{Tr}_g(\hat{O} \mathrm{e}^{-\beta \hat{K}})/Z_g$. In practice, only the Green function has to be calculated, and more complicated operators are evaluated using Wick's theorem, as shown in detail below. The Green function for a given fixed configuration of classical spins is given by $G_{i,j,\sigma,\sigma'} = \langle a_{i,\sigma} a^\dagger_{j,\sigma'} \rangle$.

Consider the special case in which $\hat{O} = a_{i,\sigma} a^\dagger_{j,\sigma'}$, relevant for the Green function. In this case,

$$G_{i,j,\sigma,\sigma'} = \mathrm{Tr}_g \left(a_{i,\sigma} a^\dagger_{j,\sigma'} \mathrm{e}^{-\beta \hat{K}} \right) / Z_g \,. \tag{7.22}$$

Changing to the base in which $\hat{K}$ is diagonal through the transformation $a^\dagger_{i\sigma} = \sum_{\mu=1}^{2L} u^\dagger_\mu U^\dagger_{\mu,i_\sigma}$ (or, equivalently, $a_{i\sigma} = \sum_{\mu=1}^{2L} U_{i_\sigma,\mu} u_\mu$), where $i_\sigma = (i,\sigma)$, it can be shown that

$$\begin{aligned}
&\mathrm{Tr}_g \left(a_{i,\sigma} a^\dagger_{j,\sigma'} \mathrm{e}^{-\beta \hat{K}} \right) \\
&= \sum_{\lambda=1}^{2L} \sum_{\eta=1}^{2L} U_{i_\sigma,\lambda} U^\dagger_{\eta,j_{\sigma'}} \mathrm{Tr}_g \left[u_\lambda u^\dagger_\eta \prod_{\nu=1}^{2L} \mathrm{e}^{-\beta \epsilon_\nu n_\nu} \right] \\
&= \sum_{\lambda=1}^{2L} \sum_{\eta=1}^{2L} U_{i_\sigma,\lambda} U^\dagger_{\eta,j_{\sigma'}} \mathrm{Tr}_g \left[\prod_{\nu=1}^{2L} \{1 + (\mathrm{e}^{-\beta \epsilon_\nu} - 1) n_\nu\} u_\lambda u^\dagger_\eta \right] \\
&= \sum_{\lambda=1}^{2L} U_{i_\sigma,\lambda} U^\dagger_{\lambda,j_{\sigma'}} \mathrm{Tr}_g \left[\prod_{\nu=1}^{2L} \{1 + (\mathrm{e}^{-\beta \epsilon_\nu} - 1) n_\nu\} (1 - n_\lambda) \right] \\
&= \sum_{\lambda=1}^{2L} U_{i_\sigma,\lambda} U^\dagger_{\lambda,j_{\sigma'}} \prod_{\nu=1(\nu\neq\lambda)}^{2L} \left(\sum_{n_\nu=0}^{1} \{1 + (\mathrm{e}^{-\beta \epsilon_\nu} - 1) n_\nu\} \right) \\
&= \sum_{\lambda=1}^{2L} U_{i_\sigma,\lambda} U^\dagger_{\lambda,j_{\sigma'}} \prod_{\nu=1(\nu\neq\lambda)}^{2L} (1 + \mathrm{e}^{-\beta \epsilon_\nu}) \,,
\end{aligned} \tag{7.23}$$

where the trace is taken using the basis of eigenstates. Here we used $u_\lambda u^\dagger_\eta = (1 - \hat{n}_\lambda)\delta_{\lambda\eta} + u_\lambda u^\dagger_\eta (1 - \delta_{\lambda\eta})$. Thus, the Green function is given by

$$\begin{aligned}
G_{i,j,\sigma,\sigma'} &= \sum_{\lambda=1}^{2L} U_{i_\sigma\lambda} U^\dagger_{\lambda j_{\sigma'}} \frac{\prod_{\nu=1(\nu\neq\lambda)}^{2L} (1 + \mathrm{e}^{-\beta \epsilon_\nu})}{\prod_{\nu=1}^{2L} (1 + \mathrm{e}^{-\beta \epsilon_\nu})} \\
&= \sum_{\lambda=1}^{2L} U_{i_\sigma,\lambda} \frac{1}{1 + \mathrm{e}^{-\beta \epsilon_\lambda}} U^\dagger_{\lambda,j_{\sigma'}} \,.
\end{aligned} \tag{7.24}$$

From this expression, mean values of physically relevant operators can be obtained. For instance, the e_g-electron number is given by $\langle \hat{n} \rangle = \sum_{i,\sigma} \langle a^\dagger_{i,\sigma} a_{i,\sigma} \rangle = 2L - \sum_{i,\sigma} G_{i,i,\sigma,\sigma}$.

Wick's Theorem. If more complicated operators are needed, such as those involving four fermionic operators, then a more careful analysis is required. In such cases it can be shown that

$$\langle a_{j1,\sigma}a^\dagger_{j2,\sigma}a_{j3,\sigma}a^\dagger_{j4,\sigma}\rangle = \mathrm{Tr}_g\left[a_{j1,\sigma}a^\dagger_{j2,\sigma}a_{j3,\sigma}a^\dagger_{j4,\sigma}\mathrm{e}^{-\beta\hat{K}}\right]/Z_g$$
$$= \sum_{\mu\nu\lambda\eta} U_{j1\sigma\mu}U^\dagger_{\nu j2\sigma}U_{j3\sigma\lambda}U^\dagger_{\eta j4\sigma}\mathrm{Tr}_g\left[u_\mu u^\dagger_\nu u_\lambda u^\dagger_\eta \mathrm{e}^{-\beta\hat{K}}\right]/Z_g\,. \tag{7.25}$$

By transporting the first operator on the left, from left to right, we obtain $u_\mu u^\dagger_\nu u_\lambda u^\dagger_\eta = \delta_{\mu\nu}u_\lambda u^\dagger_\eta + \delta_{\mu\eta}u^\dagger_\nu u_\lambda - u^\dagger_\nu u_\lambda u^\dagger_\eta u_\mu$. Then,

$$\mathrm{Tr}_g\left[u_\mu u^\dagger_\nu u_\lambda u^\dagger_\eta \mathrm{e}^{-\beta\hat{K}}\right] = \delta_{\mu\nu}\mathrm{Tr}_g\left[u_\lambda u^\dagger_\eta \mathrm{e}^{-\beta\hat{K}}\right]$$
$$+\delta_{\mu\eta}\mathrm{Tr}_g\left[u^\dagger_\nu u_\lambda \mathrm{e}^{-\beta\hat{K}}\right] - \mathrm{Tr}_g\left[u^\dagger_\nu u_\lambda u^\dagger_\eta u_\mu \mathrm{e}^{-\beta\hat{K}}\right]\,. \tag{7.26}$$

In the last term of (7.26), the relation $u_\mu \mathrm{e}^{-\beta\hat{K}} = \mathrm{e}^{-\beta\hat{K}}u_\mu \mathrm{e}^{-\beta\epsilon_\mu}$ can be used, where ϵ_μ is the energy of the eigenstate μ. To prove this identity just act with both operators over the states $|0\rangle$ and $u^\dagger_\mu|0\rangle$. After this step, and recalling that $\mathrm{Tr}_g[u^\dagger_\nu u_\lambda u^\dagger_\eta \mathrm{e}^{-\beta\hat{K}}u_\mu] = \mathrm{Tr}_g[u_\mu u^\dagger_\nu u_\lambda u^\dagger_\eta \mathrm{e}^{-\beta\hat{K}}]$ due to standard properties of traces, from (7.26) it can be shown that $\mathrm{Tr}_g[u_\mu u^\dagger_\nu u_\lambda u^\dagger_\eta \mathrm{e}^{-\beta\hat{K}}](1+\mathrm{e}^{-\beta\epsilon_\mu}) = \delta_{\mu\nu}\mathrm{Tr}_g[u_\lambda u^\dagger_\eta \mathrm{e}^{-\beta\hat{K}}] + \delta_{\mu\eta}\mathrm{Tr}_g[u^\dagger_\nu u_\lambda \mathrm{e}^{-\beta\hat{K}}]$. In addition, note that for two eigenstate operators $\mathrm{Tr}_g[u_\mu u^\dagger_\nu \mathrm{e}^{-\beta\hat{K}}]/Z_g = \delta_{\mu\nu}/(1+\mathrm{e}^{-\beta\epsilon_\mu})$. This leads to

$$\frac{1}{Z_g}\mathrm{Tr}_g\left[u_\mu u^\dagger_\nu u_\lambda u^\dagger_\eta \mathrm{e}^{-\beta\hat{K}}\right] = \frac{1}{Z_g}\mathrm{Tr}_g\left[u_\mu u^\dagger_\nu \mathrm{e}^{-\beta\hat{K}}\right]\frac{1}{Z_g}\mathrm{Tr}_g\left[u_\lambda u^\dagger_\eta \mathrm{e}^{-\beta\hat{K}}\right]$$
$$+\frac{1}{Z_g}\mathrm{Tr}_g\left[u_\mu u^\dagger_\eta \mathrm{e}^{-\beta\hat{K}}\right]\frac{1}{Z_g}\mathrm{Tr}_g\left[u^\dagger_\nu u_\lambda \mathrm{e}^{-\beta\hat{K}}\right]\,. \tag{7.27}$$

Replacing this into (7.25)

$$\langle a_{j1,\sigma}a^\dagger_{j2,\sigma}a_{j3,\sigma}a^\dagger_{j4,\sigma}\rangle = \langle a_{j1,\sigma}a^\dagger_{j2,\sigma}\rangle\langle a_{j3,\sigma}a^\dagger_{j4,\sigma}\rangle$$
$$+\langle a_{j1,\sigma}a^\dagger_{j4,\sigma}\rangle\langle a^\dagger_{j2,\sigma}a_{j3,\sigma}\rangle\,, \tag{7.28}$$

which provides a simple relation between mean values of operators with four and two fermions. Many other relations can be obtained by this procedure, which are variations of Wick's theorem.

7.5.2 Calculation of Time-dependent Observables

Time-dependent observables are evaluated using time-dependent Green functions, which can be calculated numerically. The Green function is defined as

$$G^{>}_{i,j,\sigma,\sigma}(t) = \langle a_{i,\sigma}(t)a^\dagger_{j,\sigma}(0)\rangle\,, \tag{7.29}$$

where $a_{i,\sigma}(t) = \mathrm{e}^{\mathrm{i}\hat{H}t}a_{i,\sigma}\mathrm{e}^{-\mathrm{i}\hat{H}t}$. Note that $\hat{H}$ and $\hat{K}$ can be diagonalized by the same basis of eigenvectors $\{u^\dagger_m|0\rangle\}$, and the eigenvalues of $\hat{H}$ are denoted by ρ_λ. Working in this basis it is possible to write $a_{i,\sigma}(t)$ in terms of $a_{i,\sigma}$ as

$$a_{i,\sigma}(t) = \mathrm{e}^{\mathrm{i}t\sum_\nu \rho_\nu n_\nu} \sum_\eta U_{i_\sigma,\eta} u_\eta \mathrm{e}^{-\mathrm{i}t\sum_\nu \rho_\nu n_\nu}$$
$$= \sum_{\nu=1}^{2L} \left[\sum_{\lambda=1}^{2L} U_{i_\sigma,\lambda} \mathrm{e}^{-\mathrm{i}t\rho_\lambda} U^\dagger_{\lambda,\nu} \right] a_\nu \,, \tag{7.30}$$

where $a_\nu = a_{\nu,\uparrow}$ if $\nu \leq L$ and $a_\nu = a_{\nu-L,\downarrow}$ if $\nu > L$. Replacing (7.30) in (7.29), the time-dependent Green function is given by

$$G^>_{i,j,\sigma,\sigma}(t) = \sum_{\nu=1}^{2L} \left[\sum_{\lambda=1}^{2L} U_{i_\sigma,\lambda} \mathrm{e}^{-\mathrm{i}t\rho_\lambda} U^\dagger_{\lambda,\nu} \right] \langle a_\nu a^\dagger_{j,\sigma} \rangle \,. \tag{7.31}$$

It is known that $\langle a_\nu a^\dagger_{j,\sigma} \rangle = \sum_{\lambda=1}^{2L} U_{\nu,\lambda} \frac{1}{1+\mathrm{e}^{-\beta\epsilon_\lambda}} U^\dagger_{\lambda,j_\sigma}$, where $\epsilon_\lambda = \rho_\lambda - \mu$ (see (7.22) to (7.24)). Thus, replacing (7.24) in (7.31),

$$G^>_{i,j,\sigma,\sigma}(t) = \sum_{\lambda=1}^{2L} U_{i_\sigma,\lambda} \frac{\mathrm{e}^{-\mathrm{i}t\rho_\lambda}}{1+\mathrm{e}^{-\beta(\rho_\lambda-\mu)}} U^\dagger_{\lambda,j_\sigma} \,. \tag{7.32}$$

As an example, let us calculate the spectral function

$$A(k,\omega) = -\frac{1}{\pi} \mathrm{Im} G_{\mathrm{ret}}(k,\omega) \,, \tag{7.33}$$

which is of relevance to photoemission experiments (see Chap. 18). The retarded Green function $G_{\mathrm{ret}}(k,\omega)$ is given by

$$G_{\mathrm{ret}}(k,\omega) = \int_{-\infty}^{\infty} \mathrm{d}t \mathrm{e}^{\mathrm{i}\omega t} G_{\mathrm{ret}}(k,t) \,, \tag{7.34}$$

and

$$G_{\mathrm{ret}}(k,t) = -\mathrm{i}\theta(t) \sum_\sigma \langle [a_{k,\sigma}(t) a^\dagger_{k,\sigma}(0) + a^\dagger_{k,\sigma}(0) a_{k,\sigma}(t)] \rangle$$
$$= -\mathrm{i}\theta(t) \sum_\sigma (G^>_{k,\sigma} + G^<_{k,\sigma}) \,. \tag{7.35}$$

Note here that $G^>_{k,\sigma} = \langle a_{k,\sigma}(t) a^\dagger_{k,\sigma}(0) \rangle$ and $G^<_{k,\sigma} = \langle a^\dagger_{k,\sigma}(0) a_{k,\sigma}(t) \rangle$ are implicitly defined. Since the measurements are performed in coordinate space, $G^>_{k,\sigma}$ and $G^<_{k,\sigma}$ must be expressed in terms of real-space operators using $a_{k,\sigma} = \frac{1}{\sqrt{L}} \sum_j \mathrm{e}^{-\mathrm{i}kj} a_{j,\sigma}$. Then

$$G^>_{k,\sigma} = \frac{1}{L} \sum_{j,l} \mathrm{e}^{-\mathrm{i}k(j-l)} \langle a_{j,\sigma}(t) a^\dagger_{l,\sigma}(0) \rangle = \frac{1}{L} \sum_{j,l} \mathrm{e}^{-\mathrm{i}k(j-l)} G^>_{j,l,\sigma,\sigma} \,, \tag{7.36}$$

and, analogously, an expression for $G^<_{k,\sigma}$ can be obtained. Equation (7.35) becomes

$$G_{\mathrm{ret}}(k,t) = -\mathrm{i}\theta(t) \frac{1}{L} \sum_{j,l,\sigma} [\mathrm{e}^{\mathrm{i}k(l-j)} G^>_{j,l,\sigma,\sigma} + \mathrm{e}^{-\mathrm{i}k(l-j)} G^<_{j,l,\sigma,\sigma}] \,. \tag{7.37}$$

Replacing (7.37) in (7.34) it can be shown that

$$G_{\mathrm{ret}}(k,\omega) = \frac{-\mathrm{i}}{L}\int_0^\infty \mathrm{d}t\mathrm{e}^{\mathrm{i}\omega t}\sum_{j,l,\sigma}[\mathrm{e}^{\mathrm{i}k(l-j)}G^{>}_{j,l,\sigma,\sigma} + \mathrm{e}^{-\mathrm{i}k(l-j)}G^{<}_{j,l,\sigma,\sigma}]\,. \quad (7.38)$$

The next step is to evaluate the integral. Using (7.32), the first term in (7.38) becomes

$$\frac{-\mathrm{i}}{L}\sum_{j,l,\sigma}\mathrm{e}^{\mathrm{i}k(l-j)}\sum_{\lambda=1}^{2L}\frac{U_{j_\sigma,\lambda}U^\dagger_{\lambda,l_\sigma}}{1+\mathrm{e}^{-\beta(\rho_\lambda-\mu)}}\int_0^\infty \mathrm{d}t\mathrm{e}^{\mathrm{i}(\omega-\rho_\lambda)t}\,. \quad (7.39)$$

Note that the integral is equal to $\pi\delta(\omega-\rho_\lambda)$. A similar expression is obtained for the second term so that finally the spectral function can be expressed in terms of the eigenvectors and eigenvalues of $\hat{H}$:

$$\begin{aligned} A(k,\omega) &= \frac{1}{L}\mathrm{Im}\Bigg\{\mathrm{i}\sum_{j,l,\sigma,\lambda}\Bigg[\mathrm{e}^{\mathrm{i}k(l-j)}\frac{U_{j_\sigma,\lambda}U^\dagger_{\lambda,l_\sigma}}{1+\mathrm{e}^{-\beta(\rho_\lambda-\mu)}} \\ &\quad + \mathrm{e}^{-\mathrm{i}k(l-j)}\frac{U_{l_\sigma,\lambda}U^\dagger_{\lambda,j_\sigma}}{1+\mathrm{e}^{\beta(\rho_\lambda-\mu)}}\Bigg]\delta(\omega-\rho_\lambda)\Bigg\} \\ &= \frac{1}{L}\mathrm{Im}\Bigg\{\mathrm{i}\sum_{j,l,\sigma,\lambda}\Bigg[\mathrm{e}^{\mathrm{i}k(j-l)}U_{l_\sigma,\lambda}U^\dagger_{\lambda,j_\sigma}\delta(\omega-\rho_\lambda) \\ &\quad \times\left(\frac{1}{1+\mathrm{e}^{-\beta(\rho_\lambda-\mu)}}+\frac{1}{1+\mathrm{e}^{\beta(\rho_\lambda-\mu)}}\right)\Bigg]\Bigg\}\,. \end{aligned} \quad (7.40)$$

Noticing that the sum on the final line is equal to 1, the final expression is

$$A(k,\omega) = \frac{1}{L}\mathrm{Re}\left[\sum_{j,l,\sigma,\lambda}\mathrm{e}^{\mathrm{i}k(j-l)}U_{l_\sigma,\lambda}U^\dagger_{\lambda,j_\sigma}\delta(\omega-\rho_\lambda)\right]\,. \quad (7.41)$$

Similar algebraic manipulations allow us to express other dynamical observables, such as the dynamical spin-correlation functions and the optical conductivitity, in terms of the eigenvalues and eigenvectors of the Hamiltonian matrix. In particular, the dynamical quantum (namely, involving the mobile electrons) spin spin correlation function is defined by

$$S^{\gamma\gamma'}(k,\omega) = \frac{1}{2\pi}\int_{-\infty}^{+\infty} < \sigma^\gamma_k(t)\sigma^{\gamma'}_k(0) > \mathrm{e}^{\mathrm{i}\omega t}\,\mathrm{d}t\,, \quad (7.42)$$

where $\sigma^\gamma_i = \sum_{\alpha,\beta}\sigma^\gamma_{\alpha,\beta}a^\dagger_{i,\alpha}a_{i,\beta}$, the Fourier transform is defined with $1/\sqrt{N}$ (N = number of sites), σ^γ are the Pauli matrices, $\gamma,\gamma' = x,y,z$, and for $\omega \neq 0$ the dynamical function can be expressed as

$$\begin{aligned} S^{\gamma\gamma'}(k,\omega) &= \frac{1}{L}\sum_{jl}\mathrm{e}^{\mathrm{i}k(j-l)}\sum_{\lambda\neq\lambda'}\sum_{\alpha,\beta}\sum_{\alpha',\beta'}(\sigma^\gamma_{\alpha\beta}\cdot\sigma^{\gamma'}_{\alpha'\beta'}) \\ &\quad \times\frac{U^\dagger_{\lambda,l\alpha}U_{l\beta,\lambda'}U^\dagger_{\lambda',j\alpha'}U_{j\beta',\lambda}}{(1+\mathrm{e}^{\beta(\rho_\lambda-\mu)})(1+\mathrm{e}^{-\beta(\rho_{\lambda'}-\mu)})}\delta(\omega+\rho_\lambda-\rho_{\lambda'})\,. \end{aligned} \quad (7.43)$$

A generalization to dimensions larger than one is easy and left as an exercise. The full $S(k,\omega)$, involving the mobile and localized electrons, is in principle directly related to the cross section measured in magnetic neutron scattering experiments (see Chap. 11 for an introduction).

The optical conductivity is defined as

$$\sigma(\omega) = \frac{\pi(1-\mathrm{e}^{-\beta\omega})}{\omega L} \int_{-\infty}^{+\infty} \frac{\mathrm{d}t}{2\pi} \mathrm{e}^{\mathrm{i}\omega t} < j_x(t) j_x(0) > , \tag{7.44}$$

where

$$j_x = \mathrm{i}te \sum_{j\sigma} (a^\dagger_{j+x,\sigma} a_{j,\sigma} - \mathrm{H.c.}) . \tag{7.45}$$

For $\omega \neq 0$, $\sigma(\omega)$ can be written as

$$\sigma^{xx}(\omega) = \frac{\pi t^2 e^2 (1-\mathrm{e}^{-\beta\omega})}{\omega L} \tag{7.46}$$

$$\times \sum_{\lambda \neq \lambda'} \frac{|\sum_{j\sigma}(U^\dagger_{j+\hat{x}\sigma,\lambda} U_{j\sigma,\lambda'} - U^\dagger_{j\sigma,\lambda} U_{j+\hat{x}\sigma,\lambda'})|^2}{(1+\mathrm{e}^{\beta(\rho_\lambda-\mu)})(1+\mathrm{e}^{-\beta(\rho_{\lambda'}-\mu)})} \delta(\omega + \rho_\lambda - \rho_{\lambda'}) .$$

Experimental measurements related with the optical conductivity are described in Chap. 12.

7.6 Details of the Monte Carlo Programs

The Monte Carlo programs used to perform some of the simulations described in this book have an initialization stage, followed by a main stage where the system evolves and quantities of interest are calculated. For the first portion of the simulation, the initial configuration of classical spins, $\{\theta_i\}$ and $\{\phi_i\}$, is created and stored. In principle, any initial configuration can be chosen since the final results should be independent of this choice; most common initial configurations are antiferromagnetic, ferromagnetic, or simply random. Also in the initialization process the program calculates the matrix elements of the Hamiltonian (7.13) in the $a^\dagger_{1,\uparrow}|0\rangle, \ldots, a^\dagger_{L,\uparrow}|0\rangle$, $a^\dagger_{1,\downarrow}|0\rangle, \ldots, a^\dagger_{L,\downarrow}|0\rangle$ basis for the electronic configuration, and diagonalizes that matrix, storing its eigenvalues. Here, $a^\dagger_{i,\sigma}|0\rangle$ denotes an electron at site i with z-projection of the spin equal to σ.

The main part of the program is a loop through Monte Carlo iterations or steps. For each iteration, the whole lattice is "swept". At each lattice point a local change in the spin is performed. In other words, the values of the classical spin angles θ_i and ϕ_i are modified by adding or subtracting a "small" random value chosen to be within a predefined range or "window". Note that the spin change is local, affecting only the classical spin located at site i. A local spin change can be implemented as follows:

$$\theta'_i = \theta_i + (r_1 - 0.5)\Delta\theta , \quad \phi'_i = \phi_i + (r_2 - 0.5)\Delta\phi , \tag{7.47}$$

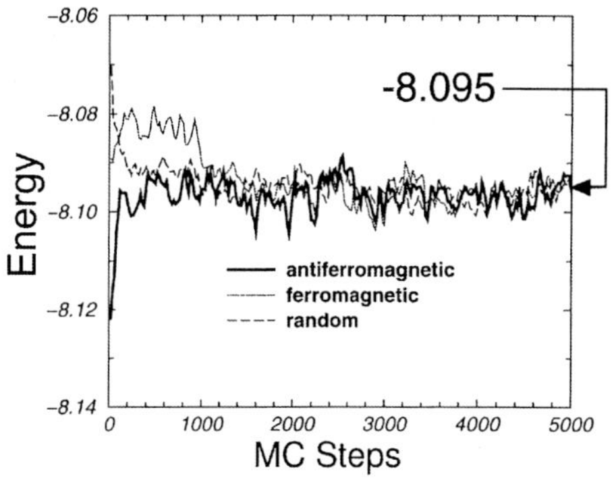

Fig. 7.2. Energy vs. Monte Carlo steps for a lattice of 16 sites at temperature $t/100$. The average shown on the *right* is over the last 2 500 steps, after equilibrium was reached. The Hund coupling is 8.0 and $J_{\mathrm{AF}} = 0$

where r_1 and r_2 are random numbers between zero and unity, and $\Delta\theta$ and $\Delta\phi$ are predefined parameters, chosen to be relatively "small" compared to the full range of the spin angles. After this new configuration is proposed, the matrix elements are calculated again and the matrix is diagonalized. At this point one must have a procedure to accept or reject the proposed configuration. Any algorithm that satisfies the detailed balance condition (for more details on this condition see the introductory portion of this chapter) can, in principle, be used, but the Metropolis [7.6] and Glauber [7.17] algorithms are the most common. For the Glauber algorithm, the program draws a random number r between zero and one and accepts the change of the spin at site i, if $r < P'/(P + P')$, where according to (7.20):

$$P = \sin\theta_i \mathrm{e}^{-\beta E_{\mathrm{H}}} \prod_{\lambda=1}^{2L} \left[1 + \mathrm{e}^{-\beta(\epsilon_\lambda - \mu)}\right] , \tag{7.48}$$

and $E_{\mathrm{H}} = J_{\mathrm{AF}} \sum_{<ij>} (\cos\theta_i \cos\theta_j + \sin\theta_i \sin\theta_j \cos(\phi_i - \phi_j))$. P' is obtained from P by replacing $\theta_i \to \theta'_i$ and $\phi_i \to \phi'_i$. If condition $r < P'/(P + P')$ is not satisfied, the program rejects the proposed configuration and the classical spins remain the same. In any case, the new configuration becomes the current configuration and the program goes on to the next lattice point and repeats the procedure. When all lattice points have been swept, the observables of interest are calculated for this Monte Carlo step and added to an accumulator. When all Monte Carlo steps have been performed, the program divides the accumulators by the number of Monte Carlo steps, yielding the averages of the observables. In actual simulations, a number of Monte Carlo steps at the beginning are not used to compute observables, to let the system thermalize. Note also that, since one must work with finite size lattices, the boundary conditions must be specified through the neighbors of each side.

Initial Configuration and the Angle "Window" Size. In Fig. 7.2, the energy is plotted against the Monte Carlo steps, to analyze the dependence of results with the initial configuration. Starting with an antiferromagnetic ordered system, it is clear that the energy is initially smaller than for ferromag-

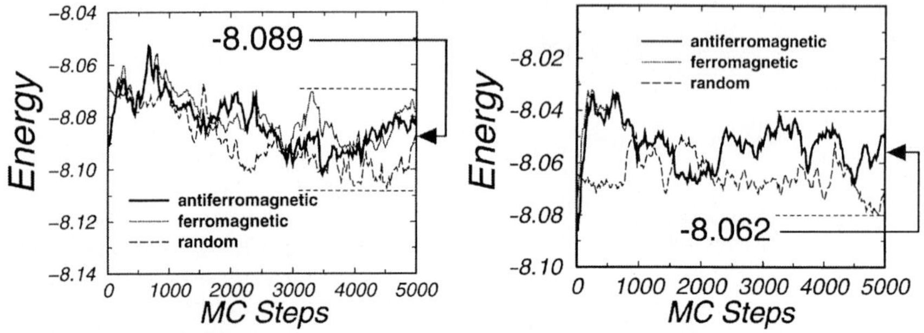

Fig. 7.3. (**a**) Energy vs. Monte Carlo steps for a 1D lattice of 16 sites at a temperature of $t/25$. The average shown on the *right* is over the last 2 500 steps. (**b**) Energy vs. Monte Carlo steps for a 1D lattice of 16 sites at a temperature of $t/2.5$. The average shown on the *right* is over the last 2 500 steps

netic or random ones. However, as the system continues evolving the initial configuration dependence disappears, and the average gives approximately the same value of -8.095 for the energy in the three cases. In Fig. 7.3a the same is plotted for a higher temperature of $T = t/75$. Due to thermal fluctuations, the instantaneous value of the energy fluctuates around a certain value more strongly than before. The same happens even at higher temperatures such as $T = t/2.5$, see Fig. 7.3b, with the fluctuations even stronger. Figure 7.4a shows the initial configuration dependence of the classical spin spin correlation. The correlation distance was chosen to be half the lattice length side, which is its maximum value since the simulation was performed using periodic boundary conditions. This quantity oscillates substantially more than the energy because it is a non local observable. The perfect ferromagnetic state has correlation equal to 1 and the antiferromagnetic one has correlation equal to -1. The initial configuration influences the first portion of the run, but then the observable starts oscillating around their proper mean values. In Fig. 7.4b, the "window" dependence of the simulation is analyzed. The windows $\Delta\theta$ and $\Delta\phi$ are the parameters that control the change in the classical spin angles θ_i and ϕ_i in the program. The newly proposed values of θ_i and ϕ_i in every Monte Carlo step are given by (7.47). In order to achieve a good sampling, one has to correctly choose the values for $\Delta\theta$ and $\Delta\phi$. If these numbers are too large, then the acceptance could also be large, and the system may never reach proper equilibrium. On the other hand, if these numbers are too small then the acceptance will be correspondingly small, and the system will take a long time to evolve. In addition, it can also occur that very small values of $\Delta\theta$ and $\Delta\phi$ make the system remain in a meta stable state and never evolve away from it. Also, if the system is initially far from equilibrium very small values for $\Delta\theta$ and $\Delta\phi$ will cause the simulation to take a huge amount of time to reach equilibrium.

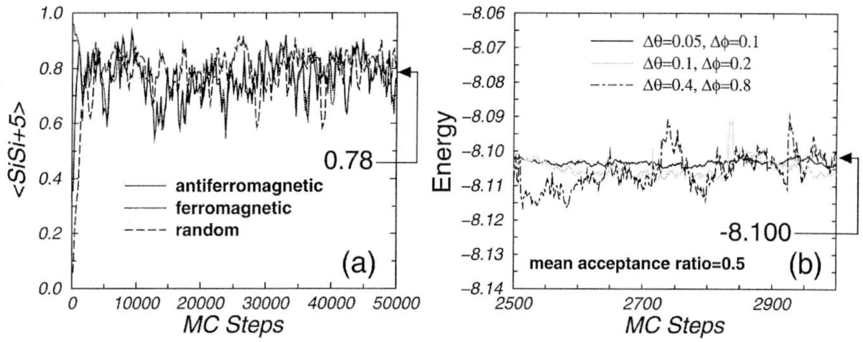

Fig. 7.4. (**a**) Classical spin correlation at a distance of 5 lattice spacings vs. Monte Carlo steps using a 1D lattice of 10 sites. The average shown on the *right* is over the last 2 500 steps. (**b**) Energy vs. Monte Carlo steps for three different choices of the "windows" $\Delta\theta$ and $\Delta\phi$. The average on the *right* is over all iterations shown

Figure 7.5a shows the electronic mean-density versus the chemical potential for a lattice of sixteen sites at a temperature of 1/75, in units of the hopping parameter t. The "forward" calculation was done by increasing the chemical potential, and each new point was calculated starting with the configuration obtained with the previous chemical potential. The same was done "backwards", i.e., decreasing μ. In both cases an abrupt jump is observed at $\mu \approx -3.1$, indicating a first-order transition for the one-orbital model in one dimension. The points that lie very close to the transition were recalculated using better statistics (shown with arrows in the figure).

Figure 7.5b shows the electronic mean density versus the chemical potential for a lattice of twenty sites at a temperature of 1/75 in units of the hopping parameter t. Again the simulation was performed by increasing and also decreasing the chemical potential. Now a "hysteresis loop" can be observed.

Monte Carlo Simulations and Boundary Conditions. In this section we want to study the properties of the ground state of the Hamiltonian (7.13) with $J_{\mathrm{AF}} = 0$ in one dimension, and its dependence with respect to the boundary conditions (BC) used. We remind the reader that on a lattice of L sites, for L electrons, the ground state is antiferromagnetic [7.18]. On the other hand, for $n < L$, where n is the number of electrons, the energy of the ferromagnetic configuration is smaller than that of the antiferromagnetic. However, it is a well-known fact [7.19–7.21] that the ground state is not always fully ferromagnetic for $n < L$, as Table 7.1 shows. We present now calculations of the ground-state energy that illustrate these facts. The calculations were performed by substituting properly selected classical spin configurations into (7.13) and calculating the energy of the ground state exactly. For the ferromagnetic regime ($n < L$), we have chosen two configurations: the FM ($\theta_i = 0$ and $\phi_i = 0$ for all sites i) and the single twist ($\theta_i = 2\pi i/L$ and $\phi_i = 0$

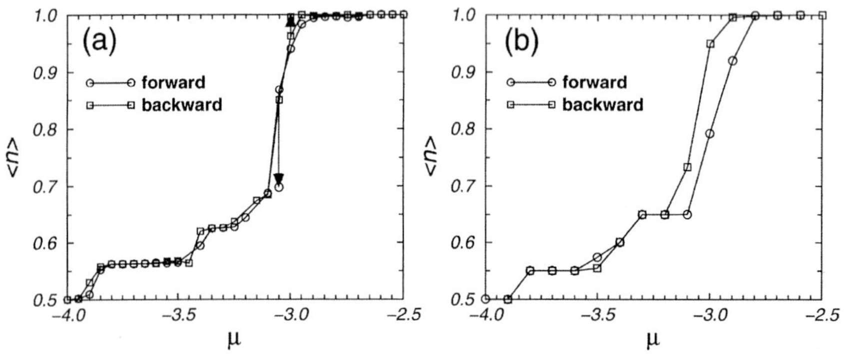

Fig. 7.5. (**a**) Electronic density vs. μ for a lattice of 16 sites calculated with increasing and decreasing μ. *Arrows* show points calculated using better statistics. (**b**) Electronic density vs. μ for a lattice of 20 sites calculated for increasing and decreasing μ. A hysteresis cycle is clearly seen

for all sites i). Note that lattice points are labeled by the index i, which runs from 0 to $L-1$, where L is the side of the lattice. The energies of these two configurations are shown in Table 7.2 for PBC and different fillings using a 10-site lattice. These results verify the rule of thumb that for an even number of electrons the ground state is FM at short distances, but with a global spin twist. It can also be seen from Table 7.2 that the states with the lowest energy have "closed-shell" configurations [7.19]. Table 7.3 shows the energies of the same configurations (FM and single twist) in the case of antiperiodic BC (APBC) for different fillings on a one-dimensional lattice with 10 sites. In this case the ground state is FM when the number of electrons is even. Results for open boundary conditions (OBC) are shown in Table 7.4 verifying the fact that the FM states are more robust: in all cases the energy of the FM state is less than that of the twisted configuration. Overall it is clear that the tendency toward ferromagnetism at short distances is very robust, but at large distances it is often confusing to find that the system may not be globally polarized due to the stabilization in some cases of a spin twist, which leads to the cancellation of the magnetization. This is a spurious effect, due to boundary conditions and finite clusters, and should not confuse the readers away from the ferromagnetic tendencies in the ground state. To fully convince the reader that antiperiodic boundary conditions (APBC) indeed reverse the twist that appears in the ground state with PBC in some cases, Fig. 7.6 shows the spin spin correlation function versus d (distance between sites) for 8 electrons on a lattice of 20 sites using APBC. This system has a twist in the ferro arrangement if PBC are used. However, as expected the figure shows that the state is truly FM and no "twisted" configuration is observed when APBC are used.

Table 7.1. Properties of the (stable) ground state of the one-orbital model, on one-dimensional lattices for $n < L$ (n is the number of electrons). The "ferromagnetic" state is considered to be fully spin polarized. The "nonferromagnetic" state shows ferromagnetism at short distances, but not globally due to a twist in the spin arrangement

Boundary conditions (BC)	n odd	n even
periodic (PBC)	ferromagnetic (FM)	nonferromagnetic
antiperiodic (APBC)	nonferromagnetic	ferromagnetic (FM)
open (OPC)	FM	FM

Table 7.2. Energies for the FM configuration (E(FM)) and the twisted FM (E(TFM)) configuration with PBC on a 10-site lattice

n	Shell (FM)	E(FM)	Shell(TFM)	E(TFM)
1	closed	−14	open	−13.8105
2	open	−27.618	closed	−27.621
3	closed	−41.236	open	−40.7494
4	open	−53.854	closed	−53.8778
5	closed	−66.472	open	−65.8937
6	open	−77.854	closed	−77.9096
7	closed	−89.236	open	−88.802
8	open	−99.618	closed	−99.6944
9	closed	−110	open	−109.8869

Table 7.3. Energies for the FM configuration (E(FM)) and the twisted FM (E(TFM)) configuration with APBC on a 10-site lattice

n	Shell(FM)	E(FM)	Shell(TFM)	E(TFM)
1	open	−13.9021	closed	−13.9021
2	closed	−27.8042	open	−27.4464
3	open	−40.9798	closed	−40.9907
4	closed	−54.1554	open	−53.5929
5	open	−66.1554	closed	−66.1951
6	closed	−78.1554	open	−77.6217
7	open	−88.9798	closed	−89.0483
8	closed	−99.8042	open	−99.515
9	open	−109.9021	closed	−109.9817

We also constructed a twisted AFM configuration as follows: $\theta_i = 2\pi i/L$ for i even and $\theta_i = 2\pi i/L + \pi$ for i odd. The energy of the twisted AFM configuration was calculated and compared to the standard AFM configuration ($\theta_i = 0$ for i even and $\theta_i = \pi$ for i odd) for electronic density equal to unity. Results (Table 7.5) show that the AFM energy at density 1 is always less than the twisted AFM, for all boundary conditions studied. This is in

Table 7.4. Energies for the FM configuration (E(FM)) and the twisted FM (E(TFM)) configuration with OBC on a 10-site lattice

n	Shell(FM)	E(FM)	Shell(TFM)	E(TFM)
1	closed	−13.919	closed	−13.8261
2	closed	−27.6015	closed	−27.43
3	closed	−40.9112	closed	−40.683
4	closed	−53.742	closed	−53.4839
5	closed	−66.0266	closed	−65.7673
6	closed	−77.742	closed	−77.5094
7	closed	−88.9112	closed	−88.7301
8	closed	−99.6015	closed	−99.492
9	closed	−109.919	closed	−109.8957

agreement with what was stated before, namely, that for electronic density equal to 1 the ground state is antiferromagnetic.

Table 7.5. Energies for the AFM configuration (E(AFM)) and the twisted AFM (E(ATFM)) configuration on a 10-site lattice with 10 electrons. C(AF) and C(TAFM) indicate if the shell is closed (0) or open (1)

BC	C(AF)	E(AF)	C(TAF)	E(TAF)
periodic	0	−120.829	0	−120.7504
antiperiodic	0	−120.8288	0	−120.7502
open	0	−120.7466	0	−120.6755

Finally, it is also important to remark that both periodic and antiperiodic boundary conditions preserve the traslational invariance of the Hamiltonian. The hopping term can be written as:

$$H_{\mathrm{K}} = -t \sum_{j=1}^{L-1} a_{j+1}^{\dagger} a_j + a_1^{\dagger} a_L \exp(\mathrm{i}\phi) + \mathrm{H.c.}\,,$$

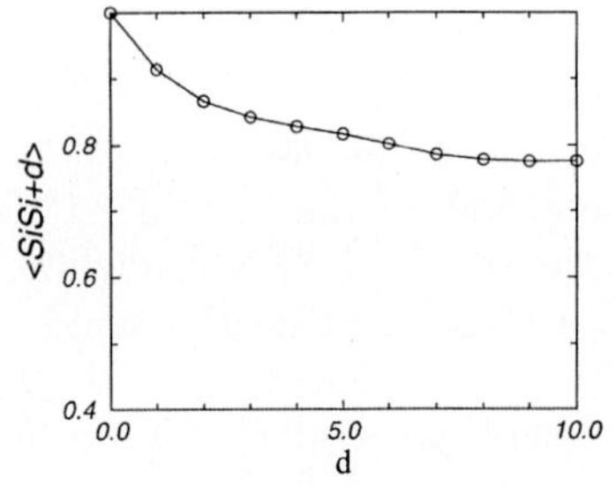

Fig. 7.6. Spin spin correlation $S(d)$ versus d (distance) obtained with MC techniques at $J_{\mathrm{H}} = 12t$, $T = t/100$, and $< n >= 8$, with antiperiodic boundary conditions

where ϕ is defined in such a way that $\exp(\mathrm{i}\phi) = 1$ for PBC and $\exp(\mathrm{i}\phi) = -1$ for APBC. To show that this Hamiltonian is translationally invariant, we define the transformation $a_j^\dagger \to a_j^\dagger \exp(\mathrm{i}\phi j/L)$. In terms of the new variables, the Hamiltonian becomes:

$$H_{\mathrm{K}} = -t \exp(\mathrm{i}\phi/L) \sum_{j=1}^{L} a_{j+1}^\dagger a_j + \mathrm{H.c.},$$

where $a_{L+1} = a_1$, which is explicitly invariant under translations. Note, however, that when working with "open" boundary conditions this symmetry is no longer preserved.

Summarizing, this chapter has provided an introduction to MC techniques, both in general and specialized to the one-orbital model that applies to manganites. The extension to more orbitals is straightforward.

7.7 Appendix: Summary of Rigorous Results for the Heisenberg Model

In the literature, exact theorems that address the presence or absence of phase transitions in some models under general circumstances are quite often invoked. This is of relevance for MC simulations of manganites, since knowing exact results allows us to judge the accuracy of the techniques employed. For example, quite frequently it is argued that the double-exchange or one-orbital model in 1D and 2D cannot have a phase transition at a finite temperature. However, the exact theorems invoked are based on particular assumptions regarding the Hamiltonian, such as short-range interactions. In the double-exchange model, the effective Hamiltonian for the localized spins can be obtained by expanding the fermionic trace in powers of those localized spins. Interactions at all distances will likely arise, as well as terms with multiple spin interactions, introducing doubts on the applicability of the theorems. Actually, the presence of ferromagnetism in one-dimensional systems has been unveiled in recent experiments [7.22], showing that the subject is still not settled. In this Appendix, the exact theorems are briefly reviewed with emphasis on their assumptions, information available here for the reader to judge their applicability to the cases discussed in this book.

In practice, these issues are not crucial, since a 1D or 2D system with a finite but huge correlation length, behaves at low temperatures for many practical purposes as an ordered ferromagnet. In addition, in real materials couplings between the dominant 1D or 2D substructures are inevitable, leading to order at finite temperatures. Nevertheless, for completeness here is a list of some relevant theorems.

7.7.1 A. Short-Range Interactions

A1. The one-dimensional and two-dimensional *quantum* Heisenberg models with short-range interactions have no finite-temperature phase transitions [7.23,7.24]. This is the Mermin and Wagner's theorem, which more specifically reads as follows [7.23]. Consider the quantum Heisenberg model defined as

$$\mathcal{H} = \frac{1}{2}\sum_{\mathbf{ij}} J_{\mathbf{ij}}\mathbf{S}_{\mathbf{i}} \cdot \mathbf{S}_{\mathbf{j}} - hS^z_{\mathbf{q}} \tag{7.49}$$

with short-range interactions that obey

$$\bar{J} = \frac{1}{2N}\sum_{\mathbf{ij}} |J_{\mathbf{ij}}|\,|\mathbf{x}_{\mathbf{i}} - \mathbf{x}_{\mathbf{j}}|^2 < \infty\,. \tag{7.50}$$

In this case, there can be no spontaneously broken symmetry at finite temperatures in one and two dimensions, namely $\lim_{h\to 0^+}\lim_{N\to\infty} m_{\mathbf{q}}(h,N) = 0$, where $S^z_{\mathbf{q}} = \sum_i \mathrm{e}^{\mathrm{i}\mathbf{q}\cdot\mathbf{x}_i} S^z_{\mathbf{i}}$ and $m_{\mathbf{q}}(h,N) = (1/NZ)\mathrm{Tr}\left[\mathrm{e}^{-\mathcal{H}(h)/T} S^z_{\mathbf{q}}\right]$. N is the number of sites. The Hamiltonian (7.49) has a continuous rotational symmetry. The theorem is valid for both, ferromagnets and antiferromagnets, and for any value of S. Ferromagnetism is ruled out by taking $\mathbf{q} = 0$, and antiferromagnetism by choosing $\mathrm{e}^{\mathrm{i}\mathbf{q}\cdot\mathbf{x}} = 1$ when $\mathbf{x}$ connects sites in the same sublattice, and -1 when it connects sites in different sublattices. When $J_{\mathbf{ij}} = |\mathbf{x}_{\mathbf{i}} - \mathbf{x}_{\mathbf{j}}|^{-p}$, condition (7.50) holds if $p > 3$ in 1 dimension or $p > 4$ in 2 dimensions.

A2. The one-dimensional and two-dimensional *classical* Heisenberg models with short-range interactions have no finite-temperature phase transition [7.23, 7.24]. Since the theorem **A1** holds for all S, we can deduce that it also holds for the classical (S-independent) limit of the Heisenberg model: $\mathcal{H}^{\mathrm{cl}} = \frac{1}{2}\sum_{\mathbf{ij}} J^{\mathrm{cl}}_{\mathbf{ij}}\Omega_{\mathbf{i}} \cdot \Omega_{\mathbf{j}} - h^{\mathrm{cl}}\Omega^z_{\mathbf{q}}$, where Ω are c-number unit vectors. The correspondence between the quantum and classical models is given by scaling the parameters: $J_{\mathbf{ij}}S(S+1) \to J^{\mathrm{cl}}_{\mathbf{ij}}$, $hS \to h^{\mathrm{cl}}$ and $m_{\mathbf{q}}/S \to m^{\mathrm{cl}}_{\mathbf{q}}$. Thus, the same result as in 1 applies to the classical model, and it is valid for both, ferromagnets and antiferromagnets.

A3. The quantum and classical Heisenberg models with short-range interactions and the additional condition that there exists a gap in the excitation spectrum, have no phase transition, even at $T = 0$, in one and two dimensions [7.23]. In principle, the Mermin and Wagner's theorem does not apply at $T = 0$. However, if $E_m - E_0 > \Delta$ for all states m, where Δ is independent of h and N, then the ground state of the Heisenberg model must be disordered. As an example, the antiferromagnetic integer spin chains and two-leg ladders are known to exhibit a spin-gap in their spectrum, and thus a disordered ground state. All the spin excitations have a gap. However, the presence of gapless excitations do not imply long-range order: for example, the spin-half Heisenberg antiferromagnet in one dimension has gapless excitations but no long-range order at $T = 0$ (although it is critical, i.e., nearly ordered).

A4. The ground state of the one-dimensional Heisenberg antiferromagnet with nearest-neighbour interactions, has no long-range order for any value of the spin [7.25]. The ground state of the two-dimensional *quantum* Heisenberg antiferromagnet can be shown to be ordered for $S \geq 1$ [7.26]. A variety of numerical simulations have strongly suggested that this theorem is valid for spin 1/2 as well.

7.7.2 B. Long-Range Interactions

A possible way to avoid the theorems expressed above is to consider long-range interactions, as follows:

B1. The *classical* 1D Heisenberg model with long-range interactions of the form r^{-p} has a finite-temperature phase transition for $1 < p < 2$ [7.27]. Analogously, the *classical* 2D Heisenberg model with long-range interactions of the form r^{-p} has a finite-temperature phase transition for $2 < p < 4$ [7.27].

B2. The *quantum* Heisenberg ferromagnet in d dimensions has a finite-temperature phase transition in the region $d < p < 2d$, and no finite-temperature phase transition for $p \geq 2d$ [7.28]. In this case, the Hamiltonian is assumed to be [7.28]:

$$\mathcal{H} = -\frac{1}{2}\sum_{\mathbf{j}}\sum_{\rho} J(\rho)\mathbf{S}_{\mathbf{j}} \cdot \mathbf{S}_{\mathbf{j}+\rho} - H\sum_{\mathbf{j}} S^z_{\mathbf{j}}$$

and the strength of the coupling is defined by

$$\lim_{N\to\infty} \frac{J(\rho)}{J_0} = \begin{cases} 0 & (|\rho| = 0) \\ |\rho|^{-p} & (\text{otherwise}) \end{cases},$$

where N is the number of sites, and $J_0 > 0$ is the nearest-neighbor interaction. The ground state for $p = 2$ was found exactly by Shastry [7.29] and Haldane [7.30], independently.

B3. Suppose $J(n) = n^{-\alpha}$. Then, the 1D *quantum* Heisenberg antiferromagnet

$$H = -\sum_{n\neq m} (-1)^{n-m} J(|n-m|)\mathbf{S}_n \cdot \mathbf{S}_m$$

of spin S has a first-order phase transition if S is sufficiently large, at some finite temperature, for $1 < \alpha < 2$ [7.27].

7.7.3 C. Ising Interactions

The case of Ising variables is important since many theorems depend on the continuous vs. discrete nature of the symmetries of the Hamiltonian. In this case it is known that:

The 1D Ising model with interactions r^{-p} has a finite temperature phase transition when $1 < p \leq 2$, and no long-range order when $p > 2$ [7.31, 7.32]. Also it is worth remembering that the 2D Ising model is FM [7.33], and in the absence of a magnetic field it was solved exactly [7.34].

7.8 Appendix: MC Code for the Ising Model

The following Fortran90 code performs a single-walker Monte Carlo for the 2D Ising model. The code employs the Fortran90 random number generator, but we strongly suggest to use a more efficient one (see for instance [7.7] or [7.35]).

```
!--------------------------------------------------------------------
! Monte Carlo algorithm for the Ising model on a 2D square lattice
!--------------------------------------------------------------------
module tables
 integer, parameter :: nsmax=499
 integer :: ns ! Linear size of the lattice
 real :: xj, xt ! J and T
 integer, dimension (0:nsmax, 0:nsmax) :: si ! spins
 integer, dimension (0:nsmax, 0:3) :: nnx, nny ! tables for neighbors
 real(kind(0.d0)), dimension (-1:1,-4:4) :: site_e ! table for energies
 real(kind(0.d0)), dimension (-1:1,-4:4) :: w ! table for probabilities
end module tables
!--------------------------------------------------------------------
program ising
use tables
implicit none
integer :: i, j, ix, iy, n ! auxiliary variables
integer :: niter, nterm ! Constants for the run
real(kind(0.d0)) :: ran ! Random number
real(kind(0.d0)) :: beta ! 1/T
real(kind(0.d0)) :: ener, e0, e2, se ! Average of energy
real(kind(0.d0)) :: magnet, m0, m2, sm ! Average of magnetization
write(*,*)'Linear size of the lattice (ns):'
read(*,*) ns
write(*,*)'Number of thermalization steps (nwarmup):'
read(*,*) nterm
write(*,*)'Number of measurements (niter):'
read(*,*) niter
write(*,*)'Temperature (T):'
read(*,*) xt
xj = 1.d0
!--------------------------------------------------------------------
! Tables for neighbors
!--------------------------------------------------------------------
!          1
!          |
! 2 - (x,y) - 0
!          |
!          3
!
! Coordinates of neighbors in any direction n are given by
! (nnx(x,n),nny(y,n)) n=0,..,3
!--------------------------------------------------------------------
do ix=0,ns-1
 nnx(ix,0)=ix+1
 nnx(ix,1)=ix
 nnx(ix,2)=ix-1
 nnx(ix,3)=ix
enddo
do iy=0,ns-1
  nny(iy,0)=iy
  nny(iy,1)=iy+1
  nny(iy,2)=iy
  nny(iy,3)=iy-1
enddo
! Periodic boundary conditions
nnx(ns-1,0) = 0
```

```
nnx(0,2) = ns-1
nny(ns-1,1) = 0
nny(0,3) = ns-1
!-----------------------------------------------------
! Build transition probabilities and local energies
!-----------------------------------------------------
beta = 1.d0/xt
do i=-1,1,2 ! Spin in the site can have value -1 or 1
do n=-4,4 ! Sum of the spins in neighboring sites -4,...,0,...,4
   site_e(i,n)=-xj*dble(i*n)
  w(i,n)=exp(-beta*xj*2*i*n)
 enddo
enddo
!-----------------------------------------------------
! Random starting configuration
!-----------------------------------------------------
do ix=0,ns-1
 do iy=0,ns-1
  si(ix,iy)=1
!we recommend using a more efficient random number generator (see Knuth)
   call random_number(ran)
 if(ran > .5d0) si(ix,iy)=-1
 enddo
enddo
!-----------------------------------------------------
! Warmup
!-----------------------------------------------------
print *,'--- WARMUP ---'
do i=1,nterm !nterm steps of metropolis for thermalization
 call metropolis()
enddo
!-----------------------------------------------------
! Start measurement
!-----------------------------------------------------
print *,'--- MEASUREMENT ---'
e0=0.d0
e2=0.d0
m0=0.d0
m2=0.d0
do i=1,niter
 do n=1,5 !decorrelation steps, arbitrarily set to 5
 call metropolis()
 enddo
 call metropolis()
! Calculate observables and add to averages
 ener = 0.d0
 magnet = 0.d0
 do ix=0,ns-1
 do iy=0,ns-1
 n = 0
 do j=0,3 !Sum of 4 first neighbor spins of site (ix,iy)
 n = n + si(nnx(ix,j), nny(iy,j))
 enddo
 ener=ener+site_e(si(ix,iy),n)
 magnet = magnet + si(ix, iy)
 enddo
 enddo
 ener=ener/2.d0 !Tables count spins twice
 e0=e0+ener
 e2=e2+ener*ener
 m0=m0+abs(magnet)
 m2=m2+magnet*magnet
enddo
! Averages and statistical errors
e0=e0/dble(niter)
e2=e2/dble(niter)
```

```
se=sqrt((e2-e0*e0)/dble(niter))
m0=m0/dble(niter)
m2=m2/dble(niter)
sm=sqrt((m2-m0*m0)/dble(niter))
! Output
print *,xt,'ENER=',e0,' Err=',se
print *,xt,'M=',m0,' Err=',sm
open(20,file='evst.dat',status='unknown',position='append')
write(20,*) xt,e0,se
close(20)
open(20,file='mvst.dat',status='unknown',position='append')
write(20,*) xt,m0,sm
close(20)
end
!--------------------------------------------------
subroutine metropolis()
use tables
implicit none
integer :: i, i0, ix, iy, n
real(kind(0.d0)) :: ran
do ix=0,ns-1
 do iy=0,ns-1 ! Sweep the lattice
 i0 = si(ix,iy)
 n = 0
 do i=0,3 ! Sum the spins in the neighborhood
 n = n + si(nnx(ix,i), nny(iy,i))
 enddo
 call random_number(ran)
 if(ran < w(i0,n)) si(ix, iy) = -i0 ! Flip Spin ?
 enddo
enddo
return
end
```

8. Mean-Field Approximation

with T. Hotta

As emphasized in previous chapters, one of the most severe problems in the field of strongly correlated electrons, including the manganites, is the difficulty theorists encounter to find accurate approximations to handle the many-body Hamiltonians that they themselves write. This leads to the proliferation of a variety of fairly different proposed ground states for the *same* model Hamiltonian, an issue that often perplexes experimentalists. This problem is not only typical of correlated electrons, say manganites and cuprates, but it appears in other areas of physics where perturbative calculations do not work. For instance, in the context of high-energy physics, this is also a notorious problem, since quarks are confined to form hadrons; an effect that cannot be considered as a perturbation over the kinetic energy portion of the quantum chromodynamics (QCD) Hamiltonian. To attempt to handle these non perturbative QCD effects the area of research known as lattice gauge theories was developed. In general, whenever many particles interact strongly, finding the properties of models associated to those systems is complicated. In previous chapters, we described how numerical techniques can help in alleviating this complication. In this chapter, another quite flexible approach is presented. This time it is mainly analytical, and often works fairly well in establishing the low-temperature phase diagrams, although typically fails in finding critical exponents and other more subtle information. The method is the *mean-field* approximation, where the influence over a given particle of the rest of the system (the other particles) is in the form of an effective external field that the particle under study feels. Self-consistency allows us to obtain equations for that field, and the mean-field problem can usually be handled exactly or almost exactly. A key assumption of the mean-field technique when applied to electrons in transition-metal oxides is that quantum fluctuations are not too strong.

8.1 Brief Introduction

As a quick reminder of the mean-field approximation, which is usually taught in elementary solid-state physics, consider the standard ferromagnetic Ising model, with Hamiltonian $H = -J\sum_{\langle \mathbf{lm}\rangle} S^z_{\mathbf{l}} S^z_{\mathbf{m}}$, where $S^z_{\mathbf{l}} = \pm 1$ is an Ising variable at site $\mathbf{l}$ of a D-dimensional hypercubic lattice. The notation $\langle \mathbf{lm}\rangle$

denotes nearest-neighbor sites, J is a positive coupling constant, and k_B is the Boltzmann constant. This problem would be trivial at zero temperature (namely all spins are either up or down), but it becomes interesting at finite temperature T since the model develops ferromagnetism at a critical Curie temperature T_C. In principle, one should calculate the expectation value of any spin as order parameter, say $S^z_\mathbf{i}$, and monitor its behavior as a function of temperature. Above T_C the expectation value is zero, and below it is nonzero. This mean value is defined as

$$\langle S^z \rangle = \frac{\sum_{\{S^z\}} S^z_\mathbf{i} \mathrm{e}^{\frac{J}{k_\mathrm{B}T}\sum_{\langle \mathbf{l},\mathbf{m}\rangle} S^z_\mathbf{l} S^z_\mathbf{m}}}{\sum_{\{S^z\}} \mathrm{e}^{\frac{J}{k_\mathrm{B}T}\sum_{\langle \mathbf{l},\mathbf{m}\rangle} S^z_\mathbf{l} S^z_\mathbf{m}}} . \tag{8.1}$$

Note that the site index of the order parameter on the left was not indicated, since translational invariance in a ferromagnet or paramagnet makes this value the same for all sites. On the right, an arbitrary site $\mathbf{i}$ was selected. The mean value can be calculated using computational techniques. Actually, the most accurate information known about non exactly solvable Ising models (such as in $D = 3$ dimensions) has been obtained by such procedures. Note that this is not an easy task, since the sums in the equation above are over all possible arrangements of spins. For an N-site lattice, there are 2^N spin configurations and the problem depends exponentially on the lattice size, if finite clusters are solved exactly. Monte Carlo methods alleviate considerably this effort, but they cannot be applied under any circumstances in quantum cases. However, there is a simple approximation that allows us to obtain a rough order of magnitude estimation of the Curie temperature. Simply focus on a given site $\mathbf{i}$, and assume that the other degrees of freedom are "frozen" to a mean value. Replace the interaction of the studied spin with the rest by an interaction with an effective external field. This is the mean-field approximation, and in this case the equation to be solved is much simpler, namely,

$$\langle S^z \rangle_\mathrm{MF} = \frac{\sum_{S^z=\pm 1} S^z_\mathbf{i} \mathrm{e}^{\frac{2JD}{k_\mathrm{B}T} S^z_\mathbf{i} \langle S^z \rangle_\mathrm{MF}}}{\sum_{S^z \pm 1} \mathrm{e}^{\frac{2JD}{k_\mathrm{B}T} S^z_\mathbf{i} \langle S^z \rangle_\mathrm{MF}}} = \tanh\left(\frac{2DJ}{k_\mathrm{B}T} \langle S^z \rangle_\mathrm{MF} \right) . \tag{8.2}$$

Using $\tanh(x) \approx x$, we arrive at a nontrivial solution for $T < T_\mathrm{C}^\mathrm{MF} = 2DJ/k_\mathrm{B}$, the main result of the mean-field approximation for the Ising model. In general, we assume that the product of two operators or degrees of freedom that appear in typical Hamiltonians is replaced by the product of one still active variable, times another one frozen to a mean value. This mean value is found self-consistently, and most mean-field approximations follow this simple idea. Of course, the technique is as accurate as the assumptions made for the decoupling of the product of operators. Sometimes we assume that there will be a particular phase transition in a system, with a concomitant order parameter, and the mean-field approach will find a finite critical temperature or coupling for the stabilization of such phase, but in reality that phase may never be stable. Thus, one has to be very careful and always compare mean-field approximation results with other more powerful techniques. But when the approach works, it is so simple that its use considerably illuminates the physics of the problem under discussion.

8.2 Application to Manganite Models

Let us return to the problem of manganese oxides. In this section, the mean-field approximation (MFA) is developed for the $J_{\rm H} = \infty$ Hamiltonian H^∞ (see Chap. 5), in an attempt to grasp the essential physics of models for manganites [3.32]. The procedure described here is very general. The details are similar to mean-field approximations for electronic systems employed in other contexts, such as in the half-filled Hubbard Hamiltonians used for the antiferromagnetic state of copper oxides [8.1] and alternative mean-field approximations for manganites (discussed briefly below). The effort focuses on the zero-temperature limit, and various transitions are found as couplings are changed (while in the Ising toy example, the transition takes place as the temperature is varied). More specifically, the MFA for manganites is based on the following extra assumptions: (i) The background t_{2g}-spin structure is fixed through the calculation by assuming that the nearest-neighbor t_{2g} spins can only be in the parallel or antiparallel configuration. This is not to be confused with the assumption that the state of the entire lattice is either FM or AF. The actual assumption is only that each individual nearest-neighbor spin pair is parallel or antiparallel. Out of individual FM or AF bonds, complicated patterns can emerge, including the famous CE-phase of half-doped manganites, discussed later. (ii) The JT- and breathing-mode distortions are assumed to be non cooperative. Thus, the mean-field approximation is carried out in the $\{Q\}$ variables, instead of the oxygen coordinates. These assumptions are discussed in more detail later in this section.

Before moving into the fine details of the manganite mean-field approximation, the reader should be aware of the existence of other important mean-field approaches since there is no unique procedure to carry out a mean-field approximation. A given problem can typically be rewritten in a variety of ways, using different formulations and variables and, in each case, a mean-field approximation is obtained. Mean-field results of a variety different from those described here can be found, e.g., in Maezono et al. [9.17] and references therein. Alonso et al. [6.50] adapted the variational formulation of the mean-field method to situations where fermions are quadratically coupled to classical degrees of freedom. In this way, fermions are treated exactly (including Berry phases), and the mean-field approximation is done only for the spins. This method yields an excellent mean-field approximation to the Curie temperature, as compared to Monte Carlo calculations (Alonso et al. [6.4]).

Returning to the mean-field approach including Jahn Teller phonons, let us first rewrite the electron phonon portion of H^∞ by applying a standard mean-field decoupling procedure. In this approximation, a given operator O that appears in the Hamiltonian is first rewritten exactly as $O = \langle O \rangle + \delta O$, where $\langle O \rangle$ is the mean value of the operator in the ground state. $\delta O = O - \langle O \rangle$ is a measure of the deviation of the operator from the mean value, namely of the strength of the fluctuations. Thus far no approximations have actually been made. The mean-field assumption arises when we consider a product of operators $O_1 O_2$. The approximation consists of discarding terms of order $\delta O_1 \delta O_2$ in $O_1 O_2$, namely $O_1 O_2 \approx \langle O_1 \rangle \langle O_2 \rangle + \delta O_1 \langle O_2 \rangle + \langle O_1 \rangle \delta O_2 = -\langle O_1 \rangle \langle O_2 \rangle + O_1 \langle O_2 \rangle + \langle O_1 \rangle O_2$.

Let us first apply this to the portion of the $J_{\mathrm{H}} = \infty$ Hamiltonian that has the phononic coordinates, namely to

$$H^{\infty}_{\mathrm{el-ph}} = E_{\mathrm{JT}} \sum_{\mathbf{i}} \left[2(q_{1\mathbf{i}} n_{\mathbf{i}} + q_{2\mathbf{i}} \tau_{x\mathbf{i}} + q_{3\mathbf{i}} \tau_{z\mathbf{i}}) + \beta q_{1\mathbf{i}}^2 + q_{2\mathbf{i}}^2 + q_{3\mathbf{i}}^2 \right] , \tag{8.3}$$

where $\beta = E_{\mathrm{JT}}/E_{\mathrm{br}}$ (this should not be confused with the inverse temperature). To arrive at the equation above from the original (5.11) of Chap. 5, a few redefinitions were introduced. To start with, the pseudospins include now a factor 2 through $T^z_{\mathbf{i}} = (1/2)\tau_{z\mathbf{i}}$, and a similar definition for the x-component. Another redefinition is $Q_{2\mathbf{i}} = (g/k_{\mathrm{JT}}) q_{2\mathbf{i}}$, and $Q_{3\mathbf{i}} = (g/k_{\mathrm{JT}}) q_{3\mathbf{i}}$. In addition, $E_{\mathrm{JT}} = (g^2/2k_{\mathrm{JT}})$, and a coupling with the breathing mode $Q_{1\mathbf{i}}$ was introduced involving a different spring constant k_{br}, and $E_{\mathrm{br}} = (g^2/2k_{\mathrm{br}})$. With all these redefinitions, the Hamiltonian (8.3) can be derived. It can be easily shown that the mean-field approach leads to the replacements

$$\begin{cases} q_{1\mathbf{i}} n_{\mathbf{i}} \approx \langle q_{1\mathbf{i}} \rangle n_{\mathbf{i}} + q_{1\mathbf{i}} \langle n_{\mathbf{i}} \rangle - \langle q_{1\mathbf{i}} \rangle \langle n_{\mathbf{i}} \rangle , \\ q_{2\mathbf{i}} \tau_{x\mathbf{i}} \approx \langle q_{2\mathbf{i}} \rangle \tau_{x\mathbf{i}} + q_{2\mathbf{i}} \langle \tau_{x\mathbf{i}} \rangle - \langle q_{2\mathbf{i}} \rangle \langle \tau_{x\mathbf{i}} \rangle , \\ q_{3\mathbf{i}} \tau_{z\mathbf{i}} \approx \langle q_{3\mathbf{i}} \rangle \tau_{z\mathbf{i}} + q_{3\mathbf{i}} \langle \tau_{z\mathbf{i}} \rangle - \langle q_{3\mathbf{i}} \rangle \langle \tau_{z\mathbf{i}} \rangle , \\ q_{\alpha\mathbf{i}}^2 \approx 2\langle q_{\alpha\mathbf{i}} \rangle q_{\alpha\mathbf{i}} - \langle q_{\alpha\mathbf{i}} \rangle^2 \ (\alpha = 1,2,3) , \end{cases} \tag{8.4}$$

where the mean values will be found self-consistently by solving the mean-field Hamiltonian defined below. Introducing the approximation (8.4) explicitly into (8.3), we have

$$\begin{aligned} H^{MF}_{\mathrm{el-ph}} = E_{\mathrm{JT}} \sum_{\mathbf{i}} [& 2(q_{1\mathbf{i}} \langle n_{\mathbf{i}} \rangle + \langle q_{1\mathbf{i}} \rangle n_{\mathbf{i}} - \langle q_{1\mathbf{i}} \rangle \langle n_{\mathbf{i}} \rangle \\ & + q_{2\mathbf{i}} \langle \tau_{x\mathbf{i}} \rangle + \langle q_{2\mathbf{i}} \rangle \tau_{x\mathbf{i}} - \langle q_{2\mathbf{i}} \rangle \langle \tau_{x\mathbf{i}} \rangle + q_{3\mathbf{i}} \langle \tau_{z\mathbf{i}} \rangle + \langle q_{3\mathbf{i}} \rangle \tau_{z\mathbf{i}} - \langle q_{3\mathbf{i}} \rangle \langle \tau_{z\mathbf{i}} \rangle) \\ & + 2(\beta q_{1\mathbf{i}} \langle q_{1\mathbf{i}} \rangle - \beta \langle q_{1\mathbf{i}} \rangle^2 + q_{2\mathbf{i}} \langle q_{2\mathbf{i}} \rangle - \langle q_{2\mathbf{i}} \rangle^2 + q_{3\mathbf{i}} \langle q_{3\mathbf{i}} \rangle - \langle q_{3\mathbf{i}} \rangle^2)] . \end{aligned} \tag{8.5}$$

For classical phonons, as assumed in the following calculations, their optimal values in (8.5) can be easily obtained since these degrees of freedom are now no longer coupled to the quantum electrons, but to mean values that play the role of frozen effective external fields. Thus, by a mere minimization of (8.5) with respect to the phononic variables (for classical phonons the mean values and actual optimal values in the ground state are the same due to the absence of fluctuations), we obtain

$$q_{1\mathbf{i}} = -\langle n_{\mathbf{i}} \rangle / \beta , \quad q_{2\mathbf{i}} = -\langle \tau_{x\mathbf{i}} \rangle , \quad q_{3\mathbf{i}} = -\langle \tau_{z\mathbf{i}} \rangle , \tag{8.6}$$

which reduces considerably the complexity of the problem. Replacing (8.6) into the mean-field Hamiltonian, we get

$$H^{\mathrm{MF}}_{\mathrm{el-ph}} = -2\sum_{\mathbf{i}}[E_{\mathrm{br}}\langle n_{\mathbf{i}}\rangle n_{\mathbf{i}} + E_{\mathrm{JT}}(\langle\tau_{x\mathbf{i}}\rangle\tau_{x\mathbf{i}} + \langle\tau_{z\mathbf{i}}\rangle\tau_{z\mathbf{i}})]$$
$$+\sum_{\mathbf{i}}[E_{\mathrm{br}}\langle n_{\mathbf{i}}\rangle^2 + E_{\mathrm{JT}}(\langle\tau_{x\mathbf{i}}\rangle^2 + \langle\tau_{z\mathbf{i}}\rangle^2)]\,, \tag{8.7}$$

where $E_{\mathrm{br}} = E_{\mathrm{JT}}/\beta$.

Consider now the electron electron interaction among the e_g electrons in the limit of $J_{\mathrm{H}}{=}\infty$. This term is explicitly defined later in (8.13), and it was also discussed in Chap. 5. At first glance, it appears sufficient to follow a similar decoupling procedure for $H_{\mathrm{el-el}}$, as carried out for the electron phonon term. However, such a decoupling cannot be uniquely accomplished since, as shown below, $H_{\mathrm{el-el}}$ is invariant with respect to the choice of e_g-electron orbitals due to the presence of a local SU(2) symmetry in the orbital space of that portion of the Hamiltonian. Thus, it is necessary to find the optimal orbital set by determining the relevant e_g-electron orbital self-consistently at each site. For this purpose, it is convenient to use the expressions $q_{2\mathbf{i}} = q_{\mathbf{i}}\sin(\xi_{\mathbf{i}})$ and $q_{3\mathbf{i}} = q_{\mathbf{i}}\cos(\xi_{\mathbf{i}})$. Using the mean-field results (8.6), the amplitude $q_{\mathbf{i}}$ and phase $\xi_{\mathbf{i}}$ are

$$q_{\mathbf{i}} = \sqrt{\langle\tau_{x\mathbf{i}}\rangle^2 + \langle\tau_{z\mathbf{i}}\rangle^2}\,,\quad \xi_{\mathbf{i}} = \pi + \tan^{-1}(\langle\tau_{x\mathbf{i}}\rangle/\langle\tau_{z\mathbf{i}}\rangle)\,, \tag{8.8}$$

where "π" is added to $\xi_{\mathbf{i}}$ motivated by the mean-field results of (8.6). $\xi_{\mathbf{i}}$ could also be $\xi_{\mathbf{i}} = \tan^{-1}(q_{2\mathbf{i}}/q_{3\mathbf{i}})$, but the minus signs in front of $\langle\tau_{x\mathbf{i}}\rangle$ and $\langle\tau_{z\mathbf{i}}\rangle$ indicate that an additional phase π in $\xi_{\mathbf{i}}$ is needed for consistency.

The phase $\xi_{\mathbf{i}}$ is very important since it fixes the actual combination of orbitals that is favored energetically at each site, namely it fixes the relative values of $q_{2\mathbf{i}}$ and $q_{3\mathbf{i}}$. Only for special cases will the "optimal" combination have 100% x^2-y^2 or $3z^2-r^2$ character (see Table 8.1). In general, $\xi_{\mathbf{i}}$ will take values such that neither of the two orbitals are totally favored. Then, the fermionic operators $c_{\mathbf{i}\mathrm{a}}$ and $c_{\mathbf{i}\mathrm{b}}$, and their corresponding states $c^{\dagger}_{\mathbf{i}\mathrm{a}}|0\rangle$ and $c^{\dagger}_{\mathbf{i}\mathrm{b}}|0\rangle$, are not intrinsically important. Using $\xi_{\mathbf{i}}$, it is more convenient to transform $c_{\mathbf{i}\mathrm{a}}$ and $c_{\mathbf{i}\mathrm{b}}$ into the so-called phase-dressed operators, $\tilde{c}_{\mathbf{i}\mathrm{a}}$ and $\tilde{c}_{\mathbf{i}\mathrm{b}}$, through

$$\begin{pmatrix}\tilde{c}_{\mathbf{i}\mathrm{a}}\\ \tilde{c}_{\mathbf{i}\mathrm{b}}\end{pmatrix} = \mathrm{e}^{\mathrm{i}\xi_{\mathbf{i}}/2}\begin{pmatrix}\cos(\xi_{\mathbf{i}}/2) & \sin(\xi_{\mathbf{i}}/2)\\ -\sin(\xi_{\mathbf{i}}/2) & \cos(\xi_{\mathbf{i}}/2)\end{pmatrix}\begin{pmatrix}c_{\mathbf{i}\mathrm{a}}\\ c_{\mathbf{i}\mathrm{b}}\end{pmatrix}, \tag{8.9}$$

where the 2×2 matrix is SU(2) symmetric: 2×2, unitary, and of determinant one. The dressed operators *by construction* satisfy the property

$$|\text{"b"}\rangle = \tilde{c}^{\dagger}_{\mathbf{i}\mathrm{b}}|0\rangle = [-\sin(\xi_{\mathbf{i}}/2)c^{\dagger}_{\mathbf{i}\mathrm{a}} + \cos(\xi_{\mathbf{i}}/2)c^{\dagger}_{\mathbf{i}\mathrm{b}}]|0\rangle\,, \tag{8.10}$$

state referred to as the "b" orbital, which is the combination with the lowest energy at a given site. The excited state or "a" orbital is simply obtained by requesting it to be orthogonal to "b", i.e.,

Table 8.1. Phase $\xi_{\rm i}$ and the corresponding $e_{\rm g}$-electron orbitals. Note that "b" corresponds to the lowest-energy orbital for $E_{\rm JT} \neq 0$

$\xi_{\rm i}$	"a" orbital	"b" orbital
0	$x^2 - y^2$	$3z^2 - r^2$
$\pi/3$	$3y^2 - r^2$	$z^2 - x^2$
$2\pi/3$	$y^2 - z^2$	$3x^2 - r^2$
π	$3z^2 - r^2$	$x^2 - y^2$
$4\pi/3$	$z^2 - x^2$	$3y^2 - r^2$
$5\pi/3$	$3x^2 - r^2$	$y^2 - z^2$

$$|\text{"a"}\rangle = \tilde{c}^{\dagger}_{\rm ia}|0\rangle = [\cos(\xi_{\rm i}/2)c^{\dagger}_{\rm ia} + \sin(\xi_{\rm i}/2)c^{\dagger}_{\rm ib}]|0\rangle\,. \tag{8.11}$$

These dressed operators are more relevant than the "bare" ones. They create states that are the proper ground state of a given ion, and its orthogonal first excited state. Their use simplifies the mean-field analysis.

Note that if $\xi_{\rm i}$ is increased by 2π, the SU(2) matrix changes sign. To keep the transformation unchanged upon a 2π-rotation in $\xi_{\rm i}$, a phase factor $e^{i\xi_{\rm i}/2}$ was added in (8.9). In the expression for the ground state of the single JT molecule – the single-site problem discussed in previous chapters – this phase factor was not added, because the electron does not hop from site to site, and the phases do not correlate with each other. This means that it was enough to pay attention to the double-valuedness of the wave function at a single site. However, in the JT crystal in which e_g electrons move in the periodic array of the JT centers, the addition of this phase factor takes into account "Berry-phase" effects arising from the circular motion of e_g-electrons around the JT center (see [8.2]).

To understand the meaning of particular values of the phases, the procedure is simple. Start by replacing $c^{\dagger}_{\rm ia}|0\rangle$ by $(x^2 - y^2)/\sqrt{2}$ and $c^{\dagger}_{\rm ib}|0\rangle$ by $(3z^2 - r^2)/\sqrt{6}$ (note the need for a proper normalization of each function). If $\xi_{\rm i} = 0$ or π, the optimal state is "pure", namely all $(3z^2 - r^2)$ or $(x^2 - y^2)$, compatible with a distortion that is purely Q_3. However, for a phase such as $\xi_{\rm i} = 2\pi/3$, the lowest-energy state ("b" state) is found to be proportional to $(3x^2 - r^2)$, which is a linear combination of the original orbitals. These ground-state combinations occur frequently. In Table 8.1, the correspondence between $\xi_{\rm i}$ and the local orbital is summarized for important values of $\xi_{\rm i}$ (overall phase factors that may affect the orbitals are not included in Table 8.1). Note that these results are similar to those found in the discussion of Chap. 4 when the physical meaning of the angle θ was addressed. Actually, θ of an isolated MnO_6 octahedra, as studied before, and $\xi_{\rm i}$ are certainly equal. However, the discussion here focused more on the electronic degree of freedom than the phononic one.

Note also that $d_{3x^2-r^2}$ and $d_{3y^2-r^2}$ *never* appear as the local orbital set. Actually, these states are not orthogonal to each other (their overlap is 1/2).

Sometimes, these states are considered as if they were an orthogonal orbital set to reproduce experiments, but such a treatment is an approximation, since orbital ordering is a simple alternation of two arbitrary kinds of orbitals.

Using the transformations described above to the phase-dressed operators, $H^{\mathrm{MF}}_{\mathrm{el-ph}}$ and $H_{\mathrm{el-el}}$ can be rewritten, after some algebra, as

$$H^{\mathrm{MF}}_{\mathrm{el-ph}} = \sum_{\mathbf{i}} \{E_{\mathrm{br}}(-2\langle n_{\mathbf{i}}\rangle \tilde{n}_{\mathbf{i}} + \langle n_{\mathbf{i}}\rangle^2) + E_{\mathrm{JT}}[2q_{\mathbf{i}}(\tilde{n}_{\mathbf{i}\mathrm{a}} - \tilde{n}_{\mathbf{i}\mathrm{b}}) + q_{\mathbf{i}}^2]\}\,, \quad (8.12)$$

and

$$H_{\mathrm{el-el}} = U' \sum_{\mathbf{i}} n_{\mathbf{i}\mathrm{a}} n_{\mathbf{i}\mathrm{b}} + V \sum_{\langle \mathbf{i},\mathbf{j}\rangle} n_{\mathbf{i}} n_{\mathbf{j}} = U' \sum_{\mathbf{i}} \tilde{n}_{\mathbf{i}\mathrm{a}} \tilde{n}_{\mathbf{i}\mathrm{b}} + V \sum_{\langle \mathbf{i},\mathbf{j}\rangle} \tilde{n}_{\mathbf{i}} \tilde{n}_{\mathbf{j}}\,, \quad (8.13)$$

where $\tilde{n}_{\mathbf{i}\gamma} = \tilde{c}^{\dagger}_{\mathbf{i}\gamma}\tilde{c}_{\mathbf{i}\gamma}$ and $\tilde{n}_{\mathbf{i}} = \tilde{n}_{\mathbf{i}\mathrm{a}} + \tilde{n}_{\mathbf{i}\mathrm{b}}$. Note that $H_{\mathrm{el-el}}$ is invariant with respect to the choice of $\xi_{\mathbf{i}}$.

To arrive at (8.12), use (8.6) and (8.8), as well as the identity $q_{2\mathbf{i}}\tau_{x\mathbf{i}} + q_{3\mathbf{i}}\tau_{z\mathbf{i}} = q_{\mathbf{i}}(\tilde{n}_{\mathbf{i}\mathrm{a}} - \tilde{n}_{\mathbf{i}\mathrm{b}})$ (proof left as exercise to the reader). Equation (8.13) can be obtained by calculating $\tilde{c}^{\dagger}_{\mathbf{i}\mathrm{a}}\tilde{c}_{\mathbf{i}\mathrm{a}} + \tilde{c}^{\dagger}_{\mathbf{i}\mathrm{b}}\tilde{c}_{\mathbf{i}\mathrm{b}}$ using (8.9). This leads to $\tilde{n}_{\mathbf{i}} = n_{\mathbf{i}}$, due to the unitary properties of the SU(2) matrix. In addition, the last result implies that $\tilde{n}_{\mathbf{i}}^2 = n_{\mathbf{i}}^2$. Recalling that $n_{\mathbf{i}\gamma}^2 = n_{\mathbf{i}\gamma}$ and $\tilde{n}_{\mathbf{i}\gamma}^2 = \tilde{n}_{\mathbf{i}\gamma}$ for $\gamma = \mathrm{a}$ and b (the spin is no longer relevant in the $J_{\mathrm{H}} = \infty$ limit), and calculating $\sum_{\mathbf{i}} \tilde{n}_{\mathbf{i}}^2 = \sum_{\mathbf{i}} n_{\mathbf{i}}^2$ it can be shown that $\sum_{\mathbf{i}} \tilde{n}_{\mathbf{i}\mathrm{a}}\tilde{n}_{\mathbf{i}\mathrm{b}} = \sum_{\mathbf{i}} n_{\mathbf{i}\mathrm{a}} n_{\mathbf{i}\mathrm{b}}$. This completes the proof of (8.13).

Let us now use the mean-field decoupling procedure learned when discussing (8.4) to calculate the product of the two number operators that appear in the electron electron term (8.13). The approximation $\tilde{n}_{\mathbf{i}}\tilde{n}_{\mathbf{j}} \approx \langle \tilde{n}_{\mathbf{i}}\rangle \tilde{n}_{\mathbf{j}} + \tilde{n}_{\mathbf{i}}\langle \tilde{n}_{\mathbf{j}}\rangle - \langle \tilde{n}_{\mathbf{i}}\rangle\langle \tilde{n}_{\mathbf{j}}\rangle$ can be immediately applied to the nearest-neighbor Coulombic repulsion portion of the electron electron interaction, namely

$$\begin{aligned} V \sum_{\langle \mathbf{i},\mathbf{j}\rangle} \tilde{n}_{\mathbf{i}}\tilde{n}_{\mathbf{j}} &\approx V \sum_{\langle \mathbf{i},\mathbf{j}\rangle} [\tilde{n}_{\mathbf{i}}\langle \tilde{n}_{\mathbf{j}}\rangle + \langle \tilde{n}_{\mathbf{i}}\rangle \tilde{n}_{\mathbf{j}} - \langle \tilde{n}_{\mathbf{i}}\rangle\langle \tilde{n}_{\mathbf{j}}\rangle] \\ &= V \sum_{\mathbf{i},\mathbf{e}} [\tilde{n}_{\mathbf{i}}\langle n_{\mathbf{i}+\mathbf{e}}\rangle + \langle n_{\mathbf{i}}\rangle \tilde{n}_{\mathbf{i}+\mathbf{e}} - \langle n_{\mathbf{i}}\rangle\langle n_{\mathbf{i}+\mathbf{e}}\rangle] \\ &= V \sum_{\mathbf{i},\mathbf{a}} [\langle n_{\mathbf{i}+\mathbf{a}}\rangle \tilde{n}_{\mathbf{i}} - (1/2)\langle n_{\mathbf{i}+\mathbf{a}}\rangle\langle n_{\mathbf{i}}\rangle]\,, \end{aligned} \quad (8.14)$$

where we used again that $n_{\mathbf{i}} = \tilde{n}_{\mathbf{i}}$, and the sum over nearest-neighbors was replaced by a sum over sites **i**, and unit vectors **e** that point along the positive directions of the three crystal axes. The set $\{\mathbf{a}\}$ used in the last step of (8.14) are unit vectors along the axes in *both* directions, namely, they are defined as $(\pm 1, 0, 0)$, $(0, \pm 1, 0)$, and $(0, 0, \pm 1)$.

For the U' term we need to use

$$\tilde{n}_{\mathbf{i}\mathrm{a}}\tilde{n}_{\mathbf{i}\mathrm{b}} \approx \langle \tilde{n}_{\mathbf{i}\mathrm{a}}\rangle \tilde{n}_{\mathbf{i}\mathrm{b}} + \tilde{n}_{\mathbf{i}\mathrm{a}}\langle \tilde{n}_{\mathbf{i}\mathrm{b}}\rangle - \langle \tilde{n}_{\mathbf{i}\mathrm{a}}\rangle\langle \tilde{n}_{\mathbf{i}\mathrm{b}}\rangle\,, \quad (8.15)$$

as well as the relations

$$\langle \tilde{n}_{\rm ia} \rangle = (\langle n_{\rm i} \rangle - q_{\rm i})/2\,, \quad \langle \tilde{n}_{\rm ib} \rangle = (\langle n_{\rm i} \rangle + q_{\rm i})/2. \tag{8.16}$$

To show the validity of the last equations note that that the sum satisfies $\langle \tilde{n}_{\rm i} \rangle = \langle \tilde{n}_{\rm ia} + \tilde{n}_{\rm ib} \rangle = \langle n_{\rm i} \rangle$, which was derived earlier. A second equation can be obtained starting with the operators of the rotated fermions (those with a tilde) written in terms of the original operators using (8.9). By this procedure, $\tilde{n}_{\rm ia} = \cos^2(\xi_{\rm i}/2) n_{\rm ia} + \sin^2(\xi_{\rm i}/2) n_{\rm ib} + \sin(\xi_{\rm i}/2)\cos(\xi_{\rm i}/2)(c^\dagger_{\rm ia} c_{\rm ib} + c^\dagger_{\rm ib} c_{\rm ia})$. A similar expression is derived for $\tilde{n}_{\rm ib}$. From their difference, and considering mean values, it can be shown that

$$\langle \tilde{n}_{\rm ia} - \tilde{n}_{\rm ib} \rangle = \cos(\xi_{\rm i}) \langle \tau_{{\rm i}z} \rangle + \sin(\xi_{\rm i}) \langle \tau_{{\rm i}x} \rangle\,. \tag{8.17}$$

The mean-field results $\langle \tau_{z{\rm i}} \rangle = -q_{3{\rm i}} = -q_{\rm i} \cos(\xi_{\rm i})$ and $\langle \tau_{x{\rm i}} \rangle = -q_{2{\rm i}} = -q_{\rm i} \sin(\xi_{\rm i})$, lead to $\langle \tilde{n}_{\rm ia} - \tilde{n}_{\rm ib} \rangle = -q_{\rm i}$. This equation can be used to complete the proof of (8.16).

Putting all this together, the electron electron interaction term becomes

$$\begin{aligned} H^{\rm MF}_{\rm el-el} &= (U'/4) \sum_{\rm i} [2 \langle n_{\rm i} \rangle \tilde{n}_{\rm i} - \langle n_{\rm i} \rangle^2 + 2 q_{\rm i} (\tilde{n}_{\rm ia} - \tilde{n}_{\rm ib}) + q_{\rm i}^2] \\ &\quad + V \sum_{\rm ia} [\langle n_{\rm i+a} \rangle \tilde{n}_{\rm i} - (1/2) \langle n_{\rm i+a} \rangle \langle n_{\rm i} \rangle]\,. \end{aligned} \tag{8.18}$$

Remember again that the unit vectors $\{\mathbf{a}\}$ have the same meaning as in the hopping term $H_{\rm kin}$, namely, they point along both positive and negative directions for each axis. It should be noted that the type of orbital ordering would have been automatically fixed as either $(x^2 - y^2)$ or $(3z^2 - r^2)$, if the original operators c would have been used for the mean-field approximation. However, as emphasized above, the $H_{\rm el-el}$ term has rotational invariance in orbital space, and there is no reason to fix the orbital only as $x^2 - y^2$ or $3z^2 - r^2$. To discuss properly the orbital ordering, the local e_g-electron basis, i.e., the phase $\xi_{\rm i}$ must be determined self-consistently.

Combining $H^{\rm MF}_{\rm el-ph}$ with $H^{\rm MF}_{\rm el-el}$ and transforming $\tilde{c}_{\rm ia}$ and $\tilde{c}_{\rm ib}$ back into the original operators $c_{\rm ia}$ and $c_{\rm ib}$, the mean-field Hamiltonian is finally

$$\begin{aligned} H^{\infty}_{\rm MF} &= - \sum_{{\rm ia}\gamma\gamma'} t^{\rm a}_{\gamma\gamma'} c^\dagger_{{\rm i}\gamma} c_{{\rm i+a}\gamma'} + J_{\rm AF} \sum_{\langle {\rm i,j} \rangle} \mathbf{S}_{\rm i} \cdot \mathbf{S}_{\rm j} \\ &\quad + \tilde{E}_{\rm JT} \sum_{\rm i} [-2(\langle \tau_{x{\rm i}} \rangle \tau_{x{\rm i}} + \langle \tau_{z{\rm i}} \rangle \tau_{z{\rm i}}) + \langle \tau_{x{\rm i}} \rangle^2 + \langle \tau_{z{\rm i}} \rangle^2] \\ &\quad + \sum_{\rm i} [(\tilde{U}'/2) \langle n_{\rm i} \rangle + V \sum_{\rm a} \langle n_{\rm i+a} \rangle](n_{\rm i} - \langle n_{\rm i} \rangle/2)\,, \end{aligned} \tag{8.19}$$

where the renormalized JT energy is given by

$$\tilde{E}_{\rm JT} = E_{\rm JT} + \frac{U'}{4}\,, \tag{8.20}$$

and the renormalized interorbital Coulomb interaction is

$$\tilde{U}' = U' - 4E_{\rm br}\,. \tag{8.21}$$

Physically, the renormalization of the JT coupling indicates that the JT energy is effectively enhanced by U'. Hence, the strong on site interorbital Coulombic correlation plays the *same* role as the JT phonon, at least at the mean-field level, indicating that it is not necessary to include U' explicitly in the models, as emphasized by Hotta et al. [8.3]. The physical origin of this effect is clear. In the regime of large Hund coupling, electrons have their spins aligned at a given site. Then, the coupling U, which penalizes double occupancy (with spin up and down) of the same orbital, becomes redundant in that regime. However, occupancy with parallel spins of available different orbitals at a given ion is, in principle, possible. This is suppressed *both* by a large U' Coulombic repulsion, as well as by a large JT coupling, since the latter tends to spread charge to distort as many octahedra as possible. This is an important point that becomes mathematically clear after the mean-field approximation. At large $J_{\rm H}$, the ground state of a model with no Coulomb interactions but large JT coupling becomes qualitatively *equivalent* to the ground state of a model with large Coulomb and no JT coupling. For Monte Carlo simulations studying a nonzero JT and zero Coulomb couplings is technically the simplest. Fortunately, the analysis of this section shows that the results should be similar to those obtained with Coulombic interactions. Of course, whether the similarity between the two approaches goes beyond the ground state, and it is present for excitations and in dynamical and temperature-dependent observables as well, still remains to be investigated.

The equation for the renormalized $\tilde{U}'$ indicates that the interorbital Coulomb interaction is effectively reduced by the breathing-mode phonon, since this phonon provides an effective attraction between electrons. The expected overall positive value of $\tilde{U}'$ indicates that e_g electrons dislike double occupancy of a given site. Thus, to exploit the gain due to the static JT energy and avoid the loss due to the on site repulsion, an e_g electron will singly occupy a given site.

8.3 Solution of the Mean-Field Equations

Let us discuss how to solve the mean-field Hamiltonian derived in the previous section on the background of fixed t_{2g} spins. For such a fixed-spin pattern, by using appropriate initial values for the local densities $\langle n_{\rm i}\rangle^{(0)}$, $\langle \tau_{x{\rm i}}\rangle^{(0)}$, and $\langle \tau_{z{\rm i}}\rangle^{(0)}$ (the only three mean-field parameters that appear in (8.19)), the mean-field Hamiltonian $H_{\rm MF}^{\infty(0)}$ is constructed, where the superscript number (j) indicates the iteration step. $H_{\rm MF}^{\infty(0)}$ can be easily diagonalized on finite clusters since it is *quadratic* in the fermionic operators. Using a variety of boundary conditions depending on the problem, the improved local densities, $\langle n_{\rm i}\rangle^{(1)}$, $\langle \tau_{x{\rm i}}\rangle^{(1)}$, and $\langle \tau_{z{\rm i}}\rangle^{(1)}$, are obtained (namely, the ground state of iteration "0" is known, and we can calculate any expectation value in that state). This procedure is repeated such that in the j-th iteration step, the local densities $\langle n_{\mathbf{i}} \rangle^{(j+1)}$, $\langle \tau_{x\mathbf{i}} \rangle^{(j+1)}$, and $\langle \tau_{z\mathbf{i}} \rangle^{(j+1)}$ are obtained by using $H_{\mathrm{MF}}^{\infty(j)}$. The iterations can be terminated if $|\langle n_{\mathbf{i}} \rangle^{(j+1)} - \langle n_{\mathbf{i}} \rangle^{(j)}| < \delta$, $|\langle \tau_{x\mathbf{i}} \rangle^{(j+1)} - \langle \tau_{x\mathbf{i}} \rangle^{(j)}| < \delta$, and $\langle \tau_{z\mathbf{i}} \rangle^{(j+1)} - \langle \tau_{z\mathbf{i}} \rangle^{(j)}| < \delta$ are satisfied, where δ is a small number that controls the convergence.

As for the cluster choice, to obtain the charge- and orbital-ordering patterns of relevance in the insulating phase, it is usually enough to study a finite-size cluster with periodic boundary conditions. The cluster size should be large enough to reproduce the periodicity in the spin, charge, and orbital ordering under investigation. Sometimes, some clusters are more "optimal" than others, since extended complicated patterns, as frequently observed in charge-ordered phases, fit better into particular lattices than others.

Finally, let us address some of the assumptions employed when the mean-field equations are solved in practice. As a first approximation for t_{2g} spins, their pattern is fixed throughout the mean-field calculation and the nearest-neighbor spin configuration is assumed to be only parallel or antiparallel. As already explained, this assumption does not indicate only the fully FM phase or three-dimensional G-type AFM spin pattern, but it includes more complicated patterns, including the CE-type AFM phase. Note that, under this assumption, several possible phases such as the spin-canted phase and the spin-flux phase, in which neighboring spins are neither FM nor AFM, are in principle neglected. This assumption for the fixed t_{2g}-spin pattern needs extra tests. Thus, it is a fact that other methods are needed to confirm the mean-field results. For this check, unbiased numerical calculations such as the Monte Carlo simulations and relaxation techniques can determine the local distortions, and the local spin directions. The optimization technique was found to work quite well in this type of problems, especially for a fixed electron number.

Our strategy to complete the mean-field analysis is as follows: (i) For some electron density and working on a small-size cluster, the mean-field calculations are carried out for several fixed configurations of t_{2g} spins. (ii) For the same electron density in the same size of cluster as in (i), both local distortions (JT phonons) and t_{2g}-spin directions are optimized using computational techniques. (iii) Results obtained in (i) are compared to those in (ii). If there occurs a serious disagreement between them, go back to step (i) and/or (ii) to do the calculations again by changing the initial inputs. In this retrial, by comparing the energies between the cases (i) and (ii), the initial condition for the case with higher energy should be replaced with that for the lower energy. To save CPU time it is quite effective to combine analytic mean-field and numerical techniques. (iv) After several iterations, if a satisfactory agreement between (i) and (ii) is obtained, the mean-field on a larger-size cluster is used to improve the results in (i). Combining the mean-field and optimization technique, it is possible to reach interesting results in a fairly reliable way.

Regarding the use of non cooperative phonons, this assumption can also be checked by comparing the non cooperative mean-field results with the optimized ones for cooperative distortions. Here note that, due to CPU and memory restrictions, the optimization technique cannot treat large-size clusters. However, the technique has the clear advantage that it is easily extended to include cooperative effects, simply by changing coordinates from $\{Q\}$ to $\{u\}$, where $\{u\}$ indicates oxygen displacements, while $\{Q\}$ denotes local distortions of the MnO_6 octahedron. The effect of the cooperative phonons should be discussed separately for several values of the hole density. In cases such as in the search for stripes at hole density $x = 1/4$ (Chap. 10), cooperative effects are very important, and care must be taken when non cooperative phonons are used. Nevertheless, on several occasions cooperative and non cooperative phonons leads to similar physics.

In other chapters of this book, results using the mean-field approximation constructed here are used to obtain properties of realistic models for manganites, comparing in several cases the data against Monte Carlo simulations. In several cases the agreement between mean-field and numerics is quite reasonable, and the mean-field methods appear to work very well to describe the individual phases of manganite models. Unfortunately the competition among phases and concomitant inhomogeneities, crucial for understanding the CMR phenomenon (Chap. 17), are quite difficult to handle with mean-field methods, and computational techniques are mandatory.

9. Two-Orbitals Model and Orbital Order

In this chapter, we describe Monte Carlo and mean-field results for the two-orbitals model with Jahn Teller phonons or Coulombic interactions. Multiorbital models are more realistic than the one-orbital model described in previous chapters. Here we report a rich phase diagram, with a variety of phases, that is in good qualitative agreement with experiments.

The present chapter also contains an introduction to the experimental X-ray resonant technique. This method has been very important in the study of manganites, particularly to experimentally confirm the presence of orbital order in these compounds. This type of order is an important ingredient of the theoretical results described in this chapter. The authors of the experimental review of the resonant X-ray method are Nelson, Hill and Gibbs. The associated theoretical review was written by Ishihara and Maekawa with emphasis on the relevance of Coulomb interactions (for another view, giving more weight to the Jahn–Teller distortions, see [9.1, 9.2]).

9.1 Monte Carlo Results and Phase Diagram

Any theoretical study of manganites that attempts to be quantitative needs to take into account the *orbital* degrees of freedom, in addition to the charge and the spin. Experiments clearly show that orbital order exists in these materials, and that this order manifests itself through distortions of the lattice (for a review see [1.5]). For this reason, a careful analysis of the two-orbitals model discussed in Chap. 5, with electrons coupled to Jahn Teller phonons, is very important for a quantitative understanding of the Mn oxides. In principle, we also need to include explicitly the Coulomb interaction (we do so in the last section of this chapter). Fortunately, a large Hund coupling between e_g and t_{2g} spins plays in many respects the role of a Coulomb interaction among the e_g electrons since it penalizes the double occupancy of an orbital. As discussed in previous chapters, this occurs because one e_g spin can be parallel to the localized t_{2g} spin, but the other e_g spin in the same orbital must be antiparallel due to the Pauli principle, increasing the energy. In addition, an intermediate or large electron phonon coupling λ favors the spreading of charge (activating as many JT octahedra as possible to reduce the energy), penalizing double occupancy of the same ion with parallel spins in different

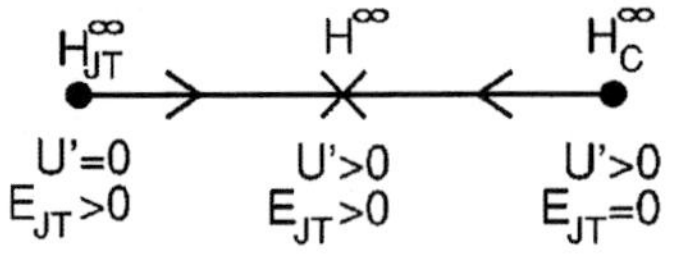

Fig. 9.1. Schematic representation of the two main approaches followed in theoretical descriptions of manganites, from Hotta et al. [6.23]. On the *left*, $E_{\rm JT} > 0$ represents a non zero Jahn Teller electron phonon coupling, while $U' = 0$ is a vanishing Coulombic interaction. On the *right*, the opposite case is represented. The real manganites are expected to be in the middle, with both couplings nonzero. The Hamiltonians are represented with an "∞" superscript to remind the reader that in both cases the Hund coupling e_g–t_{2g} is assumed large

orbitals. This case is not suppressed by a large Hund coupling alone. As a consequence, it is possible to carry out mean-field and computational studies of the two-orbital model with phonons without Coulombic terms, and expect qualitative agreement with more realistic cases. As remarked by Hotta et al. [6.23], the real materials may need a combination of both Coulombic and Jahn Teller electron phonon couplings. But this middle ground can in principle be approached from the extreme purely Coulombic (see for instance [9.3]) or purely phononic cases, as sketched in Fig. 9.1, as long as no "singularities" are found in between. Current theoretical work suggests that it is indeed possible to learn much about the real materials from one of the extreme cases of the figure.

Besides neglecting Coulombic interactions, other approximations are still needed to be able to simulate properly the model with JT phonons: (i) The localized spins must be assumed classical, as already done in the analysis of the one-orbital model described in Chap. 6. This approximation has been tested in one-dimensional cases and it appears to work, at least regarding the dominant phases at low temperatures. The t_{2g} classical spin approximation is indeed widely accepted. (ii) The phonons must also be assumed classical. This approximation is also widely expected to capture the essential properties of the phase diagram of the two-orbital model. However, a systematic comparison against results of models with quantum phonons has not been carried out even in 1D. A pragmatic reason for accepting classical phonons is rapidly emerging: the results are simply in very good agreement with experiments, as discussed in this chapter.

Before addressing the complex phases with spin, orbital, and charge ordering of the two-orbital model, a brief comment on the results for the one-orbital model of previous chapters is important. While it is certainly correct that two orbitals are needed to be accurate in the theoretical studies, many interesting results can be obtained from the one-orbital model. Of primary importance is the competition between FM and AF tendencies, which is present with both one and two orbitals. Since, technically, the one-orbital is far simpler than the two-orbital model, considerable intuition can be gathered from studies with just one orbital. In future chapters, we will gather qualitative aspects of the physics under study using both models. The one-orbital model is certainly not useless.

A second disclaimer is important at this point. The study reported in the following is far from complete, and considerable progress is still needed to fully understand the phase diagram of models with two orbitals. Careful studies in one dimension have been carried out, but, in higher dimensions, only some regions of parameter space have been analyzed in detail. In addition, most of the reported results are for the so-called "non cooperative" phonons, while the phase diagram of the more realistic "cooperative" case has only recently begun to be seriously addressed. The reader should expect considerable work to be reported in this area in the near future.

9.1.1 Phase Diagram in the Hole-Undoped Case

We start the analysis by studying the limit of zero hole density (i.e., $\langle n \rangle = 1 - x = 1.0$ in electronic language). This case of one electron per Mn ion corresponds to the situation present, for instance, in $LaMnO_3$. To study the properties of the two-orbital model, a variety of correlations have been calculated with the Monte Carlo technique, including the spin and orbital structure factors $S(q)$ and $T(q)$, respectively. The latter is defined as $T(q) = (1/N)\sum_{mj} e^{iq(m-j)}\langle \boldsymbol{T}_m \cdot \boldsymbol{T}_j \rangle$, where $\boldsymbol{T}_m = (1/2)\sum_{\sigma ab} c^{\dagger}_{ma\sigma}\boldsymbol{\sigma}_{ab}c_{mb\sigma}$. An analogous definition holds for $S(q)$.

Typical results are shown in Fig. 9.2, for one, two, and three dimensions. In the study of Fig. 9.2 the main goal was to analyze the influence of the coupling λ on the phase diagram, which is the main new ingredient compared with the case of just one orbital. As a consequence, the Hund coupling and the antiferromagnetic coupling between localized spins J_{AF} are simply fixed to numbers believed to be realistic (see caption of Fig. 9.2). The results for the spin structure factor indicate that the signal for zero momentum is robust at small and intermediate λ while, at a value $\lambda \sim 2$, it is replaced by $q{=}\pi$. Zero momentum is indicative of a spin-uniform state, and the robust value of $S(0)$ indicates the dominance of ferromagnetism. This correlates with the observed spin spin correlations of ferromagnetic character extending over distances that are as large as the studied lattices. At large λ, the dominance of momentum π indicates a sudden change to an antiferromagnetic state, again with staggered correlations extending over long distances. Other momenta are not as important as $q = 0$ and π. In the orbital sector, $T(q)$ is not substantially large at small λ for any momentum, but at intermediate coupling λ a robust value exists at $q = \pi$. Further increasing λ, the dominant momentum switches to $q = 0$.

The results of the simulations can be easily understood. At small λ, the orbital degrees of freedom appear to be disordered, while the spin is ordered ferromagnetically. These are properties of a standard ferromagnet [9.5]. The reader should not confuse the physics of $\langle n \rangle = 1.0$ for the one-orbital model,

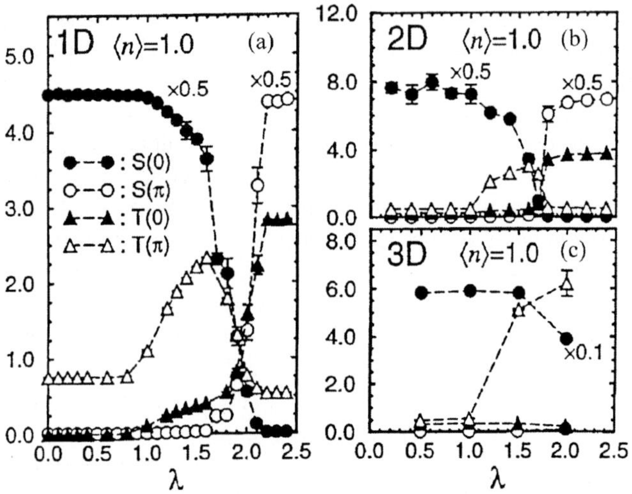

Fig. 9.2. (**a**) Orbital and spin structure factors, $S(q)$ and $T(q)$, for the two-orbital model with Jahn Teller classical phonons in one dimension. λ is the electron phonon coupling. Results, reproduced from [9.4], are shown at density $\langle n \rangle = 1.0$, $J_{\text{H}} = 8$, $J_{\text{AF}} = 0.05$, and using $t_{aa} = 1$ as the scale. In the early studies of [9.4] the hoppings were set to $t_{aa} = t_{bb} = 2t_{ab} = 2t_{ba}$, which are only close to realistic values. Further investigations with other sets of variables have shown that the results do not change dramatically. (**b**) Same as (**a**) but using a 4×4 cluster and realistic hoppings. (**c**) Same as (**a**) but on a 4^3 cluster with $J_{\text{H}} = \infty$

where antiferromagnetism dominates due to the total freezing of a ferromagnetic state by the Pauli principle, and that for the two-orbital case where only one of the two orbitals per site is populated on average and even a ferromagnetic state can exhibit electronic mobility. For two orbitals, we expect that $\langle n \rangle = 2.0$ should present antiferromagnetic tendencies. Note that experiments at $\langle n \rangle = 1.0$ do not reveal an orbital-disordered ferromagnet, but a more complex state with orbital ordering. Thus, the regime of small λ is not realistic for our purposes.

At intermediate values of λ, a situation similar to that found in experiments is observed. The spin is still ferromagnetic but now the orbital degree of freedom is active, with a *staggered* pattern, since the important momentum is π in 1D, (π, π) in 2D, and (π, π, π) in 3D. A staggered pattern in the orbital context means that a particular combination of orbitals is favored (lowers the energy) in the even sites of the lattice, while odd sites favor a different combination. The orbitals that dominate are linear combinations of the two orbitals of the original problem, in the same way as spin ordering can occur with moments pointing along the x- or y-axis instead of the quantization axis z. There is nothing special about $x^2 - y^2$ and $3z^2 - r^2$, and linear combinations can be better energetically.

Why is orbital order present in this case? An argument quite similar to the procedure followed by Kugel and Khomskii [9.6] in one of the first seminal papers on orbital order can be adapted to the case at hand. Consider for simplicity a perturbative calculation in the hopping amplitudes involving nearest-neighbor manganese ions, and focus only on the case of a diagonal hopping matrix (i.e., neglect hopping between different orbitals). In Fig. 9.3 the four possible cases are shown (i.e., spin and orbital, uniform or staggered). In case (a), the spin is FM and the orbital is staggered. In this case, the e_g electrons can move between Mn ions but paying the energy difference between orbitals, which is denoted by the positive number E_{JT}. The energy gained (negative) by this movement is then proportional to t^2/E_{JT} in second-order (the Hund coupling energy is unchanged in the intermediate virtual state of the calculation). In case (b), with a uniform spin and orbital arrangement, the system is frozen in the case of diagonal hopping. In (c), with orbital uniform but spin staggered, the virtual movement indicated by the dashed line does not change the Jahn Teller energy, but it affects the Hund energy. This means that the spin arriving to the right ion has its z-projection spin antiparallel to the localized spin at that site. The energy gain by this virtual hopping is proportional to $t^2/(2J_{\mathrm{H}})$. Finally, in (d) both the spin and the orbital are staggered, and the energy gained by virtual hopping is proportional to $t^2/(2J_{\mathrm{H}} + E_{\mathrm{JT}})$. As a consequence, cases (b) and (d) can never be the ground states, since there is always another state that shows smaller energy. The true competition is between cases (a) and (c) (Note that this calculation is effectively set up in a regime where the Jahn Teller degrees of freedom are active, namely, intermediate to large λ coupling. The case of an orbital disordered ferromagnet with extended e_g states is not considered in the present discussion). If the Hund coupling is large compared with E_{JT}, then the energy is lowest for case (a), which corresponds to a spin ferromagnetic, orbital "antiferro" or staggered state. As argued before, this state, or rather its cousin in 3D, is the state that matches experiments for $LaMnO_3$. In the other limit where the coupling λ dominates, rendering E_{JT} much larger than J_{H}, case (c) dominates and the system becomes spin antiferromagnetic but orbitally uniform. The numerical results shown in Fig. 9.2 are clearly in agreement with this simple picture.

Intermediate values of λ correspond to a ground state with staggered orbital ordering (the most novel feature) and concomitant spin ferromagnetic ordering. However, experiments for the hole-undoped manganites (i.e., $x = 0$) show that the actual spin order is of the A-variety, as explained in Chap. 3. This A-type order corresponds to a ferromagnetic arrangement in planes, with antiferromagnetic ordering between planes. Note that overall moment cancels when averaged between subsequent planes. How can we reproduce this effect using models for manganites? Recent investigations have shown the important role played by the antiferromagnetic coupling J_{AF} between localized spins. In Chap. 5, it was argued that J_{AF} is the smallest of the

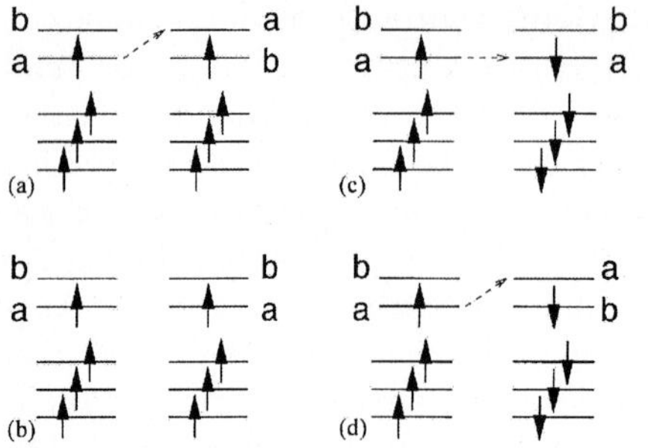

Fig. 9.3. The four possible cases of spin and orbital order involving two nearest-neighbor Mn ions. The discussion in the text shows that (**a**) and (**c**) can be the ground state depending on parameters, with (**a**) being the case closest to experiments for undoped manganites. The notation in the figure is standard, with "a" and "b" denoting e_g levels. The other three per site are t_{2g} levels

couplings of the one- and two-orbital models. However, even a small coupling becomes relevant when competing states are close in energy. Hotta et al. [9.7] showed that an A-type AF state can be stabilized at intermediate λ couplings, if $J_{\rm AF}$ is increased from the value 0.05 used in Fig. 9.2 to larger but still reasonable values. Others have reached similar conclusions [9.8, 9.9] using JT phonons. In fact, the relevance of $J_{\rm AF}$ in the context of manganites has been addressed also in the context of models with robust Coulombic interactions instead of robust electron phonon couplings (see for example [9.10–9.16]). As remarked before, the author believes that models with dominant Coulombic and dominant JT interactions lead to similar physics, simply because the large Hund coupling present in both cases heavily penalizes double occupancy. Of course, the actual location of the various phases in coupling space will change depending on the relative strength of the Coulombic and electron JT-phonon coupling, but the qualitative aspects will remain the same. This is a *conjecture*, and further work is needed to confirm it, but the evidence gathered thus far supports this conjecture strongly. From this perspective, the Coulomb vs JT battle in the literature among different theory groups does not seem to be of much relevance.

The effect of $J_{\rm AF}$ is illustrated in Fig. 9.4a where the energy of the two-orbital model ground state corresponding to a small cluster is shown as a function of J' (which is another often-used notation for $J_{\rm AF}$). At small values of this coupling the system is ferromagnetic in all three directions. There is no reason to become antiferromagnetic at hole-density zero for the two-orbital model (it would be different at higher electronic densities, such as 2.0). However, as J' increases, since this coupling favors antiferromagnetic ordering, then the FM state is penalized and its energy increases linearly with J'. At some finite, but still fairly small, J' critical coupling an abrupt transition to a state that is at least partially more favorable for ferromagnetism occurs: the A-type state is stabilized. Upon further increase of J', the tendency to AF order must be more and more favored. For this reason, beyond the A-type phase, a second transition to a C-type order is found (C-order shows ferromagnetism in one direction, and antiferromagnetism in the other two). Finally, for sufficiently large, and perhaps already unrealis-

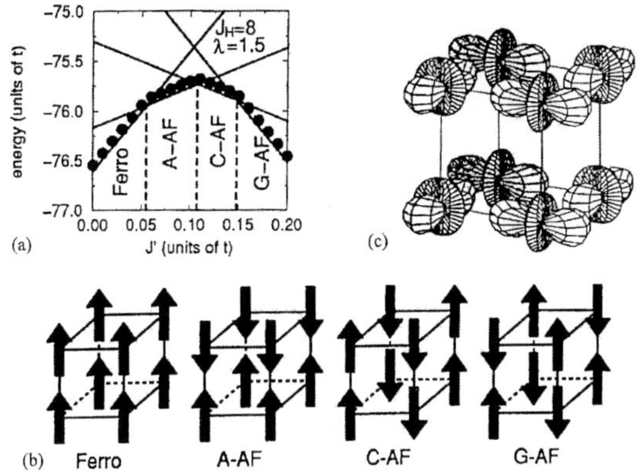

Fig. 9.4. (**a**) Energy of the ground state of a 2^3 cluster, using the two-orbital model for manganites at the indicated Hund and electron phonon couplings. The electronic density is one particle per site. The results from Hotta et al. [9.7] were obtained with Monte Carlo (*points*) and mean-field (*solid lines*) approximations. (**b**) Spin arrangements corresponding to the four states in part (**a**). (**c**) Orbital order corresponding to the A-type AF state, according to [9.7]

tic, J' couplings the system turns into a G-type AF, namely AF in all three directions. The several spin arrangements are shown in Fig. 9.4b. Of course, it could have happened that an abrupt transition separate FM from G-AF directly, but apparently it is better energetically to generate intermediate A- and C-phases. In addition, it is interesting to observe that the results in Fig. 9.4a are similar to those obtained with mean-field approximations in the context of Coulombic models [9.17]. This illustrates again the analogies between Coulomb and JT approaches. However, others [9.18] believe that the JT coupling is truly crucial to stabilize the A-type phase. The last word has not yet been said on this issue.

Concomitant with the spin states discussed in Fig. 9.4, there are particular orbital arrangements. The orbital order shown in Fig. 9.4c was reported by Hotta et al. [9.7] after perturbations away from the ideal two-orbital model were introduced. The results are in good agreement with experiments. Other authors arrived at similar conclusions (see [9.19]).

9.1.2 New Phases in the Undoped Limit?

Even in the undoped limit $x = 0$, with typical representative materials such as $LaMnO_3$, surprises can still be found. Recent experimental work by Muñoz et al. [9.20] and Kimura et al. [9.21] replacing La by Ho and Tb (both with smaller ionic radius than La) has revealed the presence of a *new phase*. This new phase may still have the same staggered arrangement of orbitals (more experiments are needed to confirm this assumption), but the spins now form "zigzag" chains. This is an alternating pattern, one lattice spacing along the

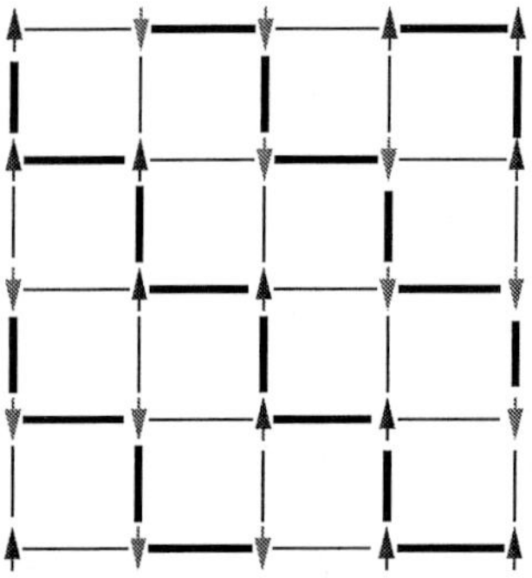

Fig. 9.5. Schematic representation of the E-state at $x = 0$, involving zigzag chains of parallel spins, coupled antiparallel to each other. For more details see [9.22]. *Thick lines* indicate the chains, the *thin lines* are the antiferromagnetic links between them

x-axis, one along the y-axis, with parallel spins along those zigzag chains, antiferromagnetically coupled between chains (see Fig. 9.5).

Very recently, it has been shown that this new phase, called the E-phase, appears in numerical studies of models with Jahn–Teller phonons [9.22]. Figure 9.6 (left) contains a rough qualitative phase diagram at $x = 0$, in the plane λ–J' (the latter being the AF coupling between localized spins, and the former the electron JT-phonon coupling). The standard $x = 0$ phase appears on the left, a G-type AF phase on the right, and the "new phase" or E-phase in the middle. Figure 9.6 (right) is the energy vs J', showing once again the presence of an intermediate regime between FM and AF. This new phase has clear similarity with the CE-state of the half-doped system, where the zigzag pattern involves two lattice spacing in both directions, also alternating. In fact, the studies by Hotta et al. reported later in Chap. 10 and [3.32] already contain the pattern of the E-phase as a possible state. Experiments [9.21] suggest that the critical temperatures of the A-AF- and E-phases are very small when they meet, varying the A-site ionic radius. This may correspond to a quantum-critical-point like behavior, similar to that discussed with toy models in Chap. 17. Clearly, even the x=0 limit still contains many surprises!

In addition to these results involving a new phase, spin-glass behavior can be generated at x=0 from $LaMnO_3$ by replacing Mn by Ga, as was shown by Zhou et al. [9.23]. Even adding hole carriers, it appears that $La_{2/3}Ca_{1/3}Mn_{1-x}Ga_xO_3$ also generates a quantum-critical-point-like behavior with increasing Ga concentration, according to recent investigations by De Teresa et al. [9.24]. Theoretical studies by Alonso et al. [9.25] suggest that as more Ga is introduced, a transition to an Anderson localized state may be induced. More work should be devoted to these interesting "QCP" materials. Studies of impurity effects in double-exchange systems have also been recently reported by Motome and Furukawa [9.26].

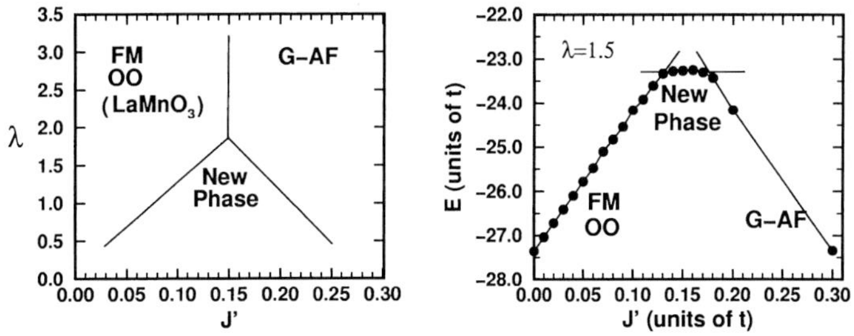

Fig. 9.6. Numerical evidence for the E-phase obtained with Monte Carlo simulations of the two-orbital model with Jahn Teller phonons [9.22]. The phase boundaries are not accurate, only qualitative. *(Left)* Schematic phase diagram in the λ–J' plane. The "new phase" is the E-phase. *(Right)* Energy vs J' at fixed and large λ, indicating an intermediate regime between FM and AF

9.1.3 Phase Diagram at Hole Density $x \neq 0$

The previous results are not difficult to obtain using Monte Carlo and mean-field techniques. They corresponded to zero hole-density, or the undoped limit, in which uniform order tends to occur, forming insulating states with a periodic arrangement of spin and orbital degrees of freedom. The situation is far more complicated when we move away from zero hole-density. In particular, at low hole-density, the individual holes distort their immediate vicinity creating "polarons" that can involve spin, orbital, and lattice degrees of freedom. Inside the polaron, deviations from the undoped limit can be found in all those variables. When many polarons are present, they may decide to produce periodic arrangements, or to join forces into a giant polaron (which should be better called a "hole cluster"). The study of this physics is complicated and, in the context of manganites with two active orbitals, only preliminary work has been carried out and documented in the literature. Thus, the reader should stay alert for further developments beyond those reported here, since there is plenty still to be done.

The phase diagram in 1D for the two-orbital model with non cooperative phonons is presented in Fig. 9.7c, reproduced from Yunoki et al. [9.4]. Plots of density vs. chemical potential are shown in Figs. 9.7a–b for 1D and 2D. The discontinuities in the latter are very reminiscent of those found in the one-orbital model (Chap. 6). Its meaning has been discussed already: the densities that cannot be accessed by tuning the chemical potential are simply unstable, and the regime corresponds to "electronic phase separation". Upon further introduction of long-range Coulombic forces (or disorder, or stress), this regime is widely expected to produce some complex periodic arrangement of charge, at the nanometer-scale level. It appears that phase separation is a robust property of models for manganites, as already discussed for the one-orbital case [9.27]. Note that the phase separation close to $x = 0$ in Fig. 9.7c is *induced by the orbital degree of freedom*, since the two phases in competition

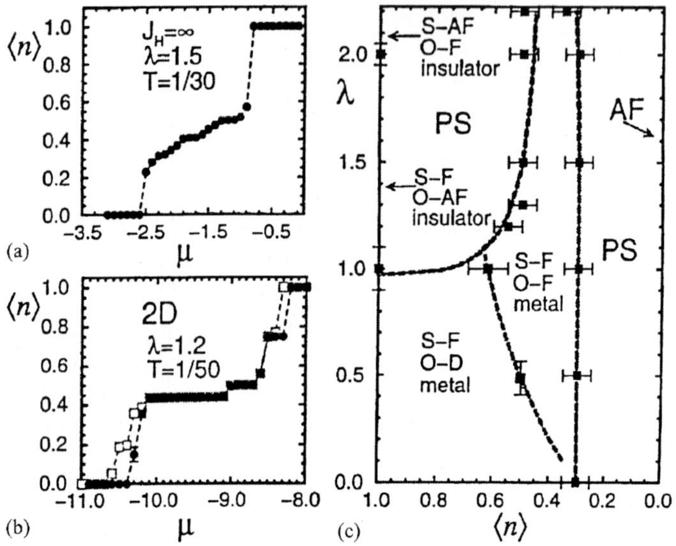

Fig. 9.7. Electronic density $\langle n \rangle$ vs. chemical potential μ for the two-orbital model with Jahn–Teller phonons at the couplings shown. (**a**) Corresponds to a finite length 1D system, while (**b**) is for a small 2D cluster (details can be found in [9.4]). The discontinuities correspond to phase separation. (**c**) Phase diagram of the two-orbital model in one dimension, at $J_{\mathrm{H}} = 8$ and $J' = J_{\mathrm{AF}} = 0.05$. Details and notation are explained in [9.4]

at fixed λ are spin ferromagnetic (as explained below). This is an interesting novel concept. Similar results were also reported by Okamoto et al. [9.28].

The phase diagram shown in Fig. 9.7c emerges by combining the results for $\langle n \rangle$ vs. μ to search for phase separation, with measurement of spin, charge, and orbital correlations as well as the study of the Drude weight in the optical conductivity (see Chap. 12). Clearly, this diagram is far from trivial, and many phases in competition are present, as widely observed experimentally. At values of λ less than one and small hole density $x = 1 - \langle n \rangle$, a robust ferromagnetic phase is observed at low temperature. The system is metallic, based on conductivity calculations, and orbitally disordered. In the same range of densities but increasing λ, a robust and fairly large regime of phase separation is obtained. Again, we expect that this regime will turn into a complex distribution of spin, charge, and orbital degrees of freedom upon the introduction of Coulombic forces, or the inevitable disorder present in any compound. It is interesting to remark, once again, that calculations with Coulombic models (without JT phonons) have also found phase-separated regimes in a variety of situations and approximations (see [6.11,6.12,8.3,9.29, 9.30]). Figure 9.8 shows density vs. chemical potential for a model with two orbitals (and no localized spins, simply two orbitals), showing a discontinuity characteristic of phase separation. The presence of electronic phase separation

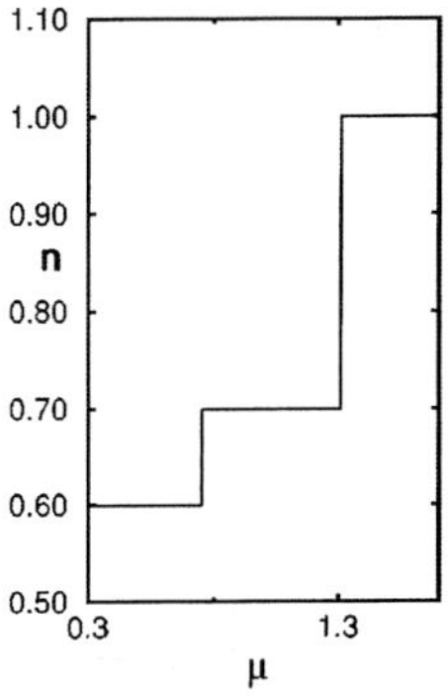

Fig. 9.8. Density vs. μ obtained by Malvezzi in the numerical study reported by Hotta et al. [8.3]. The result shows a discontinuity characteristic of phase separation. The model used involves only Coulombic interactions, two orbitals, and no localized spins. This illustrates how general this phenomenon actually is

appears to be a very robust property, at least in models for manganites. In the context of models for cuprates, the situation in realistic regimes of couplings is still unclear.

In the regime of hole densities close to 0.5, Fig. 9.7c shows another stable phase that is spin ferromagnetic, orbitally uniformly ordered, and metallic (at least in low dimension). This phase is reminiscent of the "A-type" phase found in experiments (see Chap. 3) at similar hole densities, for example, in Sr-based manganites. In real manganites, the $x = 0.5$ A-type phase is believed to be composed of metallic FM planes with uniform orbital order, which are coupled antiferromagnetically between planes leading to a poor metal or insulting behavior overall. In the reduced dimensionality of the computational study Fig. 9.7c, the latter effect is not observed, but the state found is clearly a "precursor" of the one observed experimentally. At the same density, and further increasing the electron phonon coupling (or the J_{AF} coupling at constant λ), a charge-ordered state, precursor of the famous CE-state (see Chap. 3) was found in the simulations of Yunoki et al. These results will be discussed in detail in Chap. 10. Recent mean-field studies in bilayers have also reported interesting results, which agree with experiments [9.31].

Finally, upon increasing the hole density to large values, again a regime of phase separation is obtained, which is fairly independent of the value of λ (it is present even at $\lambda = 0$). This is the straightforward translation to the two-orbital case, of the phase separation found in the one-orbital model (Chap. 6), both, at very small x and $x \approx 1$ (which translates into very small x and $x \approx 2$ in the two-orbital language, an electronic density regime not much explored with two orbitals, that may well have interesting surprises awaiting to be unveiled).

The results shown in Fig. 9.7c combined with information about the CE-state (which is stable at half-doping and large λ if working at fixed J', see Chap. 10) provides a first glimpse into the overall phase diagram of models for manganites. At the risk of sounding too optimistic, the comparison with the overall experimental phase diagram of Fig. 3.19 (Chap. 3) combining results for many manganites, seems to suggest that theory and experiments are in fairly good agreement. For instance, note that at $x = 0.5$ and large bandwidth the A-phase is found in experiments. Large bandwidth means large hopping "t", and, thus, small λ if measured in units of t. Then, small λ in Fig. 9.7c matches properly with large bandwidth in Fig. 3.19 regarding the $x = 0.5$ A-phase. The same happens for the CE-phase but now inverting small for large (i.e., small bandwidth – large λ). In the theoretical phase diagram, the FM metallic and orbitally disorder phase occurs in the correct regime of densities and couplings, namely in between $x \sim 0$ and 0.5 and more prominent for large bandwidth. In addition, the limits of A-type $x = 0$ and G-type $x = 1$ have been found in both theory and experiments. The C-type phase of Fig. 3.19 has not yet been obtained in the theoretical study. However, this type was found in the previous subsection tuning $J_{\rm AF}$ at $x = 0.0$, thus, it is nearby in parameter space. Then, with not much effort it will likely appear in the correct regime as well. In Chap. 10, even FM insulating phases will be reported, and similar phases have been found in experiments as well. Finally, note that the experimental phase diagram Fig. 3.19 does not explicitly show the many regimes of inhomogeneities in manganites that may match the phase separation found theoretically. For all these reasons, this author believes that the theoretical studies are moving in the right direction, although certainly plenty of work still remains to be performed.

As remarked at the beginning of this subsection, the computational effort shown here is only the tip of an emerging iceberg. The model in 2D and 3D has not been studied in as much detail as in 1D. Exotic phases such as the "island" phases reported in Chap. 6 for one orbital, may exist for two orbitals as well. The systematic use of "cooperative" phonons is only at the early stages, and plenty of surprises may still be found. Even studies using non cooperative phonons in regions of parameter space not explored, such as intermediate or small Hund coupling, are certainly interesting. There are still plenty of projects to be carried out.

9.2 Effective t–J-like Hamiltonians and Orbitons

The results of the previous section were obtained using JT phonons. Here, related results are described that were obtained with the two-orbital model including now Coulombic interactions, but without phonons. The results were obtained with a t–J-like effective model, in the limit of strong Coulomb repulsion. The starting point for the derivation of the model, sometimes called the "orbital t–J" model, is naturally the multiband Hubbard model already introduced in Chap. 4. The first step is to exclude from the Hilbert space states with double occupancy of the e_g band of a Mn ion, since the electron electron repulsions U and U' are the largest couplings in the multiband model. This is achieved by redefining the original fermionic operators [9.14, 9.32, 9.33], namely, multiplying them by projector operators that avoid double occupancy. This leads to new operators $\tilde{d}_{i\gamma\sigma}$ that create electrons at site i, with orbital γ (taking two possible values in the e_g band), and spin projection σ. Note that these steps are similar to those used to derive the famous t–J model for the cuprates starting from the one-band Hubbard model (see [1.20]).

The resulting (second-order perturbation theory) effective Hamiltonian in the subspace with no double occupancy of the e_g levels has a complicated expression, with several terms, too complex to derive explicitly here. However, the leading term (still long to write but manageable) is [9.14]

$$
\begin{aligned}
H_{U'-J} = &-\frac{2}{U'-J}\sum_{\langle ij\rangle}\left(\frac{3}{4}n_i n_j + \boldsymbol{S}_i\cdot\boldsymbol{S}_j\right)\\
&\times\Bigg[\left({t_{ij}^{aa}}^2 + {t_{ij}^{bb}}^2\right)\left(\frac{1}{4}n_i n_j - T_{iz}T_{jz}\right)\\
&-t_{ij}^{aa}t_{ij}^{bb}\left(T_{i+}T_{j-} + T_{i-}T_{j+}\right) + \left({t_{ij}^{ab}}^2 + {t_{ij}^{ba}}^2\right)\left(\frac{1}{4}n_i n_j + T_{iz}T_{jz}\right)\\
&-t_{ij}^{ab}t_{ij}^{ba}\left(T_{i+}T_{j+} + T_{i-}T_{j-}\right)\\
&-\left(t_{ij}^{ab}t_{ij}^{aa} - t_{ij}^{ba}t_{ij}^{bb}\right)\left(T_{iz}T_{j+} + T_{iz}T_{j-}\right)\\
&-\left(t_{ij}^{ba}t_{ij}^{aa} - t_{ij}^{ab}t_{ij}^{bb}\right)\left(T_{i+}T_{jz} + T_{i-}T_{jz}\right)\Bigg]\,,
\end{aligned}
\tag{9.1}
$$

where n_i is the electron number operator at site i, $\boldsymbol{S}_i$ is the spin operator of the e_g electron at site i defined as $\boldsymbol{S}_i = \frac{1}{2}\sum_{\gamma\alpha\beta}\tilde{d}^{\dagger}_{i\gamma\alpha}\boldsymbol{\sigma}_{\alpha\beta}\tilde{d}_{i\gamma\beta}$, and $\boldsymbol{T}_i$ is the pseudospin operator for the orbital degree of freedom at site i defined as $\boldsymbol{T}_i = \frac{1}{2}\sum_{\alpha\gamma\gamma'}\tilde{d}^{\dagger}_{i\gamma\alpha}\boldsymbol{\sigma}_{\gamma\gamma'}\tilde{d}_{i\gamma'\alpha}$. This last operator was already used elsewhere in this chapter. Its physical interpretation is simple. For example, the states with $T_{iz} = 1/2$ or $-1/2$ correspond to having one electron at site i at orbitals x^2-y^2 or $3z^2-r^2$, respectively. The operators T_{i+} and T_{i-} are defined similarly as the usual raising and lowering operator of spins. The hopping term t_{ij}^{ab} links orbitals a–b of the nearest-neighbor sites i–j.

Equation (9.1) contains the product of spin and orbital operators, since both degrees of freedom are mixed by perturbation theory. Then, it is likely that the spin arrangement in the ground state will depend on the orbital arrangement, and vice versa. The complicated form for the orbital contribution in (9.1) arises from the lack of a continuous symmetry in the orbital sector, unlike the spin sector, which has a simple form due to rotational invariance. To gain intuition on what sort of states are favored, consider just the first term of this Hamiltonian for the case $n_i = n_j = 1$ and with spin operators along the z-axis, i.e., the product $-(\frac{3}{4}+S_{iz}S_{jz})(\frac{1}{4}-T_{iz}T_{jz})$. To minimize its value the two spins must be parallel to each other. Then, this term favors spin *ferromagnetism.* The second must be as large as possible, and that is achieved by an orbital *staggered* distribution of electrons, in orthogonal orbitals from site to site. (An even simpler explanation for the appearance of spin-FM and staggered orbital order is shown in Fig. 9.9). These orthogonal orbitals do not need to be the original two used to construct the multiband Hubbard model. Linear combinations at each site could be favored as well. This is not

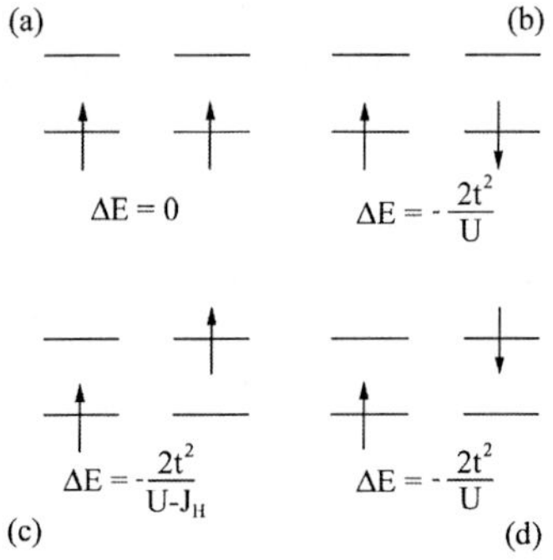

Fig. 9.9. Discussion of possible configurations of nearest-neighbor spins in a two-orbital model with Coulomb interactions, considering degenerate orbitals (split in the figure for visualization purposes), from [9.33]. For simplicity, assume a diagonal hopping matrix. (**a**) is a spin-FM state with the same orbital populated. Its energy is zero, since the Pauli principle prevents any movement. (**b**) is a spin-AF state. Since electrons can now move within the same orbital, the energy is reduced compared with (**a**), but a same-orbital repulsion energy U must be paid. (**c**) is spin-FM and orbitally staggered. The electrons populate different orbitals in different sites. Both electrons can move but, in doing so, they pay on site Coulomb repulsion while gaining the Hund energy. Finally, (**d**) is spin-AF, and orbitally staggered. Here, energy is lost both due to Coulombic repulsion and Hund coupling. Configuration (**c**) is the one that reduces the energy the most, leading to staggered orbital order in a spin-polarized background. Note that the U in (**b**) correctly stands for the intraorbital Coulomb element, while in (**c**) and (**d**) it stands for the interorbital Coulomb element usually represented by U'. Note also that $U = U' + 2J_{\rm H}$, as explained in Chap. 4. The results in this figure are quite similar to those sketched in the first section of this chapter, using electrons coupled to Jahn Teller phonons. However, in the case described in this figure there are no localized spins

obvious from the term considered with only T_{iz} active but the other two x and y components are needed as well (similarly as in spins, where staggered order can be established along the z-axis, but also along other axes). Results similar to those reported in [9.14] have also been presented in [9.33], where additional technical information on the effective Hamiltonians derived from the multiband Hubbard model can be found.

The same type of ordered states found in Coulombic-based models in [9.14, 9.32, 9.33] have been found in Monte Carlo simulations of the two-orbital model with Jahn–Teller phonons (see first section of this chapter). This illustrates again the similarities between approaches based on Coulombic models and those with strong electron Jahn–Teller phonon coupling. At the heart of this similarity is the large value of the Hund coupling, preventing double occupancy. This analogy is even stronger after an antiferromagnetic coupling $J_{\rm AF}$ between localized spins is introduced. As discussed by Ishihara et al. [9.14], this term is crucial in both approaches to generate the experimentally observed A-type antiferromagnetic phase of manganites. An

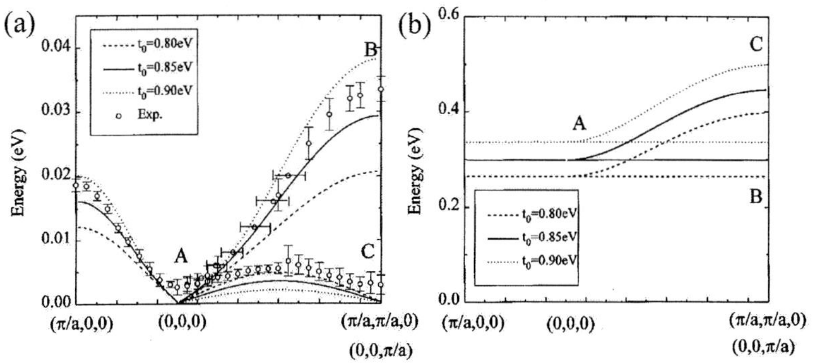

Fig. 9.10. (**a**) Dispersion relation of the spin-wave excitation, reproduced from Ishihara et al. [9.14]. The *open circles* are experimental neutron scattering data for $LaMnO_3$ (from [9.34]). Line AB joins (0,0,0) with $(\pi/a,\ \pi/a,\ 0)$. It corresponds to excitations in the ferromagnetic planes. AC joins (0,0,0) with $(0,\ 0,\ \pi/a)$, corresponding to the direction perpendicular to those planes, i.e., in the antiferromagnetic direction. (**b**) Dispersion relation of the orbital excitations, from [9.14]. Results for several hoppings are shown. The rest of the couplings and the convention for labels ABC are as in (**a**). Note the presence of a clear gap in the dispersion

analogous relevance of J_{AF} for A-type AF stabilization also appeared in the Monte Carlo/mean-field calculations detailed at the beginning of this chapter, using a strong electron JT-phonon coupling. Since both starting points are expected to lead to very similar physics, one must select the framework that allows for more flexibility in the many-body studies. Given that in the Jahn Teller approach it is possible to carry out systematic computational investigations beyond the mean-field approximation, allowing for the analysis of finite-temperature effects, cluster formation, influence of disorder, and other features of much relevance, it is then natural that the theoretical results in this book are mainly based on that JT approach.

Very interesting information about *excitations* has been gathered [9.14, 9.32, 9.33] using the Coulombic-based approach and mean-field approximations. These results are briefly discussed here. For instance, in [9.14] the spin-wave dispersion was calculated, assuming a realistic $(3x^2-r^2/3y^2-r^2)$ orbital ordering, selecting couplings to be in the A-type antiferromagnetic state, working in the mean-field approximation, and applying the conventional Holstein–Primakoff transformation. The result is shown in Fig. 9.10a, where it is compared with experiments. There is good agreement once the appropriate set of couplings is used (details can be found in [9.14]). Note that along the z-axis the dispersion is AF-like with a node at π, while the dispersion in the plane is ferromagnetic-like, as expected from an A-type AF state.

The formal similarity between the spin and pseudospin (or orbital) operators in the Hamiltonian shows that "orbital excitations" (or orbitons) – excitations somewhat analogous to the spin-waves – can also be created. The calculation in [9.14] uses the same approximations and couplings as needed to produce the data in Fig. 9.10a. The result is shown in Fig. 9.10b (note that other authors have also contributed to calculations in the context of orbitons. See, for instance, the work of Khaliullin and Oudovenko and others [9.35]). The main qualitative difference between the spin-wave and orbital-wave dispersions is the presence in the latter of a clear *gap*. In addition, the dispersion is flat in the ferromagnetic plane, and this is due to the particular orbital order of the undoped limit. The reason for the existence of a gap is the lack of rotational invariance in the pseudospin sector, which prevents the existence of a low-energy excitation when spontaneous symmetry breaking occurs in the arrangement of occupied orbitals. In some respects, the orbital excitations are analogous to spin-flip excitations in Ising models of spins and, like the latter, they cost a finite amount of energy. However, there are models of coupled spins and orbitals, the so-called SU(4) model, that have a continuous symmetry in the orbital sector as described in the review of Oleś et al. [9.33], where references can be found. This model is not realistic – it is not expected that in real systems the orbital sector will induce gapless excitations in manganites – but the SU(4) symmetric approach allows for interesting formal studies, which could be applicable to other systems where low-energy orbitons may be found. The review [9.33] also contains interesting information about the possibility of disordered ground states in multiband Hubbard models. The existence of orbitons is compatible with recent Raman scattering results for $LaMnO_3$ [9.36]. Note that the agreement between theory and experiment becomes quantitative once the theoretical calculations include Jahn–Teller distortions [9.36].

In summary, it can be safely expressed that studies of two-orbital models (either with JT phonons or Coulombic interactions) leads to states that are in excellent agreement with experiments. The A-type AF phase of the hole undoped $x = 0$ system, as well as other phases identified experimentally (C-type AF, A-type AF at half-doping, FM metallic, G-type at full doping, and charge ordering at large JT couplings) can all be found in some region of parameter space, especially for models with JT phonons. This is a significant triumph of the theoretical description of manganites. However, the key issue of the CMR effect remains unclear at this point, and further work (described in subsequent chapters) is needed to better understand real Mn oxides. The important new concept to be discussed is the *competition* among the many phases individually analyzed in this chapter.

9.3 Resonant X-ray Scattering as a Probe of Orbital and Charge Ordering

by C.S. Nelson, J.P. Hill, and D. Gibbs

Resonant X-ray scattering is a powerful experimental technique for probing orbital and charge ordering. It involves tuning the incident photon energy to an absorption edge of the relevant ion and observing scattering at previously "forbidden" Bragg peaks, and it allows high-resolution, quantitative studies of orbital and charge order – even from small samples. Further, resonant X-ray scattering from orbitally ordered systems exhibits polarization- and azimuthal-dependent properties that provide additional information about the details of the orbital order that is difficult, or impossible, to obtain with any other technique. To date, resonant X-ray scattering has been applied to studies of $La_{1-x}Sr_xMnO_4$ [9.37, 9.38], $LaMnO_3$ [9.39], $La_{1-x}Sr_xMnO_3$ [9.40, 9.41], V_2O_3 [9.2], $Nd_{0.5}Sr_{0.5}MnO_3$ [9.42], DyB_2C_2 [9.43, 9.44], $Pr_{1-x}Ca_xMnO_3$ [9.45, 9.46], Fe_3O_4 [9.47], $LaSr_2Mn_2O_7$ [9.48], YVO_3 [9.49], $LaTiO_3$ [9.50], NaV_2O_5 [9.51], $YTiO_3$ [9.52], CeB_6 [9.53], and UPd_3 [9.54] – and this long list of materials continues to grow.

In the manganites, the sensitivity to charge and orbital ordering is enhanced when the incident photon energy is tuned near the Mn K absorption edge (6.539 keV), which is the lowest energy at which a 1s electron can be excited into an unoccupied state. In this process, the core electron is promoted to an intermediate excited state, which decays with the emission of a photon. The sensitivity to charge ordering is believed to be due to the small difference in K absorption edges of the Mn^{3+} and Mn^{4+} sites. For orbital ordering, the sensitivity arises from a splitting – or difference in the weight of the density of states [9.55] – of the orbitals occupied by the excited electron in the intermediate state. In the absence of such a splitting, there is no resonant enhancement of the scattering intensity.

In principle, other absorption edges in which the intermediate state is anisotropic could be utilized, but the strong dipole transition to the Mn 4p levels – and their convenient energies for X-ray diffraction – make the K edge well suited to studies of manganites. The Mn 4p levels are affected by the symmetry of the orbital ordering, which makes the technique sensitive to the orbital degree of freedom. Therefore resonant X-ray scattering can be used to obtain important quantitative information concerning the details of this electronic order.

Two mechanisms for splitting of the 4p states have been proposed: a Coulomb coupling of the Mn 3d and 4p states – either directly, or indirectly through the hybridization with O 2p states [9.56] – and the motion of oxygen ions due to the Jahn–Teller interaction. These two mechanisms have opposite signs in terms of the direction of the splitting, but both are consistent with experimental results to date. Which mechanism is dominant therefore remains an open question; however, in either case, the resonant scattering reflects the symmetry of the orbital ordering through the perturbation of the local electronic states at the Mn^{3+} sites, and the peak positions and widths determined in the X-ray experiments measure the orbital periodicity and correlation lengths, respectively. In addition, as is described below, detailed analysis of the resonant scattering provides information about the occupation

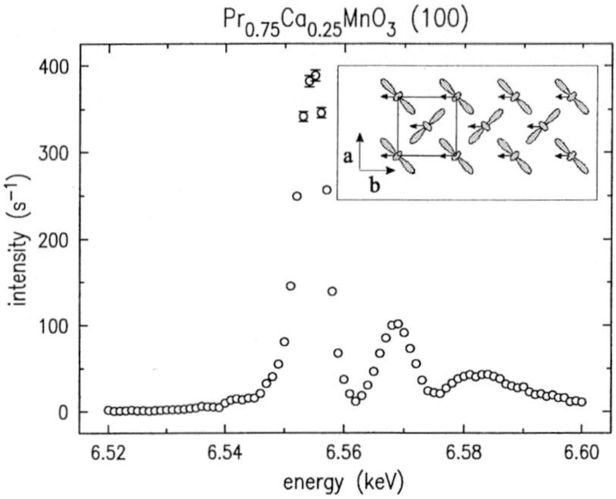

Fig. 9.11. Peak intensity versus incident photon energy of the orbital (100) reflection of the $Pr_{0.75}Ca_{0.25}MnO_3$ sample. *Inset* displays a schematic of the orbital and magnetic ordering believed to occur in this material. The *shaded figure*-8s represent the occupied e_g orbitals of the Mn^{3+} ions, and the *arrows* indicate the in-plane components of the magnetic ordering. In the absence of orbital order – e.g., if the materials were in an orbitally disordered, ferromagnetic, metallic state – no peak would be observed. For short-range orbital order, a very weak, very broad (in q-space) peak would be observed

of the ordered orbitals – a determination that cannot be made through the use of nonresonant scattering.

An example of resonant scattering from orbital order, in $Pr_{0.75}Ca_{0.25}MnO_3$, is shown in Fig. 9.11. This material is believed to exhibit $(3x^2 - r^2/3y^2 - r^2)$-type ordering of the e_g orbitals, as sketched in the inset. The data show the energy dependence of the (100) orbital order peak, which is forbidden by the crystallographic space group, measured at room temperature. A large resonant signal is visible just above the Mn K absorption edge, at $\hbar\omega = 6.555\,\text{keV}$. In addition, a pre-edge feature and two smaller peaks at higher energies reflecting the $4p$ density of states, can be seen. At energies more than $\sim 50\,\text{eV}$ away from the Mn K absorption edge, the scattering intensity is zero.

The data in Fig. 9.11 indicate the key requirement for resonant X-ray scattering measurements – tunability of the incident photon energy. Synchrotron radiation sources meet this requirement, and additionally provide high intensities and well-defined polarization states. This last property can be utilized in order to probe the polarization dependence exhibited by resonant X-ray scattering from orbital order. Working in the linear polarization basis, which is appropriate for the polarization of X-rays produced at conventional bending magnet and undulator beamlines, it can be shown that for a σ-polarized (perpendicular to the scattering plane) beam, the resonant cross section for a (100) reflection of the orbital structure shown in the inset to Fig. 9.11 results in a rotation of the polarization to a π'-polarized (parallel to the scattering

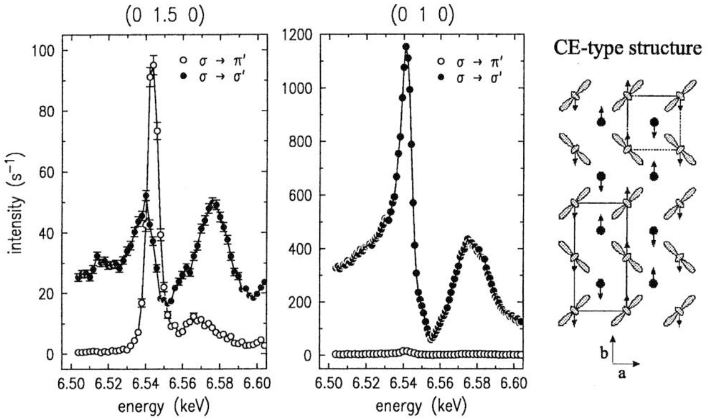

Fig. 9.12. Polarization-resolved scans of intensity versus incident photon energy of the orbital (0 1.5 0) and charge (0 1 0) reflections of the $Pr_{0.6}Ca_{0.4}MnO_3$ sample [9.46]. A schematic of the orbital, charge, and magnetic ordering of the CE-type structure is shown on the *right*. The *shaded figure*-8s represent the occupied e_g orbitals of the Mn^{3+} ions, the *filled circles* represent Mn^{4+} ions, and the *arrows* indicate the in-plane components of the magnetic ordering. The *rectangle* in the lower left shows the orbital order unit cell; the *square* in the *upper right* shows the charge order unit cell

plane) state. Since ordinary charge scattering results in a σ'-polarized beam, polarization analysis of the scattered beam is a useful technique for selecting the scattering contribution arising from orbital ordering.

In addition to this polarization dependence, resonant X-ray scattering from orbital and charge ordering also exhibits an azimuthal dependence. While nonresonant scattering is independent of the azimuthal angle (this angle characterizes the rotation of the sample about the scattering wavevector), resonant scattering from orbital and charge order exhibits characteristic azimuthal oscillations with intensity and periodicity that depend upon the type of ordering. These two properties of the scattering – the azimuthal and polarization dependences – are intimately connected in the cross section, and measurements of both can allow additional information, such as which orbital is occupied, as well as its periodicity (see, e.g., [9.52]), to be obtained.

Examples of the polarization dependence of the scattering from orbital and charge ordering are shown in Fig. 9.12. These data were collected from measurements of $Pr_{0.6}Ca_{0.4}MnO_3$, which is believed to exhibit CE-type charge and orbital ordering below a temperature of $\sim 240\,K$. In this structure, charge and orbital order peaks occur at $(0\ 2k \pm 1\ 0)$ and $(0\ 2k \pm \frac{1}{2}\ 0)$, respectively, for integer k and using orthorhombic notation.

At the (0 1.5 0) orbital order peak, the energy dependences of the two scattered polarization states are clearly different: the π' component has a peak near the Mn K absorption edge, while the σ' component decreases as the incident photon energy is tuned through the edge. This latter behavior is consistent with ordinary charge scattering, suggesting that orbital ordering is accompanied by a longitudinal lattice distortion. At the (010) charge-order peak, the scattering occurs entirely in the $\sigma \rightarrow \sigma'$ channel, to within the experiment detection limits, as expected for such charge-order scattering (see [9.46] for a discussion of the lineshapes displayed in Fig. 9.12).

In summary, resonant X-ray scattering brings new strengths, such as enhanced scattering intensity and the additional information contained in the polarization and azimuthal dependences, to studies of orbital and charge ordering. Measurements are currently being carried out at many of the synchrotron radiation facilities throughout the world – e.g., the Photon Factory in Japan, the National Synchrotron Light Source at Brookhaven National Laboratory in New York, the European Synchrotron Radiation Facility in France, and the Advanced Photon Source at Argonne National Laboratory in Illinois – in order to investigate, for example, ordering wavevectors and correlation lengths as a function of temperature and applied field. Resonant X-ray scattering complements conventional X-ray and neutron crystallographic studies, and promises to be a widely used probe of the fascinating ordering phenomena exhibited by the manganites and other materials.

We acknowledge helpful conversations with our colleagues and support by the U.S. Department of Energy, Division of Materials Science, under Contract No. DE-AC02-98CH10886.

9.4 Orbital Ordering and Resonant X-ray Scattering in Colossal Magnetoresistive Manganites

by S. Ishihara and S. Maekawa

Manganites with the perovskite crystal structure exhibit the colossal magnetoresistance (CMR) and have revived special attention on the subject. More than 40 years ago, the strong correlation between magnetism (spin) and electrical transport (charge) in these compounds was stressed, i.e., the double-exchange interaction. However, recent intensive experimental activity has revealed that this interaction cannot explain CMR. In addition to spin and charge, a Mn^{3+} ion has the orbital degree of freedom. Now it is widely accepted that a number of exotic and dramatic properties observed in manganites are ascribed to strong interplay of spin, charge, and orbital degrees of freedom of an electron, as well as the lattice. The parent compound $LaMnO_3$ of CMR manganites is a Mott insulator. Through systematic crystal and magnetic structural experiments, it has been believed for a long time that the long-range orbital ordering, where $3d_{3x^2-r^2}$ and $3d_{3y^2-r^2}$ orbitals are aligned alternately in the *ab* plane, is realized below 780 K. The A-type antiferromagnetic structure and two-dimensional-like spin-wave are predicted theoretically in this orbital ordered state [9.57]. However, the orbital degree of freedom has until recently remained hidden, since they do not couple directly to most of experimental probes.

Recent development of synchrotron light sources has changed the situation; Murakami et al. [9.58] applied the resonant X-ray scattering (RXS) technique to one of the orbital ordered CMR manganites $La_{0.5}Sr_{1.5}MnO_4$ and have successfully observed the orbital ordering. When the energy of the incident X-ray is tuned to the Mn K-edge, the X-ray scattering intensity was observed at (3/4 3/4 0), which corresponds to the periodicity of the orbital superlattice. By rotating a sample around the scattering vector, the scattering intensity shows a characteristic oscillation as a function of the rotation angle. This azimuthal-angle dependence is caused by the anisotropic distribution of the electron cloud. After the first observation, this technique was applied to several $3d$ transition-metal oxides, as well as $4f$ electron systems [9.59–9.62]. Now, through the experimental and theoretical studies of RXS, this technique is recognized as a tool for the detection of orbital ordering, equivalent to neutron and X-ray scatterings for the spin and charge degrees of freedom, respectively.

The differential scattering cross section for RXS is given by

$$\frac{\mathrm{d}^2\sigma}{\mathrm{d}\Omega \mathrm{d}\omega_f} = \sigma_{\mathrm{T}} \frac{\omega_f}{\omega_i} \sum_f \left| \sum_{\alpha\beta} (e_{k_i\lambda_i})_\alpha S_{\alpha\beta} (e_{k_f\lambda_f})_\beta \right|^2 \times \delta(\varepsilon_f + \omega_f - \varepsilon_i - \omega_i) \,, \tag{9.2}$$

where

$$S_{\alpha\beta} = \frac{m}{e^2} \sum_m \frac{\langle f | \boldsymbol{j}(\boldsymbol{k}_f)_\beta | m \rangle \langle m | \boldsymbol{j}(-\boldsymbol{k}_i)_\alpha | i \rangle}{\varepsilon_i - \varepsilon_m + \omega_i + i\eta} \,. \tag{9.3}$$

We consider the scattering of an X-ray with momentum $\boldsymbol{k}_i$, energy $\omega_i = |\boldsymbol{k}_i|$ and polarization λ_i, to $\boldsymbol{k}_f$, ω_f and λ_f. $\boldsymbol{e}_{k_l\lambda_l}$ $(l = i, f)$ is the polarization vector of the X-ray, and $|i\rangle$, $|m\rangle$ and $|f\rangle$ indicate the electronic states in the initial, intermediate, and final scattering states with energies ε_i, ε_m, and ε_f, respectively. A factor $\sigma_{\mathrm{T}} = (e^2/mc^2)^2$ is the total scattering cross section of the Thomson scattering, and η indicates an inverse of the lifetime of the intermediate scattering states. $\boldsymbol{j}(\boldsymbol{k})$ describes the electronic excitation between Mn $1s$ and $4p$ orbitals at the same site. In comparison with other experimental methods, RXS has the following advantages to detect the orbital ordering in transition-metal oxides: (1) The wavelength of the X-ray, which is tuned near the K absorption edge of a transition-metal ion, is shorter than a lattice constant of the perovskite unit cell. Thus, the diffraction experiments can be carried out for the orbital superlattice, unlike Raman scattering and optical reflection/absorption experiments. (2) S in (9.3) is a tensor with respect to the incident and scattered X-ray polarizations. On the contrary, the atomic scattering factor for the normal X-ray scattering is a scalar. (3) Tuning the X-ray energy to the absorption edge, the scattering from a specific element is identified. Note that in this case – tuning of energy to the absorption edge – the anomalous part of the atomic scattering factor dominates the scattering (the denominator in $S_{\alpha\beta}$ tends to zero). The tensor character of $S_{\alpha\beta}$ is caused by the aspherical shape of the orbitals.

One of the main issues in RXS as a probe to detect orbital ordering is its microscopic mechanism for scattering out of an orbitally ordered state, i.e. the mechanism leading to nontrivial anisotropic tensor elements of the atomic scattering factor [9.63]. This is because the incident X-ray induces the Mn $1s \to 4p$ transition, which does not access the $3d$ orbital directly. One of the promising candidates, which bring about the anisotropy of the atomic scattering factor, is the Coulomb interaction between $3d$ and $4p$ electrons [9.63–9.65]. The Coulomb interaction between electrons in the $3d_\gamma$ and $4p_\alpha$ orbitals is represented by

$$V_{\gamma\alpha} = F_0(3d,4p) + \frac{4}{35} F_2(3d,4p) \cos\left(\theta_\gamma - \frac{2\pi}{3} m_\alpha\right) , \tag{9.4}$$

where $(m_x, m_y, m_z) = (1, 2, 3)$ and $F_{0(2)}(3d, 4p)$ is the Slater integral between $3d$ and $4p$ electrons. θ_γ characterizes the wave function of the occupied orbital as $|d_\gamma\rangle = \cos(\theta_\gamma/2)|d_{3z^2-r^2}\rangle - \sin(\theta_\gamma/2)|d_{x^2-y^2}\rangle$. Due to this interaction, the three $4p$ orbitals split and the energy levels of the $4p_y$ and $4p_z$ orbitals relatively decrease at the site where the $d_{3x^2-r^2}$ orbital is occupied. As a result, the anisotropy of the atomic scattering factor $\Delta f_{\alpha\beta}(= S_{\alpha\beta})$ is brought about such that $|\Delta f_{lyy}| = |\Delta f_{lzz}| \equiv f_s$, $|\Delta f_{lxx}| \equiv f_l$ and $f_l < f_s$. At a site where $d_{3y^2-r^2}$ orbital is occupied, the conditions $|\Delta f_{lxx}| = |\Delta f_{lzz}| \equiv f_s$ and $|\Delta f_{lyy}| \equiv f_l$ are derived. A RXS intensity at the orbital superlattice reflection point is proportional to $|\Delta f_{3x^2-r^2\alpha\alpha} - \Delta f_{3y^2-r^2\alpha\alpha}|^2$ and is finite in the cases of $\alpha = x$ and y. These are consistent with the experimental results [9.58–9.62].

Another mechanism of the anisotropy of the atomic scattering factor is proposed from the viewpoint of the Jahn Teller-type lattice distortion [9.66–9.68]. Consider the $d_{3x^2-r^2}$ orbital occupied site. A lattice distortion of an O_6 octahedron along the x-direction is often observed around this Mn site. This is termed the Jahn Teller-type lattice distortion. Since the Mn O bond lengths along the x directions are elongated, the hybridization between O $2p_x$ and Mn $4p_x$ orbitals in these bonds is weaker than others. As a result, the three $4p$ orbitals split so that the energy level of the $4p_x$ orbital is lower than those of the $4p_y$ and $4p_z$ orbitals. Until now, the dominant mechanism of the anisotropy of the atomic scattering factor is still controversial. The measurements of the interference term of the scattering from the orbital, and that from the core electrons, may be available to determine relative position of the three $4p$ orbitals.

Here, we note the following experiments that suggest that the Coulomb mechanism may dominate in these compounds: (1) Neutron scattering experiments in $La_{0.88}Sr_{0.12}MnO_3$ show that the Jahn Teller-type lattice distortion exists among $291\,K > T > 145\,K$, and below 145 K, this distortion is almost quenched. On the other hand, RXS intensity is not observed in the region of $291\,K > T > 145\,K$ but appears below 145 K [9.61]. That is, the region where the Jahn Teller distortion appears and that where the RXS intensity is observed are different. (2) RXS experiments in $YTiO_3$ are recently carried out at several orbital superlattice reflection points and the polarization configurations. Through the quantitative analysis, it is concluded that the relative scattering intensities can not be explained by the atomic scattering factor expected from the lattice distortion [9.69]. To clarify the mechanism of the anisotropy of the atomic scattering factor, further studies are required.

RXS is not only applied to detection of the orbital orderings but also, recently, to that of the orbital fluctuations and excitations by utilizing the diffuse scattering and inelastic scattering techniques [9.70–9.73]. Through the rapid progress of the synchrotron radiation source and the systematic experimental and theoretical study of RXS, the origin of exotic electronic phenomena, such as CMR, observed in spin-charge-orbital coupled systems will be uncovered.

10. Charge Ordering: CE-States, Stripes, and Bi-Stripes

with T. Hotta

10.1 Early Considerations

The intuitive physical explanation for the stabilization of the fairly complex CE-state at hole density $x = 0.5$ is still under discussion. There are at least two different approaches: (i) calculations based mainly on elastic energies, à la Goodenough, and strong Coulomb repulsion, and (ii) more recent calculations where the electron kinetic energy and the J_{AF} coupling play a far more important role. Regarding (i), very early in the study of manganites, Goodenough [2.24] already discussed the CE-state, after it was reported in the neutron scattering experimental work of Wollan and Koehler [2.5]. Working at $x = 0.5$, he proposed the CE-state as a strong candidate, together with other possible competing states, including one with the charge spread as much as possible in a Na Cl lattice arrangement (sometimes known in the literature as the "Wigner crystal" state). We recall that experiments have shown [2.5] that the charge *stacks* along the z-axis in the CE-state, i.e., the charge distribution is repeated from plane to plane, instead of being shifted by one lattice spacing as in the Wigner crystal state. Using Madelung-energy considerations, Goodenough observed that all the candidate states he considered had similar electrostatic energy, but quite different elastic energies. In such analysis, the Mn O Mn bonds have different lengths depending on whether they favor an antiferromagnetic or ferromagnetic arrangement among the manganese spins. It is important to know how many couplings of one type or another are present in a given state. Qualitative considerations of this nature led to the conclusion that the CE-state is stable (note that Goodenough also considered the possible movement of the Mn^{4+} cations to relax even more the lattice strain, thus lowering the energy of the CE-state further). Note that the kinetic energy of electrons does not play an important role, and the Hund coupling is not mentioned explicitly. In Goodenough's view, elastic energies and the so-called "semicovalent" bonds (see discussion in [2.24]) are sufficient to explain the CE-state.

However, it is clear that more quantitative considerations of the competition between electrostatic, elastic, and electronic kinetic energies are still lacking in this context. Once all the effects are taken into account, will the CE-state still be the one that minimizes the energy? In addition, it is crucial to carry out truly quantitative calculations where the states *competing* with the insulating CE-state are properly considered (for example, the metallic ferromagnetic state). As shown in the experimental introduction, there are manganites that have ground states that are metallic at $x = 0.5$, and it is important to explain why this occurs. Thus, a more robust "proof" of the stability of the CE-state must necessarily involve a model Hamiltonian calculation where different half-doped states become stable in different regions of parameter space. It is only in very recent times that the stability of the CE-state has been revisited following such an elaborate approach, many years after the pioneering work of Goodenough. In fact, recent computational and mean-field studies have allowed for a comparison among competing insulating and metallic states at $x = 0.5$. The results have interesting consequences for the physics of the colossal magnetoresistance, which appears to emerge from the competition among metallic and insulating states in a sort of mixed-phase combined state, as described extensively in this book. The approach discussed mainly in this chapter is item (ii) at the beginning of this section.

10.2 Monte Carlo and Mean-Field Approximations

The CE-state, shown in detail in the introductory chapters of this book, contains "zigzag" FM arrays of t_{2g} spins, coupled antiferromagnetically perpendicular to the zigzag direction. There is a checkerboard charge ordering in the x–y plane, and charge stacking along the z-axis, together with orbital ordering involving the $3x^2$–r^2 and $3y^2$–r^2 orbitals. Can we arrive at this state using an unbiased many-body approach? To obtain the checkerboard charge distribution it appears quite natural to use a large Coulomb repulsion, beyond the on site Hubbard term. With a nearest-neighbor repulsion V of sufficiently large strength, charge certainly spreads. However, this approach does not explain the distribution of charge in the three directions, since a large V would favor a three-dimensional Wigner crystal over the CE-state found experimentally, with charge stacking along the z-axis (see next section). In addition, a very large V will quite likely make a metallic ferromagnetic state with uniform charge distribution very costly in energy. However, since this state is found experimentally in some half-doped manganites, then it must have an energy close to that of the CE-state. A large V is not the best way to explain the phenomenology of half-doped manganites!

Recently, considerable progress has been made on this front. The stability of the CE-state has been reported in realistic calculations using unbiased techniques and fairly realistic models. The calculations include Monte Carlo simulations and mean-field studies, to be reviewed here, and band insulator considerations, discussed later in this chapter. The simulations were carried out by Yunoki et al. [6.42], and typical results are shown in Fig. 10.1. In (a) the energy for a half-doped $x = 0.5$ cluster is shown at very low temperature as a function of J_{AF}, the direct coupling between localized spins. The calculation is carried out with a robust electron JT-phonon coupling, using the two-orbital model. Note the excellent agreement between Monte Carlo and mean-field results. At small values of J_{AF} the state is ferromagnetic. Part (b) of the figure shows that as a function of the strength of the electron phonon interaction λ, the FM character of the state persists if J_{AF} is small, while charge ordering and orbital ordering emerge. This is an interesting result, in that it appears that the relatively small coupling J_{AF} plays a key role for antiferromagnetism (see actual numbers in Chap. 5). In fact, upon further increasing this coupling to values of the order of a tenth of the hopping, a first-order transition is observed to a state with charge, orbital, and spin correlations characteristic of the CE-state (the first-order character of these transition will play a key role in studies to be described later in Chap. 17, when the CMR effect is analyzed). In two dimensions, the CE-state is independent of J_{AF}, since there is an equal number of FM and AF bonds (see part (a) of the figure), and eventually upon further increase of J_{AF} a truly antiferromagnetic regime is reached, perhaps already in an unrealistic region of parameter space. All these are much-welcomed results, since the calculation is unbiased, uses a fairly realistic Hamiltonian, and the results capture the competition between the FM and CE configurations, a feature that was absent in the early considerations of Goodenough. Now it is possible to address the metal insulator competition using robust techniques.

Figure 10.1b, showing the phase diagram in the $\lambda - J_{\mathrm{AF}}$ plane, is quite instructive. It shows the CE-state at intermediate values of λ, a state that can be stable even at smaller couplings (at least using non cooperative phonons; calculations with cooperative phonons were in progress when this book was being prepared, and they will clarify the true range of stability of the CE phase. Some results for cooperative JT phonons can already be found in [10.1]). At intermediate or small λs, the distribution of charge in the checkerboard is far from the extreme [3+/4+] that is often used to describe the CE-state, and it is more of the form $[(3.5+\delta)+/(3.5-\delta)+]$, with δ small. This region appears phenomenologically promising since the FM metallic orbital- and charge-disorder state is stable, and this state competes with the CE-state, at least in $La_{1-x}Ca_xMnO_3$. Could it be that the distribution of charge in the CE-state is not as extreme as widely believed? More experiments and theory are needed to clarify this important matter.

The shape of the phase diagram suggests that the coupling J_{AF} plays an even more important role than λ in the stabilization of the CE-phase. This issue will be extensively discussed in Sect. 10.5, where the zigzag chains of the CE-phase are obtained even without electron phonon coupling.

Finally, it is worth remarking that the calculations reported in this section are far from being complete. The vast parameter space of the two-orbital model has not been systematically explored. The study of cooperative phonons, as opposed to non cooperative ones as used in Fig. 10.1, is currently in progress. Differences between one, two, and three dimensions have not been studied systematically. Studies at densities different from 0.0 and 0.5 are few,

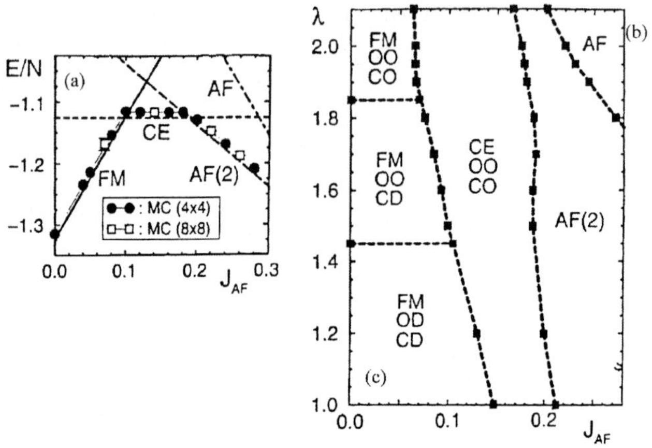

Fig. 10.1. (**a**) Energy obtained using Monte Carlo techniques on the clusters indicated (*points*), as well as mean-field approximations shown with *solid lines*. The results were obtained at density $x = 0.5$, a robust electron phonon coupling $\lambda = 1.5$, low temperature $T = 1/100$ in units of the hopping, and $J_{\mathrm{H}} = \infty$ for simplicity. The model is the two-orbital model with noncooperative Jahn Teller phonons. The FM, CE, and AF states were identified measuring spin, charge, and orbital correlations. The AF(2) is an antiferromagnetic state with a different period from the standard checkerboard pattern of up and down spins. For details, the reader can consult [6.42]. (**b**) Phase diagram in the $\lambda - J_{\mathrm{AF}}$ plane at density $x = 0.5$, obtained on 8×8 clusters. The transitions are all first order. The FM, AF, and CE states are indicated. The notation is CD = charge disorder, CO = charge order, OO = orbital order, and OD = orbital disorder. Reproduced from [6.42]

and they should be carried out to better understand the rich behavior of models for manganites. Surely, interesting results will be uncovered in the near future. In closing, we note that several groups have contributed much to the study of the CE-phase. A brief list of related work contains [10.2–10.5]. Other references will be mentioned in Sect. 10.5.

10.3 Charge Stacking

As described in the previous section, the use of a large Coulomb repulsion V to stabilize a charge-ordered state leads to a Wigner crystal without charge stacking along the z-axis, contrary to experiments. In the previous section, results of 2D simulations and mean-field approximations were described and the CE-pattern was found to be stable even without a large V. However, the issue of charge stacking needs further work. Fortunately, in [6.42] Monte Carlo studies in small 3D clusters revealed the stability of the CE-state with charge stacking. The rationale is the following: in each plane the CE-state dominates, as found in the previous section. In adjacent planes, it is convenient to stack the charge since by this procedure the J_{AF} interaction between planes is optimized the most, creating an antiferromagnetic arrangement between those planes. In such an arrangement, the energy of the 3D CE-state is now dependent on J_{AF}, decreasing as J_{AF} increases. On the other hand, the charge-shifted CE-state – adjacent planes shifted by one lattice spacing – can be shown to have no dependence on J_{AF} (equal number of FM and AF bonds). This advantage of the charge-stacked configuration survives the introduction of a small repulsion V between charges, but it is eventually destabilized by this effect. Then, to argue that the charge-stacked state is stable, it is necessary to invoke a large dielectric constant or screening effect such that V is reduced substantially from its bare value. Details can be found in [6.42]. As in the previous section, more work is still needed to fully understand charge stacking. Also note that serious studies with on site Coulombic effects are needed, although the mean-field analysis of Hotta et al. [6.23] suggests that large λ and J_{H} are qualitatively similar to large on site Coulombic interactions.

10.4 Stripes

Let us now consider hole densities other than 0.5. One of the most challenging and interesting open problems in the theoretical study of manganites are the properties of the ground states in the regime of charge ordering corresponding to hole densities different from 0.0 and 0.5. Some work in this direction has been carried out in [10.6–10.8]. The emphasis in this section will be on recent studies that suggested the presence of stripes in models for manganites [10.9]. A typical result is shown in Fig. 10.2, where the configurations that minimize the energy within the spin ferromagnetic phase at hole densities $x = 1/2, 1/3$, and $1/4$ are indicated. The result is obtained using relaxation techniques to find the configuration with the lowest energy, and for the *cooperative* Jahn–Teller phonons. The results clearly show that stripes of charge are obtained in the ground state, although a plethora of competing low energy states have also been found (implying that these stripes can likely melt very easily with temperature or disorder). The stripes are *diagonal*, and the spin does not play an important role in its stabilization, since the system is fully polarized. The driving force for stripe stabilization in Fig. 10.2 appears to be the orbital degree of freedom, which has a π-shift in its arrangement across the stripe. This has to be contrasted with the phenomenology of high-temperature superconductors, where the stripes tend to be horizontal or vertical, the spin is believed to play a key role, and a π-shift also appears across the stripes but in the antiferromagnetic order parameter (references to work in cuprates can be found at the end of this section). Note that in between the stripes the order is the same as for the $x = 0$ undoped case. Since the distance between the stripes can only be an integer number of rows, densities such as $1/2, 1/3, 1/4, \ldots$ [in general $1/(1 + \mathit{integer})$] acquire a particular relevance. It

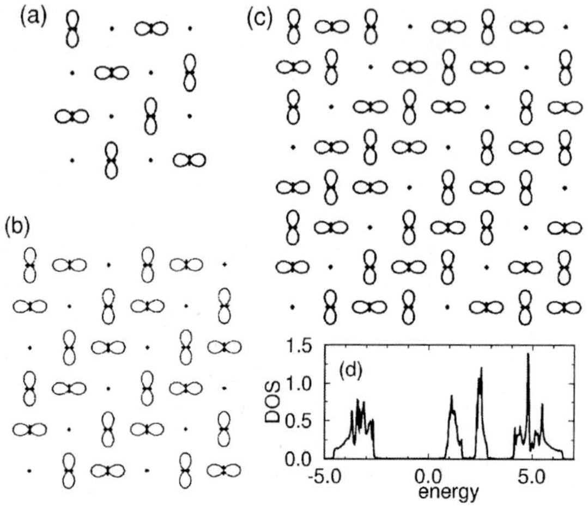

Fig. 10.2. States stabilized in simulations of the two-orbital model with robust electron JT-phonon coupling. (**a–c**), are at densities $x = 1/2$, $1/3$, and $1/4$, respectively. The spin is not shown since the systems are perfectly polarized into a FM arrangement. (**d**) is the density of states, showing the presence of a gap. Details can be found in [10.9]

is unclear what sort of state would be recovered at intermediate densities, perhaps a mixture of the two limiting states.

In the study by Hotta et al. [10.9], other interesting arrangements of spin, charge, and orbital degrees of freedom were discovered. For example, working on an 8×8 cluster at $x = 1/4$, and increasing J_{AF} to intermediate values similar to those needed to stabilize the CE-state at $x = 0.5$ (previous section), a state with diagonal stripes similar to Fig. 10.2c was observed, but with antiferromagnetic spin coupling between the zigzag chains that run perpendicular to the charge stripes. This state resembles the CE state at $x = 0.5$. Once again, other states of comparable energy were found in the simulations, and probably a glassy state is stabilized under realistic conditions.

An interesting point about the zero-temperature studies is that the charge-ordered phases need a robust electron JT-phonon coupling for stabilization, while J_{AF} is much smaller. This suggests that upon heating the system, the charge-ordering critical temperature T_{CO} could be much larger than the spin-ordering temperature (T_{N} for the CE-state). However, as described before, some of the CE-like states appear in simulations even at small λ. In this case, T_{CO} should be smaller than T_{N}, but once the zigzag paths are formed at T_{N}, due to their 1D character, any small λ will lead to charge ordering (Peierls instability argument). Then, in this case it is more likely to expect $T_{\mathrm{CO}} = T_{\mathrm{N}}$.

Recent experimental results [3.31] have been interpreted as providing evidence for stripe charge ordering in the metallic A-type antiferromagnetic phase of $Pr_{0.5}Sr_{0.5}MnO_3$, a very interesting result indeed. Other results for bilayers [15.11,15.15] can be interpreted in a similar way, and more experimental results with possible indications of stripes will be discussed in Chap. 19. Theoretical work trying to produce stripes in the A-type phase of manganites would be quite interesting. In striped regimes, the mechanism for CMR may remain the same as described before. In this case competition would be between metals and striped insulators, since in the results of Chap. 17 it will be shown that the nature of the competing phases is not very important.

Note. Stripes appear in many other contexts besides manganites. In fact, the possible existence of stripes in high-temperature superconductors and nickelates is much discussed at present. The actual experimental evidence that has led to the stripe theoretical description (see [1.3]) is based on the presence of spin incommensurability in the cuprates. In the hole-undoped case of, e.g., La_2CuO_4 the peaks in neutron scattering experiments are at momentum (π, π) in the square lattice notation, denoting an antiferromagnetic state. As the hole doping x in $La_{2-x}Sr_xCuO_4$ increases from zero, the position of the peak deviates from π by an amount δ, which is linearly proportional to x, at small x [10.10]. Other compounds, such as YBCO [10.11], also show the same effect, although it is fair to say that there is still no consensus on the universality of spin incommensurability in all cuprates. As already emphasized, theoretical studies have proposed stripes as an explanation of the spin incommensurability [1.3]. If a bundle of stripes populated by holes are aligned in one of the crystalline directions, equally spaced, with antiferromagnetism in the copper spins in the regions between stripes, and a π-shift in the AF order parameter across the stripes linking consecutive undoped domains, then a pattern of spin incommensurability similar to that found experimentally emerges [1.3]. It is still too early to say whether these stripes are indeed the explanation of the effect found in neutron scattering, but it is a serious possibility. On the other hand, in Chap. 22 it will be shown that some cuprates such as BSCCO present nanometer-scale inhomogeneities, which at least at first sight do not seem related to stripes in any obvious way. Thus, it appears that inhomogeneities are present in cuprates, but it is still unclear what is its origin. Perhaps some of the ideas proposed for inhomogeneities in manganites apply to cuprates as well.

10.5 The Band-Insulator Picture

The "band-insulator picture" for the CE-state stability, appears to be a promising concept. In the previous section, with the use of Monte Carlo techniques it was already shown that the CE-state is stable and competes with other states in a realistic region of the phase diagram. But among the many insulating charge-ordered states, it would be desirable to find a simple intuitive explanation for the stability of the CE phase. Why is this complex pattern of spin, orbitals, and charge energetically better than others?

10.5.1 A Reminder of the Band-Insulator Concept

As a preliminary exercise, consider noninteracting spinless electrons moving along a one-dimensional chain with $2N$ sites and a hopping Hamiltonian

$$H = -\sum_{\langle ij\rangle} t_{ij}(c_i^\dagger c_j + c_j^\dagger c_i)\,, \tag{10.1}$$

where $\langle ij\rangle$ denote nearest-neighbor sites, with an associated nearest-neighbor hopping t_{ij}. If all hopping amplitudes are the same, namely, for $t_{ij} = t$, the problem can be easily solved in momentum space leading to the energy dispersion $e_k = -2t\cos(k)$. Consider now the case where the hopping amplitudes corresponding to the "odd" links are still t, but those on the "even" ones are t', as shown in Fig. 10.3a. In this case, the unit cell now has two sites. A procedure to handle the problem is to define a cell index ℓ to enumerate the cells, as indicated in the figure, while an internal index $a = 1, 2$ distinguishes between the two sites of the cell. Then, the Hamiltonian can be rewritten as

$$H = -\sum_{\ell}(tc_{\ell,1}^\dagger c_{\ell,2} + t'c_{\ell,2}^\dagger c_{\ell+1,1} + \text{h.c.})\,. \tag{10.2}$$

Defining the Fourier transform through $c_{\ell,a} = \frac{1}{\sqrt{N}}\sum_p e^{ip\ell}c_{p,a}$ $(a = 1, 2)$, the Hamiltonian becomes

$$\begin{aligned} H &= -\frac{1}{N}\sum_{\ell,p,k} e^{i(k-p)\ell}(tc_{p,1}^\dagger c_{k,2} + t'c_{p,2}^\dagger c_{k,1}e^{-ik} + \text{h.c.}) \\ &= -\sum_k (tc_{k,1}^\dagger c_{k,2} + t'c_{k,2}^\dagger c_{k,1}e^{-ik} + \text{h.c.})\,, \end{aligned} \tag{10.3}$$

where in the last step we used that $\frac{1}{N}\sum_\ell e^{i(k-p)\ell} = \delta_{k,p}$. Although the Hamiltonian in k-space is diagonal in that index, a simple 2×2 extra diagonalization must be carried out in the a-index to complete the solution of the problem. The final result for the eigenenergies is

$$e_k = \pm\sqrt{t^2 + t'^2 + 2tt'\cos(k)}\,, \tag{10.4}$$

showing that there are two solutions. To recover the result of the uniform chain as a special case we must set $t = t'$, obtaining $e_k = \pm 2t\cos(k/2)$. The difference of a factor 2 in the cosine originates from the difference by the same factor in the length of the unit cell between the uniform problem and that defined for $t \neq t'$. The energy (10.4) is plotted in Fig. 10.3b. At $k = \pi$, a *gap* opens when $t \neq t'$. This originates in the well-known concept of a *band insulator*, which we learn in elementary classes of solid-state physics. The reason for this effect in the case at hand lies in the breaking of the original translational invariance properties of the system through the change in the hopping from t to t' on half the links. Then, destructive interference between incoming and reflected waves can occur at every site where the traveling electrons change their hopping amplitudes, leading to forbidden energies.

In the one-dimensional model, a possible physical reason for the alternation in the hopping between t and t' may reside in distortions of the lattice. For instance, if the ions move alternatively to left or right as sketched in Fig. 10.3c, then the hoppings will be altered and, as a result, the energy of the electrons is reduced due to the opening of the gap at $k = \pi$, at

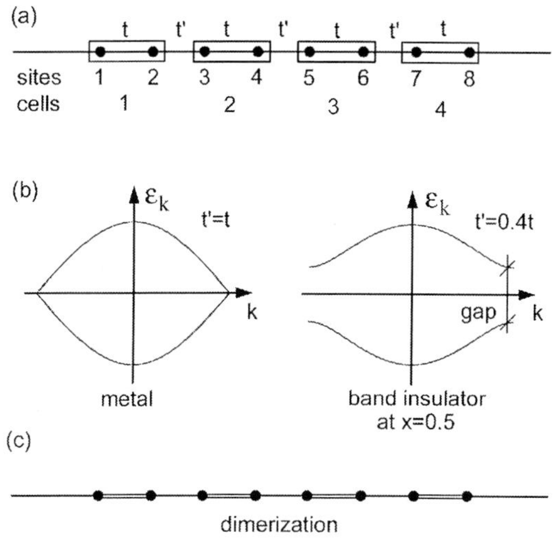

Fig. 10.3. (**a**) Schematic representation of the hopping Hamiltonian that illustrates the notion of band-insulator. (**b**) Energies for the cases $t = t'$ (metal) and $t \neq t'$ (band-insulator at half-filling). (**c**) Dimerization process that induces the pattern of hoppings used in (**a**)

least in the vicinity of that momentum. However, to properly account for this *dimerization*, we must also consider the energy penalization caused by the dimerization itself. In other words, the effective "springs" linking the ions, are stretched or expanded in the arrangement of Fig. 10.3c, leading to a quadratic penalization in the displacement from equilibrium. Once both effects are taken into account it can be shown that a spontaneous dimerization indeed occurs, at least in the one-dimensional toy problem described thus far. This is an example of a phenomenon called *Peierls instability.* The effect is not of immediate consequence to the manganites for which sign changes of the hopping amplitudes may arise by a quite different procedure, without necessarily involving displacements of the ions that form the lattice. As discussed next, these sign changes emerge from signs already contained in the original hopping amplitudes when many orbitals are considered.

10.5.2 Band-Insulator in the Context of Half-Doped Manganites

A possible explanation for the stability of the CE-state is based upon the notion that the key driving force is the *shape* of the zigzag chains that form the CE-state. Here, λ, the electron phonon coupling, does not necessarily play the most crucial role in this effect, but the argument relies on kinetic-energy considerations for zigzag chains of different forms, concluding that the one found in the CE-state is actually energetically the best at density $x = 0.5$. As shown by Hotta et al. [3.32], the zigzag ferromagnetic chains with antifer-

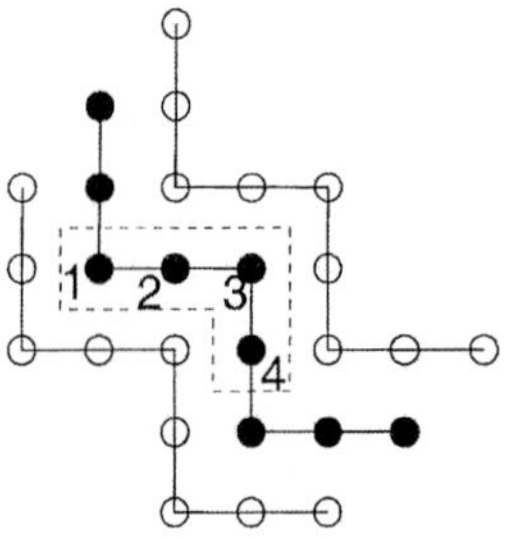

Fig. 10.4. The unit cell for the zigzag ferromagnetic chain in the half-doped CE-type phase. Note that the hopping direction is changed periodically as $\{\cdots, x, x, y, y, \cdots\}$. Reproduced from [1.9]

romagnetic interchain coupling emerge as the dominant configuration *even at* $\lambda = 0$ and large J_{H}, as a consequence of the "band-insulator" character of those chains. Similar conclusions were independently obtained by Solovyev and Terakura [10.12], and by van den Brink et al. [10.13]). The role of a band-insulator in this context was described in [8.2]. In this section, we do not prove that the band-insulator idea is necessarily the key concept to understand the CE-state in real systems. Phonons and Coulomb interaction cannot be neglected. However, the discussion here provides an alternative to ideas based on elasticity and strong Coulomb repulsion. It is only the experiments that can determine which concepts are more fundamental.

To understand the essential physics present in half-doped manganites, it is convenient to study the complex CE-structure (including the highly nontrivial stacking of charge along the z-axis) in Hamiltonians simpler than those analyzed in previous sections. Here, we start considering $H = H^{\infty}(E_{\mathrm{JT}} = U' = V = 0)$ (for a discussion of models see Chap. 5), namely, a two-orbital model where both the Coulomb interaction as well as the coupling with JT phonons are turned *off*. Only the strong Hund coupling $J_{\mathrm{H}} = \infty$ remains, with the double-exchange mechanism active. In spite of this apparent drastic simplification, we will learn that the zigzag chains of the CE-phase are already stable at this level of approximation, which is a remarkable result.

As is described in previous chapters, the CE-type antiferromagnetic phase is composed of a bundle of spin ferromagnetic chains, each with the zigzag geometry, and with antiferromagnetic interchain coupling. The reason for the stabilization of this special zigzag structure will be clarified later. For the time being, let us discuss what happens if this zigzag geometry is assumed, and how it compares with a natural competitor in energy such as a bundle of straight chains. The e_g electrons can move only along the zigzag ferromagnetic path, since the hopping perpendicular to the zigzag direction vanishes due to the standard DE mechanism and the antiferromagnetism between chains, indicating that the spin degree of freedom can be effectively neglected. Thus, the problem is reduced to the analysis of the e_g-electron motion, using noninteracting spinless particles along a one-dimensional zigzag chain. While this is conceptually simple, the math is somewhat involved. To solve this effectively "one-body" problem, the unit cell is defined as in Fig. 10.4, where the hopping amplitudes change with a period of four lattice spacings, since the hopping direction varies as $\{\cdots, x, x, y, y, \cdots\}$ along the zigzag chain. The actual hoppings along x and y are

$$t^{\boldsymbol{x}} = \frac{t_0}{4}\begin{pmatrix} 3 & -\sqrt{3} \\ -\sqrt{3} & 1 \end{pmatrix}, \quad t^{\boldsymbol{y}} = \frac{t_0}{4}\begin{pmatrix} 3 & \sqrt{3} \\ \sqrt{3} & 1 \end{pmatrix}, \tag{10.5}$$

according to the results discussed in Chap. 5. The sign difference in the non-diagonal elements of the two matrices is essential for this problem. At this point, it is useful to transform the spinless e_g-electron operators by using a unitary matrix as (Koizumi et al., [8.2]),

$$\begin{pmatrix} \alpha_{\boldsymbol{i}} \\ \beta_{\boldsymbol{i}} \end{pmatrix} = \frac{1}{\sqrt{2}}\begin{pmatrix} 1 & i \\ 1 & -i \end{pmatrix}\begin{pmatrix} c_{\boldsymbol{i}\mathrm{a}} \\ c_{\boldsymbol{i}\mathrm{b}} \end{pmatrix}. \tag{10.6}$$

After this transformation, the kinetic energy H_{kin} can be rewritten as

$$H_{\mathrm{kin}} = -t_0/2\sum_{\boldsymbol{i},\boldsymbol{a}}(\alpha_{\boldsymbol{i}}^{\dagger}\alpha_{\boldsymbol{i}+\boldsymbol{a}} + \beta_{\boldsymbol{i}}^{\dagger}\beta_{\boldsymbol{i}+\boldsymbol{a}} + \mathrm{e}^{\mathrm{i}\phi_{\boldsymbol{a}}}\alpha_{\boldsymbol{i}}^{\dagger}\beta_{\boldsymbol{i}+\boldsymbol{a}} + \mathrm{e}^{-\mathrm{i}\phi_{\boldsymbol{a}}}\beta_{\boldsymbol{i}}^{\dagger}\alpha_{\boldsymbol{i}+\boldsymbol{a}}) \tag{10.7}$$

where the phase $\phi_{\boldsymbol{a}}$ depends only on the hopping direction, and it is given by $\phi_{\boldsymbol{x}} = -\phi$, $\phi_{\boldsymbol{y}} = \phi$, with $\phi = \pi/3$. Note that the e_g electron effectively picks up a phase change when it moves between different neighboring orbitals.

To introduce the momentum k *along the zigzag chain*, the usual Bloch's phase $\mathrm{e}^{\pm\mathrm{i}k}$ is added to the hopping term (do not confuse this k with the vector momentum (k_x, k_y) corresponding to a truly two-dimensional system). Then, the problem for one electron is reduced to finding the eigenvalues of an 8×8 matrix (an electron can be in 4 possible sites per cell, and at each site either in an α or β state), given by

$$\hat{h}\begin{pmatrix} \psi_{\alpha 1} \\ \psi_{\beta 1} \\ \psi_{\alpha 2} \\ \psi_{\beta 2} \\ \psi_{\alpha 3} \\ \psi_{\beta 3} \\ \psi_{\alpha 4} \\ \psi_{\beta 4} \end{pmatrix} = \varepsilon_k \begin{pmatrix} \psi_{\alpha 1} \\ \psi_{\beta 1} \\ \psi_{\alpha 2} \\ \psi_{\beta 2} \\ \psi_{\alpha 3} \\ \psi_{\beta 3} \\ \psi_{\alpha 4} \\ \psi_{\beta 4} \end{pmatrix}, \tag{10.8}$$

where $\psi_{\alpha j}$ and $\psi_{\beta j}$ are coefficients that determine the weight of the states $\alpha_j^{\dagger}|0\rangle$ and $\beta_j^{\dagger}|0\rangle$. The Hamiltonian matrix $\hat{h}$ is given by

$$\hat{h} = -\frac{t_0}{2}\begin{pmatrix} \hat{O} & \hat{T}_k^{\boldsymbol{x}} & \hat{O} & \hat{T}_k^{\boldsymbol{y}*} \\ \hat{T}_k^{\boldsymbol{x}*} & \hat{O} & \hat{T}_k^{\boldsymbol{x}} & \hat{O} \\ \hat{O} & \hat{T}_k^{\boldsymbol{x}*} & \hat{O} & \hat{T}_k^{\boldsymbol{y}} \\ \hat{T}_k^{\boldsymbol{y}} & \hat{O} & \hat{T}_k^{\boldsymbol{y}*} & \hat{O} \end{pmatrix}. \tag{10.9}$$

Here, $\hat{O}$ is the 2×2 matrix for which all the components are zeros, the $*$ operation denotes adjoint, and the hopping matrix $\hat{T}_k^{\boldsymbol{a}}$ along the $\boldsymbol{a}$-direction is defined by

$$\hat{T}_k^{\boldsymbol{a}} = \mathrm{e}^{\mathrm{i}k} \begin{pmatrix} 1 & \mathrm{e}^{\mathrm{i}\phi_{\boldsymbol{a}}} \\ \mathrm{e}^{-\mathrm{i}\phi_{\boldsymbol{a}}} & 1 \end{pmatrix}, \tag{10.10}$$

where, again, $-\phi_{\boldsymbol{x}} = \phi_{\boldsymbol{y}} = \phi = \pi/3$.

Although it is easy to solve the present eigenvalue problem using a computer diagonalization routine, it is instructive to find the solution analytically. In the process of finding this solution, several interesting points will be clarified. First, note that there are two eigenfunctions of the 8×8 matrix that have a "localized" character (and zero energy), satisfying

$$\text{(a)}\, \hat{h}(\psi_{\alpha 2} - \mathrm{e}^{-\mathrm{i}\phi}\psi_{\beta 2}) = 0\,, \quad \text{and} \quad \text{(b)}\, \hat{h}(\psi_{\alpha 4} - \mathrm{e}^{+\mathrm{i}\phi}\psi_{\beta 4}) = 0\,. \tag{10.11}$$

This result can be checked as follows. Equation (10.11a) defines an eigenstate of the cell Hamiltonian with nonzero coefficients only for two states of the eight possible used to construct the Hamiltonian. Written in a second-quantized form, the state is $(\alpha_2^\dagger - \mathrm{e}^{\mathrm{i}\phi}\beta_2^\dagger)|0\rangle$, where "2" denotes the site according to the convention of Fig. 10.4. Transforming back to the original "c" basis, the state is $[\frac{1}{\sqrt{2}}(1 - \mathrm{e}^{\mathrm{i}\phi})c_{2\mathrm{a}}^\dagger - \frac{\mathrm{i}}{\sqrt{2}}(1 + \mathrm{e}^{\mathrm{i}\phi})c_{2\mathrm{b}}^\dagger]|0\rangle$. Transforming now to a real-space representation where $c_{2\mathrm{a}}^\dagger|0\rangle \to (x^2 - y^2)/\sqrt{2}$ and $c_{2\mathrm{b}}^\dagger|0\rangle \to (3z^2 - r^2)/\sqrt{6}$, the eigenstate can be shown to be proportional to the orbital $y^2 - z^2$ at site "2". By a similar procedure, the eigenstate defined through (10.11)(b) is proportional to $x^2 - z^2$ at site "4".

By orthogonality, the active orbitals are then $3x^2 - r^2$ and $3y^2 - r^2$ at sites 2 and 4, respectively. This suggests that, if some additional potential acts over the e_g electrons along the chains, the $(3x^2 - r^2/3y^2 - r^2)$-type orbital ordering will immediately occur due to the standard Peierls instability. This point will be discussed again later in the context of charge-orbital ordering due to the JT distortion.

Having found two eigenstates, the Hamiltonian that still needs to be diagonalized is 6×6. To find the other extended eigenstates, consider the active basis functions at sites 2 and 4, given by $(\psi_{\alpha 2} + \mathrm{e}^{-\mathrm{i}\phi}\psi_{\beta 2})/\sqrt{2}$ and $(\psi_{\alpha 4} + \mathrm{e}^{+\mathrm{i}\phi}\psi_{\beta 4})/\sqrt{2}$, respectively. After some algebra, two 3×3 block Hamiltonians $\tilde{h}^{\pm}$ can be constructed using the new basis functions defined as

$$\Phi_1^{\pm} = (1/2)[(\mathrm{e}^{-\mathrm{i}\phi/2}\psi_{\alpha 4} + \mathrm{e}^{\mathrm{i}\phi/2}\psi_{\beta 4}) \pm (\mathrm{e}^{\mathrm{i}\phi/2}\psi_{\alpha 2} + \mathrm{e}^{-\mathrm{i}\phi/2}\psi_{\beta 2})]\,, \tag{10.12}$$

$$\Phi_2^{\pm} = (\psi_{\alpha 1} \pm \psi_{\beta 3})/\sqrt{2}\,, \quad \text{and} \quad \Phi_3^{\pm} = (\psi_{\beta 1} \pm \psi_{\alpha 3})/\sqrt{2}\,. \tag{10.13}$$

The block Hamiltonian defined using the "+" states is the one that includes the ground state. The Hamiltonian in this subspace is

$$\tilde{h}^{+} \begin{pmatrix} \Phi_1^{+} \\ \Phi_2^{+} \\ \Phi_3^{+} \end{pmatrix} = \varepsilon_k^{+} \begin{pmatrix} \Phi_1^{+} \\ \Phi_2^{+} \\ \Phi_3^{+} \end{pmatrix}, \tag{10.14}$$

where the 3×3 matrix $\tilde{h}^{+}$ is given by

$$\tilde{h}^{+} = -\sqrt{2}t_0 \begin{pmatrix} 0 & \cos k_- & \cos k_+ \\ \cos k_- & 0 & 0 \\ \cos k_+ & 0 & 0 \end{pmatrix}, \tag{10.15}$$

with $k_\pm = k \pm \phi/2$. By solving this eigenvalue problem, it can be easily shown that the eigenenergies are

$$\varepsilon_k^+ = 0\,, \quad \pm t_0\sqrt{2+\cos(2k)}\,. \tag{10.16}$$

Here note that the momentum k is restricted to the reduced zone, $-\pi/4 \le k \le \pi/4$. As expected, the lowest-energy band has a minimum at $k = 0$, indicating that this block correctly includes the ground state of the one e_g-electron problem. The block Hamiltonian for the other sector is given by

$$\tilde{h}^- \begin{pmatrix} \Phi_1^- \\ \Phi_2^- \\ \Phi_3^- \end{pmatrix} = \varepsilon_k^- \begin{pmatrix} \Phi_1^- \\ \Phi_2^- \\ \Phi_3^- \end{pmatrix}, \tag{10.17}$$

with

$$\tilde{h}^- = -\sqrt{2}t_0 \begin{pmatrix} 0 & \mathrm{i}\sin k_- & \mathrm{i}\sin k_+ \\ -\mathrm{i}\sin k_- & 0 & 0 \\ -\mathrm{i}\sin k_+ & 0 & 0 \end{pmatrix}. \tag{10.18}$$

The eigenenergies are given by

$$\varepsilon_k^- = 0\,, \quad \pm t_0\sqrt{2-\cos(2k)}\,. \tag{10.19}$$

Summarizing, we obtain the following eight eigenenergies

$$\varepsilon_k = 0\,, \quad \pm t_0\sqrt{2+\cos(2k)}\,, \quad \pm t_0\sqrt{2-\cos(2k)}\,, \tag{10.20}$$

where the zero-energy state exhibits four-fold degeneracy.

The band structure is shown in Fig. 10.5. The most remarkable feature is that the system at quarter filling, i.e., $x = 0.5$, is *band insulating*, with a bandgap of value t_0. This is the key result of this section. Below, we argue that other shapes of the zigzag chain lead to higher energies. The opening of the gap reduces the energy of the system in this case. The band-insulating state, without interactions between the electrons moving along the zigzag chains, is caused by the phase difference between $t_{\mu\nu}^{\boldsymbol{x}}$ and $t_{\mu\nu}^{\boldsymbol{y}}$ ($\mu \neq \nu$). Intuitively, such a band-insulator originates in the presence of a standing-wave state due to the interference between two traveling waves running along the x- and y-directions. In this interference picture, the nodes of the wave function exist on the "corner" of the zigzag structure, and the probability amplitude becomes larger in the "straight" segment of the path. Thus, based on this band-insulating phase even a weak potential can produce the proper charge and orbital ordering. Since t_0 is at least of the order of 1000 K, this state is very robust. If a weak potential is included into such an insulating phase, the system maintains its "insulating" properties, and a modulation in the orbital density will appear.

Let us now discuss the stabilization of the zigzag structure in the CE-type phase. Although it is true that the zigzag one-dimensional spin ferromagnetic chain has a large band gap, this fact alone does not guarantee that this band-insulating phase is the lowest-energy state. To prove that the phase composed

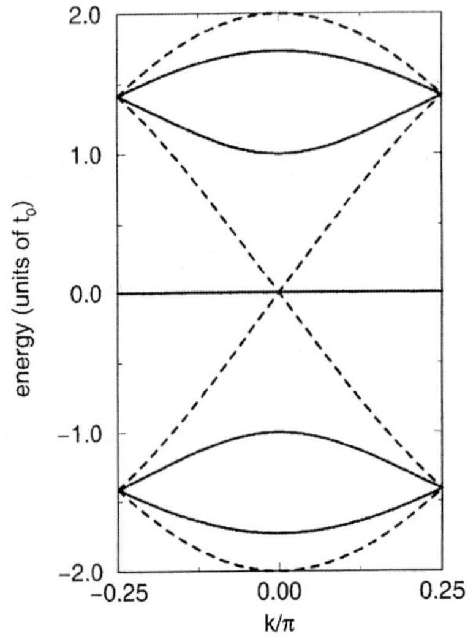

Fig. 10.5. Band structure for the zigzag 1D chain (*solid curve*) in the reduced zone $-\pi/4 \le k \le \pi/4$. For reference, the band structure $-2t_0 \cos k$ for the straight 1D path is also shown (*broken curve*). The line at zero energy indicates the four-fold degenerate flat-band present for the zigzag 1D path. From [1.9]

of these zigzag chains is truly the ground state, at least the following issues should be clarified: (i) Does this zigzag structure have the lowest energy compared to other zigzag paths with the same four-sites periodicity and compared with the straight path? (ii) Does the periodicity with four lattice spacings produce the *global* ground state? In other words, can zigzag structures with another periodicity reduce the energy? (iii) Is the energy of the zigzag phase with its particular arrangement of spins lower than uniformly ferromagnetic or antiferromagnetic phases? These points have been already clarified [3.32], and a summary of the main ideas is described below.

The first point can be checked by comparing directly the energies for all possible zigzag structures with the periodicity of four lattice spacings. Due to translational invariance, there exist four types of zigzag states that are classified by the sequence of the hopping directions, namely, $\{x, x, x, x\}$, $\{x, y, x, y\}$, $\{x, x, x, y\}$, and $\{x, x, y, y\}$. For the doping of interest, which is "quarter-filling" if one recalls there are two orbitals per site or $x = 0.5$ in a more standard notation, explicit calculations show that the zigzag pattern denoted by $\{x, x, y, y\}$ has the lowest energy [3.32]. This result can be understood as follows. The state characterized by $\{x, x, x, x\}$ is, of course, the one-dimensional metal with a dispersion relation given simply by $-2t_0 \cos k$. Note that in this straight ferromagnetic chain, the active orbital at every site is $3x^2 - r^2$, if the hopping direction is restricted to be along the x-direction. As emphasized in the discussion above, a periodic change in the hopping amplitude produces a band gap, indicating that metallic states have energies *higher* than band-insulating states. Among the latter, the state with the largest bandgap will likely be the ground state. Several calculations at quarter-filling show that the zigzag state with $\{x, x, y, y\}$ has the lowest energy. This result can also be deduced based on the interference effect. As argued above, due to interference between the two traveling waves along the x- and y-directions, a standing-wave state occurs with nodes at the corner sites, and a large probability amplitude at the sites included in the straight segments. To contain two e_g electrons in the unit cell with four lattice spacings, at least two sites in the straight segment are needed. Moreover, as long as no external potential is applied the nodes are distributed with equal spacing in the wave function. Thus, it is clear that the lowest-energy state corresponds to the zigzag structure with $\{x, x, y, y\}$.

As for the second point (ii) regarding the periodicity, it is difficult to carry out a direct comparison among the energies for all possible states, since there are infinite possibilities for the combinations of hopping directions. Koizumi et al. [8.2] developed an interesting calculation that makes the pattern of the CE-state the most reasonable energetically. They based their considerations on the use of a one-dimensional e_g-electron system with a phase factor ξ added to the hopping process, similar to the phases that appear in the realistic model due to the population of different orbitals along the chains, as discussed before. They assumed that this phase is a uniform smooth twist with a period of M lattice spacings, and solve the problem exactly. From the energies obtained in the calculation, they showed that the optimal periodicity is 2 for the half-filled situation considered here (for details, see the review [1.9]). The unconvinced reader can try solving exactly a variety of chains of noninteracting spinless electrons with different shapes by a brute force approach. The result is that the chain with period $M = 2$ is the one that lowers the energy at $x = 0.5$.

To further elaborate on the stability of the CE state (part (iii) above), it is necessary to include the effect of the magnetic coupling between adjacent t_{2g} spins. The appearance of the antiferromagnetic phase with the zigzag geometry can be understood as due to the competition between the kinetic energy of e_g electrons and the magnetic energy gain of t_{2g} spins. Namely, if $J_{\rm AF}$ is very small, for instance equal to zero, the uniform ferromagnetic phase best optimizes the kinetic energy of the e_g electrons. On the other hand, when $J_{\rm AF}$ is as large as t_0, the system stabilizes G-type antiferromagnetism to exploit the magnetic energy of the t_{2g} spins. For intermediate values of $J_{\rm AF}$, the phase with a zigzag structure takes advantage, at least partially, of both interactions. The e_g electrons can move easily along the zigzag chain with alignment of t_{2g} spins, optimizing the kinetic energy and, at the same time, there is a magnetic energy gain due to the antiferromagnetic coupling between adjacent zigzag chains. Overall, in the 2D CE-state the number of ferromagnetic and antiferromagnetic nearest-neighbor spins is the same. In Monte Carlo simulations and the mean-field approximation, a finite "window" in $J_{\rm AF}$ in which the zigzag AFM phase is stabilized is found around $J_{\rm AF} \approx 0.1t_0$.

In summary, at half-filling, the zigzag chains of the CE-type AFM phase can be stabilized even *without* Coulombic and/or JT-phononic interactions, using only large Hund and finite J_{AF} couplings. Of course, the neglected interactions are still needed to reproduce the proper charge and orbital ordering but, as already mentioned in the above discussion. Due to the special geometry of the one-dimensional zigzag ferromagnetic chain, it is easy to imagine that the checkerboard-type charge ordering and $(3x^2 - r^2/3y^2 - r^2)$ orbital-ordering pattern will be stabilized. Furthermore, the charge confinement in the straight segment (sites 2 and 4 in Fig. 10.4), naturally leads to charge stacking along the z-axis, with stability caused by the special geometry of the zigzag structure and the influence of J_{AF}.

Thus, the complex spin-charge-orbital structure for half-doped manganites can be understood intuitively simply as the result of its band-insulating nature, a quite different approach from that followed by Goodenough [2.24] in his pioneering work. Of course, more work is needed to further clarify whether the CE-state stability emerges from kinetic energy, or purely elastic considerations. Probably a combination of both effects is at work in the real system, but one may be more dominant than the other. Further work in this area is important since recent experimental results question the presence of a 3+/4+ charge-ordered state at half-doping [10.14,10.15]. If these potentially important experimental results are confirmed, the picture of Goodenough constructed around a charge-ordered state would no longer be valid, and the band-insulator picture could acquire more relevance.

10.5.3 Bi-stripe Structure at $x > 0.5$

In the previous subsection, the discussion focused on the half-doped CE-type phase. It may be expected that similar arguments can be applied to the regime $x > 1/2$, since in the phase diagram for $La_{1-x}Ca_xMnO_3$, the antiferromagnetic phase is stable at low temperatures for $0.50 < x < 0.88$ (see experimental discussion earlier in the text). Then, let us try to extend the band-insulating picture to the density $x = 2/3$ based on the infinite Hund coupling model H^∞, and, again, without considering the JT-phononic and Coulombic interactions, as for half-filling. Density 2/3 is important due to the observation of bi-stripe structures in experiments, as already discussed in previous chapters. However, calculations already reported by Hotta et al. [3.32] in the absence of JT phonons indicate that the lowest-energy state is characterized by the *straight* path, not the zigzag one, leading to the C-type antiferromagnetic phase, which was also discussed in previous chapters. At first glance, the zigzag structure, for instance the $\{x,x,x,y,y,y\}$-type path, could be the ground state for the same reason as in the case of $x = 0.5$, namely due to the opening of a gap. However, while it is true that the state with such a zigzag structure is a band-insulator, the energy gain near the chemical potential due to the opening of the bandgap is not always the most dominant effect. In fact, even in the case of x=0.5, the energy of the *bottom* of the band for the straight-path configuration is $-2t_0$ while, for the zigzag path, is $-\sqrt{3}t_0$. Then, the bottom is penalized by the formation of zigzag paths. For $x = 1/2$, the energy gain due to the gap opening overcomes the energy difference at the bottom of the band, leading to the band-insulating ground state. However, even if a band gap opens for $x = 2/3$, the energy of the zigzag structure is not lower than that of the metallic straight-line phase. Qualitatively, this can be understood in the limit of just one electron as follows: An electron can move smoothly along the one-dimensional path if it is straight. However, if the path is zigzag, "reflection" of the wave function occurs at the corner, and then a smooth movement of one electron is no longer possible. Thus, for a *small* number of carriers, where only the bottom of the band matters, it is natural that the ground state is characterized by the straight path to optimize the kinetic energy of the e_g electrons. The band-insulator picture is expected to work better at intermediate electronic densities.

In neutron scattering experiments, a spin pattern similar to the CE-type antiferromagnetic phase has been observed (Radaelli et al. [3.2]), suggesting that the "zigzag-path" approach to the problem may still be useful. In order to stabilize a zigzag antiferromagnetic phase it is necessary to include the JT distortion, i.e., it is important to study the case $\lambda \neq 0$ (otherwise straight lines are stable, as already explained). As noted by Hotta et al. [3.32], a variety of zigzag paths could be stabilized in principle when the JT phonons are included, and their energies may be close to each other. In such a case, the classification of zigzag paths is an important issue to understand the competing "bi-stripe" vs. "Wigner-crystal" structures. The former has been proposed by Mori et al. [3.8], while the latter was claimed to be stable by Radaelli et al. [3.2]. The experimental situation is still unclear, i.e., both states appear to be very close in energy in real manganites. In the scenario of [3.32], the shape of the zigzag structure is characterized by the "winding number" w associated with the Berry-phase ξ of an e_g electron parallel-transported through Jahn Teller centers, along zigzag one-dimensional paths. The Berry phase was also discussed in Chap. 8 with mean-field approximations for the two-orbital model. The winding number is defined as

$$w = \oint \frac{\mathrm{d}\boldsymbol{r}}{2\pi} \nabla \xi . \tag{10.21}$$

This quantity can be shown to be an integer (see, e.g., Hotta et al., [8.2]). Note that the integral indicates an accumulation of the phase difference along the one-dimensional ferromagnetic path in the unit length M (the periodicity of the structure). This quantity can be shown to be equal to half the number of corners included in the unit path. Thus, the phase does not change for a straight segment and, therefore, $w = 0$ for the straight-line path.

After several attempts to include effectively the JT phonons, mainly within the mean-field approximation, it was found [3.32] that the bi-stripe phase and the Wigner crystal phase universally appear for $w = x/(1 - x)$ and $w = 1$, respectively. The winding number for the bi-stripe structure has a remarkable dependence on x, reflecting the fact that the distance between

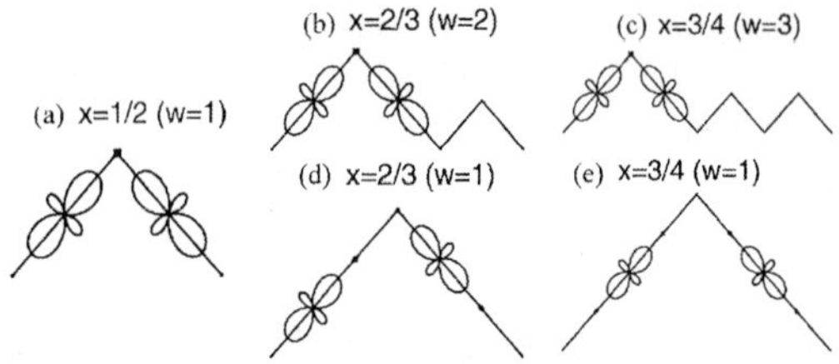

Fig. 10.6. (**a**) Path with $w = 1$ at $x = 1/2$. At each site, the orbital shape is shown with its size scaling with the orbital density, after a mean-field calculation [1.9]. (**b**) The BS-structure path with $w = 2$ at $x = 2/3$ (BS = bi-stripe). (**c**) The BS-structure path with $w = 3$ at $x = 3/4$. (**d**) The WC-structure path with $w = 1$ at $x = 2/3$ (WC = Wigner crystal. (**e**) The WC-structure path with $w = 1$ at $x = 3/4$

adjacent bi-stripes changes with x. This x dependence of the modulation vector of the lattice distortion has been observed in electron microscopy experiments (Mori et al., [3.8]). The corresponding zigzag paths with the charge and orbital ordering are shown in Fig. 10.6. In the bi-stripe structure, the charge is confined to the short straight segment, as for the CE-type structure at $x = 0.5$. On the other hand, in the Wigner-crystal structure, the straight segment includes two sites, indicating that the charge prefers to occupy either one or the other of those sites. Then, to minimize the JT energy and/or the Coulomb repulsion, the e_g electrons are distributed with equal spacing. The corresponding spin structure is shown in Fig. 10.7. A small difference in the zigzag geometry can produce a significant different in the spin structure. Following the definitions for the C- and E-type structures (see Wollan and Koehler [2.5] and Chap. 2), the bi-stripe and Wigner-crystal structures have $C_{1-x}E_x$-type and C_xE_{1-x}-type spin arrangements, respectively. Note that at $x = 1/2$, half of the plane is filled by the C-type, while another half is covered by the E-type, clearly illustrating the meaning of "CE" in the spin structure of half-doped manganites. Note that both states considered here are candidates for the ground state, but the actual calculation deciding which one wins is complicated and it may depend on details of models and couplings used. It is an interesting open problem to clarify whether bi-stripes or Wigner-crystal states are stable at $x > 1/2$. Reference [3.32] provided arguments to understand why these states may be stable, but they are likely very close in energy.

For the charge structure along the z-axis at $x = 2/3$ shown in Fig. 10.8, a remarkable feature can be observed. Due to the confinement of charge in the short straight segments, the charge *stacking* is suggested from the argument expressed above. The reason why this arrangement is preferred is that each zigzag chain must be coupled antiferromagnetically to the next chain along the z-axis to improve the magnetic portion of the energy. This cannot be achieved if we shift the zigzag chains by one lattice spacing from one plane to the next. Then, as discussed already, it is believed that the nonzero value of

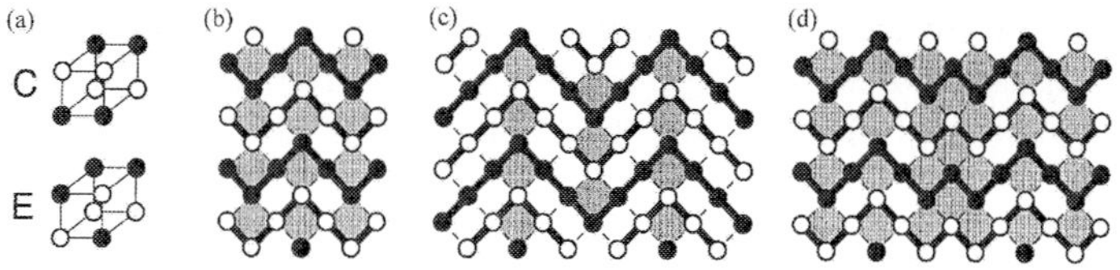

Fig. 10.7. (**a**) C- and E-type unit cell (Wollan and Koehler, [2.5]). (**b**) The spin structure in the a–b plane at $x = 1/2$. *Open* and *solid circle* denote the spin up and down, respectively. The *thick line* indicates the zigzag FM path. The *open* and *shaded squares* denote the C- and E-type unit cells. At $x = 1/2$, C-type unit cell occupies half of the 2D plane, clearly indicating the "CE"-type phase. (**c**) The spin structure at $x = 2/3$ for Wigner-crystal-type phase. Note that 66% of the 2D lattice is occupied by C-type unit cell. Thus, it is called "C_2E"-type AFM phase. (**d**) The spin structure at $x = 2/3$ for bi-stripe-type phase. Note that 33% of the 2D lattice is occupied by C-type unit cell. Thus, it is called "CE_2"-type AFM phase

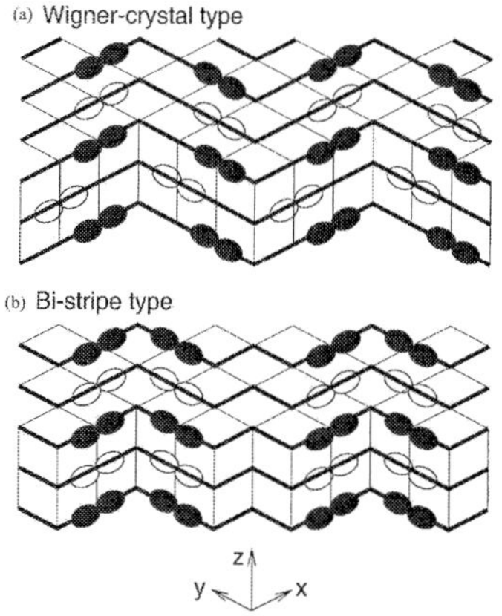

Fig. 10.8. Schematic spin, charge, and orbital ordering for (**a**) WC and (**b**) BS structures at $x = 2/3$. The *open* and *solid symbols* indicate spin up and down, respectively. The FM 1D path is denoted by the *thick line*. The empty sites denote Mn^{4+} ions, while the lobes indicate the Mn^{3+} ions in which $3x^2-r^2$ or $3y^2-r^2$ orbitals are occupied

J_{AF} plays a *key role* in favoring the formation of charge-stacked structures in manganites. On the other hand, in the Wigner-crystal-type structure, charge is not stacked, but it can be shifted by one lattice constant to avoid the Coulomb repulsion, while keeping the J_{AF} portion of the Hamiltonian satisfied. Thus, charge stacking is not natural for the Wigner crystal. Indeed, if charge stacking is actually observed in experiments at $x = 2/3$, the topological scenario suggests that the bi-stripe phase is the actual ground state. This is a key observation. The experimental discovery of charge stacking in manganites (not only at $x = 2/3$, but at $x = 1/2$ and other densities as well) clearly indicates that *mechanisms other than the strong long-range Coulomb repulsion must be at work to stabilize such a structure.* Hotta et al. [3.32] proposed that the kinetic energy of electrons in combination with the magnetic energy is the missing factor, and at half-filling robust calculations can be carried out confirming this picture. On the other hand, to firmly establish the final "winner" in the bi-stripe and Wigner-crystal competition at

$x = 2/3$, more precise experiments, as well as quantitative calculations, are needed. Note that recently Khomskii and Kugel [10.16] arrived at similar conclusions starting from the quite different point of view of considering elastic interactions.

In summary, the charge-ordered state of models for manganites is extremely interesting and nontrivial. The CE-state can be obtained from Monte Carlo simulations, together with the competing states. It is possible to attribute the existence of charge stacking in these compounds as due to the influence of J_{AF}. In addition, stripes and bi-stripes can be studied at special densities. However, plenty of work remains to be done. For example, the charge-ordered states associated with the broad charge-ordered regime of PCMO around 40% hole doping remains to be clarified. What states are stabilized in this regime? Are stripes a possibility or the state keeps its CE-like $x = 0.50$ characteristics? Hopefully these issues will be clarified in the near future.

11. Inhomogeneities in Manganites: The Case of $La_{1-x}Ca_xMnO_3$

The theoretical analysis presented in previous chapters has clearly unveiled a previously unknown feature of simple models for manganites: there is an intrinsic tendency to electronic phase separation. Once the long-range Coulombic interaction is considered, as well as the effect of pinning centers and disorder, the macroscopic phase separation becomes a microscopic inhomogeneous state. Similar ideas have been analyzed before in the context of high-temperature superconductors (see [1.3] and Chap. 1), and it is rapidly becoming a unifying theme in the study of transition-metal oxides, and even in some non-oxide compounds. In this chapter, we describe experimental results for manganites focusing on the issue of inhomogeneities in $La_{1-x}Ca_xMnO_3$ (LCMO). The reader will find that the evidence for inhomogeneities in LCMO is simply overwhelming. Figure 11.1, reproduced from Moreo et al. [1.10], shows the phase diagram of LCMO containing words reproduced from the experimental literature, at the temperature and hole density of those experiments. Clearly, in all the phases, including the ferromagnetic metallic one, and even at high temperatures, inhomogeneities have been unveiled. Experimentalists use a variety of names to describe their results, such as "droplets", "domains", "clusters", "polarons", and others, but they all refer to the same notion: the system is inhomogeneous at small length scales.

Equally convincing evidence for similar features in other manganites is presented in future chapters as well. Many of those features are in agreement with the theory presented thus far, but others will need further refinement. In Chap. 17, theoretical descriptions alternative to the electronic phase separation are presented. The presentation of results in this chapter is mainly ordered by experimental method. The chronological order of discoveries is not necessarily the order in which the results are presented below. The contributions by J.A. Fernandez-Baca and G. Papavassiliou will provide the readers with introductions and additional information about the very important neutron scattering and NMR techniques, widely used in manganites and other compounds. Another contribution by J.J. Neumeier and A.L. Cornelius illustrates the use of heat-capacity measurements to study phase separation in Mn oxides.

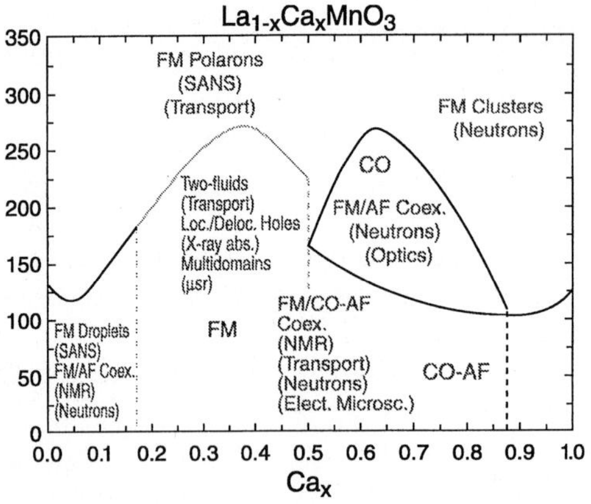

Fig. 11.1. Schematic phase diagram of $La_{1-x}Ca_xMnO_3$, from Moreo et al. [1.10]. The words added refer to statements reproduced from the experimental literature, at the density and temperature of those experiments, as well as the technique used (NMR, neutrons, etc.). The use of words such as droplets, domains, polarons, clusters, and others indicate a tendency toward inhomogeneous behavior, present in most of the phase diagram

11.1 Inhomogeneities in $La_{1-x}Ca_xMnO_3$

11.1.1 Brief Introduction to Neutron Scattering

by J. A. Fernandez-Baca

Neutrons have unique properties that make them a powerful probe of condensed matter. They have no charge and, unlike electromagnetic radiation or electrons, neutrons interact with atoms mainly via short-range nuclear interactions. For this reason they are weakly interacting in most materials and, thus, they penetrate matter far better than charged particles or X-rays. Also unlike other common probes, neutrons can be produced to simultaneously have wavelengths and energies that are in the ideal range for the study of condensed matter. Neutrons with wavelengths of a few angstroms, a length scale comparable to the interatomic distances in solids and liquids, are ideal for structural studies. These neutrons have energies (about 25 meV) comparable to those of thermally induced fluctuations in solids, and are a natural choice to study their dynamics. Finally, neutrons have a spin that couples strongly to electronic spins in magnetic materials. This makes them ideally suited to study the magnetic structures as well as the magnetic excitations, such as magnons. The neutron spin can also couple to nuclear spins but this interaction is much weaker and, in most cases, negligible.

Neutrons for scattering experiments can be produced in two ways: by fission in a nuclear reactor, or by a process called spallation at an accelerator-based facility. The term "spallation" describes the process that occurs when an energetic beam of ions (generally protons) strikes a heavy metal target, knocking neutrons from it. In both processes, fission and spallation, high-energy (i.e., MeV) neutrons are produced. These are subsequently slowed-down in a moderator by multiple collisions until they approach thermal equilibrium. Neutrons in equilibrium with a moderator at room temperature are referred to as "thermal" neutrons, while those thermalized in moderators at low temperatures (like liquid hydrogen at approximately 20 K) are referred to as "cold" neutrons. Reactors produce a continuous stream of neutrons and, for most experiments, only those in a narrow wavelength band are selected by a "monochromator" crystal. Although a continuous supply of neutrons can be produced by spallation, most of the existing sources produce pulsed neutron beams. In these cases "monochromator" crystals are not normally used and the energy of each neutron is determined from its time-of-flight.

The foundations of the neutron scattering technique were developed in the 1940s and 1950s, when researchers in the United States and Canada gained access to the significant neutron fluxes that the first reactors provided. It was between 1948 and 1955, at Oak Ridge's Graphite Reactor, that E.O. Wollan and C. Shull laid the basis of modern neutron diffraction [11.1]. In particular, inspired by the classic papers of Halpern and Johnston [11.2], they investigated systematically the scattering of neutrons by magnetic materials and demonstrated the existence of the ferrimagnetic state in Fe_3O_4 [11.3]. In this state, first proposed by Néel [11.4], the ferrous and ferric ions couple antiferromagnetically but the balancing of parallel and antiparallel moments is not complete. They extensively studied the scattering of neutrons by paramagnetic and antiferromagnetic materials [11.5], as well as by ferro- and ferrimagnets. They also determined the magnetic structures of various alloys, investigated the magnetic density of ferromagnetic materials, and the critical scattering near the Curie point of Fe. In the mid-1950s Bertram Brockhouse developed the inelastic neutron scattering technique (the process in which neutrons gain or lose energy) at Chalk River's NRX reactor. He designed the first triple-axis spectrometer to analyze the spectrum of neutrons after they have been scattered. After Brockhouse's work inelastic neutron scattering became a useful tool to study elementary excitations in solids, such us phonons and magnons, diffusion in liquids, and fluctuations in magnetic materials. In 1994, Clifford Shull and Bertram Brockhouse were awarded the Nobel Prize in Physics for their "pioneering contributions to the development of neutron scattering techniques for studies of condensed matter" (Wollan died in 1967 and could not share this honor).

From the early neutron scattering work, of particular interest to the readers of this book is the study of the magnetic structures of $La_{1-x}Ca_xMnO_3$ by E.O. Wollan and W.C. Koehler in the 1950s. The now familiar magnetic structures of the perovskite manganese oxides (A- and CE-type antiferromagnetic, ferromagnetic, etc.) were established in this study published in 1955 in the Physical Review [11.6]. This work, which has become one of the most frequently cited papers in the CMR field, was performed on polycrystalline specimens at Oak Ridge's Graphite Reactor.

Since the early days at Oak Ridge and Chalk River, significant advances have been made in the development of intense neutron sources, as well as in neutron scattering instrumentation. Neutron scattering has now become a very popular technique to study the statics (structures) and dynamics (excitations) of condensed-matter systems throughout the world. Since the construc-

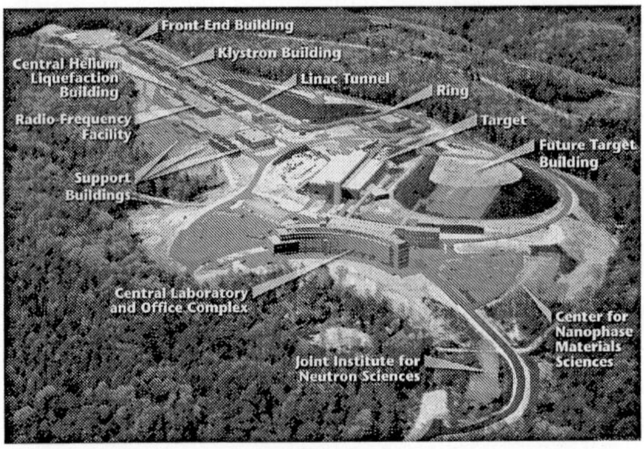

Fig. 11.2. Artist conception of the spallation neutron source (SNS). This facility is being constructed in Oak Ridge (USA) by the US Department of Energy. When the construction of the SNS is completed in the year 2006, it will provide the world's most intense pulsed neutron beams for scientific research

tion of reactors or spallation sources requires enormous investments, most of the neutron scattering experiments are concentrated around the major user facilities such as the Institut Laue-Langevin (Grenoble, France), the ISIS spallation source (Rutherford Appleton Laboratory, Didcot, UK), the NIST Center for Neutron Research (NIST, Gaithersburg, USA), the High Flux Isotope Reactor (Oak Ridge National Laboratory, Oak Ridge, USA) and the JRR-3m reactor (JAERI, Tokai, Japan). New-generation neutron sources are also being planned in Europe (the European Spallation Source) and in Japan (the High Intensity Proton Accelerator Facility), and the Spallation Neutron Source (SNS) is being constructed in Oak Ridge (USA) (Fig. 11.2).

Considering the special focus of this book, in this section we will concentrate on the use of neutron scattering to study the dynamics of spins in a solid. As this has been reviewed extensively elsewhere [11.7–11.9] we will only summarize the most relevant features of this technique. We start with an incident beam of neutrons of linear momentum $\hbar\mathbf{k}_\mathrm{i}$ and energy $\hbar^2 {k_\mathrm{i}}^2 2m_\mathrm{n}$ (the subscript "i" stands for "initial" or prior to the scattering process, and m_n is the neutron mass) that strikes a sample of the material we want to study. The energy and momentum of this beam are selected by a "monochromator" crystal (in a reactor) or by time-of-flight techniques (in a spallation source). As a result of the interaction of the incident neutrons with the nuclei and the electronic spins of the sample, the neutrons exchange momentum and energy with the sample and are scattered. By measuring the momentum and energy spectra of the scattered neutrons we can obtain important information about the sample's atomic structure and dynamics. In a typical experiment we can measure both the momentum $\hbar\mathbf{k}_\mathrm{f}$ and energy $\hbar^2 {k_\mathrm{f}}^2 2m_\mathrm{n}$ of the scattered neutrons (here the subscript "f" means "final" or after the scattering process), and we can then define the scattering wavevector and energy transfer $\hbar\omega$ of this process as:

$$\boldsymbol{Q} = \boldsymbol{k}_{\mathrm{i}} - \boldsymbol{k}_{\mathrm{f}}\,, \quad \hbar\omega = \frac{\hbar^2}{2m_{\mathrm{n}}}\left({k_{\mathrm{i}}}^2 - {k_{\mathrm{f}}}^2\right), \tag{11.1}$$

which denote the momentum and energy imparted to the system under study by the incident neutrons. *Elastic* neutron scattering is the process in which there is no energy imparted to the sample at the atomic level ($\hbar\omega = 0$). *Inelastic* scattering is the process that results in energy transfer to the sample ($\hbar\omega \neq 0$), as in the creation or annihilation of an elementary excitation. Inelastic neutron scattering is the ideal probe of the dynamical behavior of magnetic systems because it provides a direct measurement of the imaginary part of the generalized susceptibility $\chi(\boldsymbol{Q},\omega)$, which contains all the information of the spin correlations of the system under study. The partial differential cross section for the scattering of unpolarized neutrons from a system of electron spins, located at the coordinates $\boldsymbol{r}_n$, is [11.7]:

$$\frac{\mathrm{d}^2\sigma}{\mathrm{d}\Omega\mathrm{d}\omega} = \left(\frac{\gamma e^2}{m_e c^2}\right)^2 \left[\frac{1}{2}gf(\boldsymbol{Q})\right]^2 \frac{k_{\mathrm{f}}}{k_{\mathrm{i}}} \sum_{\alpha,\beta}(\delta_{\alpha\beta} - \hat{Q}_\alpha\hat{Q}_\beta)S_{\alpha\beta}(\boldsymbol{Q},\omega)\,, \tag{11.2}$$

where the scattering function $S_{\alpha\beta}(\boldsymbol{Q},\omega)$ is the space and time Fourier transform of the time-dependent spin-pair correlation function, i.e.,

$$S_{\alpha\beta}(\boldsymbol{Q},\omega) = \int \frac{\mathrm{d}t}{2\pi}\mathrm{e}^{-\mathrm{i}\omega t}\sum_{m,n}\exp[\mathrm{i}\boldsymbol{Q}\cdot(\boldsymbol{r}_m - \boldsymbol{r}_n)]\langle S_\alpha(\boldsymbol{r}_m,0)S_\beta(\boldsymbol{r}_n,t)\rangle\,. \tag{11.3}$$

In the above equations $\left(\frac{\gamma e^2}{m_e c^2}\right)^2 = 0.291$ barns is the coupling constant of the neutron to unpaired electron spins; g is the Landé factor and $f(\boldsymbol{Q})$ is the magnetic form factor, which is related to the Fourier transform of the radial part of the electronic wave function. $\boldsymbol{Q}$ and $\hbar\omega$ are the scattering wavevector and the energy transfer defined above (11.1), and $\hat{\boldsymbol{Q}} = \boldsymbol{Q}/|\boldsymbol{Q}|$. The indices α and β denote Cartesian coordinates, thus $\hat{Q}_\alpha$ and S_α are the α components of the unit vector $\hat{\boldsymbol{Q}}$ and the magnetic moment $\boldsymbol{S}$.

The essential information contained in (11.2) is that neutrons scattered with momentum and energy transfer $\boldsymbol{Q}$ and $\hbar\omega$ directly probe a single Fourier component of the spin-pair correlation function. As we are mainly concerned with dynamical processes, we will use $S_{\alpha\beta}(\boldsymbol{Q},\omega)$ to denote the inelastic ($\hbar\omega \neq 0$) part of the scattering function, unless specified otherwise. By using the fluctuation-dissipation theorem we can relate $S_{\alpha\beta}(\boldsymbol{Q},\omega)$ to the imaginary part of the generalized susceptibility:

$$S_{\alpha\beta}(\boldsymbol{Q},\omega) = [n(\omega)+1]\mathrm{Im}\chi_{\alpha\beta}(\boldsymbol{Q},\omega)\,, \tag{11.4}$$

where $n(\omega) = [\exp(\hbar\omega/k_{\mathrm{B}}T) - 1]^{-1}$ is the Bose thermal factor. As noted above, the neutron scattering technique provides a direct measurement of $\mathrm{Im}\chi_{\alpha\beta}(\boldsymbol{Q},\omega)$. Another important feature is that $S_{\alpha\beta}(\boldsymbol{Q},\omega)$ can also be expressed in terms of the isothermal susceptibility $\chi_{\alpha\beta}(\boldsymbol{Q})$ and the spectral weight function $F_{\alpha\beta}(\boldsymbol{Q},\omega)$, i.e.,

$$S_{\alpha\beta}(\boldsymbol{Q},\omega) = [n(\omega)+1]\omega\chi_{\alpha\beta}(\boldsymbol{Q})F_{\alpha\beta}(\boldsymbol{Q},\omega)\,. \tag{11.5}$$

The isothermal susceptibility $\chi_{\alpha\beta}(\boldsymbol{Q})$ is the equilibrium response function of the α-th component of the magnetization to an external magnetic field in the direction β with a spatial dependence of the form $H[\exp(\mathrm{i}\boldsymbol{Q}\cdot\boldsymbol{r})]$. The spectral weight function $F_{\alpha\beta}(\boldsymbol{Q},\omega)$ is the normalized temporal Fourier transform of the magnetization response when the same external field is suddenly switched off. This function, describing the shape of the scattering as a function of energy, satisfies the normalization condition

$$\int_{-\infty}^{\infty} F_{\alpha\beta}(\boldsymbol{Q},\omega)\mathrm{d}\omega = 1\,. \tag{11.6}$$

When the total z-component of the spin is a constant of the motion, as in the case of a Heisenberg ferromagnet, the terms with $\alpha \neq \beta$ in the cross section vanish and (11.2) simplifies to:

$$\frac{\mathrm{d}^2\sigma}{\mathrm{d}\Omega\mathrm{d}\omega} = \left(\frac{\gamma e^2}{m_e c^2}\right)^2 \left[\frac{1}{2}gf(\boldsymbol{Q})\right]^2 \frac{k_\mathrm{f}}{k_\mathrm{i}} \times\left\{\left(1+\hat{Q}_z^{\,2}\right)S_{xx}(\boldsymbol{Q},\omega)+\left(1-\hat{Q}_z^{\,2}\right)S_{zz}(\boldsymbol{Q},\omega)\right\}. \tag{11.7}$$

This cross section is the sum of two terms, which we denote as transverse (i.e., perpendicular to the z-axis) and longitudinal (i.e., along the z-axis). The longitudinal term is expected to lead to elastic or quasielastic scattering only. The transverse term corresponds to the creation and annihilation of magnons or spin waves. This is the term in which we are mostly interested. In particular, the transverse term of the scattering function is

$$S_{xx}(\boldsymbol{Q},\omega) = [n(\omega)+1]\omega\chi_{\alpha\beta}(\boldsymbol{Q})F_{xx}(\boldsymbol{Q},\omega)\,. \tag{11.8}$$

The exact form of the shape of the spectral weight function $F_{xx}(\boldsymbol{Q},\omega)$ depends on the specific processes involved. For a periodic lattice at low temperatures, where the spin-wave interactions are negligible, the spin waves are approximate eigenstates of the system and we have the familiar result

$$F_{xx}(\boldsymbol{Q},\omega) = \frac{1}{2}\left[\delta(\hbar\omega-\hbar\omega_{\boldsymbol{Q}})+\delta(\hbar\omega+\hbar\omega_{\boldsymbol{Q}})\right], \tag{11.9}$$

where δ is Dirac's delta function and $\hbar\omega_{\boldsymbol{Q}}$ is the spin-wave energy. The two terms in this spectral weight function correspond to the creation and annihilation of spin-waves. The presence of disorder or interactions between spin waves reduces the lifetime of these excitations. This has to be taken into account in the spectral weight function. For this purpose $F_{xx}(\boldsymbol{Q},\omega)$ is often approximated by a double Lorentzian function or other forms.

The static part of the scattering function (11.3) can be written as:

$$S_{\alpha\beta}(\boldsymbol{Q},\omega=0) = \delta(\omega)S_{\alpha\beta}(\boldsymbol{Q})\,, \tag{11.10}$$

where $S_{\alpha\beta}(\boldsymbol{Q})$ is the "infinite-time" spin-pair correlation function

$$S_{\alpha\beta}(\boldsymbol{Q}) = \sum_{m,n} \mathrm{e}^{[\mathrm{i}\boldsymbol{Q}\cdot(\boldsymbol{r}_m-\boldsymbol{r}_n)]}\langle S_\alpha(\boldsymbol{r}_m,0)S_\beta(\boldsymbol{r}_n,\infty)\rangle\,. \tag{11.11}$$

This corresponds to the (static) magnetic structure of the system and gives rise to the magnetic Bragg peaks. Here, the use of times 0 and ∞ provides a definition of static correlations, an alternative to the equal-time correlations more frequently used in textbooks.

Another important relationship can be derived by performing the energy integration of (11.5) and using the normalization of the spectral weight function $F_{\alpha\beta}(\boldsymbol{Q},\omega)$ (11.6):

$$\int_{-\infty}^{\infty} \frac{1}{[n(\omega)+1]\omega} S_{\alpha\beta}(\boldsymbol{Q},\omega)\,\mathrm{d}\omega = \chi_{\alpha\beta}(\boldsymbol{Q})\,. \tag{11.12}$$

Of particular interest is the condition when the "inelasticity" is small (i.e., when only small energies are involved) and $1/([n(\omega)+1]\omega) \to \hbar/k_{\mathrm{B}}T$, in this case:

$$\hbar \int_{-\infty}^{\infty} S_{\alpha\beta}(\boldsymbol{Q},\omega)\mathrm{d}\omega = k_{\mathrm{B}}T\chi_{\alpha\beta}(\boldsymbol{Q})\,. \tag{11.13}$$

In this limit (often called the *quasistatic* approximation) the energy integrated scattering function is proportional to the isothermal susceptibility $\chi_{\alpha\beta}(\boldsymbol{Q})$. In practice, this energy integration can be approximately realized in an experiment with no energy analysis, in which one measures all the neutrons scattered at a fixed angle. This technique is frequently used to determine the "Q-dependent" susceptibility $\chi_{\alpha\beta}(\boldsymbol{Q})$ in magnetic systems, and has also been used in the CMR manganites. At this point it is important to note that the scattering wavevector $\boldsymbol{Q}$ defined in (11.1) can be expressed as $\boldsymbol{Q} = \boldsymbol{\tau} + \boldsymbol{q}$, where $\boldsymbol{\tau}$ is a reciprocal lattice vector and $\boldsymbol{q}$ is the reduced wavevector (see Fig. 11.3). This is a consequence of Bloch's theorem which requires that the energy of any (wave-like) eigenstate in a periodic lattice must be periodic in $\boldsymbol{\tau}$ and, thus, the relevant wavevector is $\boldsymbol{q}$. Then, what is measured in this fixed-angle technique (with no energy analysis) is $\chi_{\alpha\beta}(\boldsymbol{q})$, which is frequently approximated by the Ornstein Zernike form:

$$S_{\alpha\beta}(\boldsymbol{q}) \propto \frac{1}{q^2+\kappa^2}\,, \tag{11.14}$$

where $\kappa = 1/\xi$ and ξ is the "correlation length" of the electronic spins. It is important to perform this measurement at small values of the wavevector $\boldsymbol{q}$, in the vicinity of a reciprocal lattice vector $\boldsymbol{\tau}$. Frequently the experiments are performed near $\boldsymbol{\tau} = 0$. In this case $\boldsymbol{Q} = \boldsymbol{q}$ and these become "low-Q" measurements. Current instrumentation allows the measurement of wavevectors as low as $Q = 0.001\,\text{Å}^{-1}$. This technique is often referred to as "*small-angle neutron scattering*" (SANS) because the measurement of small Q values need to be performed at small angles. The use of neutron scattering techniques with no energy analysis is much broader than the measurement of magnetic susceptibilities or measurements at low Qs. Neutron scattering studies at arbitrary values of $\boldsymbol{Q}$ with no energy analysis, are generally referred to as "*neutron diffraction*".

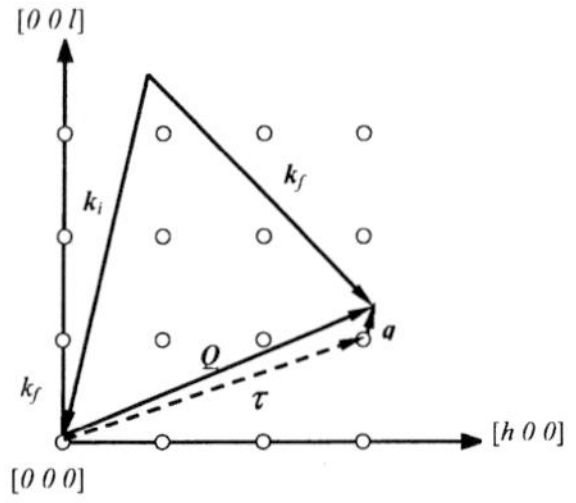

Fig. 11.3. Schematic scattering diagram. $\boldsymbol{Q}$ is the scattering wavevector, $\boldsymbol{\tau}$ is a reciprocal lattice vector, and $\boldsymbol{q}$ is the reduced wavevector

We can now discuss some practical considerations related to the neutron scattering experiments. The samples can be single crystals or polycrystalline specimens placed in a suitable environment (low/high temperature, magnetic field, high pressure, etc.). Most inelastic scattering experiments are performed on single crystals, and usually a few grams of material are needed. For example, measurement of spin-waves in $Pr_{0.63}Sr_{0.37}MnO_3$ and $Nd_{0.70}Sr_{0.30}MnO_3$ [11.10] were performed in high-quality single crystals of about 2 g. This kind of measurement, however, can also be performed on polycrystalline specimens with some limitations. The first measurements of spin-waves in $La_{0.70}Ca_{0.30}MnO_3$ [11.12] were indeed performed on polycrystalline specimens at a time when large single crystals of this compound were not available. What is measured in these cases is the average over many crystallites oriented in random directions, the only practical lattice vector is then $\boldsymbol{\tau} = 0$ and the scattering wavevector is averaged over all possible directions. This is useful only if the system under study is isotropic (or nearly isotropic). In general, neutron diffraction experiments do not require samples as large as those needed for inelastic scattering measurements, and normally a fraction of a gram is sufficient for these measurements.

Another practical consideration is the instrumental resolution. The discussions above assume "perfect" resolution, in which all the energies and wavevectors are determined exactly. In a real experiment, however, there is a finite instrumental-resolution driven by the need to integrate over a region of $(\boldsymbol{Q}, \omega)$ space in order to measure a significant signal. All measurements in $(\boldsymbol{Q}, \omega)$ space are then smeared by the instrumental resolution function $R(\boldsymbol{Q}, \omega)$, which is normally approximated by a 4-dimensional ellipsoid that defines the extent of the smearing. Thus, when we measure the scattering function for a particular point in $(\boldsymbol{Q}, \omega)$ space, what we really measure is the convolution of $S_{\alpha\beta}(\boldsymbol{Q}, \omega)$ and the resolution function:

$$I_{\alpha\beta}(\boldsymbol{Q}, \omega) = \int S_{\alpha\beta}(\boldsymbol{Q}', \omega') R(\boldsymbol{Q} - \boldsymbol{Q}', \omega - \omega') \mathrm{d}\omega' \mathrm{d}^3 Q' \,. \tag{11.15}$$

Figure 11.4 illustrates the effect of the instrumental resolution in the measurement of the low-q spin-waves in $La_{0.70}Ca_{0.30}MnO_3$ at $T = 80\,\mathrm{K}$ [11.11]. The peaks at the positive and negative side of the energy axis correspond to

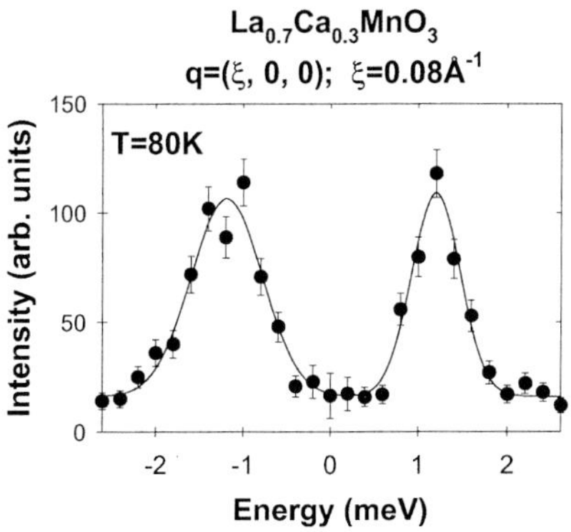

Fig. 11.4. Spin-waves in $La_{0.70}Ca_{0.30}MnO_3$ at $T = 80\,K$. This corresponds to a double Dirac's delta spectral weight function broadened by the instrumental energy resolution. The intrinsic broadening of the spin-waves is negligible. A temperature-independent non magnetic component at $\hbar\omega = 0$ has been subtracted. From [11.11]

the creation and the annihilation of spin-waves. At this low temperature the spin wave peaks are expected to have a spectral-weight function consisting of nearly Dirac's delta functions (11.9). The broadening of the peaks observed in this figure is just that of the instrumental resolution, confirming that the intrinsic broadening of spin-waves is negligible.

The finite instrumental resolution implies uncertainities in the determination of the angles and energies. Thus, when we measure $S_{\alpha\beta}(\boldsymbol{Q})$ – the "infinite-time" (i.e., static) spin-pair correlation function – we are limited to how well we can measure $\hbar\omega = 0$. If our energy resolution (uncertainty in the measurement of the energy) is $\Delta(\hbar\omega)$, the time uncertainty of our measurements is determined by the Heisenberg uncertainty principle "$\Delta(\hbar\omega).\Delta t \geq \hbar$", i.e. any process with a lifetime of $\Delta t \geq \hbar/\Delta(\hbar\omega)$ will appear to be static within the experimental energy resolution. Likewise, any length scale $\Delta x \geq 1/\Delta(Q)$ (where $\Delta(Q)$ is the wavevector Q resolution) will appear to be infinite. The values of the energy or wavevector Q resolution depend both on the instrument design and configuration and vary greatly. A typical value of the energy resolution is $\Delta(\hbar\omega) \approx 100\,\mu eV$ (at $\hbar\omega = 0$), and to this instrument any process with a lifetime greater than $\approx 7\,ps$ will appear to be static. A typical wavevector resolution in the same instrument is $\Delta Q \approx 0.002\,Å^{-1}$, which means that any length scale greater than $\approx 500\,Å$ will appear to be infinite.

ORNL is managed by UT-Batelle, LLC for the US DOE under contract DE-AC05-00 OR22725.

11.1.2 Neutron Scattering in $La_{1-x}Ca_xMnO_3$

Neutron scattering techniques have been used for the study of LCMO by several groups. Early in the analysis of manganites and shortly after the discovery of the CMR effect, Lynn et al. [11.12] provided key evidence of unconventional behavior in the ferromagnetic phase of LCMO $x = 0.33$ ($T_C = 250\,K$), at temperatures as low as 200 K. What is anomalous is the observation of a strongly field-dependent *diffusive* component (zero frequency) in the neutron spectrum below T_C, while at low temperatures the system behaves as an ideal isotropic ferromagnet. In their seminal work, Lynn et al. [11.12] concluded that the anomalous quasielastic component is associated with electrons that are localized within the FM metallic phase, in agreement with a variety of other more recent results [11.13].

Figure 11.5 shows inelastic neutron scattering spectra for LCMO $x = 0.33$ at two different temperatures [11.12]. As in Fig. 11.4, the peaks at the positive and negative sides of the energy axis correspond to the creation and annihilation of spin-wave excitations. Note that the excitation energies in Fig. 11.5 are lower than those in Fig. 11.4. The reason for the difference is that the data of Fig. 11.5 are for a different wavevector and at different temperatures from that in Fig. 11.4. It is known that ferromagnets have spin-1 excitations with a dispersion $E = Dq^2$, where D is the stiffness coefficient (the quadratic dispersion is characteristic of ferromagnets, while in antiferromagnets the relation would be linear, after a shift in momentum space from zero to the wavevector characteristic of the antiferromagnetic order). It is also known that as the temperature is raised to approach T_C (i.e., $T \to T_C$) the effects of spin-wave interactions lead to a reduction of D (renormalization of the spin-wave stiffness coefficient) in such a way that $D \to 0$ as $T \to T_C$. It is then expected that the excitation energies of Fig. 11.5 are smaller than in Fig. 11.4. The spin-wave peaks of Fig. 11.5 indeed follow a Dq^2 relation, and, as the temperature is reduced, they become the dominant feature in the spectra.

Note, however, that Fig. 11.5 shows unusual features not expected in conventional ferromagnets. There is a large peak at zero energy (not present in Fig. 11.4), believed to originate in a diffusion mechanism. Later studies by Fernandez-Baca et al. [11.10] have shown that as the value of T_C decreases, the spin dynamics become more anomalous. From the analysis of the data, Lynn et al. [11.12] conclude that the diffusion has a length scale of about 12 Å, a size that appears in other independent studies as well (to be described below). The momentum independence of the quasielastic peak shows that the associated excitation is likely to be localized. An inhomogeneous picture of LCMO near T_C with localized and delocalized electrons is compatible with the neutron scattering results [11.12]. Signs of irreversibility were observed at T_C, showing that the transition is not a conventional second-order one, it probably is weakly first order.

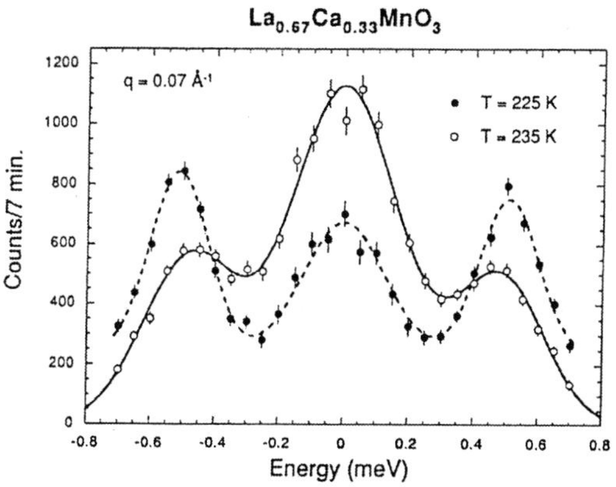

Fig. 11.5. Inelastic spectrum of LCMO $x = 0.33$, at two temperatures below T_C and fixed momentum, reproduced from Lynn et al. [11.12]. The peaks on the *left* and *right* are due to spin-waves, while the central quasielastic peak has a diffusive origin. This result is compatible with an inhomogeneous picture of manganites, even within the FM metallic phase

The results of [11.12] were found to be representative of manganites with a substantial MR effect. Fernandez-Baca et al. [11.10] studied $x = 0.30$ $Nd_{1-x}Sr_xMnO_3$, with a Curie temperature $\sim 200\,K$, and $Pr_{1-x}Sr_xMnO_3$ at $x = 0.37$ with $T_C = 300\,K$. These two compounds have a T_C below and above the ordering temperature of LCMO $x = 0.30$. It was found that the compound with the lowest T_C displayed a diffusive central peak, with characteristic length 20 Å, while the sample with the largest T_C exhibited a conventional behavior. Results for these compounds, in Fig. 11.6, show that NSMO develops a prominent central peak as $T \to T_C$. Results for LSMO with similar features will be shown in Chap. 14.

The low-energy magnons, however, show a universal behavior with similar values of D, independently of the actual T_C. Recently, it was proposed that the values of D and its doping dependence originate from a superposition of the double exchange for correlated e_g electrons and the superexchange, in a model that includes orbital degeneracy [11.14].

Note on Spin Diffusion. The analysis of the inelastic neutron scattering results discussed above relies on the concept of spin diffusion, briefly reviewed here. In a pioneering paper, Halperin and Hohenberg [11.15] developed a hydrodynamic theory of spin excitations for magnetic systems. "Hydrodynamic considerations" refer to a theory based only on conservation laws of the microscopic Hamiltonian, and the symmetry properties of the ground state – which in a spontaneously symmetry-broken state can be of lower symmetry than the Hamiltonian. Other assumptions involve the expansion of quantities in powers of the gradients and magnitudes of the deviations from equilibrium. This approach does not rely on microscopic effects at the lattice spacing level, but on the behavior at large distances. A microscopic relaxation time τ is assumed to exist. This is the time it takes to return back to local equilibrium after a disturbance is created. For an isotropic ferromagnet, the

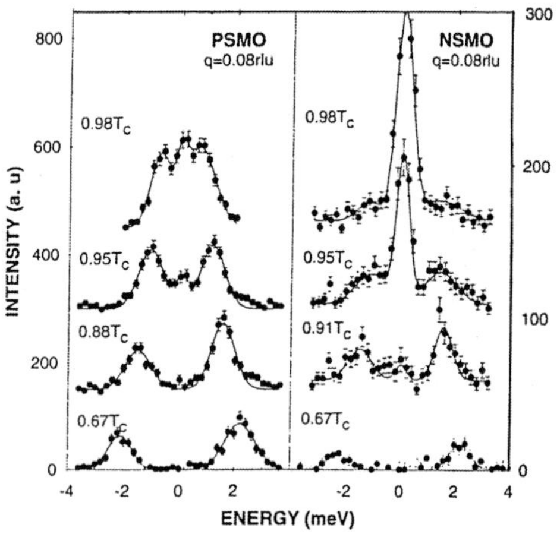

Fig. 11.6. Energy scans for (*left*) $Pr_{1-x}Sr_xMnO_3$ $x = 0.37$ and (*right*) $Nd_{1-x}Sr_xMnO_3$ $x = 0.30$ at fixed momentum q, from Fernandez-Baca et al. [11.10]. The development of a prominent central peak in NSMO (*right*) denotes unconventional behavior

hydrodynamic approach predicts [11.15] the existence of spin-waves below the ordering temperature T_C, with frequency $\omega(\boldsymbol{k}) \propto k^2$, and a "damping" of order k^4, at least for small k. For this reason spin-waves are particularly stable excitations of a ferromagnet (and also antiferromagnets, in this case with a linear k dependence of the spin-wave energy, and a damping proportional to k^2). These spin-waves are observed at low temperature in the ferromagnetic regimes of manganites using neutron scattering. The damping term depends on the spin diffusion coefficient Λ, which can be expressed in terms of microscopic properties of the system [11.15].

Above the Curie temperature, the system is presumed to be disordered, i.e., in a paramagnetic state. In such a case, excitations do not cost energy and a peak at very low energy is expected. In the ordered state below T_C, flipping a spin must cost energy since the flipped spin is not aligned with the rest. However, in a very disordered state (say $T = \infty$ as an extreme limit), flipping a spin does not cost energy since locally the spins can point in any direction. Whether a given spin is up or down is basically irrelevant. Then, in the disordered limit a peak at zero frequency should be observed. This peak must have a width determined by Λ. This is the regime of "diffusive spin dynamics". The peak should not have a momentum dependence since the excitations created by disturbances of the equilibrium state do not propagate over large distances, but they are relaxed by spin diffusion.

Explicit calculations of Λ based on microscopic Hamiltonians are very difficult. They involve the large-time behavior of correlations, namely dynamical information, which is far more difficult to obtain than static information. However, in some limits, such as infinite temperature, results can be obtained for the one-orbital or double-exchange model analytically (see [11.16]). As emphasized in [11.16], the results are not in quantitative agreement with experiments perhaps due to the large temperature used in the calculation, but they provide conceptual information on how the relaxation process acts in these systems. In addition, effects such as the

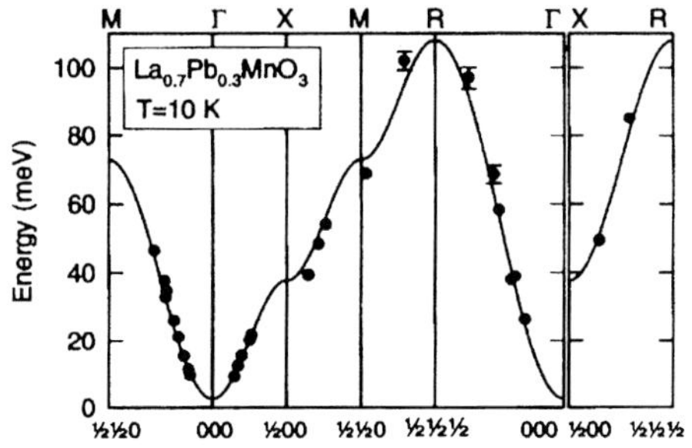

Fig. 11.7. Measured spin-wave dispersion for $La_{0.7}Pb_{0.3}MnO_3$ along major symmetry directions at 10 K, much smaller than T_C [11.17]. The *solid lines* show the dispersion relation of a nearest-neighbor Heisenberg ferromagnet that best fits the data

formation of polarons and mixed-phase tendencies should be incorporated into a more realistic calculation.

11.1.3 Anomalous Softening of Zone-Boundary Magnons

The spin-wave dispersion of $La_{0.7}Pb_{0.3}MnO_3$, a material that has a high Curie temperature of 355 K, was obtained from inelastic neutron scattering studies [11.17]. It was observed that a simple nearest-neighbors Heisenberg Hamiltonian accounted for the entire dispersion, an unexpectedly simple result considering the complexity of models usually applied to manganites, far more involved than a mere Heisenberg model (see Fig. 11.7). This suggests that $La_{0.7}Pb_{0.3}MnO_3$ is a very conventional ferromagnet, particularly at low temperatures [11.18]. However, as the temperature increases to 290 K, still within the ferromagnetic phase, the high-energy magnons acquire a large width to the point of virtually disappearing, while the low-energy magnons are weakly affected. These results are compatible with several others described in the previous section, showing that manganites have unusual properties near T_C, even within the ferromagnetic phase and even in the case of a large bandwidth material with large Curie temperature. The studies of $La_{0.85}Sr_{0.15}MnO_3$ in [11.19] arrived at similar conclusions as in [11.17], with a good fit of the peak positions by the simple Heisenberg model, but with an anomalously large broadening at 10 K, which could not be explained by such model. Studies of $La_{0.7}Sr_{0.3}MnO_3$ with a large Curie temperature $T_C = 378$ K have also revealed results similar to those of typical ferromagnets [11.20].

As the Curie temperature is reduced, increasing the strength of the CMR effect, clear deviations from a simple ferromagnet have been observed. This is to be expected in view of the many anomalous properties of low- or intermediate-bandwidth manganites, already discussed in this book. In particular, the agreement with the Heisenberg model no longer holds as T_C decreases, and it is known that a prominent diffusive central peak appears in some compounds (previous section). In [11.21], $Pr_{0.63}Sr_{0.37}MnO_3$ was studied. This material has a $T_C = 301$ K, not much different from $La_{0.7}Pb_{0.3}MnO_3$ [11.17]. However, this reduction is sufficient to induce, at 10 K, significant de-

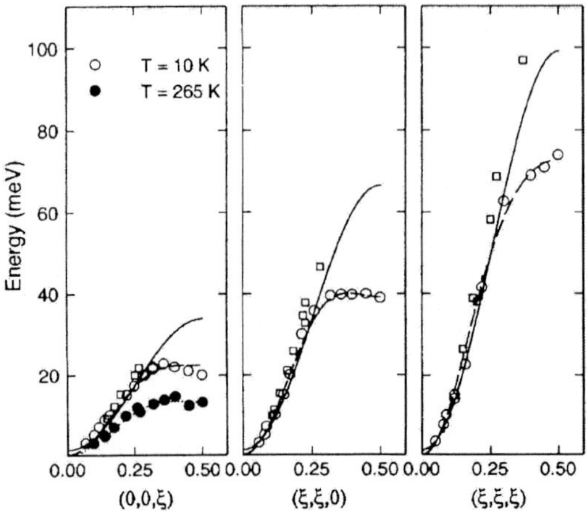

Fig. 11.8. Magnon dispersion of $Pr_{0.63}Sr_{0.37}MnO_3$ from [11.21]. The *solid line* is a fit using a nearest-neighbor Heisenberg model. The *dashed line* is a more complicated fit, with up to fourth-order nearest-neighbors. The *squares* are results from [11.17], for a higher T_C compound

viations in magnons from the simple Heisenberg description, as shown in Fig. 11.8. The deviation is the more significant at high momenta (or high energy), while at low momenta the dispersion can be fitted with a simple model. This phenomenon is sometimes described as *zone-boundary softening.* At 265 K, once again significant broadening of the high-energy magnons is observed. Slightly above T_C, the spin-wave peaks have disappeared, which is also unusual since conventional ferromagnets show a magnon-like peak in neutron scattering experiments even far above T_C (see discussion in [11.21]). This may be indicative of an abrupt, first-order-like change from the high-temperature insulating to the low-temperature metallic phases in manganites, as suggested by other techniques as well.

The results in Figs. 11.7 and 11.8 suggest that deviations from the canonical Heisenberg model may become more prominent as T_C decreases. However, this is not quite the case. In [11.22], results for $Nd_{0.7}Sr_{0.3}MnO_3$ ($T_C = 301$ K), $La_{0.7}Ca_{0.3}MnO_3$ ($T_C = 238$ K), and $Pr_{0.63}Sr_{0.37}MnO_3$ ($T_C = 198$ K) obtained at 10 K were found to be nearly identical, with zone-boundary magnon-softening and -broadening not compatible with a simple double-exchange picture. Typical results for LCMO $x = 0.30$ are reproduced in Fig. 11.9 where the spin-wave peak is sharp at small momentum and its location increases with increasing momentum, while at large momentum its energy is approximately constant and the width is large.

What are the reasons for the anomalous behavior found near the zone boundary in the spin-wave dispersion? Several proposals have been made. For

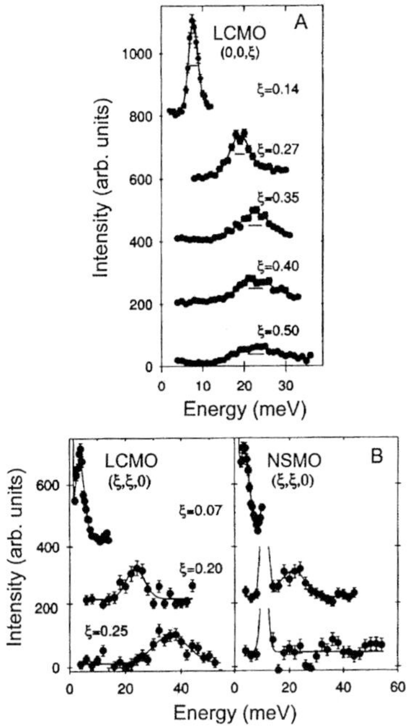

Fig. 11.9. Inelastic neutron scattering results corresponding to LCMO and NSMO at $x = 0.30$. The results shown are constant momentum scans at 10 K along the $[0,0,\xi]$ direction, with $\xi = 0.5$ denoting π/a in a more standard notation ($a = 3.86$ Å). From [11.22]

instance, magnon phonon coupling has been discussed as a possible source for the anomalous behavior. Optical phonons have been identified [11.22] with the energy of the approximately energy-independent portion of the spin-wave dispersion near the zone boundary. Spin-wave broadening can indeed be caused by phonons, which modulate the exchange coupling [11.23]. Charge and coupled orbital-lattice fluctuations were proposed as the origin of softening in the magnetic spectrum in [11.24]. On the other hand, a purely magnetic spin origin of the softening has been described in [10.12] (see also [11.25]). Calculations within the one-orbital model have also been used to analyze the broadening of spin-waves in experiments [11.26], revealing the existence of continuum states in the single-spin-flip channel that overlap with magnons at low energy. Competing FM AF states have been proposed as a possible explanation of the spin-wave softening as well [11.27]. Calculations of spin-waves and many other observables using the one-orbital model coupled to local phonons were reported in [11.28]. Explaining the anomalies near the zone boundary is a theoretical issue of much relevance at the present time.

The influence of the prominent phase-separation or mixed-phase tendencies on the spin-wave spectra has not been investigated in any detail yet. Near such regimes the compressibility is large, phonons are active, and small islands of charge ordering may be present within the ferromagnetic phase. It is expected that the inhomogeneous background may affect the magnon behavior. Indeed, experimental results [11.29] in regimes where two states are in competition show that mixed regimes severely influence the spin-wave dispersion. Recent calculations [11.30] with a novel spin-wave expansion highlight the relevance of competing tendencies in this context.

Overall, the softening described here provides more evidence that even the ferromagnetic phase of low-T_C manganites deviates from simple DE behavior (other experimental results lead to similar conclusions, as shown in the rest of the chapter, and other chapters). It is likely that degrees of freedom other than the spin must be incorporated in realistic descriptions of manganites.

The low-temperature spin dynamics of *bilayer* manganites has also been studied experimentally [11.31]. Within the planes, the spin-wave dispersion is almost perfectly two-dimensional in the regime where ferromagnetic correlations are observed at low temperatures. The interplane coupling within the double layer changes from ferromagnetic to antiferromagnetic as the density of holes x increases from 0.3 to 0.48. Anomalous broadening near the zone boundary was also found in bilayers [11.31]. Materials with an *A-type* antiferromagnetic ground state have also been analyzed using inelastic neutron scattering tecniques. For example, $Nd_{0.45}Sr_{0.55}MnO_3$ exhibits a very anisotropic spin-wave dispersion characteristic of the A-type AF state [11.32].

11.1.4 Volume Thermal Expansion

Let us consider now other experimental techniques and results that have contributed to the widely held view that manganites are inhomogeneous. We start with the important results by Ibarra et al. [11.33], which were among the first ones to properly identify those inhomogeneities.

Through the measurement of the lattice parameters as a function of temperature, the volume thermal expansion $\Delta V/V$ can be studied. A clear anomalous behavior was observed in this quantity by Ibarra et al. [11.33] studying $La_{0.60}Y_{0.07}Ca_{0.33}MnO_3$. The result for the linear thermal expansion is shown in Fig. 11.10 (left). The Curie temperature is indicated. A higher-temperature scale T_p exists where the coefficient deviates from the expected result arising from standard phononic contributions, which is shown by a thin solid line in Fig. 11.10 (left). Note that the sample's volume must increase with increasing temperature, but the issue under discussion is whether this occurs by a standard or anomalous procedure. In [11.33], the new temperature scale T_p was tentatively associated with polaron formation, but mixed-phase theories would also explain this scale as caused by the formation of clusters of competing phases, namely the T^* scale to be described in Chap. 17. The high-volume regime between T_C and T_p may be caused by a charge-localized high-volume PM insulating state. Even the inverse susceptibility (Fig. 11.10 (right)) shows the existence of a high-temperature new scale $\sim 1.6T_C$ where deviations from standard behavior are observed.

Similar effects were also found in $(La_{0.5}Nd_{0.5})_{2/3}Ca_{1/3}MnO_3$ [11.34]. They appear in $Pr_{2/3}Ca_{1/3}MnO_3$ [11.35] as well (showing that T_p also exists for

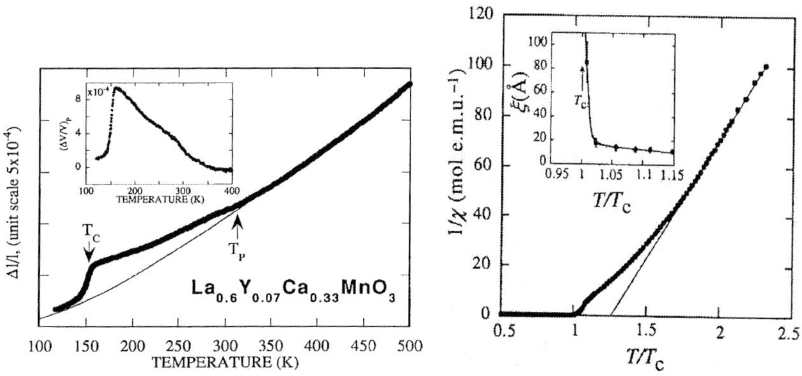

Fig. 11.10. (*Left*) Linear thermal expansion of $La_{0.60}Y_{0.07}Ca_{0.33}MnO_3$, reproduced from Ibarra et al. [11.33]. The extra contribution, i.e., the deviation from the *solid line* that contains the standard phonon induced expansion prediction, is shown in the *inset*. (*Right*) Even without Y, the $x = 0.33$ LCMO compound shows a similar anomalous behavior, which appears in the thermal expansion coefficient and in the thermal dependence of the inverse susceptibility, the latter reproduced here from De Teresa et al. [11.39]. The *solid line* is a canonical Curie Weiss law result

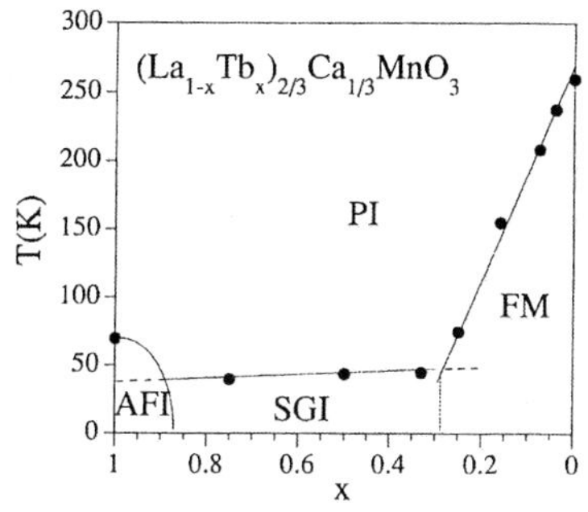

Fig. 11.11. Magnetic and electrical phase diagram of the series $(La_{1-x}Tb_x)_{2/3}Ca_{1/3}MnO_3$, from [11.36]. SGI is a spin-glass insulator phase

CO materials at low temperatures), and in $(La_{1-x}Tb_x)_{2/3}Ca_{1/3}MnO_3$ [11.36]. The phase diagram of the latter is shown in Fig. 11.11. It contains FM, AF, and spin-glass insulator (SGI) phases at low temperature. The thermal expansion coefficient shows a behavior comparable to that of other compounds. At $x = 0.25$ a mixture of two limits was observed, perhaps signaling the coexistence of metallic and insulating clusters. In the neighborhood of this composition, the resistivity vs. temperature [11.37] has the characteristic behavior of manganites with a large peak at finite temperature and a large $T = 0$ metallic resistivity, which are properties of mixed-phase states as described later in Chap. 17. Indications of mixed-phase behavior have also

would indicate smaller nanometer-scale inhomogeneities. More experimental work is needed to confirm this wide range of cluster sizes.

A very spectacular way to visualize the phase-separation phenomenon, reported by Uehara et al. [11.41], is with the use of transmission electron microscopy techniques. The results are shown in Fig. 11.13, where the dark regions denote the FM metallic portions of the sample and the gray and white regions correspond to the insulator. Clearly the structure is submicrometer in its length scale.

Note: What is Transmission Electron Microscopy? In transmission electron microscopy (TEM), a beam of electrons is directed toward a thinned sample. These (usually highly energetic) electrons interact with the atoms in the sample producing characteristic radiation and particles, which provide information about the material properties of the sample. Since the electron beam goes through the sample, it reveals properties of the interior of the specimen, such as size, shape, providing distribution of phases, and information on chemical composition and the crystal structure. Electron microscopes overcome the limitations of conventional microscopes, which are constrained by the physics of light to a resolution of a fraction of a micrometer. With electron microscopy, a far better resolution can be achieved exploiting the fact that electrons have wave properties that make them quite similar to light.

The electrons of the beam may scatter when crossing the sample. If the unscattered electrons are removed from the final image, then the result is called a *dark-field image*. This imaging mode is called dark-field because the result is dark unless the electrons are scattered, and the bright regions correspond to strongly scattered electrons (as in the case of charge-ordered insulating phases of manganites, which substantially scatter incoming electrons).

11.1.6 Brief Introduction to STM

The scanning tunneling microscope (STM) was invented in 1981 by Binnig and Rohrer at IBM Zürich. This is the first instrument that generated real-space images of surfaces with atomic resolution, allowing researchers to actually "see" atoms. In these devices a sharp conducting tip is used in a setup schematically shown in Fig. 11.14, with a bias voltage between the tip and the sample. When the tip is brought within a few angstroms of the sample, electrons can tunnel or jump from the tip to the sample or vice versa, depending on the sign of the voltage. The chances of this tunneling occurring depend on the properties of the sample, namely whether it has good or bad conductivity, and available or occupied states. By this procedure, the electronic properties of the surface can be analyzed.

STM devices are commonly considered as tools to generate images of the surface. However, they can also be used to study in detail the dynamical properties at a single point (x,y) on that surface. In particular, STM are spectroscopy tools, probing the electronic properties of a material with atomic resolution. Other spectroscopies, such as photoemission, average data over a large area of size in the scale of micrometers or millimeters. The study of the dependence of an STM signal upon the local electronic structure of the

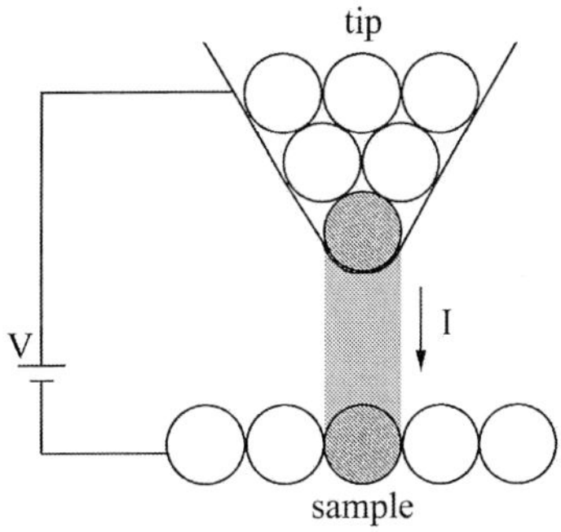

Fig. 11.14. Schematic description of a STM device, showing the sample, the tip, and bias voltage

surface is known as "scanning tunneling spectroscopy". Many methods can be used for these spectroscopic studies. These include the constant-current mode in which the tip-sample distance is adjusted every time the tip is moved to another location. Another option is the constant-height mode in which the tip is kept at a constant distance from the plane. In this situation, the tunneling current varies from place to place. In addition, it is very useful to keep the tip positioned over a feature of interest and change the bias voltage recording the tunneling current. This allows the determination of current vs. voltage (I–V) curves that characterize the electronic structure at the specific location (x,y). Also it is possible to collect $\mathrm{d}I/\mathrm{d}V$ (conductivity) vs. V curves. These are all ways to probe the local electronic properties of the surface of the material under investigation. The results of Fäth et al. and Renner et al., described in this section, have been collected with these techniques. For more details on STM techniques, there are several books on the subject as well as information on-line at, e.g., www.topometrix.com/spmguide/contents.htm from where information for this brief introduction was gathered.

11.1.7 Scanning Tunneling Spectroscopy in $La_{1-x}Ca_xMnO_3$

Scanning tunneling spectroscopy (STS) was used by Fäth et al. [11.45] to study single crystals and thin films of $La_{1-x}Ca_xMnO_3$ with $x \sim 0.30$. Below T_C, phase separation was observed where inhomogeneous structures of metallic and insulating areas coexist. The cluster size was found to be as large as a fraction of a micrometer, and that size and structure depended strongly on external fields. The picture of a percolative transition emerges clearly from this study. These results are very important since many probes used to test the mixed-phase character of the manganite state provide information averaged over the sample, while STS gives real-space pictures at a microscopic level. On the other hand, STS probes mainly the surface and, in this respect, there is always some uncertainty as to whether or not the results are representative of the bulk. In view of the overwhelming amount of experimental literature suggesting mixed-phase tendencies in manganese

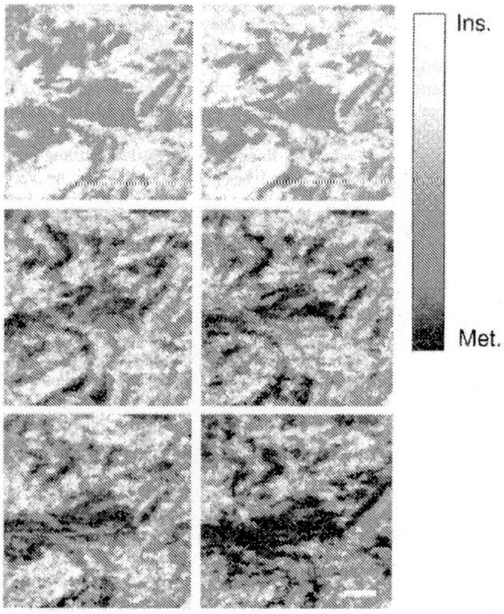

Fig. 11.15. Scanning tunneling spectroscopy images of the local electronic structure of LCMO, obtained just below T_C in magnetic fields of 0, 0.3, 1, 3, 5, and 9 T (from *left* to *right* and *top* to *bottom*). The data was taken on a thin-film sample with $x \sim 0.3$. From [11.45]

oxides, it is conceivable that the STS results of [11.45] are close to those of the bulk.

The results of Fäth et al. [11.45] are reproduced in Fig. 11.15. Note the presence of "clouds", which can be metallic or insulating. As these authors remark, this picture is not at all compatible with the picture of homogeneously distributed small polarons, which is a competing theory to the mixed-phase scenario. Note also the rapid percolation that occurs as the external field is turned on. Other STS results are reported in [11.46]. One of the main conclusions of the latter work is that the large variation in resistivity near T_C is a consequence of a variable density of states at the Fermi level, with a smaller contribution from the temperature-dependent mobility. This is in agreement with photoemission results on bilayers and theoretical calculations reporting the presence of a pseudogap in the density of states in mixed-phase regimes (see Chap. 18). This type of result is strongly against the picture of Anderson localization for the insulator above T_C since, in this mechanism, no changes in the density of states are expected to create insulating behavior. This conclusion is in excellent agreement with the analysis of Smolyaninova et al. [11.47] who studied the resistivity of LCMO thin films.

Recently, nuclear magnetic resonance experiments were reported by Bibes et al. [11.48] for a series of fully strained epitaxial $La_{2/3}Ca_{1/3}MnO_3$ thin films grown on $SrTiO_3$. This work found evidence of multiple phase segregation into ferromagnetic metallic, ferromagnetic insulating, and nonferromagnetic insulating regions. The authors believe that this phenomenon reflects an intrinsic property of LCMO/STO interfaces. Since the interaction with the substrate is strong, this experiment suggests that the phase separation found by Fäth et al. could also be strongly influenced by the substrate. As a consequence the relevance of the thin-film results for bulk materials remains unclear.

Even more recently, Renner et al. [11.49] reported remarkable scanning tunneling microscopy (STM) data obtained on analyzing $Bi_{1-x}Ca_xMnO_3$ with $x = 0.76$. This is the first experiment in a transition-metal oxide to image charge ordering and phase separation in real-space with *atomic-scale* resolution. The results provide an atomic-scale basis for descriptions of the manganites as mixtures of electronically and structurally distinct phases, in excellent agreement with modern theoretical studies [1.9] and a wide range of experiments described in this chapter and elsewhere in the book. Figure 11.16, reproduced from [11.49], provides an example of nanoscale phase separation. There is a charge-ordered region (upper part) and a more metallic homogeneous region (lower part). It is remarkable that the results were obtained at *room* temperature, above the charge-ordering temperature. Theoretical studies (Chap. 17) suggest that clusters should be formed above the ordering temperature, up to a temperature scale denoted by T^*. The local density of states also shows the mixed-phase character, with one phase insulating and the other more metallic (although not a very good metal). The lattice spacings deduce from STM are very similar to those found with X-ray crystallography, indicating that the surface is representative of the bulk. This is a remarkable experiment, with remarkable results. The mixed-phase character of manganites at the nanoscale can now be directly observed!

11.1.8 PDF, X-Ray, and Small-Angle Neutron Scattering Experiments

The PDF (pair distribution function) technique, which will be explained in more detail in the discussion of inhomogeneities in LSMO (Chap. 14), can also be applied to LCMO. In fact, PDF results for LCMO have been recently reported by Billinge et al. [11.50] using X-rays instead of neutrons, due to the high flux of X-rays from modern synchrotron sources. Among the main results emerging from PDF experiments is that the local lattice structure is *significantly different* from that observed crystallographically, i.e., the MnO_6 octahedra can have a JT distortion locally, even if globally the average JT distortion is zero or negligible. Another interesting PDF result is the presence of lattice distortions deep into the ferromagnetic phase that, according to naive double-exchange arguments, should be uniform. In the vicinity of the metal insulator transition, either by changing temperature or doping, the FM state is not in a pure homogeneous state. Both these results will appear also in the study of LSMO, as described in Chap. 14.

In Fig. 11.17a, PDF results for LCMO are summarized. The phase boundaries are taken from experiments that are sensitive to long-range order, typically magnetization and transport. The dark regions are those where "polarons" are observed, in the language of [11.50], while light shading indicates regions where localized and delocalized phases coexist (inferred from

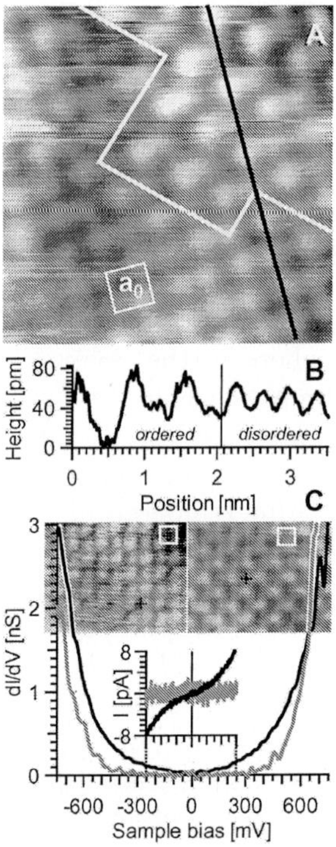

Fig. 11.16. (**A**) Atomic-scale image of phase segregation at 299 K. From Renner et al. [11.49]. The *white broken line* separates an insulating charge-ordering region from a more metallic and homogeneous phase. The compound is $Bi_{1-x}Ca_xMnO_3$ with $x = 0.76$. (**B**) Intensity profile extracted along the *straight line* shown in (**A**). Note the difference in amplitude over the Mn sites in the ordered region due to charge ordering. (**C**) Spectroscopic signature of phase separation. The spectra were acquired in the location of the *crosses*. Two clearly distinct results are obtained

the length of the Mn O bonds). The white region is the only one that can be considered truly uniform. For reference, Fig. 11.17b contains the more standard elastic neutron scattering phase diagram, including a separation between orthorhombic and pseudocubic regimes.

Note that very often in the PDF literature the results are interpreted in terms of small JT polarons. However, the technique by itself does not distinguish between small JT polarons and larger clusters of a JT active phase. For this reason, this author believes that it is risky to state that polarons are found using PDF methods, considering the standard definition of a polaron as a hole with a surrounding lattice distortion. Larger entities involving several holes inside a lattice-distorted cluster may also explain the data. This observation is compatible with neutron scattering experiments, carried out in similar regions of parameter space as the PDF experiments, which reported *correlated* polaron behavior, implying that the polarons do not form a weakly interacting "gas" (see Chap. 19).

X-ray absorption experiments by Booth et al. [11.52] led to conclusions similar to those of PDF experiments, namely that the number of delocalized holes $n_{\rm dh}$ continuously changes even well below the Curie temperature in the FM phase of LCMO. In fact, $\ln(n_{\rm dh}) \sim M$, where M is the magnetization.

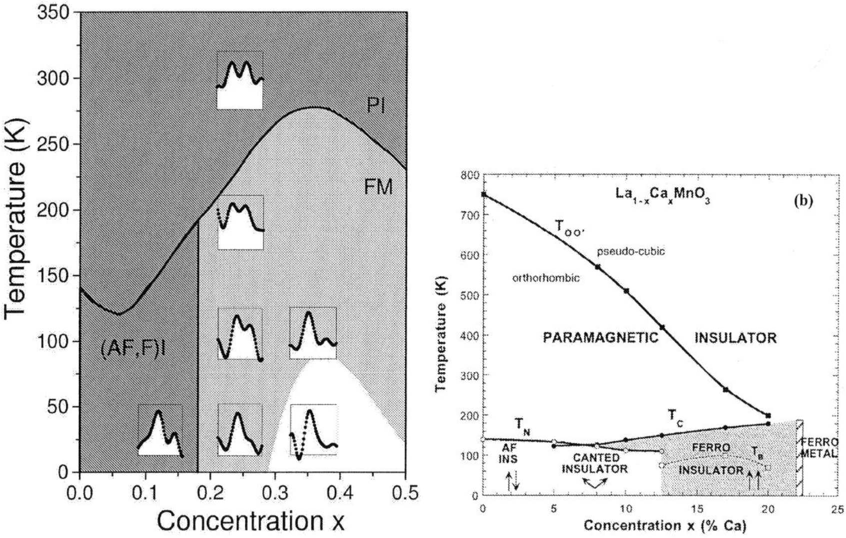

Fig. 11.17. (**a**) Schematic phase diagram of $La_{1-x}Ca_xMnO_3$. *Solid lines* are phase boundaries taken from experiments mentioned in [11.50]. The PDF results at small distance, i.e., related to the Mn O bonds, are provided in the *insets*. The *dark region* is JT distorted, while the gray regime shows a coexistence of JT distorted and undistorted phases. The white region is the homogeneous ferromagnet. The boundaries between *shaded regions* are diffuse and continuous. Result reproduced from [11.50]. (**b**) Phase diagram of $La_{1-x}Ca_xMnO_3$ obtained using elastic neutron scattering, reproduced from [11.51]. The emphasis here is on the orthorhombic to pseudocubic transition, which in view of the PDF results, separates a static Jahn–Teller distorted region from a dynamically distorted one

Elastic neutron scattering experiments by Hennion et al. [11.53] have also revealed the presence of inhomogeneities in $La_{1-x}Ca_xMnO_3$ at small $x = 0.05$ and 0.08. These experiments contain the existence of a diffuse scattering component, together with Bragg peak contributions. The results were interpreted in terms of an assembly of magnetic clusters or droplets with radius of approximately 9 Å. The number of FM clusters is very low compared to that of holes, suggesting that each cluster may contain several holes. This leads to a picture of hole-rich droplets within a hole-poor medium, compatible with the phase separation observed in theoretical studies of manganites. Recently, results similar to those for LCMO were also reported for $x = 0.06$ LSMO [11.54].

11.1.9 Brief Introduction to Nuclear Magnetic Resonance

Nuclear magnetic resonance (NMR) is a technique widely used to characterize the properties of a variety of materials [11.74, 11.79]. It involves the interaction of a nucleus, with a nonzero spin magnitude I, with an applied magnetic field B. In the presence of this field the energy levels of the nucleus split into $2I + 1$ lines, with energy splittings $\Delta E = \hbar\gamma Bm$, that can be detected experimentally by observing resonant absorption of electromagnetic radiation. The gyromagnetic ratio of the nucleus under study is γ, and m takes the values $-I$, $-I + 1$, ..., $I - 1$ and I. The importance of the NMR technique is that γ is sensitive to the local environment of the nucleus, namely, its value changes depending on whether the studies are carried out using fairly isolated ions, or in situations where the ions are immersed in a lattice with which it interacts. Analyzing these changes in γ allows for local (microscopic) studies of the properties of a given compound. These changes in γ produce a shift ΔE in the resonant frequencies of the nuclei. This shift contains several contributions which can be analyzed separately, such as the effect of closed-shell electrons. The most important for our purposes is called the *Knight shift*, caused by the interaction of the spin of the nucleus with conduction electrons. The Knight shift can be shown to be proportional to the electron-spin susceptibility or Pauli susceptibility, also measured in practice in an applied magnetic field, from the ratio between the induced magnetic moment and the magnetic field. Formally, the Knight shift is caused by the hyperfine interaction between the spin of the nucleus $\boldsymbol{I}$ and the spin of the electron $\boldsymbol{S}$.

The NMR technique is based upon the physics learned in elementary classes of quantum mechanics regarding the behavior of a spin in a magnetic field. In practice, a static magnetic field is applied, but also an additional sinusoidally varying magnetic field (perpendicular to the main one) is used to detect the transitions. The expectation value of the spin precesses about the field at the Larmor frequency, just as it would classically. The coupling of the spin with the rest of the system introduces relaxation times ("friction") in the equations for the spins to point along the magnetic field, that can also be studied from the experimental data. These relaxation times T_1 and T_2 are associated with spin lattice and spin spin relaxations, respectively. The first one is related to the longitudinal component of the magnetization while the second one is associated with the transverse component.

Typical NMR plots contain the NMR intensity of the signal in the vertical axis, which is related to the number of nuclei contributing to the signal, and the frequency in the horizontal one, all at constant external magnetic field. These plots contain a rich structure of peaks that arise from the possible locations that a given ion (with its associated nucleus) can have in a given material. Analyzing peak positions, intensities, and widths as a function of the parameters of the problem, such as temperature, conclusions can be obtained about the microscopic properties of a given compound. From the analysis of NMR data, the microscopic spin arrangement of the compound under study can be analyzed since different states lead to different effective fields at the position of the nuclei investigated. This is particularly important for those spin states, such as the antiferromagnetic one, where the overall magnetization cancels and microscopic approaches are needed to establish the properties of the system. Actually, for spin-ordered states, NMR measurements can be carried out even in the absence of external magnetic fields, since there is already an internal field that provides the splitting among the otherwise spin-degenerate levels.

The magnetic fields typically used in NMR experiments are on the order of 10 T, and typical NMR frequencies are in the region of 100 MHz. The splitting energies caused by magnetic fields of a few Teslas are small in the natural eV units of electronic problems.

NMR has been recently widely used in the context of the high-temperature superconductors, and a brief review and plenty of references can be found on p. 530 of [11.80], as well as in the contribution of Papavassiliou in this chapter. Results for manganites are discussed in this book as well, but with emphasis on the results, not on the experimental details that the reader can find in the original references. To supplement the information provided here, [11.81] gives a brief description of *nuclear quadrupole resonance* (NQR), a related technique also widely used in the study of materials. This method applies to nuclei with I larger than 1/2. Information about other important related techniques can also be found in [11.81]. This includes (i) *electron-spin resonance* (ESR), where unpaired electrons in transition ions are studied in magnetic fields (for an introduction see chapter 11 of [11.74]); (ii) *muon spin relaxation* (μSR), where the precession is now studied using muons that are artificially implanted into the sample; (iii) *Mössbauer resonance*, which involves the study of gamma rays emitted from nuclei that undergo transitions from their ground state to excited states; and others.

11.1.10 Nuclear Magnetic Resonance in $La_{1-x}Ca_xMnO_3$

NMR investigations are important for the analysis of manganites, and especially for testing the predictions of the phase-separation theory. Very interesting NMR investigations have been presented by Papavassiliou et al. [11.56]. Typical NMR spectra reproduced from [11.56] are shown in Fig. 11.18a, obtained at low temperature $T = 3.2$ K. The analysis of the NMR spectra is the most straightforward way to analyze phase coexistence. At $x = 0.50$ a *double-peaked* spectrum is observed in Fig. 11.18a, one is AF and the other FM, signaling coexistence of phases. However, at $x = 0.33$ the state appears homogeneous, with only one peak. At $x = 0.25$, two peaks are again observed. The one located at the lowest frequency is associated with localized Mn^{4+} states. The results are summarized in the phase diagram shown in Fig. 11.18b. It is clear that even phase diagrams that are presumably well established, such as for LCMO, may need substantial revision once microscopic probes are used. In fact, recent NMR results by Kapusta et al. [11.57] revealed also the need for revision of the phase diagram of $La_{1-x}Ca_xMnO_3$. These authors found the presence of FM clusters embedded in a charge-ordered background at density $x = 0.65$, presumed to be "deep" into the charge-ordered regime.

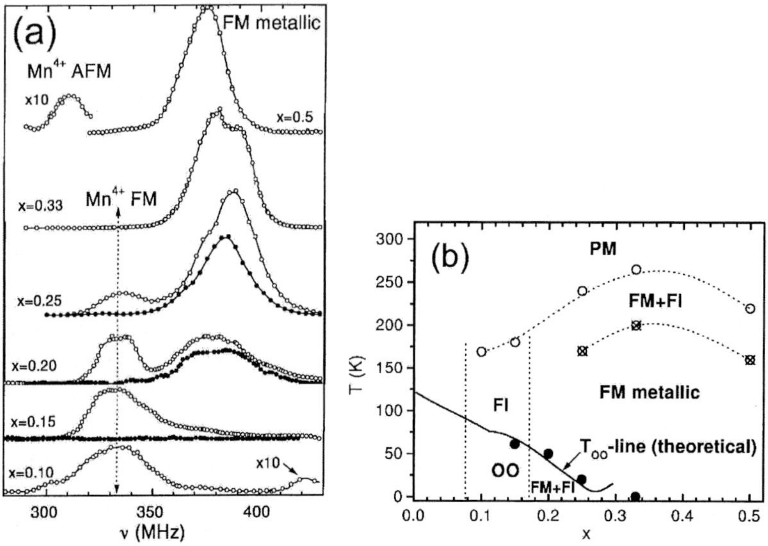

Fig. 11.18. (**a**) NMR spectra of $La_{1-x}Ca_xMnO_3$ at the hole densities indicated, in a zero external field and temperature 3.2 K (*open circles*). *Filled circles* are spectra at 62 K ($x = 0.15$), 50 K ($x = 0.20$), and 20 K ($x = 0.25$). From [11.56]. (**b**) Revision of the phase diagram of $La_{1-x}Ca_xMnO_3$ obtained using NMR techniques, reproduced from Papavassiliou et al. [11.56]. The novel regimes found by NMR investigations are the FM+FI region below the Curie temperature, and the low-temperature region next to orbital order (OO). For more details, see [11.56]

Using NMR techniques applied to LCMO, Allodi et al. [11.55] also observed the coexistence of FM and AF resonances leading to a picture of electronic inhomogeneities for this compound. Dho et al. [11.58] arrived at similar conclusions regarding phase coexistence at $x = 0.50$. Mori et al. [11.59] reported similar mixed-phase tendencies with microdomains of size 20–30 nm. More recently, Mössbauer spectroscopy results also support the mixed-phase FM-AF picture of $x = 0.5$ LCMO [11.60]. Very recent electron spin resonance (ESR) results for the same compound by Rivadulla et al. [11.61] found indications of phase separation as well, showing that ESR can be a useful tool to investigate mixed-phase states. In addition, NMR results for $La_{1-x}Ca_xMnO_3$ in agreement with the predictions of the phase-separation theory, were reported in [11.62]. In particular, at $x = 0.5$ the system is shown to break into FM and AF regions at low temperature. A detailed study of phase separation in $(La_{0.25}Pr_{0.75})_{0.7}Ca_{0.3}MnO_3$ was also presented by Yakubovskii et al. [11.63] who concluded that their "NMR results demonstrate directly the presence of magnetic phase separation at a local level". Clearly, the evidence for phase separation using NMR methods is robust.

11.1.11 The NMR "Wipeout" Effect

by G. Papavassiliou

A common aspect in underdoped La-based cuprates and CMR manganites is the tendency to spin-freezing that has been observed in the doping range between the AFM phase at $x = 0$, and the superconductive or FM metallic phase. In the case of $La_{2-x}Sr_xCuO_4$, this intermediate phase is formed by frozen AFM clusters, i.e., a cluster spin-glass phase is formed, which coexists with superconductivity for $x \geq 0.06$ [11.64, 11.65]. On the other hand, in CMR manganites the freezing is rather related to orbital than with spin rearrangements. Competing AFM (staggered) and FM orbital configurations in a spin FM background [11.66] probably lead to the freezing dynamics, which have been recently observed with neutron scattering [11.51] and NMR techniques [11.56, 11.67, 11.68] in $La_{1-x}Ca_xMnO_3$ for $x \leq 0.20$.

Remarkably, in both these systems the tendency to freeze is accompanied by a dramatic loss of the NMR or NQR signal intensity, the so-called "wipeout effect" [11.68–11.72]. Initially, the term wipeout has been used to describe the strong decrease of the NMR signal in cubic metals, by introducing non magnetic impurities or defects [11.73]. Near such irregularities the local symmetry is no longer cubic, and the electric-field gradients induce a strong quadrupolar spread of the resonance frequencies, thus smearing out the NMR signal. On the other hand, in cuprates and manganites the wipeout effect is produced by the extreme shortening of the spin lattice ($1/T_1$) and spin spin ($1/T_2$) relaxation rates, due to coupling of the nuclear spins with slow fluctuations of the hyperfine magnetic field (magnetic relaxation) or the electric-field gradient (quadrupolar relaxation). As a consequence, the relaxation rates become so short that most of the NMR signal relaxes before it can be observed. The influence of a broad distribution of "freezing" correlation times τ_c on the relaxation rates $1/T_{1,2}$ may be followed by considering a stretched exponential autocorrelation function for the hyperfine field [11.68], $< B_{\text{hf}}(t)B_{\text{hf}}(0) >=< B^2_{\text{hf}} > \exp\left(-(t/\tau_c)^\alpha\right)$, with $0 \leq \alpha \leq 1$. The relaxation rate $1/T_1$ from such a distribution is given by, $1/T_1 \propto J(\omega) =< B^2_{\text{hf}} > \int_0^\infty \exp(-(t/\tau_c)^\alpha) \exp(-\mathrm{i}\omega t) dt$ [11.74]. In the fast motion regime, $\omega\tau_c \ll 1$, this gives $1/T_1 \propto< B^2_{hf} > \Gamma(\frac{1+\alpha}{\alpha})\tau_c$, whereas in the "slow motion" regime, $\omega\tau_c \gg 1$, $1/T_1 \propto 2\alpha/(\omega^{1+\alpha}\tau_c^\alpha)$. In the case of gradual freezing, as spectral fluctuations slow down and cross the Larmor frequency ω, $J(\omega)$ crosses over from the fast- to the slow-motion regime, and $1/T_1$ exhibits a maximum at a temperature defined as the freezing temperature T_{f}. The strongest wipeout effect is thus expected in the neighborhood of T_{f}. It is also notable that in both systems the relaxation enhancement is accompanied by the crossover of the nuclear spin echo decay from a Gaussian to a Lorenzian shape. The Gaussian decay is dominated by the indirect spin spin coupling between like nuclear spins, whereas the Lorentzian decay is produced by a Redfield T_1-process [11.75].

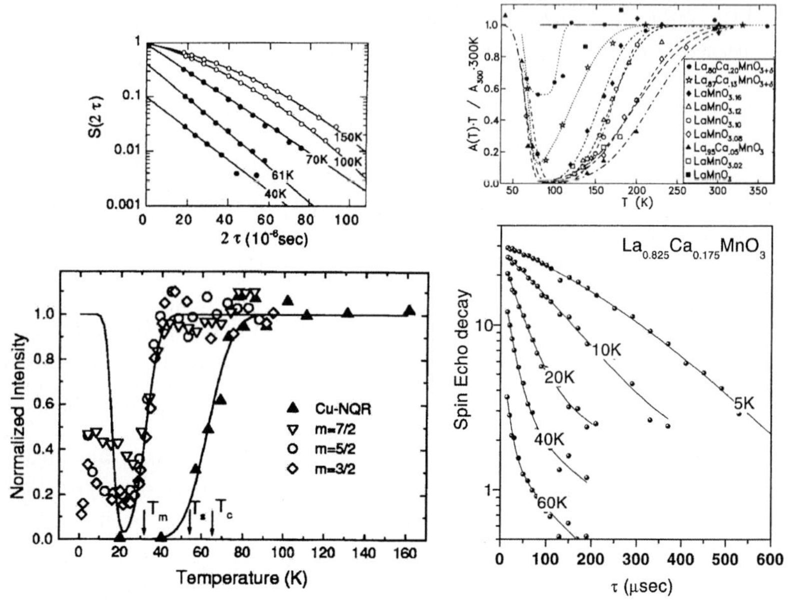

Fig. 11.19. Figure related to the discussion on the wipeout effect. For details see text

The similar freezing dynamics exhibited by these two systems (cuprates and manganites) become clear in Fig. 11.19. In thc upper-left panel of this figure, the gradual transition from a Gaussian to Lorentzian spin-echo decay, which accompanies the wipeout effect, is clearly demonstrated by ^{63}Cu NQR measurements on $La_{1.48}Nd_{0.4}Sr_{0.12}CuO_4$ [11.76]. This system exhibits a superconducting transition at 5 K. In the lower-left panel the normalized signal intensity of ^{139}La and ^{63}Cu NQR is shown as a function of temperature for the same system [11.77]. The wipeout effect is observed to be different for La and Cu nuclei. This is expected as La ions lie outside the CuO_2 planes and are coupled to Cu^{2+} electron spins through a smaller hyperfine interaction in comparison to ^{63}Cu, which leads to longer T_2 values. It has also been found that the ratio between the relaxation rates of ^{63}Cu and ^{65}Cu corresponds to a magnetic relaxation mechanism [11.77], a result also confirmed by other experiments [11.70, 11.71].

A similar phenomenon is also shown for ferromagnetic and insulating LCMO in Fig. 11.19 (right panels). In the upper-right panel, the wipeout effect of ^{139}La NMR is shown at low doping, for $La_{1-x}Ca_xMnO_{3+\delta}$ [11.72]. In the range $0.1 \leq x \leq 0.2$ these systems exhibit a transition to a low-T orbitally ordered state with strongly freezing characteristics at $T_f \approx 70$–80 K [11.67, 11.68]. It has also been observed that the increasing wipeout close to T_f is accompanied with a Gaussian to Lorentzian crossover in the spin-echo decay, as shown in the lower-right panel for $La_{0.825}Ca_{0.175}MnO_3$ [11.68].

In conclusion, charge localization, glassy spin or orbital freezing, and NMR/NQR wipeout are three mutually dependent properties of layered cuprates and CMR manganites. It is tempting to attribute these three features to uncorrelated slowly fluctuating spin – respectively orbital – clusters, which by cooling percolate to a low-T ordered state, as recently proposed theoretically [11.78].

11.1.12 Influence of Cr Doping on Charge-Ordered Manganites

Raveau et al. [11.82] found that the substitution of a small number of Mn ions by another transition metal has a strong effect on the magnetic and electronic properties of the parent compound. In particular, a few per cent of Cr replacing Mn suppresses the charge-ordered arrangement of some manganites, creating instead a ferromagnetic metallic state. The effect of a relatively small number of impurities on the charge-ordered state is clearly unusual. It would be expected that impurities disturb the movement of charge and, as a consequence, favor charge localization and the formation of an insulator. However, Cr doping in manganites induces a metallic state, a quite intriguing result. Exotic phenomena associated with impurity doping have been found in other transition-metal oxides, where this doping leads to an ordered antiferromagnetic state [11.83].

The issue of impurity doping was also investigated by Kimura et al. [11.84] for Cr-doped $Nd_{1/2}Ca_{1/2}MnO_3$, which has a charge-ordered state. A typical result is shown in Fig. 11.20a, where resistivity vs. temperature is shown at 2% Cr doping (results for the magnetization are given in Fig. 11.20b). It is clear that a small field produces drastic changes in the transport properties. The results are reminiscent of the percolative features found in studies of random-resistor networks, described in Chap. 17. Note that the peaks in resistance vs. temperature are not very pronounced in Fig. 11.20a, as found in some simulations also in Chap. 17. Kimura et al. explained their results as caused by the formation of impurity-induced FM regions, in the background of the charge-ordered insulating matrix. Long-time relaxation effects in magnetization and conductance were observed, similar to those seen in materials known as "relaxor ferroelectrics" [11.85–11.87].

Several other studies of the influence of Cr on manganites have been reported. Oshima et al. [11.88] studied Cr-doped $Pr_{0.5}Ca_{0.5}MnO_3$, observing photoinduced persistent conductance in thin films. Magnetic force microscopy studies on the same compound, by Oshima et al. [11.89], reported the existence of microstructures, consistent with phase-separation theories of manganites. Moritomo et al. [11.90] found coexistence of CO domains in the FM phase, forming a phase-separated state in Cr-doped $RE_{1/2}Ca_{1/2}(Mn,Cr)O_3$, with RE = La, Nd, Sm, or Eu. The phase diagram is reproduced in Fig. 11.21. Once again, it appears that the transition from a FM to a CO state is through

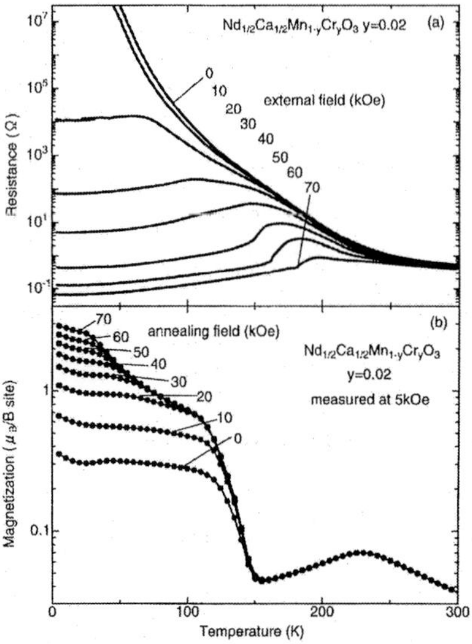

Fig. 11.20. (**a**) Resistance vs. temperature at various magnetic fields for a crystal of $Nd_{1/2}Ca_{1/2}Mn_{1-y}Cr_yO_3$ with $y = 0.02$. (**b**) Magnetization vs. temperature for the same crystal as in (**a**). From [11.84]

a mixed-phase regime. This result has similarities with the theoretical ideas discussed later in Chap. 17, where the phase diagram of two competing phases is discussed. In fact, differences between Al- and Cr-doping of CO and FM manganese perovskites have been recently explained by Takenaka et al. [11.91] based on the phase-separation scenario.

A systematic study of the effect of impurities, such as Cr, on several manganites has been presented by Katsufuji et al. [11.92]. Their main result is summarized in Fig. 11.22, showing results for (Pr,Ca)-, (La,Ca)-, and (Nd,Sr)-based manganese oxides. The black region is *charge ordered* in the absence of impurities, but becomes ferromagnetic metallic upon the replacement of a small percentage of Mn by Cr. This result shows a neat universality among the many Mn oxides, and also that Cr doping can have an effect similar to that of magnetic fields. In particular, for PCMO it is known, and puzzling, that magnetic fields destabilize the CE-state on a wide density window. Figure 11.22 shows that a similar result can be obtained with Cr doping.

The influence of the replacement of Mn by Fe was studied in [11.93], where a rapid suppression of the Curie temperature with increasing Fe concentration was observed. The shapes of the resistivity vs. temperature curves are again reminiscent of the percolative results found in the random resistor network, as described in Chap. 17. The results of [11.93] are qualitatively similar to those found for Cr doping. Both the FM state and the CO state can be rapidly destabilized after Mn is replaced by another ion. This is also similar to the rapid suppression of superconductivity with Zn doping in the high-T_c superconductors. Other studies of Cr-doped manganites were done by Hervieu et al. [11.94] using $Pr_{1-x}Ca_xMnO_3$ at $x = 0.5$ as parent compound, showing

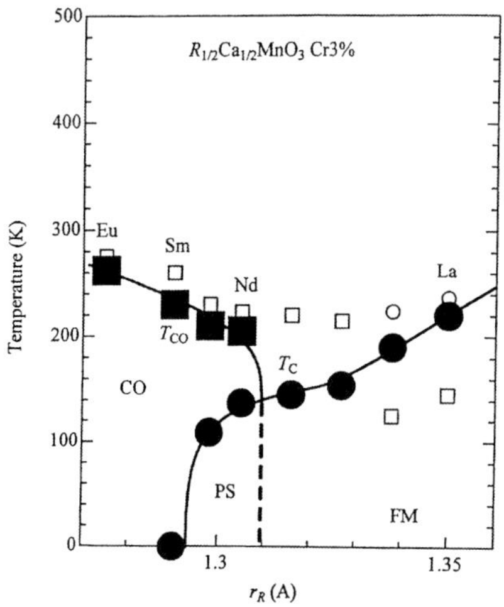

Fig. 11.21. Phase diagram of 3% Cr-doped manganites $RE_{1/2}Ca_{1/2}(Mn,Cr)O_3$ as a function of the average ionic radius $r_{R(A)}$ of the rare-earth. *Closed circles* (*squares*) are Curie temperature and charge-ordered temperatures. *Open symbols* denote results for the Cr-undoped limit. A phase-separated (PS) regime appears between the FM and CO phases. From [11.90]

that the resistivity and nanostructures are dependent on the magnetic-field history.

All these results show that doping of manganites with a tiny amount of impurities produces drastic changes in the properties of the system, as for magnetic fields, pressure, and other perturbations. It is likely that all these phenomena have a common explanation, and impurity doping may play a key role on its understanding.

11.1.13 Giant Noise in Manganites

Dramatic current inhomogeneities were observed in $La_{1-x}Ca_xMnO_3$ ($x = 0.33$) near the metal insulator transition. This manifests itself as large resistance changes in the form of steps as a function of time, as illustrated in Fig. 11.23 reproduced from Merithew et al. [11.95] (similar results were reported by Raquet et al. [11.96], where the effect is referred to as giant random telegraph noise). The voltage spectral density has a $(1/f)$ (f=frequency) form (see discussion on low-frequency noise below). The authors of [11.95, 11.96] carried out a detailed analysis, concluding that the effect arises from fluctuations between local states with different conductivities. These fluctuators are likely located along the percolating backbone expected in manganites, thus influencing substantially the conductance. The domains have a size of 10^4 to 10^6 unit cells [11.95, 11.96]. The effect does not appear to be caused by

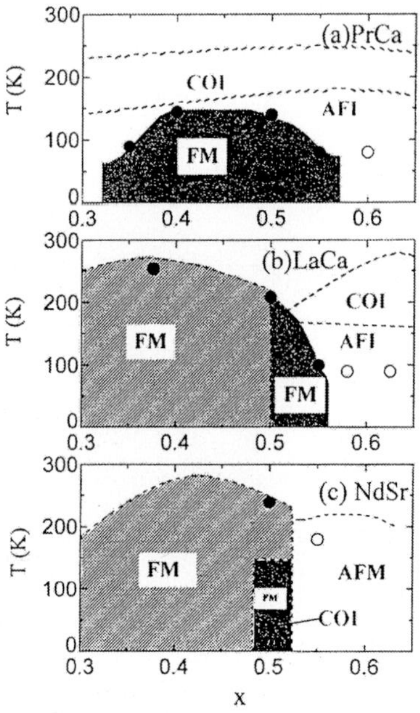

Fig. 11.22. Phase diagrams of (**a**) $Pr_{1-x}Ca_xMnO_3$, (**b**) $La_{1-x}Ca_xMnO_3$, and (**c**) $Nd_{1-x}Sr_xMnO_3$ all of them with 3% Cr replacing Mn. The *black regions* are charge ordered in the absence of Cr doping, but become FM and metallic upon such doping. From [11.92]

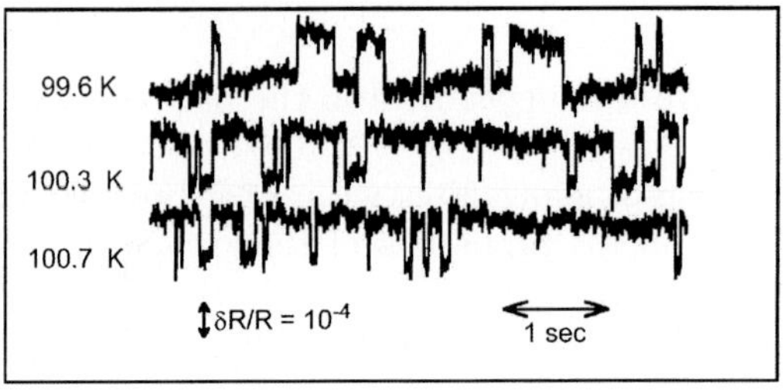

Fig. 11.23. Changes in the resistance of LCMO ($x = 0.33$) at three temperatures, reproduced from Merithew et al. [11.95]. Similar results were reported by Raquet et al. [11.96] and Podzorov et al. [11.98]

chemical inhomogeneity, but seems to be intrinsic to the material, in agreement with the vast literature discussed in this book on mixed-phase states. Similar results were reported for $Pr_{2/3}Ca_{1/3}MnO_3$ by Anane et al. [11.97], and for $(La_{1-y}Pr_y)_{1-x}Ca_xMnO_3$ in the mixed-phase regime by Podzorov et al. [11.98]. Effects related to those found in manganites appear also in so-called "relaxor ferroelectrics". Recent work by Colla et al. [11.99] addresses this issue and provides a list of references. Theoretical calculations have also supported the notion that $(1/f)$ noise is found in phase-separated regions [11.100].

Note: Low-frequency $(1/f)$ Noise. Certain materials, such as the manganites at particular compositions and temperatures, present slow changes of the resistance with time, as discussed in this subsection. In some cases, the resistance jumps between two or more locally stable values, an effect known as "telegraph noise". Resistance noise as described here occurs typically near metal insulator transitions. This time-dependent resistance induces noise in the voltage, if working at constant current. This noise is different from other sources of noise, such as the shot noise caused by the discrete nature of the charge. The latter, and other types of noise, are typically "white" in the sense of being approximately frequency independent, contrary to the "$1/f$" noise.

Consider a material where several states are in competition. For simplicity, let us imagine the sample as made out of small regions labeled with an index i, each one containing two states separated by a free-energy barrier of height W_i. The time between tunneling events from one minimum to the other is $\tau = \tau_0 e^{\beta W_i}$, where $\beta = 1/(k_B T)$, and τ_0 is a proportionality constant. The higher the barrier, the less frequent the tunneling. The barrier heights W_i are assumed to be uniformly distributed between a minimum and a maximum value. Thus, the probability $P(W_i)$ of a given height is simply a constant.

Due to tunneling-induced changes in the state at each site i of the grid, the resistance R will change (typically the two competing states are a metal and an insulator), rendering it time dependent, i.e., $R = R(t)$. This time dependence turns into a frequency dependence after a Fourier transform. For a fixed frequency ω, the tunneling rates $1/\tau$ that match that frequency are the important ones. This means that of the many barrier heights, those that satisfy $\omega = (1/\tau_0)^{-\beta W_\omega}$ contribute the most to the resistance at that frequency. Finally, the probability of tunneling at frequency ω (which regulates the resistance at the same frequency) is proportional to the probability W_ω, which is a constant as already explained, times $|dW_\omega/d\omega|$. The last factor accounts for the fact that, even with a uniform distribution of barrier heights, the probability of tunneling at a given time $\tau = 1/\omega$ is not constant. In fact, $|dW_\omega/d\omega| = 1/\omega$ using the formula a few lines above, justifying the term "$1/f$" noise. The reader can find a much more detailed description of noise in systems such as those studied here, in the book of Y. Imry, *Introduction to Mesoscopic Physics*, Oxford University Press, and references therein.

11.1.14 Other Results Related with Phase Separation

Other experimental techniques have shown indications of inhomogeneities in LCMO at $x \sim 0.3$. (1) Kida et al. [11.101] arrived at such a conclusion based on estimations of the low-energy complex dielectric constant spectrum. (2) Zhou and Goodenough [11.102] found results compatible with the coexistence of hole-rich clusters within a hole-poor matrix in the paramagnetic state, analyzing $(La_{1-x}Nd_x)_{0.7}Ca_{0.3}MnO_3$. (3) Muon spin relaxation results by Heffner et al. [11.103] suggest the existence of two spatially separated regions in LCMO with $x \sim 0.3$. (4) A "two-fluid" model for manganites was developed by Jaime, Salamon and collaborators since their early very important studies of LCMO [1.18, 11.104]. They were among the first to identify the presence of inhomogeneities in these compounds. Recent work by the same group, reported by Chun et al. [11.104], shows for LCMO $x = 0.3$ that the polaron picture holds above $T_p = 1.4T_C$, but that between T_C and T_p a mixed-phase state better describes their results. This is in agreement with the conclusions of Ibarra et al. based on the thermal expansion coefficient described in this chapter. It appears that there is a temperature scale in manganites well above T_C where mixed-phase tendencies start. This is in agreement with the theoretical discussions of Chap. 17 where a new scale T^* is introduced. (5) Photoabsorption results by Hirai et al. [11.105] for LCMO have also been interpreted as evidence of phase separation in manganites, above T_C. (6) Recent studies have shown that LCMO $x = 0.3$ also has a colossal *electroresistance*, complementary to the CMR effect [11.106]. These results were considered as strongly favoring a percolative phase-separation picture. (7) Infrared reflectivity and extended X-ray absorption fine structure measurements by Massa et al. [11.107] in LCMO $x = 0.3$ reveal inhomogeneities that can have localizing effects on the carriers. (8) Isotope effect, specific heat, and thermal expansion studies by Gordon et al. (see citation in [3.13]) led to results compatible with FM polarons, bipolarons, or clusters above T_C. (9) Recently, results for the compound $CaMn_7O_{12}$ have also been explained in terms of the percolative and phase-separated scenario [11.108].

In view of the large number of reports of inhomogeneities in LCMO, it is quite likely that other reports may have escaped the attention of this author, who apologizes in advance for possible omissions. Clearly the evidence for mixed-phase tendencies in perovskite LCMO manganites is simply overwhelming, and the reader will find similar trends in other manganites discussed in the following chapters.

11.2 Inhomogeneities in Electron-Doped Manganites

Studies in the "electron-doped" region of manganites (i.e., in the regime of large hole doping) also provide very useful information about the competition of phases. The discussion of these studies is included in this chapter since many of the electron-doped investigations in this context have been carried out on LCMO. Maignan et al. [11.109] reported resistivity vs. temperature of $Ca_{1-x}Sm_xMnO_3$, and the results are reproduced in Fig. 11.24a. A tendency toward metallicity develops as the electron density grows. Large MR has been observed in this regime [3.7]. $Ca_{1-x}La_xMnO_3$ was studied in the region of hole doping between $x = 0$ and 0.2 by Neumeier and Cohn [11.110]. These important studies will only be briefly discussed here, since Neumeier and Cornelius provide in the next section a more detailed analysis of specific-heat techniques and their use to study phase-separated materials. Figure 11.24b contains Neumeier and Cohn's magnetization vs. temperature. Clearly, a finite moment rapidly develops as the electron density increases. Figure 11.24c also shows the magnetic saturation moment at 5 K vs. La doping x. In the vicinity of 8% doping, it develops a maximum, concomitant with a peak in the conductivity (Fig. 11.24c). It is clear then that at low electron density, ferromagnetism develops with doping. The analysis of Neumeier and Cohn [11.110] also suggests that phase separation involving FM and AF regions (the latter with either G-type or C-type characteristics) dominates the physics. This is actually quite similar to results reported for low electron density bilayer

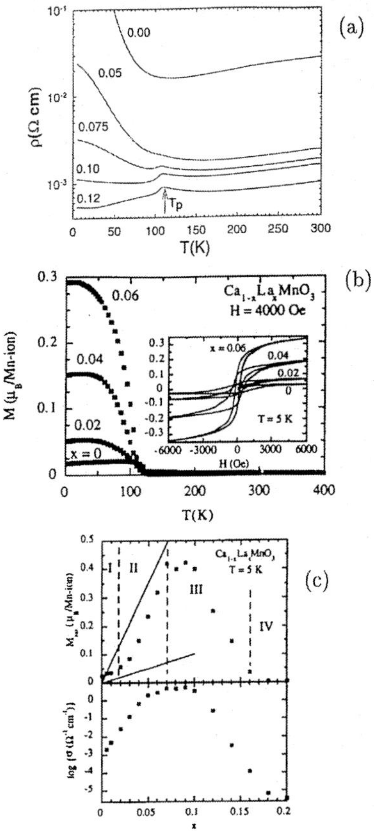

Fig. 11.24. (**a**) Resistivity vs. temperature of $Ca_{1-x}Sm_xMnO_3$, from [11.109]. (**b**) Magnetization vs. temperature for $Ca_{1-x}La_xMnO_3$, from [11.110]. (**c**) *Upper panel*: magnetic saturation at 5 K vs. La doping. *Lower panel*: Electrical conductivity at 5 K. The regions I-IV in the *upper panel* have different spin arrangements, according to [11.110]. In I a G-type AF state dominates. In region II, there are local FM regions in a G-type AF background. In III, G-type and C-type AF coexist with FM, while in IV, the C-type AF state dominates

compounds, discussed elsewhere in this book, and also to the conclusions of an analysis of $Ca_{0.85}Sm_{0.15}MnO_3$ by Mahendiran et al. [11.111]. The results are also compatible with the systematic analysis of electron-doped materials reported by Martin et al. [11.113]. Their data are, in general, described as forming a "cluster glass", compatible with the phase-separation theory. The same material $Ca_{0.85}Sm_{0.15}MnO_3$ has also been recently analyzed by Algarabel et al. [11.114] who found nano and microscale phase segregation. Experimental and theoretical studies of $Ca_{1-x}Y_xMnO_3$ by Aliaga et al. [11.115] also lead to the conclusion that the competition FM AF is the main driving force. Recent theoretical studies by Chen and Allen [11.116] have shown that upon doping $CaMnO_3$ with $x = 0.045$ electrons/site, a polaronic insulator becomes unstable against a FM metal, illustrating once again the competition between phases present in these compounds and models.

Insight into the low electron doping region has also been obtained through the study of $Bi_{1-x}Ca_xMnO_3$, with x between 0.74 and 0.82. This material is expected to behave quite similarly to LCMO. Neutron scattering results re-

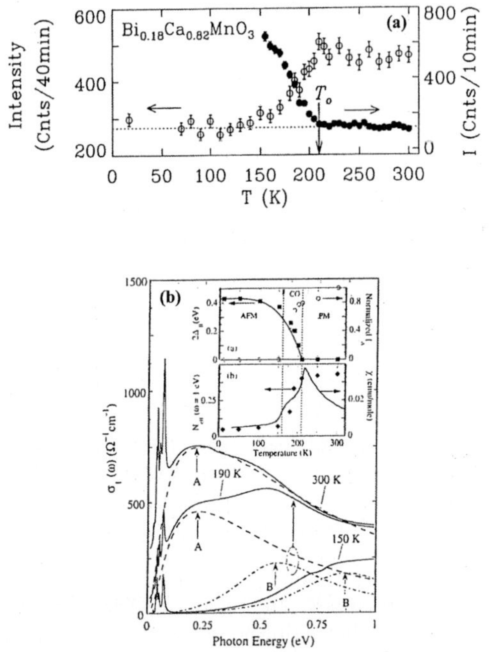

Fig. 11.25. (**a**) Temperature dependence of the intensity of the AF (*closed*) and FM (*open*) peaks in BCMO. The region between ~ 150 K and 200 K shows coexistence of both features. Reproduced from [11.117]. (**b**) Real part of the optical conductivity at three temperatures, from Liu et al. [11.119], where details of the fitting procedure (*dashed* and *dot-dashed lines*) can be found. The *upper inset* shows the temperature dependence of the energy gap (*filled squares*) and the polaron oscillator strength (*open circles*). The *lower inset* is the effective number of carriers. Peak B evolves into a clean charge-gap as the temperature decreases, while A corresponds to polarons

ported by Bao et al. [11.117] found ferromagnetic correlations at high temperature. Upon cooling, the system reaches a state with coexisting FM and AF peaks and, upon further cooling, a purely AF state is eventually reached. This information is summarized in Fig. 11.25a, reproduced from [11.117]. Note that the dominant tendency at high temperature is quite different from the properties of the actual zero-temperature ground state. This behavior is characteristic of many manganites, where many states are in close competition. Very recently, "martensitive" characteristics of the charge-ordering temperature of $Bi_{0.2}Ca_{0.8}MnO_3$ have been unveiled [11.118], and this property may explain some of the unusual phenomena found in these manganites.

Optical studies of the same compound BCMO by Liu et al. [11.119] were also interpreted in terms of a phase-separation picture involving polarons (see Fig. 11.25b). Savosta et al. [11.120] also find indications of phase separation in the regime of electron doping. More recently, inelastic light scattering results by Yoon et al. [11.121] were interpreted as being caused by magnetic fluctuations with "chiral" charge motion (broken time-reversal state), a quite interesting and challenging result. Certainly plenty of surprises are still waiting to be unveiled in the regime of electron-doped manganese oxides.

11.3 Probing of Phase-segregated Systems with Low-Temperature Heat-Capacity Measurements
by J.J. Neumeier and A.L. Cornelius

Low-temperature heat-capacity measurements ($T \leq 10\,\mathrm{K}$) provide information regarding the bulk properties of solids. This includes contributions from the conduction electrons, disorder, the lattice, and magnetic excitations due to spin waves or the nuclear hyperfine field. In the case of phase-segregated systems, such as the CMR manganese oxides, these data by themselves do not provide definitive evidence of phase segregation, especially on the microscopic scale. However, they can provide signatures of the coexistence of multiple phases, such as multiple magnetic states that possess differing magnetic excitations, or the development of disorder. This information, coupled with other bulk physical properties measurements, microscopic studies and/or phenomenological models, can then assist in constructing a physical picture for these systems.

The heat capacity of a magnetic solid at low temperature consists of numerous contributions and it is typically given by

$$C_{\mathrm{p}} = C_{\mathrm{elec}} + C_{\mathrm{hyp}} + C_{\mathrm{lat}} + C_{\mathrm{mag}}, \tag{11.16}$$

where C_{elec} is the electronic contribution, C_{hyp} arises from the hyperfine field of the nuclear moment, C_{lat} is due to the lattice, and C_{mag} is due to magnetic spin waves. The electronic contribution at low temperatures is approximately $C_{\mathrm{elec}} = \gamma T$, where γ is the Sommerfeld coefficient. In insulating CMR materials, such as $LaMnO_3$, C_{elec} is small or zero [11.122]. However, this is not always the case. For example, $CaMnO_3$ possesses a finite density of states at the Fermi level (as indicated by thermopower measurements) [11.109] and a finite value for γ [11.123]. It is also possible for disorder to lead to a linear term in the heat capacity at low temperatures [11.124]. For example, the linear term observed in $Pr_{1-x}Ca_xMnO_3$ for $0.3 \leq x \leq 0.5$ was attributed to this effect [11.125]. It is important to consider electronic transport data in order to accurately evaluate the origin of γ. This is illustrated in the case of $LaMnO_3$ where thermopower measurements [11.126] do not reveal a finite density of states at the Fermi level although a finite value for γ is observed in C_{p}; γ in this case can be attributed to disorder [11.127].

C_{hyp} is given by A/T^2 where A is related to the internal hyperfine magnetic field by the relation [11.122, 11.128]

$$A = \frac{R}{3}\left(\frac{I+1}{I}\right)\left(\frac{\mu_I H_{\text{hyp}}}{k_{\text{B}}}\right)^2 , \tag{11.17}$$

where I is the nuclear spin (5/2 for ^{55}Mn), μ_I is the nuclear magneton, and H_{hyp} is the internal field strength at the Mn site. C_{hyp} results from the existence of electrons in unfilled Mn shells and is essentially the high-temperature side of a Schottky anomaly [11.129]. It is expected to decrease with hole doping, because of increased charge-carrier mobility, which leads to a decrease in H_{hyp}. However, experiments on $La_{1-x}Sr_xMnO_3$ reveal an initial increase in A from $x = 0$ to $x = 0.2$ followed by a decrease at $x = 0.3$ [11.122]. This reflects the initial doping of holes on the oxygen sites in the region $x \leq 0.2$, followed by doping on Mn sites as x moves into the regime where $La_{1-x}Sr_xMnO_3$ is metallic. Furthermore, measurements of $CaMnO_3$ reveal a value of A of roughly 1/2 that observed in $LaMnO_3$ because of the decreased number of e_g electrons [11.123]. Thus, the magnitude of A can provide information regarding the doped charge carriers in the manganites.

The C_{latt} term is approximately $C_{\text{latt}} = \beta T^3$. In some cases, investigators use an additional term proportional to T^5 [11.122,11.130], although this is not universally necessary [11.123] in the manganites. β can be used to calculate the Debye temperature Θ_{D} through the relation

$$\Theta_{\text{D}} = \left(\frac{1.944 \times 10^6\ r}{\beta}\right)^{1/3} , \tag{11.18}$$

where β has units of mJ/mole K^4 and r is the number of atoms per unit cell. Extracting β from low-temperature heat-capacity data is difficult for AFM systems, since the AFM spin-wave excitation (see below) is also expected to behave as T^3. In such a case, heat-capacity measurements above the Néel temperature T_{N} can be used to determine Θ_{D} in the absence of AFM order. This value can then be used to estimate β for the low-temperature fit [11.123].

For the manganites, the C_{mag} term is of particular importance since it provides information regarding magnetic excitations associated to long-range FM or AFM order. It is estimated as $\sum \beta_n T^n$, where the value of the exponent n depends on the type of magnetic excitation. In the case of a simple 3D AFM (such as G-type AFM) a value of $n = 3$ is expected. This term is clearly observed [11.123] in $CaMnO_3$. For more complex systems such as $LaMnO_3$, which possesses A-type AFM order consisting of FM sheets that are antiferromagnetically coupled, the dispersion relation leads to a T^2-behavior [11.122] for C_{mag}. This occurs because of the gap in the AFM spin-wave caused by the presence of the 2D FM sheets. This example underscores the importance of inelastic neutron scattering in providing dispersion relations for magnetic materials possessing a wide rage of spin configurations, such as the manganites. For a purely FM system, an exponent $n = 3/2$ is expected [11.131].

As an example including all of the above-discussed terms, consider electron-doped $CaMnO_3$. Low-temperature heat-capacity data are in Fig. 11.26, along with data for two La-substituted samples with $x = 0.94$ and 0.97. The

Table 11.1. Summary of the fitting parameters to the data in Fig. 11.26. Definitions of the various coefficients are given in the text. The units are mJ/mol K^2 for γ and mJ/mol K^4 for β and mJ/mol K^{n-1} where n is the subscript of the coefficient. The number in parentheses is the statistical uncertainty in the last digit from the least squares fitting procedure

x	A	γ	$10^2\beta$	$10^2\beta_3$	$\beta_{3/2}$	β_2
0	4.70(3)	1.40(4)	3.57	0.83(2)	–	–
0.03	4.30(4)	2.04(9)	3.57	2.28(4)	0.73(2)	–
0.06	4.67(3)	1.93(7)	3.57	0.83(3)	–	0.72(1)

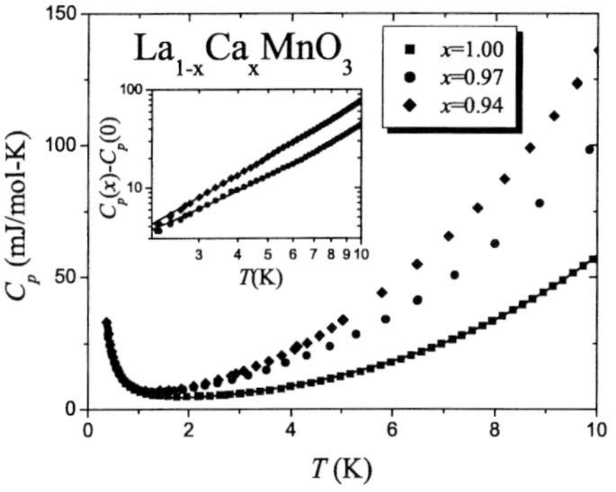

Fig. 11.26. Measured specific heat as a function of temperature for $Ca_{1-x}La_xMnO_3$. The *inset* shows the difference in the La substituted samples relative to the undoped sample. The *lines* are fits described in the text with the relevant fitting parameters summarized in Table 11.1

substitution of La^{3+} for Ca^{2+} adds electrons to this simple G-type antiferromagnet. This leads to the development of a FM saturation moment in the magnetization [11.109,11.110]. The solid line through the data illustrates the high-quality fit obtained using the parameters displayed in Table 11.1. Since the concentration $x = 1.0$ is known to possess the G-type AFM structure, the presence of a T^3 term is expected. With the lattice term also proportional to T^3, the value of $\Theta_D = 648$ K obtained through fitting at 200 K is used; this value is used as well in fitting the data for the $x = 0.94$ and 0.97 specimens discussed below, since the small substitution levels are not expected to strongly influence Θ_D. With β fixed, the fitting yields $\beta_3 = 0.083(2)$ mJ/mole K^2, which indicates the presence of AFM order. In addition, the value of γ so obtained suggests the existence of a finite density of states at the Fermi level. This is consistent with thermopower measurements [11.109]. The hyperfine term A corresponds to an internal magnetic field of 27.6 T. This value of A is only 60% of that found [11.122] for $LaMnO_3$ and a factor of 6 smaller than that found for $Pr_{0.6}Ca_{0.4}MnO_3$ [11.130]. The unusually large value found in the latter compound may be due to the fact that the fitting procedure was conducted on data at temperatures above 2 K, whereas the C_{hyp} term increases in magnitude (yielding a more reliable fit) below this temperature [11.122, 11.123].

The electron-doped samples are most interesting since they are expected to reveal changes in C_p related to phase segregation. Considering first the specimen $La_{0.03}Ca_{0.97}MnO_3$ a significant change in the heat capacity is evident in Fig. 11.26 and its inset, where $C_p(x) - C_p(0)$ is plotted versus T. Fitting the data with (11.16) reveals a ferromagnetic term proportional to $T^{3/2}$. While this suggests the existence of a long-range FM spin-wave, the magnitude of $\beta_{3/2}$ is smaller than that observed in purely FM samples [11.122] suggesting either a larger spin-wave stiffness, or partial ferromagnetism (note that $C_{\mathrm{mag}} \sim D^{-2}$ where D is the spin-wave stiffness). Interestingly, the β_3 term jumps dramatically signifying a decrease in the spin-wave stiffness for the G-type AFM component. This likely occurs as a result of inclusion of FM regions in the AFM lattice, or the existence of non collinear (or canted) AFM. Furthermore, the electronic term γ increases as would be expected because of the enhanced charge carrier concentration [11.110]. The slight decrease in A may signify a change in the charge distribution around the Mn ions due to the increased charge mobility associated with the polaronic conduction.

In the case of $x = 0.94$, the $T^{3/2}$ term is no longer observed, but a T^2 term is. This change in behavior is most clearly revealed in the inset of Fig. 11.26 where $C_p(x) - C_p(0)$ is plotted versus T on a log log scale revealing good agreement with the exponent of 2. Interestingly, the T^3 term is of similar magnitude to that of the undoped sample. The coexistence of these two terms suggests that a change of regime has occurred and that two distinct AFM phases are present. In fact, powder neutron diffraction reveals the existence of a finite fraction of C-type AFM at this doping level [11.132]. The similarity of the T^3 term may simply be a coincidence, due to a combined decrease in spin-wave stiffness, change in phase fraction, and introduction of a new AFM phase. The absence of a $T^{3/2}$ term is puzzling given the fact that neutron diffraction reveals signatures of long-range FM at this composition [11.132].

12. Optical Conductivity

In this chapter, properties of the optical conductivity in the area of manganites are reviewed, both in their experimental and theoretical aspects. The basic features of the technique itself are also reviewed here in a contribution by T.W. Noh. The reader will not only learn of results for manganites, but will obtain information to address optical experiments in many other materials as well.

12.1 $\sigma(\omega)$: Experimental Results in Manganites

We start this section with a summary of experimental results for manganites. The emphasis will be on common features among several Mn oxides, rather than anomalies of particular materials. A picture will clearly emerge showing common behavior among several compounds, suggesting that there must be a universal mechanism behind the results.

Working at room temperature with polycrystalline $La_{1-x}Ca_xMnO_3$ samples, Jung et al. [12.1] reported the optical conductivity shown in Fig. 12.1a. The region of low energy – the so-called midgap states – was investigated in detail, and effects due to Jahn–Teller couplings were identified (at higher energies a transition from the oxygen $2p$-band to the e_g band was clearly observed [12.1], but it will not be discussed here). The figure shows that the results from $x = 0$ to 1 can be fit with two Gaussians: the lower energy peak is attributed to the transition II in Fig. 12.1b, namely, from a Jahn–Teller split Mn^{3+} ion to a Mn^{4+} unoccupied electronic state (since there is no electron in that state, there is no JT distortion). The peak in Fig. 12.1a with the highest energy is attributed [12.1] to an intra-atomic transition between JT split e_g states (see I of Fig. 12.1b). Process III of Fig. 12.1b is believed to involve a large energy once Coulombic interactions are considered, thus it was discarded. Additional discussion on these interpretations can be found in [12.1]. There is no obvious $\omega = 0$ Drude weight in $\sigma(\omega)$ due to the insulating character of LCMO at the room temperature considered. The important results reported in [12.1] show that the broad features associated with JT distortions at room temperature may be active at all densities in LCMO, *regardless* of the type of state that is stabilized at low temperatures.

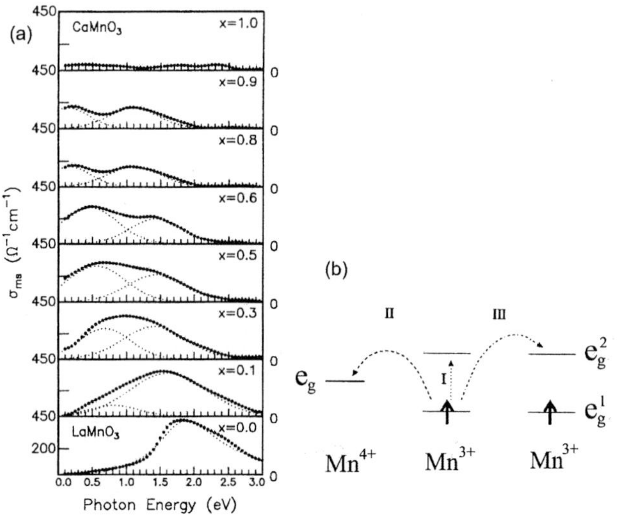

Fig. 12.1. (**a**) Optical conductivity spectra of the midgap state in $La_{1-x}Ca_xMnO_3$, from [12.1]. The *solid circles* are experimental data, while the *dotted lines* represent two Gaussian functions. (**b**) Schematic diagram of the possible intra- and interatomic processes that may lead to the peaks shown in (**a**). For details see text and [12.1]

However, note that the interpretation of these results is still under discussion. Investigations of $x = 0.175$ $La_{1-x}Sr_xMnO_3$ by Okimoto et al. [12.2], in Fig. 12.2a, observed near room temperature similar features as reported in [12.1], namely, a very broad peak. This peak was interpreted in terms of transitions between bands split by the Hund coupling, as contained in the one-orbital model, instead of invoking JT couplings. The temperature dependence of the results shows the presence of weight at low energy in the ferromagnetic phase, as expected from a metal. However, the integral of the optical conductivity up to frequency ω, usually denoted by $\tilde{N}_{\mathrm{eff}}(\omega)$, still varies substantially even when the magnetization is nearly saturated (see Fig. 12.2c). This indicates that physics beyond the double exchange must be considered, since a saturated ferromagnet should produce a nearly perfect metal, at least within the one-orbital family of theories. The missing physics may be the JT distortions. Results for $Nd_{1-x}Sr_xMnO_3$ at $x = 0.30$ have a shape and temperature dependence similar to those of LCMO and LSMO (see Fig. 12.2d, from [12.3]). The low-energy data can be fitted with a narrow Drude peak and a midinfrared absorption peak. The temperature dependence of the optical conductivity of $x = 0.30$ LCMO is shown in Fig. 12.2b, from [12.4], once again showing similarities with other compounds. Regardless of the interpretation, it is clear that (i) a very broad feature around 1.0 eV, (ii) the transfer of

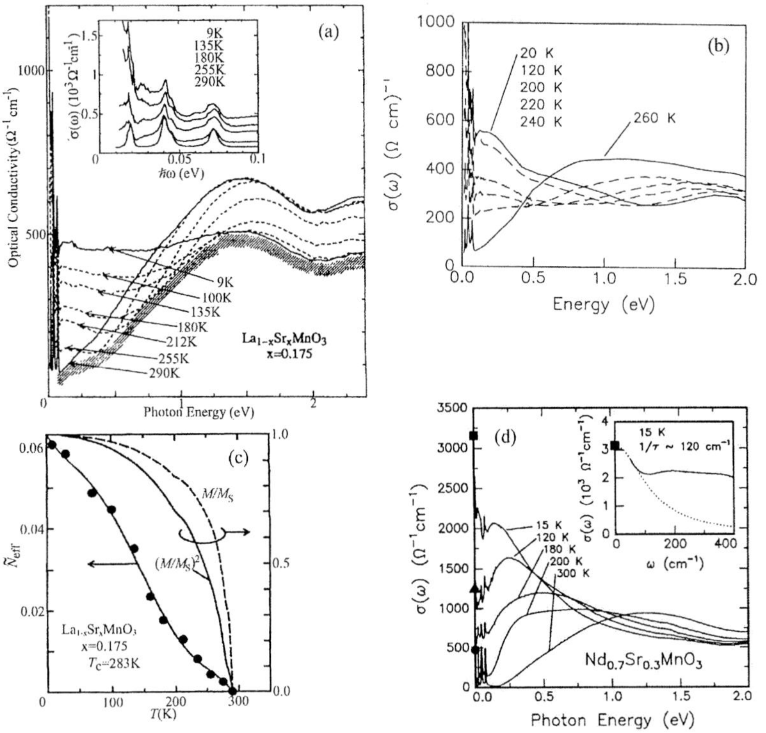

Fig. 12.2. (**a**) Optical conductivity of $La_{1-x}Sr_xMnO_3$ at $x = 0.175$, from [12.2]. The temperature dependence is indicated. The *inset* shows a magnification of the low-energy portion. (**b**) Optical conductivity of LCMO at $x = 0.30$, from [12.4]. (**c**) Temperature dependence of the effective number of electrons $\tilde{N}_{\text{eff}}(\omega)$, from [12.2]. Shown also is the magnetization of the sample. (**d**) Optical conductivity, from [12.3], corresponding to $Nd_{1-x}Sr_xMnO_3$ at $x = 0.30$. The temperature dependence is indicated. *Inset* corresponds to a simple Drude weight fitting of the data

weight from that peak to smaller energies on cooling, and (iii) a $\tilde{N}_{\text{eff}}(\omega)$ that changes substantially even within the ferromagnetic phase, are characteristic of several Mn oxides, particularly those that have a ferromagnetic metallic phase at low temperature.

To complete this brief summary of optical conductivity experimental results, data corresponding to other manganites are shown in Fig. 12.3. The relevant details of these results are the following:

(i) Mn oxides with a charge gap at low temperatures. In this family of compounds results for the single-layer $La_{1/2}Sr_{3/2}MnO_4$ are shown in Fig. 12.3a, from [12.5]. Upon cooling, weight moves to higher energies, contrary to the behavior found in materials where the low-temperature state is a metal. However, the broad feature centered near 1.0 eV is still clearly visible. A similar behavior is observed for the perovskite $Nd_{1/2}Sr_{1/2}MnO_3$, which is charge ordered at low temperatures (see Fig. 12.3b, from [12.6]). Note that results quite similar to those induced by increasing temperature were obtained by increasing magnetic fields, as shown in Fig. 12.3c, with weight being transfered from high to low energies.

(ii) Mn oxides with a metallic state at low temperatures. In addition to the results already discussed for this type of material, additional results for the compound $La_{1.2}Sr_{1.8}Mn_2O_7$ were reported in [12.7] (see Fig. 12.3d). The behavior is clearly similar to that of LCMO at $x = 0.30$.

(iii) Other cases. Materials in complex regions of the phase diagram, such as $La_{1-x}Sr_xMnO_3$ at $x = 0.125$ where phase-separation tendencies may be relevant, show a transfer of weight to low energies upon cooling, but a Drude peak does not form (see Fig. 12.3e, from [12.8]). This may correspond to a state with metallic regions that are not percolated.

Comments:

(1) From the optical experiments described above, it is difficult to address whether clusters of CO- and FM-phases compete above T_C, as claimed using a variety of other experimental probes and also studying realistic models. The reason is that the energies involved in the optical processes are compatible with local or at most nearest-neighbor electron-transfer processes, as in Fig. 12.1b. However, note that the broad character of the midgap states, both for LCMO, LSMO and others, may indicate the presence of an inhomogeneous distribution of active sites. Clearly, the origin of the broad width of midgap features should be understood in more detail. It would be highly desirable to investigate by optical means the *low-energy* portion of the spectrum trying to identify features characteristics of, e.g., a pseudo-CE phase that could be present in clustered form in the high-temperature insulating state of $x = 0.30$ LCMO, as observed in other experiments. Perhaps the low-energy phononic spectrum must be studied in more detail in future investigations. Note that, according to recent estimations [12.9], optical experiments tend to produce averages over length scales of about 10–100 nm. Therefore, if the potentially coexisting domains are larger than those lengths, signatures can be found in $\sigma(\omega)$, but for smaller nanoscale domains the effects of inhomogeneities cannot be easily observed with this technique, unfortunately.

Regarding these issues, progress was recently made through the optical investigation of $La_{5/8-y}Pr_yCa_{3/8}MnO_3$ ($y = 0.35$) [12.10], a material that clearly presents mixed-phase tendencies, with large submicrometer-size domains (see Chap. 11). With such large clusters, optical techniques should detect the coexistence of phases. Results are shown in Fig. 12.4a,b. At temperatures above the Curie temperature where FM clusters are not yet formed, the optical conductivity has features clearly reminiscent of a charge-ordered

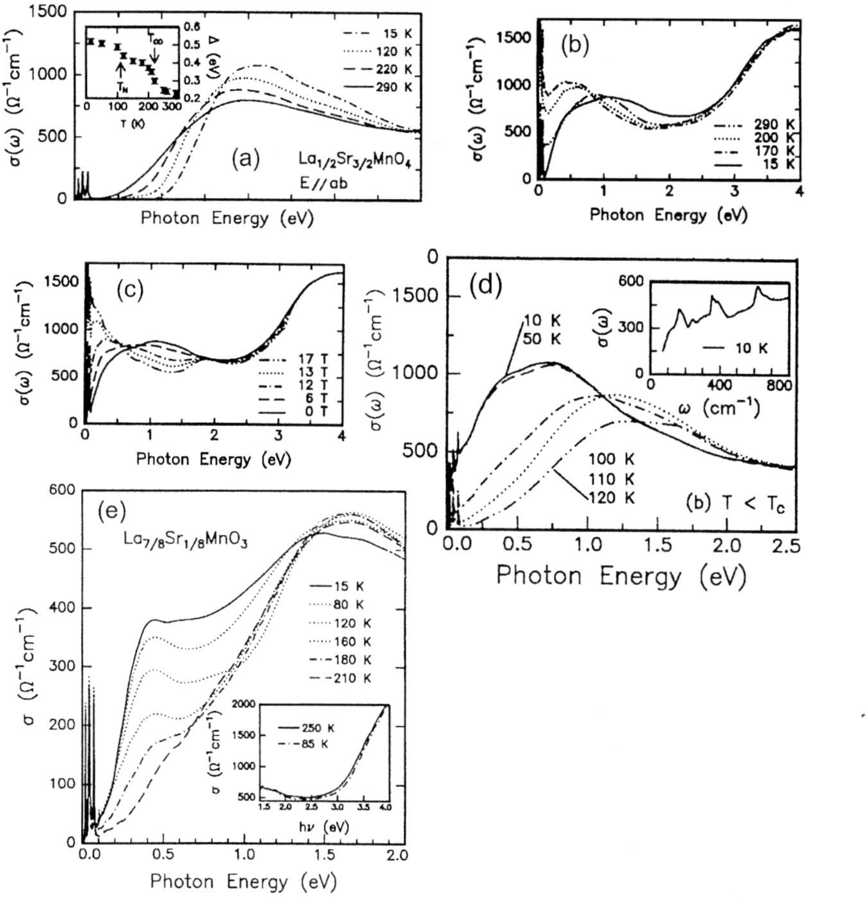

Fig. 12.3. (**a**) Optical conductivity of the single-layer $La_{1/2}Sr_{3/2}MnO_4$, from [12.5]. (**b**) Optical conductivity of $Nd_{1/2}Sr_{1/2}MnO_3$, from [12.6]. (**c**) Magnetic-field dependence of the optical conductivity of $Nd_{1/2}Sr_{1/2}MnO_3$ at low temperatures, from [12.6]. (**d**) Optical conductivity of the bilayered compound $La_{1.2}Sr_{1.8}Mn_2O_7$, from [12.7]. (**e**) Optical conductivity of LSMO at $x = 0.125$, from [12.8]

state. This occurs even at room temperature, showing that CO clusters exist in that regime. At low temperatures, panel (b) shows that the feature around 1.4 eV does not disappear (CO clusters remain), but weight appears at low energies, between 0.2 and 0.5 eV. According to the analysis carried out in [12.10], which includes a study varying magnetic fields, this weight corresponds to the contribution coming from the FM clusters formed upon cooling. The data of [12.10] clearly supports the conjectured mixed-phase tendency of $La_{5/8-y}Pr_yCa_{3/8}MnO_3$ at $y = 0.35$.

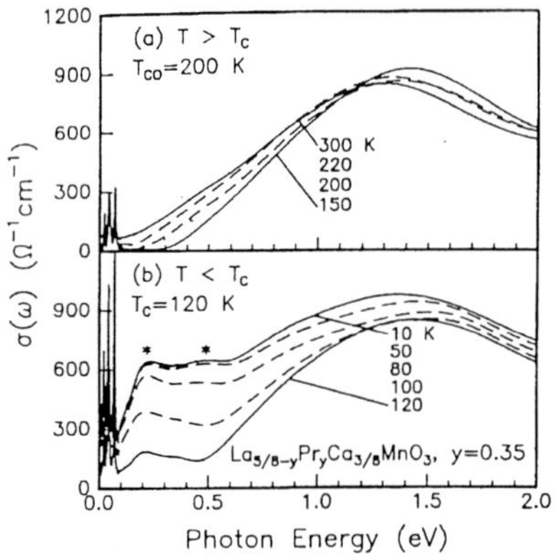

Fig. 12.4. Temperature dependence of the optical conductivity (**a**) above T_C and (**b**) below T_C, corresponding to $La_{5/8-y}Pr_yCa_{3/8}MnO_3$ with y=0.35, from [12.10]

(2) Note that Jung et al. [12.1] refer to the simple transition II shown in Fig. 12.1b, which involves just two ions, as "polaron absorption". They present their results for other materials as evidence of polaronic behavior. However, note that this type of transition can be present in states that are not dominated exclusively by polarons. For instance, states with mixed-phase tendencies containing small islands of charge-ordered phases may also have similar transitions. Thus, the reader must keep in mind that several processes identified in optical data are very local, and information about extended states is difficult to gather. For this reason the data seems quite similar among the many compounds, even if they have different ground states. Only the region below 0.5 eV appears sensitive to the actual long-range features of that ground state. In fact, recent studies [12.11] showed that FM states have an isotropic conductivity, while A-type and C-type states have anisotropic behavior among the three axes, concomitant with the type of orbital order they present. Then, the low-energy transition Mn^{3+}–Mn^{4+} can provide information about the local properties of the states under investigation.

(3) Another important detail to note is the close similarity of the results found in manganites with those in other transition-metal oxides (for an excellent review by Cooper, see [12.12]). (i) For instance, in Fig. 12.5a results for a single-layered Ni oxide are shown [12.13]. The hole concentration is $x = 1/3$, which corresponds to a charge-ordered state at low temperatures. The result, Fig. 12.5a, has a broad peak roughly centered at 1 eV, and weight moves from low energy to high energy upon cooling due to the charge gap that develops. This behavior is quite similar to the results shown before for manganites with a CE charge-ordered state. Note that usually nickelates are not associated with any kind of Jahn Teller distortion, yet the results are similar to those of Mn oxides, which is surprising. Perhaps there is a simple explanation for the existence of the broad midgap features that is common to many transition-metal oxides. (ii) Even copper-oxide compounds have features that resemble

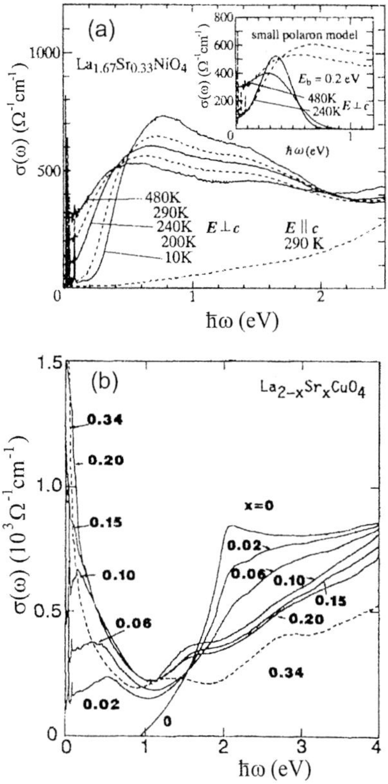

Fig. 12.5. (**a**) Optical conductivity of the single-layered nickelate $La_{1.67}Sr_{0.33}NiO_4$, from [12.13]. The *inset* shows fitting results with small polaron models. (**b**) Optical conductivity of single-layered $La_{2-x}Sr_xCuO_4$, a representative of the high-T_c family, from [12.14]. Results for both nickelates and cuprates are similar to those of manganites

those of manganites. In Fig. 12.5b the optical conductivity of LSCO is shown at several hole concentrations [12.14]. The high-energy features at 2 eV or higher are related to charge-transfer excitations and they occur also in manganites, although they were not described in the previous discussion. Here the focus should be the energy region below 1.5 eV. For small hole density x, a broad feature is centered at 0.5 eV, which moves down upon increasing the hole density until it becomes a Drude weight. This behavior resembles results found for manganites with a metallic ground state. A discussion of experimental results for $\sigma(\omega)$ in the area of Cu oxides can be found in [1.20]. Clearly, there are strong similarities in the optical conductivity between the results obtained in Mn oxides and other transition-metal oxides.

Note. A standard definition of the different ranges of energies used in optical measurements is the following. The "microwave" regime corresponds to energies smaller than 1 meV. The "far-infrared" is between 1 meV and about 20 meV. The "midinfrared" is between 20 meV and about 0.5 eV. The "near-infrared" is between 0.5 eV and 2 eV. The "visible" is a narrow range around 2–3 eV. Higher energies than the visible lead to the "ultraviolet" range. In addition, to transform from energies to inverse cm or Kelvin, use

$$1\,\mathrm{eV} = 8065\,\mathrm{cm}^{-1} = 11\,600\,\mathrm{K}\,. \tag{12.1}$$

12.2 $\sigma(\omega)$: Theoretical Results

The theoretical calculations of the optical conductivity reported in the literature can be clearly separated in two groups: (1) those using Jahn Teller phonons and a large Hund coupling as the dominant interactions, and (2) those using purely electronic Coulombic interactions. The results of these approaches are described below. However, before this it is useful to provide a brief reminder of equations widely used in the context of calculations of optical conductivities. The many models that can be used forbids a general formulation, thus here zero-temperature results corresponding to the famous one-band Hubbard model for cuprates are described. However, if the one- or two-orbital models for manganites with Coulombic or Jahn–Teller interactions is used, the formulas change only slightly. The reader interested in the fine details of these equations, and their derivation and interpretation, should consult p. 803 of [1.20], where a review of optical conductivity in the context of high-temperature superconductors was provided. The present discussion is based on that review.

Focusing on the equations for the one-band zero-temperature Hubbard model, the operator related to the response of the system to a weak electric field is the current operator [1.20] defined as

$$\hat{j}_l(\boldsymbol{r}) = \mathrm{i}t \sum_\sigma (c^\dagger_{\boldsymbol{r}\sigma} c_{\boldsymbol{r}+l\sigma} - c^\dagger_{\boldsymbol{r}+l\sigma} c_{\boldsymbol{r}\sigma})\,, \tag{12.2}$$

where the sites of the lattice are denoted by $\boldsymbol{r}$, the axes directions by l, the spin by σ, and the operators are the usual creation and destruction operators of fermions. In the case of manganites, an extra index related to orbitals must be included. The kinetic energy operator, which is used in this context to derive the Drude weight discussed below, is defined as

$$\hat{T} = -t \sum_{\boldsymbol{r}l\sigma} (c^\dagger_{\boldsymbol{r}\sigma} c_{\boldsymbol{r}+l\sigma} + c^\dagger_{\boldsymbol{r}+l\sigma} c_{\boldsymbol{r}\sigma})\,, \tag{12.3}$$

with the same notation as for the current. This operator is one of the terms of the Hubbard-model Hamiltonian. Following the derivation discussed in [1.20] and others, it can be shown that the real part of the optical conductivity can be written as the sum of a zero-frequency delta function with weight D (the Drude weight), and a "regular" or "incoherent" contribution that is present at finite frequency:

$$\sigma_1(\omega) = D\delta(\omega) + \frac{e^2\pi}{N} \sum_{n\neq 0} \frac{|\langle\phi_0|\hat{j}_x|\phi_n\rangle|^2}{E_n - E_0} \delta(\omega - (E_n - E_0))\,. \tag{12.4}$$

In the last equation the eigenstates of the problem are denoted by $|\phi_n\rangle$ and their associated energies by E_n. The number of sites is N and the charge e. The current operator is obtained from (12.2) by summing over sites. The Drude weight D is given by

$$\frac{D}{2\pi e^2} = -\frac{\langle\hat{T}\rangle}{4N} - \frac{1}{N}\sum_{n\neq 0}\frac{|\langle\phi_0|\hat{j}_x|\phi_n\rangle|^2}{E_n - E_0}, \tag{12.5}$$

and the kinetic operator expectation value is in the ground state $|\phi_0\rangle$. The zero-frequency delta function is important: when it is present the movement of carriers is "coherent". In fact, D can be used as the order parameter for metal insulator transitions (for a discussion see [1.20], and references therein). In realistic systems, for example with disorder, the delta function acquires a width and the inverse of $\sigma(\omega = 0)$ is the D.C. resistivity that can be measured experimentally. Finally, it is worth mentioning that integrating both terms of the expression for $\sigma_1(\omega)$ above, the famous "sum-rule" for the optical conductivity is derived. This expression relates the expectation value of the kinetic energy with the integral of the optical conductivity as

$$\int_0^\infty d\omega\sigma_1(\omega) = -\frac{\pi e^2}{4N}\langle\hat{T}\rangle\,. \tag{12.6}$$

This sum-rule is an important test of the accuracy of calculations since the left- and right-hand sides are evaluated independently. For unbiased numerical results, both terms must match, providing a useful way to debug programs. Once again note that the equations above are for the Hubbard model, and they should be rederived for use in other models. However, the general trends sketched here are common to most problems, and just some fine details may be different.

Let us return to the description of results in the manganite context. In the derivation of equations used to gather these results, slight modifications of the Hubbard model equations were used, particularly those that provide the Drude weight (that cannot be calculated *directly* on finite systems with periodic boundary conditions [1.20]).

For a two-dimensional system, the optical conductivity sum-rule reads:

$$\frac{D}{2} = \frac{\pi e^2\langle -\hat{T}\rangle}{4N} - \int_{0^+}^\infty \sigma_1(\omega)\, d\omega\,, \tag{12.7}$$

which is a mere rewriting of the previous formula for the Drude weight, but providing explicitly the last term as an integral of the conductivity. For the two-orbital model with J_{H} finite, the current is

$$\hat{j}_x = \mathrm{i}e \sum_{i\gamma\gamma'\sigma} (t^x_{\gamma\gamma'} a^\dagger_{i\gamma\sigma} a_{i+x\gamma'\sigma} - \mathrm{H.c.})\,,$$

and the kinetic energy operator is

$$-\hat{T} = \sum_{i\gamma\gamma' a\sigma} (t^a_{\gamma\gamma'} a^\dagger_{i\gamma\sigma} a_{i+a\gamma'\sigma} + \mathrm{H.c.})\,,$$

where the operators create or destroy electrons at site i, with orbital γ, and spin projection σ. The equivalent formulas for the infinite J_{H} model are:

$$\hat{j}_x = \mathrm{i}e \sum_{i\gamma\gamma'} (t^x_{\gamma\gamma'} S_{i,i+x} \tilde{a}^\dagger_{i\gamma} \tilde{a}_{i+x\gamma'} - \mathrm{H.c.})$$

and

$$-\hat{T} = \sum_{i\gamma\gamma' a\sigma} (t^a_{\gamma\gamma'} S_{i,i+a} \tilde{a}^\dagger_{i\gamma} \tilde{a}_{i+a\gamma'} + \mathrm{H.c.}) ,$$

where

$$S_{i,j} = \cos(\theta_i/2)\cos(\theta_j/2) + \sin(\theta_i/2)\sin(\theta_j/2)\mathrm{e}^{-\mathrm{i}(\phi_i - \phi_j)} .$$

For more details on the notation see Chap. 5.

The summary of results is the following:

(1) Calculations using Monte Carlo techniques applied to a two-orbital model with large Jahn–Teller electron phonon coupling and also a large Hund coupling have been reported in [9.4, 12.15]. Some results are given in Fig. 12.6a,b. In panel (a), the temperature dependence is presented at electronic density 0.7 (weight at very large energy of order of the Hund coupling is not shown). A one-dimensional system is used, but the results are believed to be qualitatively independent of dimensions due to the simplicity and generality of their interpretation. From the study of spin correlations, the authors of [9.4] believe that the temperature at which FM correlations become appreciable over several lattice spacings is $T = t/20$. An important point is that even in this FM phase, the Drude weight shown in the inset of (a) is still changing with temperature, as observed in experiments due to the presence of an active phononic degree of freedom even at low temperatures. In addition, it is clear that a broad peak appears at intermediate energies in $\sigma(\omega)$, as also occurs in experiments. This peak is believed to be caused by phononic Jahn–Teller excitations, in agreement with the interpretation of [12.1], and it moves to lower energies as the temperature is reduced. In Fig. 12.6b, calculations above the ordering temperature are presented, parametric with the electronic density. A clear two-peak structure is observed, quite reminiscent of the experimental results [12.1] (note that here the Drude weight is nearly negligible). The large-energy peak is caused by intersite transitions, while the low-energy peak is compatible with Mn^{3+} to Mn^{4+} transitions, once again in agreement with the experimental results of [12.1] (see also [12.2]). Overall, the theoretical calculations presented in Fig. 12.6 seem to reproduce qualitatively the main features reported in the experimental literature corresponding to materials that become metallic at low temperatures. However, more work is needed since $\sigma(\omega)$ in charge-ordered regimes has not been studied in detail.

(2) Regarding calculations using electronic Coulomb interactions instead of JT couplings, here the main rationale for the appearance of the broad peak centered at $\sim 1\,\mathrm{eV}$ in experiments is the presence of orbital fluctuations. In this context the spin degree of freedom is not important in a spin fully

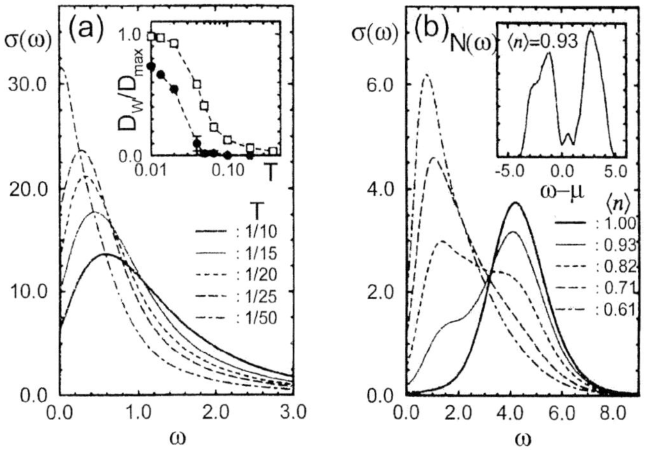

Fig. 12.6. (**a**) $\sigma(\omega)$ at $\lambda = 1.0$ (electron phonon coupling), electronic density 0.7, and several temperatures. The calculation was done using a chain of 20 sites, the Monte Carlo technique, and a two-orbital model with JT phonons and a large Hund coupling. For details the reader should consult [9.4]. The *inset* shows the Drude weight D_W vs. temperature for the two-orbital (*circles*) and one-orbital (*squares*) model. (**b**) Optical conductivity spectra parametric with the electronic density $\langle n \rangle$ at $\lambda = 1.5$, $T = t/10$ (a relatively high temperature), and using a chain of 16 sites (similar results were obtained on a 4×4 cluster). The *inset* contains the density of states $N(\omega)$ corresponding to the lower Hund coupling band, showing the splitting caused by the robust JT-phonon coupling

saturated system, but the orbital degree of freedom is active. The results of calculations in small clusters [12.16] using a two-orbital model are shown in Fig. 12.7, where only the regular part of the conductivity is presented (i.e., the Drude peak is not shown). At low temperatures, a peak at $\sim 2t$ is observed, assigned to orbital excitations. The Drude weight is shown in the inset. Even at zero temperature, it does not carry the full weight. The calculations were done in a x^2–y^2 orbitally ordered state, and the spin is assumed to be ferromagnetically aligned. It would be interesting to relax this last constraint and verify what occurs when both orbital and spin are allowed to fluctuate. It would also be important to study other phases, such as the charge-ordered CE-state.

Several other theoretical results for $\sigma(\omega)$ have been reported. Using the infinite dimension $D = \infty$ approximation, the optical conductivity of the one-orbital model has been calculated [12.17]. At low temperatures in the ferromagnetic regime the conductivity reduces to a Drude peak that carries 100% of the weight, namely, there is no incoherent structure coexisting with the Drude peak. For this reason, it is considered that other degrees of freedom such as the lattice and/or orbitals are needed to reproduce the experiments that show that the Drude peak changes substantially even in

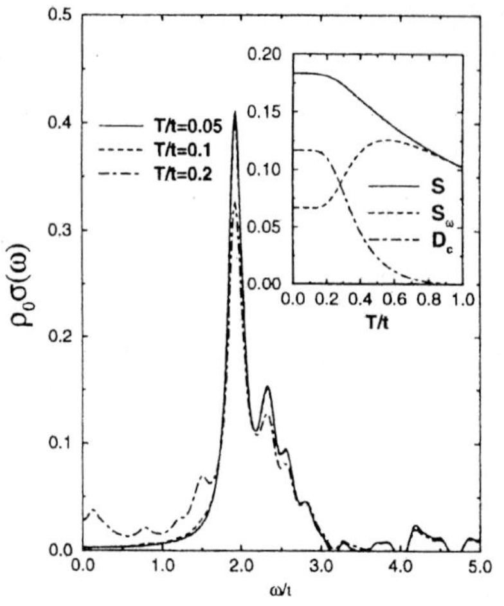

Fig. 12.7. Regular part of the optical conductivity in a low-temperature x^2-y^2 orbital-ordered regime, at the temperatures indicated. The model is the orbital t–J model (see Chap. 9). The spin is assumed to be fully saturated. The density is $x = 0.6$ and the cluster size 4×4. More details can be found in [12.16]

the FM regime. Calculations using two orbitals, in addition to those already discussed, can be found in [9.13, 12.18, 12.19]. Particularly in [9.13, 12.19], a possible "orbital liquid" character of the ferromagnetic state is analyzed. The relevance of both Coulombic and Jahn Teller interactions has been proposed in [12.20] to explain the incoherent charge dynamics of ferromagnetic manganites. Additional discussion related to theoretical calculations of $\sigma(\omega)$ using the $D = \infty$ approximation, and its comparison with experiments, can be found in [12.21]. Note that very recent experimental results [12.9] favor the interpretation of optical-conductivity data based on the JT scenario over the purely Coulombic one, but more work should be carried out to confirm those results. Theoretical investigations (reviewed elsewhere in this book) appear to suggest that the differences between the two scenarios are not as sharp as previously believed, at least regarding ground-state properties, and probably a combination of both is needed to fully account for the experimental results.

In summary, theoretical and experimental studies of the optical conductivity of Mn oxides are in reasonable agreement in the intermediate energy range of excitations. Future studies of the low-energy regime could lead to important results, particularly addressing the presence of inhomogeneities in these compounds. Clearly, optical techniques are powerful methods that have led to important progress in the study of transition-metal oxides.

12.3 Brief Introduction to Experimental Aspects of Optical Spectroscopy
by T.W. Noh

Optical spectroscopy is a simple but useful experimental method to probe numerous degrees of freedom in solids. When a light wave enters a material, its electric field can be strongly coupled with charged particles, such as electrons and ions, inside the material. The average response of the material to the electromagnetic field is represented by the frequency-dependent complex dielectric function $\widetilde{\varepsilon}$, which contains detailed dynamic information on charge, orbital, and lattice degrees of freedom. On the other hand, for most materials, the interaction of the charged particles with the magnetic field of the light wave is usually small enough to be neglected, so their spin degree of freedom is difficult to be probed by the light. However, in manganites, there is a strong interplay between spin and charge degrees of freedom, which is usually represented by the double-exchange interaction. Therefore, the optical spectroscopy can be used to probe the interplays among spin, charge, orbital, and lattice degrees of freedom in manganites.

Typical optical-spectroscopic techniques involve shining light onto a sample and measuring the reflectance or transmittance of the sample, as a function of the photon energy [12.22]. In general, transmittance measurements are much more sensitive than reflectance measurements. In the spectral regions of interest, optical absorption is generally quite high, so that often a negligibly small fraction of the incident light can be transmitted. For example, in the visible and ultraviolet regions, many materials have skin depths of the order of only 100 Å. However, it is generally not feasible to make high-quality films thin enough to obtain meaningful data. Therefore, reflectance spectra $R(\omega)$ are frequently measured using bulk samples.

When light of a frequency ω is incident at near normal incidence onto a solid, the reflectivity coefficient $\widetilde{r}$ can be defined as the ratio of the reflected electric field $\widetilde{E}_{\text{refl}}$ to the incident electric field $\widetilde{E}_{\text{inc}}$:

$$\widetilde{E}_{\text{refl}}/\widetilde{E}_{\text{inc}} \equiv \widetilde{r}(\omega) \equiv \rho(\omega)\exp[\mathrm{i}\theta(\omega)] , \tag{12.8}$$

where ρ and θ are the amplitude and phase component of the reflectivity coefficient, respectively. Using the continuity conditions of electric and magnetic fields parallel to the crystal surface, $\widetilde{r}(\omega)$ can be written in terms of $\widetilde{\varepsilon}(\omega)$ [3.3]:

$$\widetilde{r}(\omega) = \frac{\sqrt{\widetilde{\varepsilon}(\omega)} - 1}{\sqrt{\widetilde{\varepsilon}(\omega)} + 1} . \tag{12.9}$$

The quantity measured in experiments is the reflectance spectra $R(\omega)$, defined as the intensity ratio between the incoming and the reflected beams:

$$R(\omega) = \left|\widetilde{E}_{\text{refl}}\right|^2 / \left|\widetilde{E}_{\text{inc}}\right|^2 = |\widetilde{r}(\omega)|^2 = \rho(\omega)^2 . \tag{12.10}$$

It is difficult to measure $\theta(\omega)$ directly in the optical region. After reflectance measurements, we usually have only one piece of information for a given frequency, but we are interested in the full complex quantity, i.e., $\widetilde{\varepsilon}$. Fortunately, due to the Kramers Kronig (K K) relation, we can find the imaginary part of the response of a linear passive system if we know the real part of the response at all frequencies, and vice versa. Using $\ln \widetilde{r}(\omega) = \ln R^{1/2}(\omega) + \mathrm{i}\theta(\omega)$, the phase can be obtained through [12.23]

$$\theta(\omega) = -\frac{\omega}{\pi} P \int_0^\infty \frac{\ln R(s)}{s^2 - \omega^2} \mathrm{d}s\,. \tag{12.11}$$

Once we know R for all frequencies, $\widetilde{\varepsilon}(\omega)$ can be determined using the equations described above.

In any optical study, the most important first step is to obtain $R(\omega)$ accurately. Most of the samples, either in single crystalline or in polycrystalline bulk form, do not have flat and shiny surfaces, so there will be lots of scattering from those surfaces. A clean and flat cleaved surface from a single crystal is the best, however, it is quite difficult to obtain. To reduce the light-scattering problems, optical polishing, oxygen annealing, and gold-normalization processes are commonly used. Before optical measurements, the samples were polished up to a submicrometer level by mechanical polishing, using polishing paper or diamond paste. For manganites, the surface-polishing process was known to result in very large strains on the sample surface [12.24]. Due to the strong Jahn–Teller coupling, such strains will cause a serious distortion in $R(\omega)$. This surface-damage problem was found to be relieved by annealing the polished sample in an oxygen atmosphere [12.3]. Although the sample surface was polished up to a submicrometer level, the surface-scattering problem is still serious in the visible and ultraviolet regions, where the wavelength of the light becomes comparable to a typical scale of surface roughness. To obtain the correct $R(\omega)$ of the polished surface, a thin gold film is evaporated on it. Then, $R_{\mathrm{Gold}}(\omega)$ of the gold-coated surface will be measured again. The ratio between $R(\omega)$ and $R_{\mathrm{Gold}}(\omega)$ can provide information on how much of the incoming light was scattered in the original reflectance measurements and be used to determine the correction factor for the surface scattering. The polishing, oxygen annealing, and gold-normalization processes were found to be essential to obtain the correct reflectance spectra for manganites [12.3].

The spectral region of measurements will be decided by the kind of degrees of freedom to be investigated. To study the dynamics of phonons and/or electrodynamics of free carriers, one should measure $R(\omega)$ at least from far-infrared (IR) to mid-IR regions (for a definition of these and other regions, see the "Note" before the section on theoretical studies in this chapter). To investigate electronic transitions near the Fermi level, one should measure $R(\omega)$ in the frequency region from near-IR to ultraviolet (UV). For manganites, large spectral weight changes were observed from far-IR to UV regions [12.2, 12.4, 12.25], so it is important to measure $R(\omega)$ from a few meV to a few tens of eV. Moreover, the wider the measured spectral region, the smaller the errors in the calculated values of $\theta(\omega)$. To cover such a wide spectral region from far-IR to UV, three or four different spectrophotometers are commonly used with various combinations of sources, detectors, and other optical components. At Seoul National University, in the millimeter and submillimeter regions, we use a Lamellar grating spectrophotometer. To cover frequencies between 5.0 meV and 6.0 eV, we use commercial Fourier transform spectrophotometers and grating-type spectrophotometers. In the UV region above 6 eV, we use syncrotron radiation at the Pohang Light Source. Therefore, reflectance measurements on any sample in a wide frequency region is a very time-consuming process.

It is practically impossible to measure $R(\omega)$ at all frequencies. In order to use the formulas described here properly, several extrapolation techniques are commonly used to reduce the possible error in $\theta(\omega)$ due to the unmeasured spectral region [12.26]. The low-frequency extrapolation is rather simple. For a metallic sample, the Hagen–Rubens relation is used: $R(\omega) \approx 1 - \sqrt{2\omega/\sigma_{\mathrm{dc}}\pi}$, where σ_{dc} is the D.C. conductivity [12.27]. For an insulating sample, the Lorentz oscillator fitting to the lowest-lying excitation (usually, a transverse optic phonon) is assumed with $R = \left|\left(\sqrt{\varepsilon_{\mathrm{dc}}} - 1\right) / \left(\sqrt{\varepsilon_{\mathrm{dc}}} + 1\right)\right|^2$, where $\varepsilon_{\mathrm{dc}}$ is the D.C. dielectric constant. The high-frequency extrapolation is more difficult to treat accurately and leads to the largest source of error in $\theta(\omega)$. Most experiments do not include the full range of interband transitions and, thus, the measured $R(\omega)$ at even fairly high frequency can follow the behavior of free electron, i.e., $R(\omega) \propto \omega^{-4}$. In the actual K K analysis, it is useful to test various extrapolation functions that decrease in the high-frequency regions, and compare the effects of the high-frequency extrapolations.

It is quite helpful to use independent spectroscopic measurements to check the validity of the K K analysis. For example, spectroscopic ellipsometry is a powerful tool that can directly measure $\tilde{\varepsilon}(\omega)$ from the polarization-dependent reflectance changes [12.28]. Due to the lack of compensation elements in a wide frequency region, most spectroscopic ellipsometric measurements are limited to the frequencies near the visible region. However, if $\tilde{\varepsilon}(\omega)$ can be determined for some spectral region by the elliposmetry, they can be used to check the validity of the K K analysis. Since the calculated $\tilde{\varepsilon}(\omega)$ in the visible region is the most susceptible to the phase error due to the high-frequency extrapolation, this method is proven to be very useful.

The optical investigations using the above-mentioned K K analysis should be done carefully for samples with anisotropic $\tilde{\varepsilon}$. For the perovskite manganites in the paramagnetic insulating or ferromagnetic metallic states, $\tilde{\varepsilon}$ should be isotropic and the K K analysis is valid for polycrystalline samples, as well as for single-crystalline samples. However, the manganites with the layered perovskite structure have highly anisotropic $\tilde{\varepsilon}$. Some perovskite manganites show anisotropic $\tilde{\varepsilon}$ in their spin/orbital ordered states. For example, polarized optical microscope studies show that (Nd,Sr)MnO_3 single crystals are typically composed of multidomains at the spin/orbital ordered states and

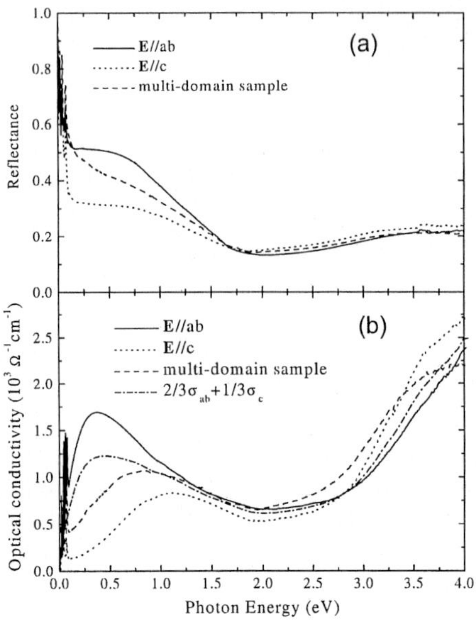

Fig. 12.8. (**a**) Reflectance and (**b**) optical-conductivity spectra of (Nd,Sr)MnO_3 in the A-type ordered state. *Solid* and *dotted lines* are data from the polarization dependent optical measurements and the K K analysis using a single-domain crystal for $E//ab$ and $E//c$, respectively. *Dashed lines* are data using a multidomain crystal whose typical domain size is $\sim$ 10 μm. Note that, in (**b**), the Kramers Kronig analysis result is different from the average conductivity, i.e., $\left[\frac{2}{3}\sigma_{ab} + \frac{1}{3}\sigma_c\right]$

their optical responses are somewhat anisotropic [12.29]. If the polarization dependent reflectance measurements are performed on a single-domain single crystal, then we will be able to use the K K analysis without any difficulty. If such a high quality sample is not available, we should be careful in the K K analysis. Simple K K analysis for manganites with anisotropic $\widetilde{\varepsilon}$ could result in some spurious effects, such as a strong midinfrared peak for polycrystalline high-T_c superconductors [12.30].

Let us consider reflectance measurements on a $Nd_{0.45}Sr_{0.55}MnO_3$ single crystal, which has an A-type antiferromagnetic ordering at low temperatures. In the long-wavelength limit, where the wavelength λ of the incoming light is larger than the domain size L, the medium should appear homogeneous with an effective conductivity. On the other hand, if $\lambda < L$, the light will see the responses of individual domains, so we will obtain an averaged optical response. If the c-axes of the domains are randomly oriented along any one of the crystallographic axes, the measured $R(\omega)$ should be $\left[\frac{2}{3}R_{ab}(\omega) + \frac{1}{3}R_c(\omega)\right]$. The solid and dotted lines in Fig. 12.8a show experimental $R_{ab}(\omega)$ and $R_c(\omega)$ for the $Nd_{0.45}Sr_{0.55}MnO_3$ single crystal in the A-type ordered state. Clearly, optical anisotropy can be observed below 1.5 eV. The corresponding real part of the optical conductivity spectra, $\sigma(\omega)$, obtained by the K K analysis is plotted in Fig. 12.8b. Now, consider the $R(\omega)$ of a multidomain $Nd_{0.45}Sr_{0.55}MnO_3$ single crystal whose typical domain size is about 10 μm. The experimental $R(\omega)$ are shown as the dashed line in Fig. 12.8a. When we perform the K K analysis, we obtain the $\sigma(\omega)$ shown as the dashed line in Fig. 12.8b. It is clear that the K K analysis result is very different from $\left[\frac{2}{3}\sigma_{ab} + \frac{1}{3}\sigma_c\right]$. In particular, the position and strength of the mid-IR peak are quite different. Therefore, more careful analyses should be used for the optically anisotropic samples [12.31].

In summary, optical spectroscopy is a simple and quite useful experimental technique to probe various degrees of freedom in manganites. Reflectance measurements in a wide frequency region on properly prepared sample surfaces and the Kramers Kronig analysis can provide $\tilde{\varepsilon}(\omega)$, which will be useful to understand the electrodynamics of charge carriers, phonon dynamics, and interband transitions. However, for anisotropic materials, the Kramers Kronig analysis is difficult to be applied except for polarization-dependent reflectance data of single-domain single crystals.

13. Glassy Behavior and Time-Dependent Phenomena

In this chapter, the evidence of glassy behavior in manganites is discussed. The results presented here supplement those shown in Chaps. 11, 12, and others, and they reinforce the notion that even very good crystals of Mn oxides are inhomogeneous. The emphasis here is in half-doped $x = 0.5$ LCMO compounds, although several others have glassy behavior as well. Also in this chapter, the issue of glassiness and time-dependent phenomena is addressed in a contribution by P. Schiffer. Overall, the main conclusion is that in manganites there are clear analogies with spin-glass systems. However, it remains to be investigated whether a one-to-one correspondence with previously known glasses can be established, or whether a new class must be defined in manganites. This is an exciting sub-area of research with much potential for interesting discoveries, and even applications.

13.1 Brief Introduction to Spin-Glasses

As an introduction to this chapter, it is important to briefly review the main concepts needed for the definition of a standard spin-glass. This information will allow us to elaborate on the glassy behavior found in Mn oxides. The book of Mydosh [13.1] on glasses is primarily used for the brief review of this section (for another excellent book on the subject see for instance [13.2]). In Mydosh's book, the reader can find most of the important references on glasses, which certainly cannot be repeated here because of lack of space. The reader will rapidly notice clear analogies between "old" glasses and some manganites, although there are also notorious differences as well.

Let us start with an intuitive description of what a spin-glass is believed to be. Following Mydosh [13.1], two important concepts are recurrent in the description of materials classified as spin-glasses. One is the existence of local correlations or "clusters" at temperatures above the so-called "freezing" temperature T_f. These clusters are believed to be the building blocks of spin-glasses. The second recurrent concept is the need to introduce more than a single relaxation time. Actually a *broad* distribution of those times is needed to explain experiments. Let us elaborate more on these items:

The clusters mentioned above are formed at temperatures $k_B T$ (energy units) similar or smaller than the typical couplings Js within those clusters.

These structures are apparently frozen, and they do not need to have a particular order: the spins inside can be mainly randomly oriented, but their orientations are frozen in time. The size of these clusters typically grow as the temperature is reduced. Note also that together with the formation of clusters in space, the relaxation times are affected. For spins that are isolated, i.e., that do not belong to a cluster, their characteristic times are fast. However, for spins belonging to a cluster, locked in a particular spin arrangement, the characteristic relaxation times are much longer. Since there are many clusters, with a variety of shapes and number of spins participating, there is a broad distribution of relaxation times. Clearly, different experimental techniques will be sensitive to different time scales and, thus, the system may appear different depending on the characteristics of the experimental method used [13.1].

As the system is cooled and $T \to T_f$, a percolative process generates an infinite cluster of rigidly frozen spins. Each spin keeps its orientation over long times, namely, $\langle S_i(t)S_i(0)\rangle$ remains finite as $t \to \infty$ for $T < T_f$ (percolation is reviewed in Chap. 16). At this temperature, certainly there are still many spins that do not belong to the so-called infinite cluster, the one that leads to percolation. As a consequence, there is still a broad distribution of relaxation times across T_f. In a percolated state, as described by Mydosh [13.1], energy barriers are natural, leading to hysteresis processes and related phenomena. Spatial inhomogeneities and a vast spectrum of relaxation times over a wide temperature range are the common features observed in a vast set of experiments. As emphasized in the previous paragraph, the freezing temperature T_f then must be a function of the time scale of the measurements.

How do we arrive at the frozen state of spin-glasses where there is no long-range order? It appears that it is important to have some source of randomness in the system (such as a random occupancy by spins of the sites of a regular lattice), mixed interactions (namely, interactions that could be, for example, ferromagnetic or antiferromagnetic), and frustration (see below). With all these ingredients present, the spin-glass system freezes at low temperatures into a *cooperative* state (cooperative in the sense that spins interact among themselves) which, due to the randomness and frustration, does not have long-range order of the standard forms. The frustration leads to many competing states at low energy. An example of systems with site disorder is $Au_{1-x}Fe_x$, where Fe has a spin, while Au does not. Other materials have bond disorder, with J couplings that can be positive or negative. A frequently quoted source of mixed interactions is the RKKY (Ruderman, Kittel, Kasuya, and Yosida) interaction where mobile electrons mediate the interaction between localized spins. This interaction has sign oscillations, and it is described in elementary text books.

Regarding the concept of frustration, a typical example of its intuitive meaning is provided by the case of four spins located at the four sites belonging to a plaquette. If the J interaction among them favors ferromagnetism

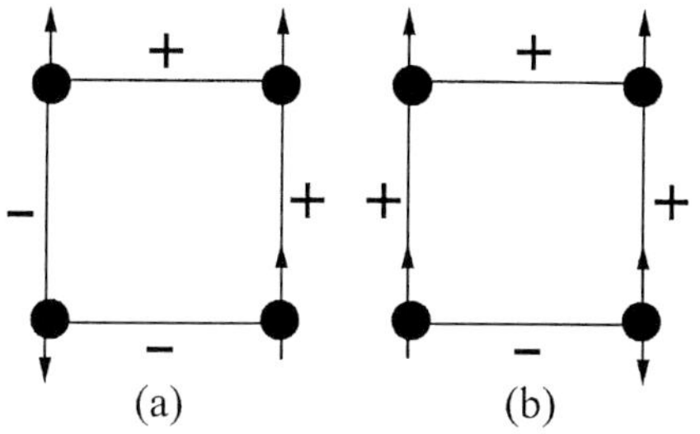

Fig. 13.1. Illustration of the concept of frustration. (**a**) is the case of an unfrustrated plaquette. The signs are those in the spin spin interaction, namely this is a mixture of FM and AF couplings. (**b**) is a frustrated plaquette. The lower-right spin could be up or down, and in both cases one of the links does not have the minimum energy, thus the plaquette is frustrated. From [13.1]

for three links, but antiferromagnetism for the other one, then it is not possible to arrange Ising spins such that all bonds are simultaneously "satisfied", i.e., with the maximum energy gained at each bond. Figure 13.1b shows that once three spins are fixed, the fourth is "frustrated" (contrary to the case of an unfrustrated system shown in a). In frustrated systems, there are many states that have the same or similar energy, and they all contribute to the low-temperature properties. Then, frustration leads to many possible ground states. These effects justify the definition of spin-glasses provided at the beginning of this section, namely, a spin glass occurs when there are randomly positioned clusters with spins that accidentally happen to have strong couplings J among them, leading to the freezing of their individual spins. As T reaches the overall freezing values, these clusters begin to interact with each other, and the system looks for the lowest-state configuration, which can be very complex. In addition, the system is often trapped into a metastable configuration. Note that the issue of whether T_{f} is a true critical temperature or just a crossover, is still under much discussion even after many years of investigations in the area of spin-glasses. The only sure result is that if T_{f} is indeed a transition, it is very unconventional [13.1]. The reason for these doubts is that there are very long times involved in these compounds, and it is difficult to judge whether a system has finally settled into an equilibrium state, or whether it is still evolving.

It is also important to note that in some spin-glasses very large ferromagnetic clusters have been found, larger than allowed by a random distribution of spins. This originates in chemical clustering or other effects. When this situation occurs, the term "cluster glass" is used. In this context, long-range magnetic order is incipient, and such systems are qualitatively similar to manganites in many respects. In fact, the term cluster glass has been repeatedly used to describe several manganites in the regime of competition of phases.

Let us now briefly discuss experiments that contribute to characterize whether a given material is a spin-glass or not:

(1) In typical glasses, the magnetic susceptibility shows that at a temperature T_{d}, which is much larger than T_{f}, ferromagnetic clusters start forming out of the paramagnetic background. According to Mydosh [13.1], specific-heat, resistivity, and neutron scattering results all suggest the existence of

these ferro-clusters at high temperature. The effect is reminiscent of experimental results in manganites, and also theoretical ideas that suggest the existence of a new temperature scale T^* for cluster formation in the same family of materials (Chap. 17).

(2) A very important type of experiment in the context of spin-glasses measures the "AC-susceptibility" $\chi(\omega)$, namely, the magnetic susceptibility in the presence of a small AC-field. The intensity of this field can be as small as 0.1 G, allowing for a study of the properties of the system without altering its behavior. Studies of $\chi(\omega)$ in the early 1970s reported unexpected sharp cusps in AuFe and CuMn, materials that are canonical spin-glasses. Some results are reproduced in Fig. 13.2a. Data for an "ideal" two-dimensional spin-glass $Rb_2Cu_{0.78}Co_{0.22}F_4$ are also shown in Fig. 13.2b. Note that using a weak field of intensity of just a few hundred gauss, the cusp is smeared out into a broad maximum. This is somewhat surprising, since an effect (cusp) that occurs at $T_f \sim 25$ K is destroyed by a field of approximately 1000 G = 0.1 T. Such a field is equivalent to heating the system by only a fraction of a degree (see discussion in Chap. 2), much smaller than the 25 K scale of T_f. The effect reminds us of similar phenomena found in Mn oxides where small perturbations can easily unbalance the state of manganites, as occurs using a magnetic field of a few Teslas in the CMR region.

(3) Another interesting property of spin-glasses is the difference at low temperatures between "field-cooled" (FC) and "zero-field-cooled" (ZFC) susceptibilities (or, equivalently, FC and ZFC low-field magnetizations). When this effect is observed in a material under study, the standard conclusion is that a spin-glass state has been found. This type of study has also been reported in the Mn oxide context, as described in this chapter. The field-cooling results are obtained when the small field h used to obtain the magnetization M (and also the susceptibility χ through the ratio M/h) is turned on at temperatures far above T_f and, then, the sample is cooled down recording the magnetization. Alternatively, the sample can be cooled in zero field below T_f and, afterwards, the field can be turned on (zero-field cooling). These two methods produce *different* results, as illustrated in Fig. 13.2c for the case of CuMn. The FC susceptibility is approximately constant below T_f, and also approximately time independent. On the other hand, the ZFC susceptibility below T_f has a value comparable with those found using an AC-field. The result in this case is time dependent, and the susceptibility increases slowly. It is expected that as time $t \to \infty$, χ_{FC} and χ_{ZFC} will eventually reach similar values. The differences between FC and ZFC results provide a widely used method to estimate T_f.

It is important to note that resistivity measurements in most canonical spin-glasses do not exhibit a transition at T_f. This is in contrast with results in CMR manganites where the resistivity is among the key measurable quantities to detect the effect. Thus, similarities and differences exist between canonical spin-glasses and manganites.

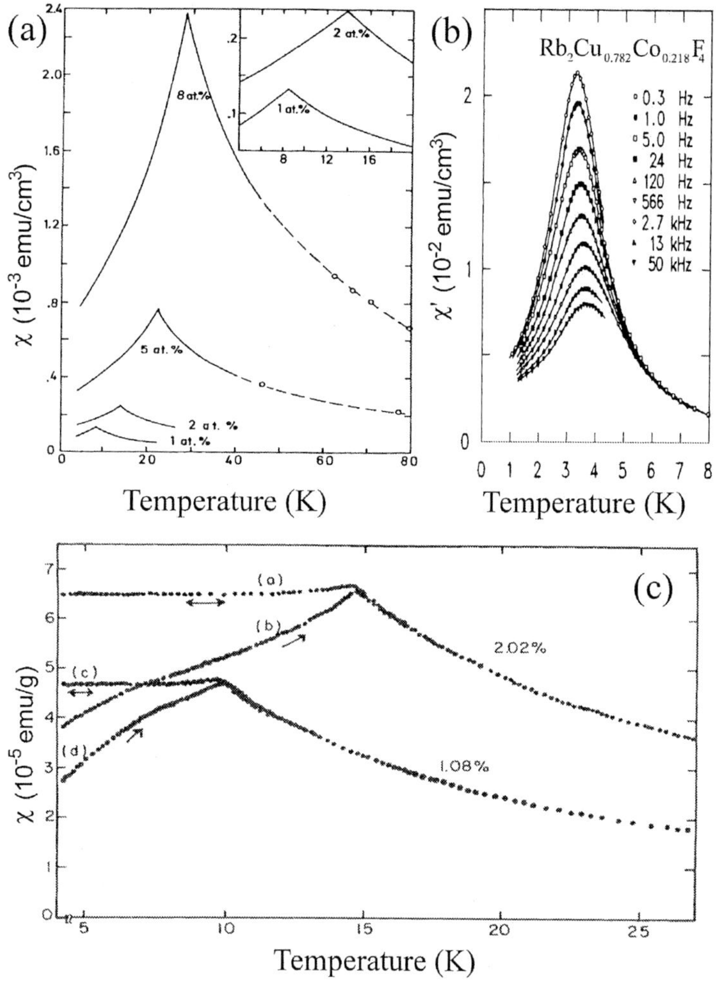

Fig. 13.2. (**a**) Low-field susceptibility of AuFe. Shown are results for different concentrations of Fe from 1 to 8%. The frequency was in the range 50 to 155 Hz. For details see Mydosh [13.1], page 65, from where this figure is reproduced. (**b**) Susceptibility of $Rb_2Cu_{0.78}Co_{0.22}F_4$ versus temperature. Data were taken at frequencies ranging from 0.3 Hz to 50 kHz. *Solids lines* are guides to the eye. From [13.1], p. 189. (**c**) Field-cooled ((**a**) and (**c**) in this frame) and zero-field-cooled ((**b**) and (**d**)) magnetizations of CuMn versus temperature. From [13.1], p. 69

13.2 Glassy Behavior in Manganites

13.2.1 Long Relaxation Times

A variety of experiments in manganites have revealed a phenomenology similar to results found in glasses. For instance, long relaxation times have been detected. Related to the mixed-phase tendencies of the half-doped $x = 0.5$ LCMO (described in Chap. 11), recently López et al. [13.22] reported the presence of *long time logarithmic* relaxation rates in the magnetization (previous studies of relaxation times are reviewed in [13.22]). Depending on the history of the switching "on" and "off" of external fields, at some temperatures the magnetization is found to continue varying even after 10^4 s. This slow relaxation was attributed to the coexistence of FM and AF regions, which produces a distribution of energy barriers. Similar effects have been found in Monte Carlo simulations in the phase-separated regime: the convergence time is anomalously large in those regimes, since the spins are trapped in configurations with slightly higher energy than the ground state.

These slow relaxations have been observed in other manganites as well, and the effects are usually associated to spin-glass behavior. The so-called *aging* effect in spin-glasses has been found in the magnetization of a sample cooled in zero field, with a value that depends on how long the sample is held in zero field before the small magnetic field needed to obtain the magnetization is applied. These aging effects have been observed recently in $La_{0.7-x}Y_xCa_{0.3}MnO_3$ by Freitas et al. [13.3]. Slow relaxation was also observed in $(La_{0.25}Pr_{0.75})_{0.7}Ca_{0.3}MnO_3$ by Voloshin et al. [13.4], and in $Pr_{0.63}Ca_{0.37}MnO_3$ by Raychaudhuri et al. [13.5]. A distribution of relaxation times has also been reported by Seiro et al. [13.6] in $La_{1-x}Ca_xMnO_3$ and $La_{1-x}Sr_xMnO_3$ with $x = 0.30$, when elastic properties were studied using a subresonant inverted pendulum.

However, care must be taken in characterizing the states of manganites as having glassy properties. Standard spin-glasses are usually described by models involving Heisenberg interactions with random couplings, and such a description cannot be directly applied to the mixed-phase state that is expected to characterize manganites. The cluster size can be as large as a fraction of a μm, and within that cluster the spins are regularly arranged at low temperature, not randomly. For this reason, several researchers prefer the use of the name "cluster glass" for manganites, as already remarked, since probably it is the moment of the FM clusters that can be considered as the degree of freedom coupled to other clusters by effectively random moment moment interactions. Note that standard procedures to characterize spin-glass, such as observing a difference between zero-field-cooled and field-cooled magnetizations vs. temperature are not sufficient to distinguish between a conventional spin-glass and an inhomogeneous clustered system.

13.2.2 Phase Separation at $x = 0.5$ LCMO

The properties of $La_{1-x}Ca_xMnO_3$ at $x = 0.50$ are complex and they are still debated. In Chap. 11, experimental evidence suggested that the system is phase separated. Here, this issue is discussed further in view of its relation with the glassy features addressed in this chapter. Results for half-doped LCMO are very sensitive to the hole density, as shown in Fig. 13.3a, where

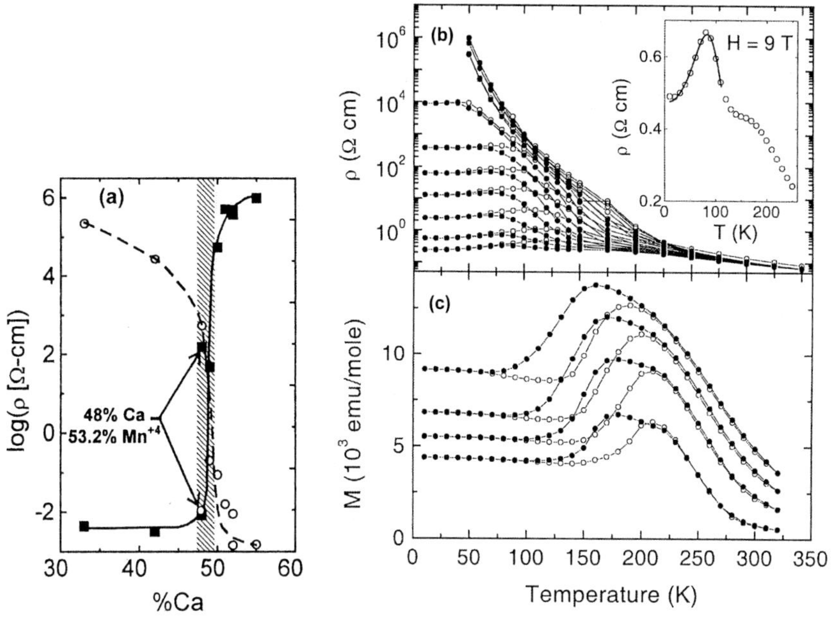

Fig. 13.3. (**a**) The metal-insulator transition near $x = 0.50$ in LCMO, from Roy et al. [13.8]. The resistivity (*solid squares*) was measured at 60 K and zero external field, while the magnetization (*open circles*) was obtained at 10 K and a field of 1 T (scale for the latter not shown). *Lines* are guides to the eye. The *arrows* indicate results for some samples that are inconsistent with the trends, showing the complications that arise in sample preparations of half-doped LCMO. (**b**) *Top panel*: resistivity vs. temperature for $x = 0.50$ LCMO on cooling (*solid*) and warming (*open*). The fields used are from 0 to 9 T, in steps of 1 T, starting from the top. The *bottom panel* shows the magnetization using fields from 1 to 7 T, in steps of 2 T. For more details the reader should consult [13.9]

the resistivity and magnetization are found to change very abruptly near $x = 0.50$. The resistivity and magnetization in external fields at $x = 0.50$ are presented in Fig. 13.3b. As in other compounds, the charge-ordered state stabilized at low temperatures is sensitive to the introduction of external fields. Note the similarity of the resistivity plots with those obtained from the random-resistor-network calculations of Chap. 17. There is plenty of experimental evidence that half-doped LCMO presents a mixed-phase state with interesting properties. Other Ca-based manganites present similar tendencies at $x = 0.5$ as well, as shown in recent studies of $Nd_{0.5}Ca_{0.5}MnO_3$ and $Gd_{0.5}Ca_{0.5}MnO_3$ [13.7].

The phase separation at $x = 0.5$ LCMO has also been studied systematically by Levy et al. [13.10] (see also [13.11]). Their results show that the mixed-phase effect can be controlled using samples with varying average grain size. These results were later confirmed in other "canonical" phase-separated compounds [11.118]. Recently, Parisi et al. [13.12] studied the variation of the metallic fraction f with applied magnetic fields, and they presented empirical laws that indicate a fast transformation into the FM state. More recent transport measurements by Levy et al. [13.13] on $La_{0.5}Ca_{0.5}Mn_{0.95}Fe_{0.05}O_3$ have revealed coexistence of ferromagnetism and charge ordering, and novel rejuvenation, persistent memory effects, and dynamical contributions to the resistivity. Rejuvenation of the resistivity after aging, and a persistent memory of low magnetic fields have also been reported by Levy et al. [13.14] in studies of $x = 0.5$ LCMO. These important investigations are unveiling the rich dynamics of mixed-phase manganites, a promising area of investigations with much potential for surprises.

13.2.3 Manganites: Canonical Glasses or a New Glassy State?

As remarked in the introduction to this chapter, there are many similarities between glasses and manganites, but it is not obvious that they necessarily have a totally analogous qualitative behavior. The key issue is whether manganites in their glassy state – usually in the regime of large MR effects – form some sort of new universality class of glasses or, more ambitiously, some new state of matter. Is the manganite glassy state a canonical glass or not? The answer to this question is unknown, and its analysis is complicated by the fact that even within standard spin-glasses there is still considerable discussion on the nature of the spin-glass state and the associated transitions (if they are true transitions). The reader can find an introductory review on spin-glasses addressing, among several issues, whether a conventional glass is a new state of matter, in [13.15] and references therein. This is a difficult issue since the glass state is not distinguished by any obvious symmetry from a liquid state with a very slow relaxation to a uniform density. In spite of considerable research over many years, the basic issue of whether a glass is or is not a new phase is still under debate. For this reason, it is likely that in the area of manganites the best we can do at present is to simply address the issue of whether "manganite glasses" are or are not the same from a phenomenological point of view as standard spin glasses. Complicating the quantitative analysis even more, it is known that the computational study of conventional spin-glass models is notoriously difficult (for a recent review see [13.16]). Similar complexity (or worse) is to be expected in studies of models for manganites in the mixed-phase region. Clearly, neither experiments nor theory are easy in the glassy regime.

In spite of these difficulties, recent studies by Deac et al. [13.35] have reported considerable progress regarding the analysis of the glassy state of manganites. These studies revealed the existence of *two* types of relaxation processes in the CMR regime: (i) One of them occurs at low magnetic field and is associated with the reorientation of the preformed ferromagnetic domains believed to be causing the large MR in these compounds. (ii) A second process occurs at higher fields, and is associated with the conversion between ferromagnetic and non ferromagnetic phases. These two processes may constitute *a new sort of glassiness* based on the macroscopic and dynamical coexistence of two phases [13.35]. From their studies of glassy-like behavior in $Y_{0.7}Ca_{0.3}MnO_3$, Mathieu et al. [13.36] also concluded that manganites are quite unusual and difficult to classify according to existing theories or phenomenology of single-phase materials with random interactions. Niazi et al. [13.37] studied a magnetic system with spin-glass characteristics, compared its behavior with manganites and concluded that it is the "cause" of the inhomogeneous behavior that makes phase-separated systems fundamentally different from spin-glasses. The possibility of a new kind of glass in manganites is a challenging and quite interesting issue that deserves further experimental and theoretical work. The potential use of the encoded information in the amount of the FM phase has recently been suggested [13.14], and as a consequence even applications can be envisioned in this context.

It is also important to remark that the study of similarities and differences between the dynamics of Mn oxides and "relaxor ferroelectrics" should be pursued. In this context, the slower kinetics of nanodomains compared with that of spins in spin-glasses has been discussed [11.85–11.87]. Are manganites like relaxor materials? More details are given in Chap. 22, where analogies between manganites and other materials are analyzed. Also in Chap. 20 several related compounds that share a common phenomenology are discussed. The lessons learned by studying the glassy behavior and strange dynamical properties of Mn oxides appear to be applicable to a variety of other materials as well.

13.3 Phase Separation, Time-Dependent Effects, and Glassy Behavior in the CMR Manganites
by P. Schiffer

Within the past few years, it has been proposed that, rather than being exclusively either conducting or charge ordered, the electronic ground state of some compositions of the CMR manganites could be a microscopically phase-separated combination of both states [13.17]. Strong experimental evidence for such phase separation has been provided by electron microscopy, tunneling microscopy, X-ray diffraction, electronic transport, noise spectra, Mössbauer spectroscopy, NMR, neutron scattering, and optical studies in a variety of manganite systems, including $La_{1-x}Ca_xMnO_3$, $Pr_{1-x}Ca_xMnO_3$, and $Nd_{1-x}Sr_xMnO_3$. These studies, not listed explicitly here since they have been extensively discussed elsewhere in this book and in reviews [13.17], suggest that the phase separation is an intrinsic property of the manganites and not attributable to poor sample quality or large-scale compositional inhomogeneity. Not surprisingly, the phase separation is most clearly observed for materials near the compositional transition between predominantly conducting and predominantly charge-ordered ground states (e.g., near $x = 0.50$

in $La_{1-x}Ca_xMnO_3$), where the two different ground states are energetically most closely balanced. There is also evidence for such coexistence of conducting ferromagnetic regions and charge-ordered regions well beyond the immediate vicinity of the compositional metal insulator transition. Phase separation can also extend into the predominantly charge-ordered states, and recent work suggests phase separation in almost the full range of doping from near $x = 0$ to near $x = 1$. These issues have been discussed in this book in detail.

Phase transitions between the predominantly charge-ordered antiferromagnetic phases and predominantly conducting ferromagnetic phases in the manganites are typically first order. This is expected theoretically [6.42] and has been confirmed experimentally by observations of hysteresis and irreversibilities. The first-order difference between the two phases is consistent with phase separation, and also suggests that we should expect interesting dynamics as the relative fractions of the two phases are altered by a change in either temperature or magnetic field. Indeed there are numerous reports in the literature of logarithmic time dependence in the relaxation of both resistivity and magnetization after sudden changes in field [11.84, 13.4, 13.18–13.24], as well as memory effects [2.16], large electrical fluctuations (noise) [13.25], an unusual frequency dependence to the effects of applied electric fields [13.26], and unusual relaxation dynamics in both optical [13.27] and μSR studies [13.28]. While many of these observations have not been explicitly connected to phase separation by the workers making the measurements, and some of them could be due to other effects, such as domain-wall motion [13.29], these are exactly the sorts of effects one might expect to observe from the dynamics of the phase separation between two energetically similar states. In fact, slow equilibration consistent with the observed logarithmic relaxation is often seen in Monte Carlo studies of phase-separation models [13.30], and experimentally the logarithmic relaxation effects seem to be enhanced in materials near $x = 0.50$ where the phase separation is most pronounced [11.84]. These relaxation effects are easily observable and can have a strong nontrivial temperature dependence [11.84, 13.21, 13.22], as demonstrated in Fig. 13.4 for $La_{0.5}Ca_{0.5}MnO_3$.

The relaxation effects are not limited to the magnetization and resistivity. In Fig. 13.5 we show the low-temperature specific heat of $La_{0.5}Ca_{0.5}MnO_3$ measured as the field is changed in 2500 Oe intervals every 10 min from $0 \to 9 \to -9 \to 9$ T. We see that the specific heat shows a rather sharp decrease in the initial sweep up in field (open circles) when the ferromagnetic domains are presumably increasing in size, but then the specific heat rises almost to the initial zero-field value after waiting an additional 10 min at the maximum field of 9 T. Upon decreasing the magnetic field to -9 T (solid line), the specific heat shows non monotonic behavior that is rather different from the initial field sweep, but which is reproduced upon raising the field back to $+9$ T (dashed line). The resistivity and the magnetization of the same

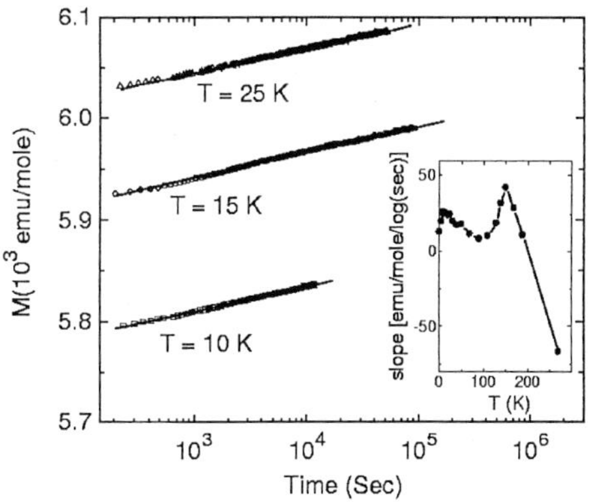

Fig. 13.4. The time dependence of the magnetization of $La_{0.50}Ca_{0.50}MnO_3$. The sample was zero-field-cooled and then the field was raised to 7 T, after which the magnetization was measured as a function of time. The *inset* shows the slope of the logarithmic time dependence, which has maxima at both $T \sim 150$ K, where the material undergoes charge ordering but also at $T \sim 20$ K where there are no obvious features in the bulk properties. (M. Roy et al., unpublished)

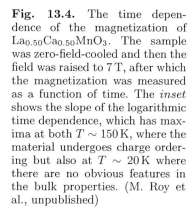

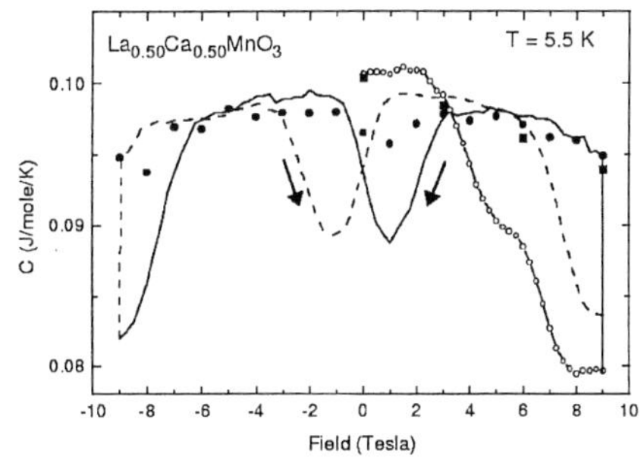

Fig. 13.5. Field dependence of the specific heat of $La_{0.50}Ca_{0.50}MnO_3$ showing non equilibrium behavior attributable to the dynamics of phase separation. *Open circles* (initial sweep up in field after zero-field cooling); *Solid line* (subsequent sweep down in field); *Dashed line* (subsequent sweep up in field). The *solid squares* and *circles* are data taken in equilibrium while increasing and decreasing the field respectively (waiting for an hour after changing the magnetic field before measuring) [13.31]

sample do not show any such non monotonic behavior during field-sweeps, and the equilibrium specific heat (as measured after waiting an hour at each field) is almost flat, as shown by the solid circles in the figure. We attribute the non equilibrium behavior of the specific heat (not seen in $La_{1-x}Ca_xMnO_3$ except near $x = 0.50$) to changes in the nature of the phase separation, i.e., the changing morphology of domains of the different phases with increasing and decreasing field [13.31] and resultant finite-size effects on the excitation spectra.

The observation of time-dependent behavior in the manganites, whether or not associated with phase separation, suggests a comparison with traditional spin-glasses [13.32], which originate from a frustration of ferromagnetic and antiferromagnetic interactions, albeit at the atomic scale [13.1]. Indeed there have been several manganite materials that have been identified as having spin-glass phases [13.33, 13.34, 13.36] based on irreversibilities in the magnetization- or frequency-dependent peaks in the real or imaginary parts of the A.C. susceptibility (see Figs. 13.6 and 13.7). These same characteristics are, however, typical of a variety of systems that are not spin-glasses (e.g.,

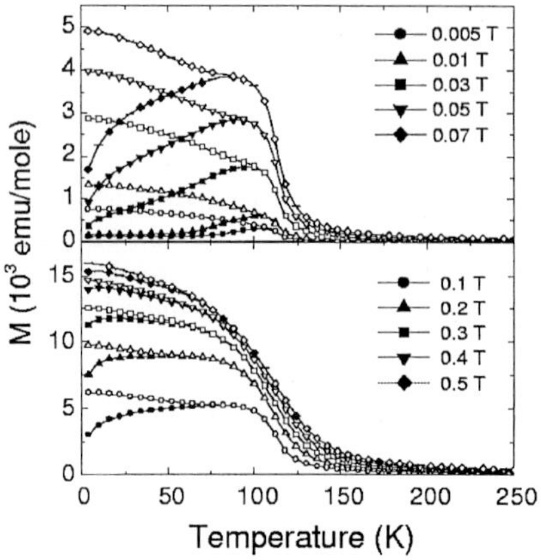

Fig. 13.6. The field-cooled and zero-field-cooled magnetization of $Pr_{0.7}Ca_{0.3}MnO_3$ measured in different applied magnetic fields. The difference between the two curves is qualitatively equivalent to that expected for a spin glass, although neutron scattering experiments show the development of phase separation is responsible. The irreversibility is suppressed by the applied field, which can be attributed to the growth of ferromagnetic clusters [13.39]

superparamagnets) and the question arises as to whether the glassy phases of the manganites constitute spin-glasses in the traditional sense or whether they are rather some sort of "cluster glass" [11.109,13.38] in which correlated groups of spins (whose size can change with temperature) undergo a glassy freezing. Indeed, a detailed examination of the susceptibility in one phase-separated compound shows that there are qualitative features of the behavior that are inconsistent with ordinary spin-glasses [13.39], strongly suggesting that phase separation leads to a new form of glassiness that remains to be well defined.

The exquisite sensitivity of the manganites to magnetic fields suggests that this glassiness associated with phase separation will have unique characteristics since the ferromagnetic fraction is tunable by application of a field, a degree of freedom that is not as accessible in other glassy magnets. The effects of applied field on the phase separation apparently leads to a growth in the ferromagnetic clusters, which has been hypothesized to result in percolation phenomena in the electronic transport and has been shown to have a significant effect on the glassy properties, as demonstrated in Figs. 13.6 and 13.7. A difficulty in analyzing this behavior is that other glassy magnets also have nontrivial field dependence to their properties, but there are clear features in the evolution of manganite behavior with magnetic field that can be attributed to the growth of ferromagnetic clusters at the expense of the charge-ordered phase [13.39].

In summary, the existing experimental data demonstrate that the manganites display glassy behavior and that this behavior is common, if not universal, among those compounds in which there is phase separation. A challenge for future work is to determine if this glassy behavior associated with phase separation has common characteristics that distinguish it from glassiness in other magnets. Since the phase separation is also strongly affected by doping and pressure, the manganites may not only provide a new

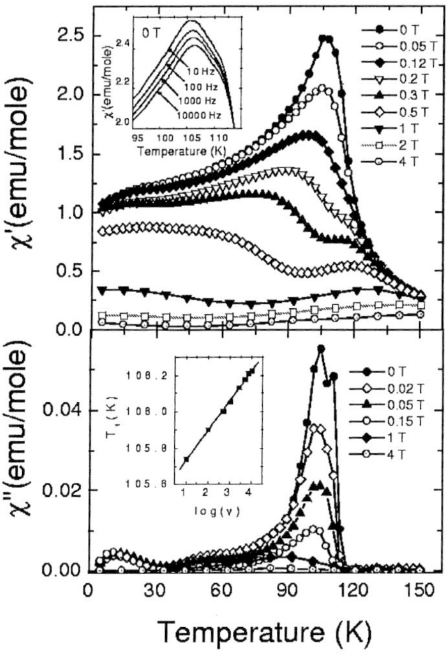

Fig. 13.7. The real and imaginary parts of the A.C. susceptibility of $Pr_{0.7}Ca_{0.3}MnO_3$ measured in applied static magnetic fields. The *insets* show the variation of the zero-field peak with frequency, which is logarithmic, as is typical for a glassy system [13.39]

type of glassines, but also a testbed in which the factors resulting in glassiness (i.e., the level of frustration among the competing interactions, the size of the ferromagnetic clusters, etc.) can be tuned to study the nature of the glass-like phases.

We gratefully acknowledge support from NSF grant DMR-0101318 and helpful discussions with E. Dagotto.

14. Inhomogeneities in $La_{1-x}Sr_xMnO_3$ and $Pr_{1-x}Ca_xMnO_3$

14.1 Results for $La_{1-x}Sr_xMnO_3$

14.1.1 Low Hole Density

We start the analysis of inhomogeneities in LSMO by describing experimental results obtained using the pair-density-functional (PDF) technique. To understand this method, consider the density correlation function $\rho(r)$ that describes the probability of finding an atom at a distance r from another atom, in the material under study. This function contains information about the atomic structure of the material expressed in real space. For a perfectly periodic crystal, $\rho(r)$ peaks sharply at a, $2a$, ... where a is the lattice spacing. However, many real materials present substantial deviations from periodicity, and this occurs in manganites as well. In these materials, $\rho(r)$ is a richer function, with peaks of finite width at several locations. This correlation function is important since it reveals features of the local structure, such as small deviations from average crystallographic positions due to tendencies to order at short distances that are not reflected in the dominant long-range order of atoms. In other words, not only are Bragg peaks revealed in this type of studies, but also those arising from diffuse scattering. The function $\rho(r)$ can be obtained from X-ray or neutron diffraction techniques without assumptions about the periodicity of the lattice. The data is usually gathered in terms of momentum-space variables, and the results must be Fourier transformed to real space for their analysis. Details about this procedure and a more detailed introduction to PDF can be found in [14.1], and references therein.

Typical PDF data for $La_{1-x}Sr_xMnO_3$ are reproduced in Fig. 14.1. Part a contains results for the undoped limit. The first minimum (peaks can be negative or positive, see [14.1]) is located at ~ 1.93 Å and the second at ~ 2.2 Å. These peaks correspond to Mn O distances, and there are two values due to the static Jahn–Teller distortions. These distortions elongate in one direction the octahedron of oxygens around manganeses, reducing the Mn O distances in the other two directions. The second large peak in Fig. 14.1a, this time positive, represents the La–O and O–O distances, split by distortions. Thus, each peak has a fairly well-defined rationale behind it, and it is interesting to observe how they evolve as the hole density is increased. In Fig. 14.1b results at several densities are shown, in the paramagnetic insulating phase. These

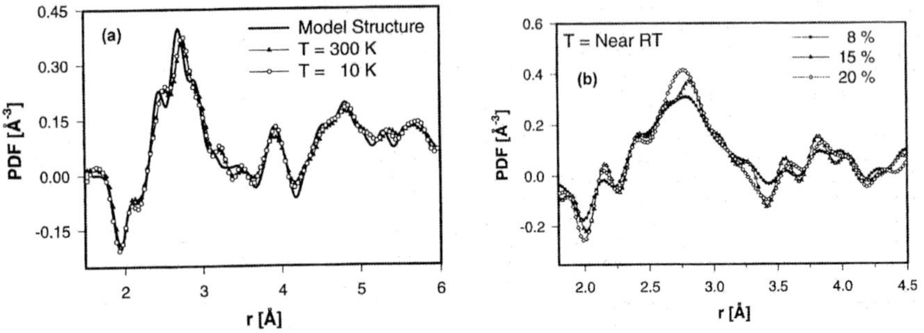

Fig. 14.1. (**a**) PDF results for $LaMnO_3$, both at room temperature and at 10 K, from [14.2]. (**b**) PDF results for $La_{1-x}Sr_xMnO_3$ at the three hole densities shown, near room temperature, from [14.2]. All samples are paramagnetic insulators. The composition dependence of the local structure is relatively small. These results indicate that local JT distortions must be present in all cases

results are similar to those found for the undoped material, even at density $x = 0.20$, where X-ray analysis would suggest that the static JT distortion has disappeared (incorrectly suggesting that the lattice spacings are uniform in the three directions). This interesting PDF result suggest that JT distortions persist at a *local* level even at high densities. At long distances, the distortions are not uniform but they randomly orient in different directions, leading *on average* to lattice spacings that are the same in the three directions. This is conceptually very important: physics at short distances may be governing the behavior of manganites, and it is crucial to investigate at a microscopic level the properties of these compounds. The results reported in [14.2] were analyzed in terms of polarons, and the picture of percolation was already presented as an explanation of the metal insulator transition of $La_{1-x}Sr_xMnO_3$. More recent PDF results [14.3] are compatible with microscopic segregation of charge-rich and -poor regions, in agreement with theoretical calculations.

The PDF evidence supporting the existence of local JT distortions in these systems appears robust even within the low-temperature FM phase, as shown in [14.2, 14.3]. It is only at $T = 10\,\mathrm{K}$ and $x = 0.40$ that the system is truly uniform, with no local distortions. The PDF results suggest that, in general, the homogeneous picture of the metallic state of manganites is incorrect. This is in agreement with a variety of other experimental results for other manganites, as described elsewhere in this and other chapters. However, it is important to mention that issues related to chemical homogeneity are still under discussion for LSMO at small hole density [14.4]. It is certainly of much relevance to clarify whether chemical inhomogeneities exist in manganites. This potential problem could explain, in part, the mixed-phase tendencies. However, the majority of manganite experimentalists are convinced of the homogeneity and quality of their single-crystal samples, and they believe the observed phase-separation tendencies are intrinsic.

14.1.2 Density $x \approx 1/8$

The region of the LSMO phase diagram in the vicinity of $x = 1/8 = 0.125$ has attracted much attention, and its properties are still not understood. Early neutron scattering studies reported charge ordering at this density and low temperatures [14.5], although the results have not been confirmed [11.19]. If charge ordering exists, a close analogy with other manganites with similar complex behavior, such as LCMO, would be unveiled (even high-T_c superconductors have interesting behavior at this density). Other groups have also investigated the region around $x = 1/8$ [14.11, 14.12]. Ivanshin et al. [14.6] and Pimenov et al. [14.8] used electron spin resonance techniques to study LSMO at dopings between 0.0 and 0.2 (see also Ivannikov et al. [14.9]). These authors reported evidence of spin canting, which appears at odds with phase separation predictions. However, an analysis by Foss and Mayr [14.10] using models that explicitly incorporate impurities to mimic Sr dopants, reproduced part of Pimenov et al. data [14.8], with a state containing pinned ferromagnetic polarons in an antiferromagnetic background. The spins near the FM polarons that locally would prefer an AF arrangement, instead form spin canting due to the influence of the large moment of that nearby polaron. While evidence accumulates that a uniform spin-canted state is not stable in manganite models, inhomogeneous states that have some sort of spin-canting characteristic can be constructed. In addition, a clustered state with nanostructures has a rapidly changing, but still continuous, density vs. chemical potential curve, as in recent photoemission experiments for Sr-based Mn oxides [14.7].

Kiryukhin et al. [14.13] reported an X-ray "photo-induced" structural transition at $x = 0.125$. Manganites appear to be very susceptible not only to small magnetic fields, but to several other perturbations in the regime of large MR. To explain their results, the authors of [14.13] concluded that the system may be phase separated, as predicted by theoretical calculations. In addition, in light-illumination studies carried out at compositions $x = 0.3$ and with oxygen deficiency, Cauro et al. [14.14] reported a persistent increase of the conductivity with light. Other neutron scattering investigations at low x in $La_{1-x}Sr_xMnO_3$ reported the existence of two ferromagnetic phases, one metallic and one insulating, that differ in their orbital ordering [14.15]. This result is interesting for at least two reasons: (1) It shows that insulating behavior can be produced with magnetic fields, instead of the field-induced metals that are usually described in the literature. This illustrates the rich properties of Mn oxides where several kinds of exotic transitions have been discovered in recent years. (2) The orbital order can be as important as the spin or charge order. The experiments of [14.15] are in good agreement with theoretical calculations that reported phase separation between two FM phases, which differ in their orbital order [9.4]. In addition, stripes in FM phases have been recently reported [10.9]. Such a state is induced by orbital order without the participation of the spin. Moreover, ferromagnetic insulators induced by a large electron JT-phonon coupling have been reported in some calculations [6.42]. Such exotic behavior – clearly beyond double-exchange ideas – can be understood if the orbital degree of freedom is incorporated into the analysis [6.42]. All this evidence suggests that the "FM insulator" phases of manganites may contain highly nontrivial hidden orders worth investigating.

A variety of additional experimental investigations have also strongly suggested that $La_{1-x}Sr_xMnO_3$ at low hole-density should not be considered as microscopically homogeneous. For example, with ultrasound techniques the elastic moduli of $x = 0.17$ LSMO were studied by Darling et al. [14.16], reporting the presence of small microstructures. Takenaka et al. [14.17] investigating a similar density with optical techniques concluded that the system is a "bad metal" at low temperatures. Also Demin et al. [14.18] found tendencies to a two-phase regime at $x = 0.10$, through the discovery of a giant red shift of the absorption edge. Jung et al. [14.19] investigated the optical conductivity of $x = 1/8$ LSMO and concluded that the results are compatible with a phase-separation percolative picture. The list of related work is simply too long to be reproduced here, and readers can find addicional references in reviews such as [1.9], and others. Overall, it can be concluded that a considerable literature suggests the lack of microscopic homogeneity in LSMO at low density. The presence of competing states induces a complex microtexture. Certainly more work is needed to fully clarify the physics of LSMO, but at least two conclusions can be stated: (1) The system at low hole-density is not homogeneous, and a plain spin-canted state is not sufficient to reproduce experiments. (2) A fully macroscopic phase-separated state is not sufficient either. The "truth" appears close to a picture with *microscopic separation of phases*, perhaps with FM clusters pinned by disorder or trapping Coulombic centers, such as the Sr dopants [14.20].

14.1.3 Large Hole Density: Double-Exchange Behavior?

According to neutron measurements, $La_{1-x}Sr_xMnO_3$ still presents anomalous behavior at carrier densities as high as 30% hole doping. For example, in Fig. 14.2 central diffusive peaks similar to those described in the neutron scattering section of Chap. 11 have been identified by Vasiliu-Doloc et al. [11.19, 14.21]. The discussion of Chap. 11 will not be repeated here, but it is clear that the results are indicative of non standard behavior since they are so similar to those found for lower-bandwidth compounds, such as LCMO, with anomalous properties. Indications of phase separation and a percolative picture in Sr-based compounds have also been discussed by Xiong et al. [14.22].

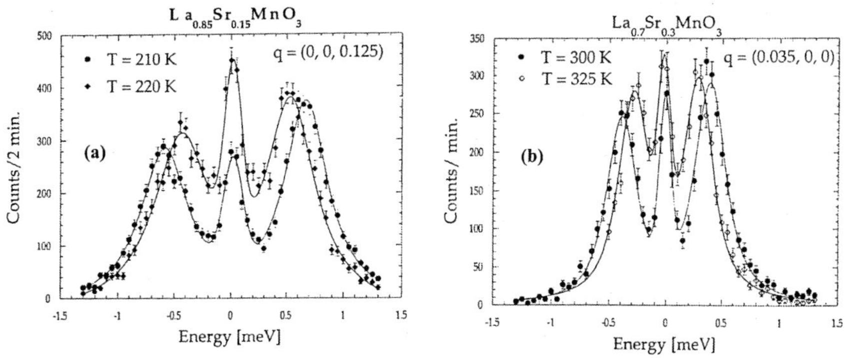

Fig. 14.2. (**a**) Neutron scattering results for $La_{1-x}Sr_xMnO_3$, similar to those described in Chap. 11 in the context of an anomalous central peak. (**a**) corresponds to 15% hole doping from Vasiliu-Doloc et al. [11.19]. (**b**) is the same as (**a**) but at 30% doping. Result from Vasiliu-Doloc et al. [14.21]

At larger density, typically $x \sim 0.4$ the resistivity above the Curie temperature is metallic, and Furukawa [14.23] has argued that the material can be described by simple double-exchange ideas. In this context there are no relevant intrinsic inhomogeneities, as at lower densities or as in other lower-bandwidth manganites. Then, it appears that as the Curie temperature increases, the behavior of the Mn oxides becomes canonical and compatible with traditional ideas on the subject. This detail is important for applications that require a large Curie temperature. On the other hand, if the main motivation for manganite investigations is the understanding of the CMR effect, then one must move in the opposite direction of decreasing T_C, as explained in the introductory chapter. In fact, recent investigations of critical exponents of the $La_{1-x}Ca_xMnO_3$ compound at $x = 0.3$ have shown that the β exponent for the spontaneous magnetization is about 0.14, quite different from the 0.37 of $La_{1-x}Sr_xMnO_3$ at the same density (the latter being compatible with the critical exponent of the 3D Heisenberg model, which is 0.365) [14.24]. Similar conclusions were reached in [14.25] through specific-heat, thermal expansion, and elastic-modulus measurements. In fact, the $La_{1-x}Ca_xMnO_3$ transition has first-order-like characteristics, an issue of much importance in manganites as explained in this book.

14.1.4 Inhomogeneities in Sr-Based Manganites at $x = 0.5$

Some Sr-based manganites at high hole-density, $x = 0.5$, also present coexistence of phases. Consider, for instance, the following results:

(1) Allodi et al. [14.26], using NMR techniques, observed that the transition from the AF to the FM state in $Pr_{1-x}Sr_xMnO_3$ at $x = 0.5$, proceeds through the nucleation of nanoscopic FM domains, suggesting a percolative

process. Evidence for phase separation in the same compound was also reported by Mahendiran et al. [14.27].

(2) Kajimoto et al. [3.28] used neutron techniques to study $Nd_{1-x}Sr_xMnO_3$, and observed that the CE-phase near $x = 0.5$ coexists with the A-type AF phase. Fukumoto et al. [14.28] arrived at similar conclusions – evidence for microscopic-scale phase separation – using transmission electron microscopy. EPR studies by Joshi et al. [14.29] also reported indications of phase separation at $x = 0.5$.

It is likely that other Sr-based manganese oxides will have similar features, in particular regions of the phase diagram. For example, $La_{1-x}Sr_xMnO_3$ at hole densities in the vicinity of the FM to A-type AF transition at low temperatures (see Chap. 3) may present microscopic phase separation between these two phases. In general, it appears that any pair of phases with sufficiently different arrangement of spin, charge, and orbital, such that the transition from one to the other should have first-order characteristics, can lead to mixed-phase states near the transition. This is compatible with the theoretical calculations described in Chap. 17, where disorder transforms a first-order transition into a continuous percolative one. Another result compatible with theoretical calculations (such as those in Chap. 10) is the competition of the CE-phase, the A-type antiferromagnet, and the FM metallic phase at $x = 0.5$. There are regions in parameter space where these *three* phases simultaneously have very close energies.

14.2 Results for $Pr_{1-x}Ca_xMnO_3$

The small-bandwidth manganite PCMO also presents regimes with inhomogeneities. Part of the evidence is in other chapters, such as Chap. 19, and for this reason the present section is brief. X-ray synchrotron and neutron diffraction studies of $Pr_{1-x}Ca_xMnO_3$ by Cox et al. [14.30] at hole density $x = 0.3$ have reported results interpreted as being produced by a phase-segregation phenomenon (see also Rivadulla et al. [14.31], and Raychaudhuri et al. [14.32]). More recent diffraction and inelastic neutron studies of the same compound by Radaelli et al. [14.33] have also been analyzed as caused by mesoscopic (500 to 2000 Å) and microscopic (5–20 Å) phase segregation. Casa et al. [14.34] studying $Pr_{1-x}(Ca_{1-y}Sr_y)_xMnO_3$ observed that X-ray illumination resulted in a microscopically phase-separated state, where charge-ordered insulating regions provide barriers against charge transport between metallic clusters.

This is similar to the pioneering results of Kiryukhin et al. [14.35] where an unusual insulator metal transition was reported when $Pr_{1-x}Ca_xMnO_3$ $x = 0.3$ was exposed to an X-ray beam. A typical result of this type of experiment is shown in Fig. 14.3, where the electrical resistance and intensity of the superlattice reflection peak that characterizes charge ordering are shown as a function of time, with and without X-ray exposure, as indicated.

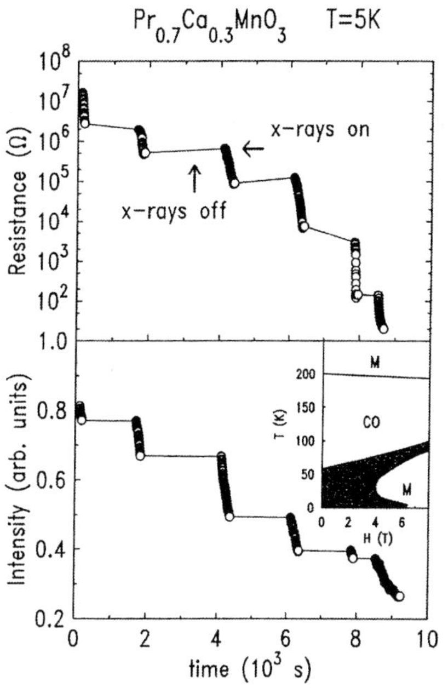

Fig. 14.3. Electrical resistance (*upper panel*) and intensity of a superlattice reflection peak characteristic of a CO state (*lower panel*) as a function of time, with and without X-ray exposure, as indicated. From [14.34]

Clearly, the X-rays reduce the resistivity. The coexistence of two phases in $Pr_{0.63}Ca_{0.37}MnO_3$ was also inferred by Guha et al. [14.36] from their analysis of nonlinear conduction in single crystals. Results by Savosta et al. [14.37], studying $Pr_{0.5}Ca_{0.2}Sr_{0.3}MnO_3$ with NMR, were also interpreted as providing evidence of magnetic phase separation. Recently, Niebieskikwiat et al. [14.38] found evidence of phase separation in $Pr_{1-x}(Ca_{1-y}Sr_y)_xMnO_3$ for a variety of compositions, through measurements of thermoelectric power, electrical resistivity, and DC magnetization in polycrystalline samples. A strong competition between FM and charge ordering was observed.

The main overall conclusion of this chapter is that both $La_{1-x}Sr_xMnO_3$ and $Pr_{1-x}Ca_xMnO_3$, and closely related compounds, present a phenomenology quite similar to that of $La_{1-x}Ca_xMnO_3$ in regimes of densities and temperatures where phases are in competition. This is an important result, and shows that the large-, intermediate-, or small-bandwidth character of the material is not necessarily the most important property for inhomogeneities to exist. The bandwidth simply regulates the properties of the individual competing phases, but not the competition itself. On the other hand, for the MR effect to be huge, materials with quite different resistivities are needed, and this occurs more often in intermediate- or low-bandwidth manganites where the "very insulating" charge-ordered state can be stabilized.

15. Inhomogeneities in Layered Manganites

In this chapter, experimental evidence of inhomogeneities in quasi-2D manganites is presented. This evidence supports the conjecture that most manganites, whether 3D perovskites or 2D layered, have mixed-phase tendencies.

15.1 Mixed-Phase Tendencies in Bilayers

The complex phase diagram of bilayer manganites shown in Fig. 3.23 of Chap. 3 clearly illustrates the competition of phases that is often present in Mn oxides, regardless of their effective dimensionality. This phase diagram presents interesting similarities with the theoretical general phase diagram found in simulations involving two competing ordered states, discussed in Chap. 17. This competition is possible in at least three regions of Fig. 3.23: (i) The "canted" region between the FM and A-type AF states, which could correspond to a mixed-phase of those two neighboring states. (ii) The region centered at $x = 0.7$ with no long-range order, which is probably a cluster-glass mixture of the two surrounding phases with x^2–y^2 and $3y^2 - r^2$ orbital ordering. This case appears to correspond to "large disorder" in the general phase diagram of competing phases, Fig. 17.10 of Chap. 17. (iii) The $x \sim 0.9$ region where there is also a competition of two states, with a reduction of both their critical temperatures as this density is approached. Future experimental work will clarify whether these conjectures related to inhomogeneities are correct or not for bilayers. In addition, in the CMR region at $x \sim 0.4$ it may occur that *three* phases compete, namely, a FM metal, an A-type AF insulator, and a charge-ordered CE-type state. This possible three-state competition was observed in computer simulations of the two-orbital model (see [6.42]). The small energy difference among a variety of states appears to be at the heart of the physics of manganese oxides.

Let us consider now the bilayers and their inhomogeneities, in particular regimes of hole densities, which have been studied by several groups.

15.1.1 Inhomogeneities at $x = 0.50$

Searching for competing phases, Kubota et al. [15.1] analyzed in detail the $x = 0.50$ bilayer manganite using neutron diffraction techniques. They studied the regime of temperatures where charge-ordering tendencies were observed in other investigations, as described in Chap. 3. These authors concluded that two types of ordering are coexisting – metallic $x^2 - y^2$ and charge-localized $(3x^2 - r^2)/(3y^2 - r^2)$ – corresponding to A-type and CE-type orderings. Similar results regarding phase coexistence were reported by Argyriou et al. [15.2]. These authors also noticed that the charge-lattice fluctuations, precursors of charge ordering at 210 K, start at temperatures as high as 340 K, or even higher. Moreover, Bewley et al. [15.3], replacing La by other ions, observed with μSR that biphasic compounds can be formed at $x = 0.50$, and in some cases spin-glass behavior is observed. These experimental observations are in qualitative agreement with the theoretical calculations reported elsewhere in this book where: (i) It has been shown that at $x = 0.50$ the A-type AF, CE-type CO, and FM metallic states are all close in energy [6.42]. (ii) As discussed in the context of two ordered phases in the presence of disorder (Chap. 17), a T^* temperature scale must exist where clusters of competing phases start forming, well above the true ordering temperatures. The work of [15.2] shows that such scale in bilayers may be as high as room temperature, in agreement with results for other manganite compounds. In Chap. 19 far more details about this issue can be found.

15.1.2 Inhomogeneities at $x = 0.40$

Results of a variety of experimental techniques have shown that $x = 0.40$ bilayers exhibit inhomogeneities as well, both below and above the Curie temperature:

(1) Optical conductivity results reported by Ishikawa et al. [15.7] for the bilayer at $x = 0.40$ are shown in Fig. 15.1. It is clear from this figure that even at temperatures as low as 10 K the weight at low energy (Drude weight) is very small (see also [15.8]). Qualitatively, such an effect indicates that electrons cannot propagate easily across the sample upon the application of a small voltage. Note that a broad midinfrared peak at ~ 0.4 eV can be observed in the optical conductivity. Strong scattering of spin-polarized carriers by phonons and orbital excitations, with energies such as those present in the midinfrared band, could be responsible for these features. It is clear that the FM state of bilayers is not a good metal, namely, the carrier motion is mainly diffusive. This is compatible with the high value of the resistivity at zero temperature, already shown in Chap. 3 when discussing bilayers.

(2) Using neutron scattering techniques applied to the $x = 0.4$ bilayer manganite in the paramagnetic state, Perring et al. [5.21] reported the presence of long-lived antiferromagnetic clusters coexisting with ferromagnetic critical fluctuations (see Fig. 15.2). This antiferromagnetism appears to be along one axis of the basal Mn O plane of the bilayers, namely it is of the A-type, with an associated peak in the spin-structure factor located at $(\pi, 0, 0)$ in the usual notation. The typical size of the AF cluster is about 2.4 lattice spacings. Perring et al. [5.21] remarked that due to the coexistence of

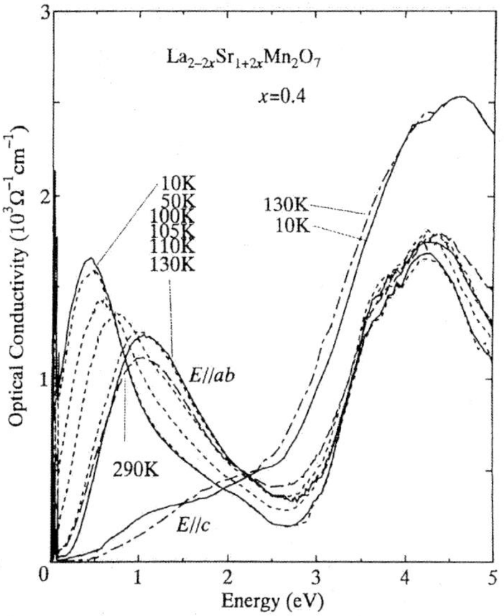

Fig. 15.1. Temperature dependence of the optical conductivity spectra of $x = 0.4$ $La_{2-2x}Sr_{1+2x}Mn_2O_7$, reproduced from Ishikawa et al. [15.7]. Results for the electric field parallel and perpendicular to the planes are presented. These results show that the material is not a good metal below the Curie temperature of 121 K, since the Drude weight is always small, and the conductivity is dominated by a feature at $\sim 1\,\mathrm{eV}$

FM and AF clusters it is not possible to describe the paramagnetic insulator state using mean-field approximations, which starts with the assumption of a uniform state. This is a recurrent conclusion of recent studies of manganites: *approximations that assume a uniform state do not capture the essence of the manganite states*, since they involve a mixture of clusters, which can be static or highly fluctuating depending on temperature and densities. Although the results of Perring et al. [5.21] were initially criticized by Millis [15.9], who proposed a uniform ferrimagnetic state, a large amount of experimental and theoretical information has shown that inhomogeneous states are the rule rather than the exception in manganites, including bilayers.

(3) Results from other techniques have also suggested that in the range between the Curie temperature and approximately room temperature, short-range ferromagnetic order can be found in the $x = 0.40$ bilayer manganite. For instance at 300 K, sharp deviations from the Curie–Weiss law start in the magnetic susceptibility, while the thermoelectric power indicates the formation of clusters in the basal plane (see Zhou et al. [15.10]).

(4) Investigations by Vasiliu-Doloc et al. [15.11] at $x = 0.4$ using X-ray and neutron scattering techniques, reported short-range polaron correlations

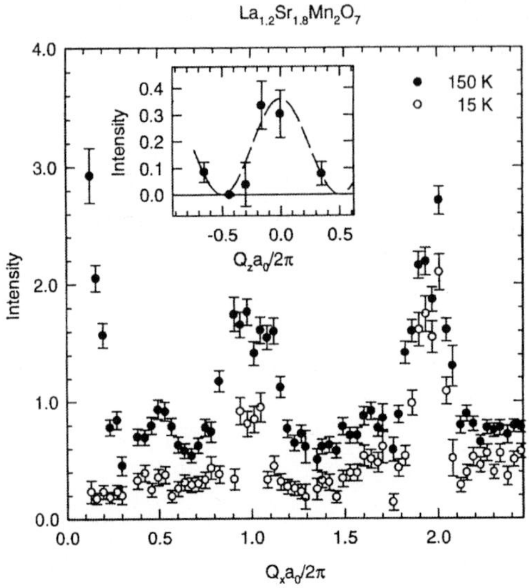

Fig. 15.2. Momentum dependence of the neutron scattering intensity, from experiments performed on $x = 0.40$ bilayers by Perring et al. [5.21]. Results for two temperatures are shown, and more details can be found in the original reference. The important point for the discussion in this section is the existence of a small peak at 0.5 in the horizontal axis (and at 150 K), which corresponds to an AF state. The more dominant peaks at other momenta indicate ferromagnetism

in the paramagnetic insulating state above T_C and up to room temperature. This issue will be discussed in more detail in Chap. 19.

(5) Studies by Kubota et al. [15.15] reported the resistivity vs. temperature for compositions $x = 0.4$, 0.45, and 0.48. The results are reminiscent of the percolative transitions discussed in Chap. 17. In particular, the zero-temperature resistivity at $x = 0.45$ is very large, a result difficult to reconcile with a homogeneous canted spin state. The spin-canted phase of bilayer manganites may be a mixture of FM and CO islands, and the authors of [15.15] arrived at similar conclusions.

15.1.3 Inhomogeneities at $x \sim 0.30$

Adding to the results of previous subsections, Heffner et al. [15.4] reported μSR data compatible with the existence of inhomogeneities in the $x = 0.3$ bilayer system as well. In addition, Chauvet et al. [15.5], using electron-spin resonance (ESR) techniques, showed that at $x = 0.325$ and in the regime of temperature between the Curie temperature and a higher temperature scale $\sim 380\,\mathrm{K}$, there are indications of FM clusters, with AF walls blocking the magnetic interaction between clusters. Indications of inhomogeneous states in $x = 0.30$ bilayers have also been reported by Hur et al. [15.6]. The experimental results described in this paragraph suggest that in the intermediate temperature regime of bilayers, a competition among many phases is to be expected, leading to cluster formation. This picture is compatible with the theoretical results of Chap. 17, where in the regime between T_C and T^*, ordered clusters of one of the competing phases are stabilized through the formation of walls of the other phase in between. By this procedure, it is possible to have an overall zero-order parameter, while being locally ordered.

15.1.4 Study of $(La_{1-z}Nd_z)_{2-2x}Sr_{1+2x}Mn_2O_7$

The "poor" metal behavior of the FM phase in $La_{2-2x}Sr_{1+2x}Mn_2O_7$ described in the previous section is suggestive of a competition between ferromagnetism and charge-ordering tendencies. Previous experience with 30% doped $La_{1-x}Ca_xMnO_3$ cubic perovskites has shown that the replacement of La by Pr induces a rapid reduction of the critical Curie temperature, and the subsequent stabilization of the charge-ordered state (see Chap. 11). In general, varying the mean value of the A-site radius, a poor-metal can be transformed into an insulator. If bilayers and cubic perovskites share similar physics, then it should be possible to alter the FM poor-metal state of $La_{2-2x}Sr_{1+2x}Mn_2O_7$, transforming it into an insulator by chemical substitution of La by other trivalent ions as well.

This expectation is in agreement with experimental data presented by Moritomo et al. [15.16] in their study of $(La_{1-z}Nd_z)_{2-2x}Sr_{1+2x}Mn_2O_7$. Typical results for the behavior of the resistivity varying the amount of Nd, at fixed $x = 0.40$, are shown in Fig. 15.3 (left). Compatible with the percolative results emerging from a random-resistor-network simulation (Chap. 17), and with many experimental studies of cubic perovskites, the resistivity has insulating behavior at room temperature, transforming into poor metal behavior at low temperature with a huge value at zero temperature. The full phase diagram in the z–x plane is shown in Fig. 15.3 (right). It would be interesting to determine the properties of the paramagnetic insulating phase. Is it charge ordered, or a cluster-glass mixture of the FM and A-type AF states, which are present in its vicinity? If it is a cluster-glass state, perhaps it is the state predicted at high values of the disorder by the theoretical arguments leading to quantum-critical-like behavior separating two order phases (Chap. 17). A cluster-glass behavior could contribute to explaining the huge MR effect of the $z = 1.0$ limit, and at $x = 0.45$ and 0.50 reported by Battle et al. [15.17]. This effect is observed *without* a transition to a low-temperature ferromagnetic state. This example shows that CMR effects do not need a metal in the process, but it can occur between two competing insulating states as well, in good agreement with theoretical expectations (Chap. 17). More work should be devoted to the interesting (La,Nd)-bilayer compounds to fully clarify their highly nontrivial properties.

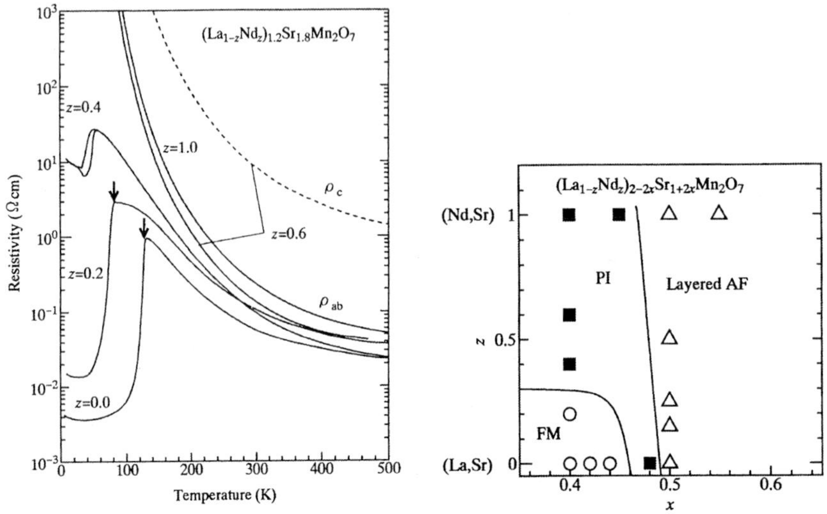

Fig. 15.3. (**a**) In-plane resistivity for single crystals of $(La_{1-z}Nd_z)_{1.2}Sr_{1.8}Mn_2O_7$. Curie temperatures are indicated with *arrows*. The out-of-plane resistivity is also shown, for one composition. A poor-metal to insulator transition occurs at low temperature, varying z. From Moritomo et al. [15.16]. (**b**) Ground states of $(La_{1-z}Nd_z)_{2-2x}Sr_{1+2x}Mn_2O_7$, from Moritomo et al. [15.18]. *Circles*, *triangles*, and *squares* represent the ferromagnetic metallic (FM), layered antiferromagnetic (layered-AF), and paramagnetic insulating (PI) states, respectively

15.2 Inhomogeneities in Single Layers

Recently, a synchrotron X-ray scattering study of the single-layer manganite $La_{1-x}Sr_{1+x}MnO_4$ in the range $0.33 \leq x \leq 0.67$ was presented by Larochelle et al. [15.19]. In this study three distinct regions were identified: (i) a disordered regime for $x < 0.4$, a (ii) mixed-phase regime for $0.40 \leq x < 0.5$, and (iii) a charge-ordered region at $x \geq 0.50$. Region (ii) was considered part of a spin-glass regime in previous investigations, as discussed for example in [3.44]. The new study of Larochelle et al. found signals of charge ordering in that density range, and these authors believed that a mixed-phase disorder/charge-ordered better describes the data. These tendencies to phase separation are in good agreement with theoretical calculations, and it is clear that many regions labeled "spin-glass" in early experimental studies of this and other manganites may correspond to more complex patterns, such as cluster-glasses or mixed-phase regimes. In addition, in the regime $x < 0.40$, the new study reports results similar to those at $x = 0.50$ above the charge-ordering temperature, where it is believed that polaron correlations must be present. Then, even at $x < 0.40$ it is possible that small-scale cluster coexistence may be a good representation of the single-layer ground state. More

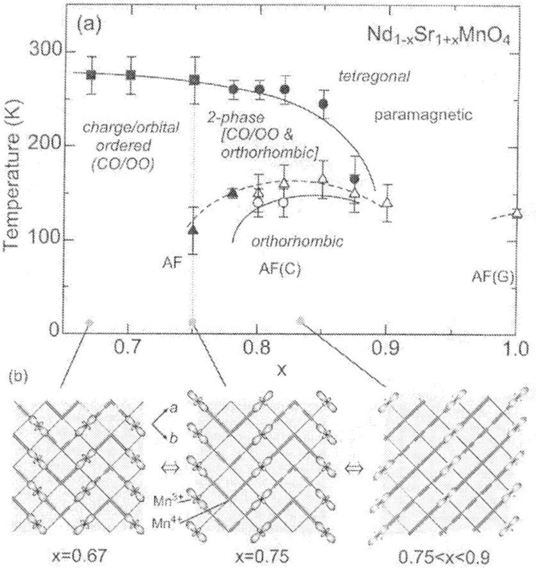

Fig. 15.4. Electronic phase diagram of $Nd_{1-x}Sr_{1+x}MnO_4$, from Kimura et al. [15.20]. The states are shown, and the region of phase coexistence is indicated. The details can be found in the original reference. Note that the region between 0.9 and 1.0 still has not been characterized

work is needed to fully clarify the very interesting and highly nontrivial properties of the layered manganites that Larochelle et al. have started to unveil.

Other recent studies of single-layer manganites have also reported regions of phase coexistence. In particular, $Nd_{1-x}Sr_{1+x}MnO_4$ was analyzed by Kimura et al. [15.20] in the density window between $x = 0.67$ and 0.90. The results of transport, magnetic, and structural measurements are summarized in Fig. 15.4. A competition between a C-type AF state and a charge-ordered state, which smoothly evolves from the CE-state at $x = 0.50$, was observed. There are regions where the two phases coexist. The states observed are in good agreement with the conclusions of Hotta et al. [3.32].

Summarizing this chapter, both for double- and single-layer manganites there is experimental evidence of mixed-phase tendencies, results compatible with those of the perovskites discussed elsewhere in this book. The results also open up the possibility for similar tendencies to occur in other transition-metal oxides, including the famous layered cuprates. More work should be devoted to the clarification of the interesting phase diagram of the single-layer Mn oxide. The results should be compared with those of other interesting two-dimensional oxide materials.

16. An Elementary Introduction to Percolation

The basic aspects of percolation theory are briefly reviewed in this chapter. The theory of percolation is important for the main issue discussed in this book, namely, the presence of nanoscale phase separation in manganites – both in theory and experiments – and the explanation of the CMR effect as arising from a percolation process (see Chap. 17). The concept of percolation is well known in physics, and it has been reviewed by Stauffer and Aharony [16.1]. Their book is much recommended to the reader to find additional information on this important subject, supplementing the elementary primer provided here.

16.1 Basic Considerations

Following Stauffer and Aharony [16.1], as a simple example to introduce the notion of percolation one can use a square lattice where each link is "occupied" randomly with probability p. The occupied links are denoted with solid lines in Fig. 16.1. Some of those occupied links are isolated from others. However, in the random distribution of these sites it is clear that "clusters" will be formed, namely, regions where several occupied links are close to each other, forming small linked structures. Percolation deals with these clusters, particularly their typical sizes and shapes, which can be very complex. As in the introductory example of Stauffer and Aharony [16.1], with thick solid lines, the "largest" cluster is marked in Fig. 16.1, namely, the cluster that has more links. At probability $p = 0.4$, even this largest cluster does not extend from one side of the sample to the other, and for that reason it is said that the system has not percolated. If one imagines the links as channels allowing for liquid or charge to flow, in the configuration of Fig. 16.1a there will be no net flow across the sample. However, results for p=0.6 are qualitatively different, see Fig. 16.1b. Now the largest cluster is sufficiently large that it extends from one side to the other, allowing for liquid flow. This suggests that there is a critical probability p_c, shown to be 0.5 below for the case under discussion in two dimensions, where the system actually *percolates*.

Note that the language introduced here, and the notion of percolation, is certainly not restricted to square lattices, but is far more general. Percolation is widely used in a variety of contexts, and in this book this notion will

(a) (b)

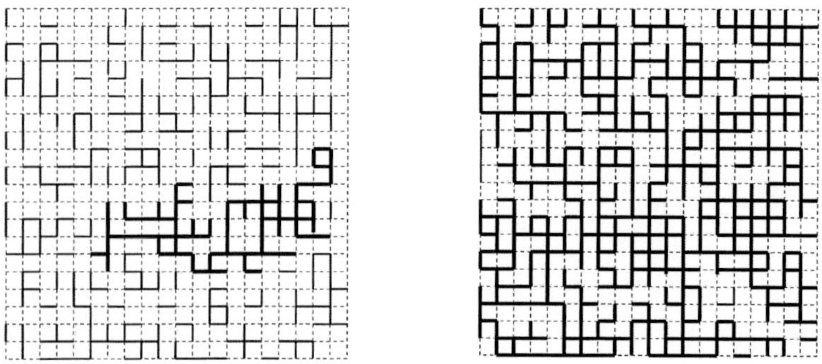

Fig. 16.1. Random distribution of occupied (*solid lines*) and unoccupied links (*dashed lines*), or, equivalently, metallic or insulating links in the random-resistor network discussed later. Results are shown for a 2D square lattice and (**a**) corresponds to probability $p = 0.4$, while (**b**) corresponds to $p = 0.6$. Marked with *thicker symbols* is the largest connected cluster at each value of p. The system (**a**) has not percolated, while (**b**) has. Figure courtesy of Matthias Mayr

be applied to the study of manganites and other transition-metal oxides. Figure 16.1 also suggests that in most cases of interest the study of percolation must be carried out numerically. This is due to the complexity of the cluster shapes involved in the process, which can be very fractalic, with a sort of effective dimensionality that can be larger than one but smaller than the dimension of the system (two in the example). In fact, the shortest path from one side of the sample to the other in Fig. 16.1b is certainly totally different from a straight line, for the case when the probability p is only slightly larger than the critical value. In addition, near criticality *many length scales* are of relevance. For instance, the largest cluster has many branches, and it contains holes, which may contain other smaller clusters. In fact, there is no obvious dominant length scale at the critical concentration. The reader can find in Stauffer and Aharony's book [16.1] far more information than provided here about these very important concepts in the context of the theory of percolation, as well as dozens of references.

Among the first studies of percolation were those by Broadbent and Hammersley [16.2], using lattice models for fluid flow in a static random medium. In their model, fluid flows through bonds that connect nearest-neighbor sites of a regular lattice. The disorder is usually introduced in the system either through modifications of the bonds or sites. In the former, a fraction of the bonds are missing from the lattice, and fluid cannot flow through them. In the latter, a fraction of sites are missing, and no current can flow through any of the bonds connected to that site. The authors of [16.2] already discussed the concept of a critical fraction of fluid-conducting bonds or sites needed to allow for the flow of the fluid from one side of the sample to the other (i.e., for percolation to occur, as already discussed). Note that percolation plays an important role in metal insulator transitions, where the "fluid" is actually the electrons moving across the sample. A review of percolation, precisely with emphasis on transport in inhomogeneous conductors, can be found in [16.3] (portions of this section are based on that review).

For a low concentration p of conducting bonds, conduction is possible within small clusters, but they are too isolated for conduction to occur through the entire system. As p increases, larger clusters occur, and their mean size increases. As p approaches the critical value p_c, the larger clusters start to merge, creating a few very large clusters, with a mean size that diverges as the sample size diverges. For $p > p_c$, in practice it usually occurs that only one large cluster of conducting medium remains, the so-called "infinite cluster", together with many small ones that are absorbed into the infinite one as p is further increased. The ratio of number of sites or bonds in the infinite cluster to the total number is denoted by $P(p)$, and it is a natural quantity to characterize transport (together with the actual conductivity or resistivity of the sample, as discussed later). Numerical studies have shown that near the percolative limit p_c, a power-law behavior is observed, namely $P(p) \sim (p - p_c)^s$, with s a positive number less than one (thus, $P(p)$ has infinite slope at p_c). This is a general result observed in lattices of different shapes and dimensionalities, even in systems without an underlying regular structure. It is also a characteristic of critical phenomena in a variety of contexts where second-order or continuous phase transitions occur. As it happens in that context, a variety of critical exponents is usually introduced, but the fine details of such discussion are beyond the scope of this elementary section (for information on the subject, see Chap. 2 of [16.1]).

16.2 Resistor Networks

Suppose the location in space of the conducting and non conducting regions of a metal-insulator mixed-phase system are known. Suppose also that the metallic regions are much larger than the electronic mean-free path, then becoming essentially classical. In this limit, the electrons are treated as a fluid of charge, and disorder, or phases that are in competition with the metallic one, enter through the boundary conditions that the electronic current must satisfy for its flow. In the limit of sufficiently large metallic regions, it is possible to use a local conductivity $\sigma(r)$ at site r to characterize charge movement, as if that region were a resistance. On the other hand, for small metallic regions this would not necessarily be correct, namely, transport could be "ballistic": electrons for sufficiently small conducting clusters may move from one side to the other without collisions, and, thus, without resistance. But in several cases a description based on $\sigma(r)$ is correct, and the goal in this context is to calculate the conductivity of the entire mixed-phase sample, which must be a function of the distribution of local conductivities.

Isolated conducting regions are clearly of no relevance for transport across the sample, namely, for calculating the DC conductivity (however, they are important for AC conductivity, where a finite frequency ω of oscillation of the external field driving the current is introduced). The regions of relevance for DC conduction are those that belong to the "infinite cluster", described by $P(p)$. However, the relation between conductivity and $P(p)$ is not obvious. For instance, the infinite cluster has many dead ends that are useless for conduction (see Fig. 16.1b). The current flows only through a fraction of the infinite cluster, the so-called "backbone", and the number of truly relevant bonds for the effective resistance is far smaller than $P(p)$.

In most of the cases studied in the literature of percolation, the conductance at a link joining sites i and j, is restricted to be either 0 (non conducting or insulating) or 1 (conducting or metallic), in the appropriate units. This is not necessarily the case in real systems, and actually in some applications to manganites described in other chapters, this "0" or "1" restriction is removed. Nevertheless, to gain intuition on the behavior of resistor networks, two-valued conductances will be used below. Also, in practice, regular lattices (square and cubic in 2D and 3D, respectively) are typically selected.

As already defined above, a variety of percolation models or resistor networks can be defined. (i) A *bond percolation model* is the one where a fraction of bonds chosen at random are removed, namely, they have zero conductance. This is very easily done in practice. To each bond a random number between 0 and 1 is first assigned. If that number is larger than the fraction of removed bonds that is to be studied, a conductance of 1 is assigned. If smaller, then the conductance is 0. The usual notation is to use p as the fraction of bonds that *remain*, i.e., those that are conducting. The fraction of removed bonds is then 1-p. The bond percolation model will be extensively used for manganites in other chapters of this book. (ii) Another popular model is the *site percolation model* where the random number is assigned to each node or site of the regular lattice. The removal of a site implies that the conductivity of all the bonds connected to that site is 0.

To calculate the overall conductance of a resistor network, the Kirchhoff equations must be solved numerically (see Appendix to this chapter). The studies of the case of a two-valued, 0 and 1, resistor have been documented extensively in the literature, and the main results are reproduced here. Conductance data for the bond percolation model are presented in Fig. 16.2 for 2D and 3D systems. The conductance is normalized to one at $p = 1$, in the limit where all bonds conduct. Results for $P(p)$ (not shown) demonstrate that this quantity and the conductance are not identical, as explained before. Although not obvious in the figures scale, typically there are two regimes found for the conductance: a critical one very near p_c, where the onset of conduction occurs with zero slope, and a linear regime at higher p.

Near p_c, the conductance $G(p)$ follows a power law, $G(p) \sim (p - p_c)^t$. The values for t found numerically in 3D are close to 1.6 (not only for bond

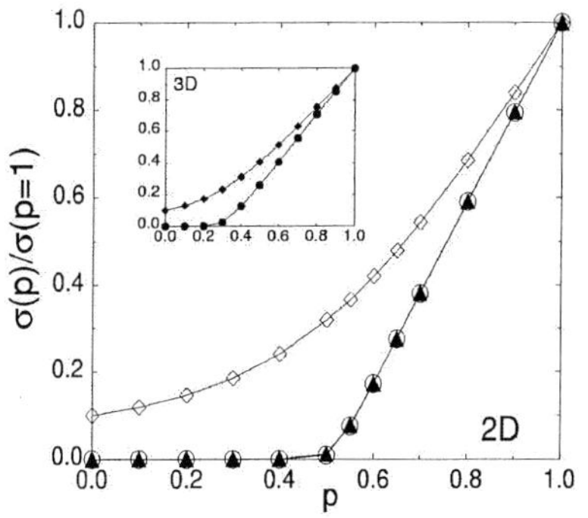

Fig. 16.2. Conductivity versus the percentage of metallic links p on a 100×100 $d = 2$ random-resistor-network model. The data are scaled such that $\sigma(p = 1) = 1$. *Open diamonds* denote results for a model with $\sigma_I = 0.1$ (still considered an insulator due to its relatively low conductivity compared with the metal) and σ_M=1 (metal). *Open circles* are earlier results from [16.3] on a 50^2 lattice. The differences from the 100^2 lattice results are very small. The *inset* shows results on a 25^3 cube for the same model, as well as for a model with a finite insulating conductivity. Data from [16.3] are not shown, but agree very well with these results. Averages over 50 configurations were made in the calculations. Figure courtesy of M. Mayr

percolation but also for site percolation), while in 2D they are closer to 1.1 (for details and references see [16.1]). Note that the fine details of the model (site vs. bond percolation, etc.) are not important near criticality, but only the dimension of the sample is relevant. This universal behavior is common in models that have critical behavior.

Note that the results of Fig. 16.2 are obtained by making an average over many distributions of links (just one distribution would be sufficient only for an infinite size lattice). For a finite lattice, it may accidentally occur that a given random-link distribution could lead to percolation, for example at $p = 0.48$ which is only slightly smaller than p_c. However, when several random-link distributions are studied, the majority will not have that property since the true critical value is 0.50. Thus, it is crucial to carry out a statistical study by generating many clusters, as many as needed until a convergence of the results to a satisfactory accuracy is achieved. The statistical analysis techniques followed in this context are similar to those of standard Monte Carlo simulations in many respects. Note that in practice, very large clusters seem to require less configurations with independent distributions of links, while small clusters need far more. A large cluster is sufficiently close to bulk that results obtained with independent random-link distributions are actually very similar in value to each other.

Table 16.1 contains information (from [16.1]) of selected percolation critical fractions p_c, for the cases of site and bond percolation. Some of the results are exact, such as for the bond model on a square lattice, and the site model on a triangular lattice.

Table 16.1. Critical probability p_c for site and bond percolation, reproduced from the book of Stauffer and Aharony

Lattice	Site	Bond
Honeycomb	0.6962	0.6527
Square	0.5927	0.5000
Triangular	0.5000	0.3473
Diamond	0.43	0.388
Simple cubic	0.3116	0.2488
BCC	0.246	0.1803
FCC	0.198	0.119

16.3 Anderson Localization

Anderson localization [16.4] is the phenomenon where the conductivity of one- or two-dimensional systems in general, and three-dimensional systems as long as disorder strengths are sufficiently strong, is shown to vanish at zero temperature due to quantum-interference effects. This occurs in models where electrons do not interact with each other, but simply scatter from static impurities. The reason for this phenomenon is in the quantum-mechanical treatment of the electronic movement, where elastic scattering from impurities produces reflected waves that interfere with the incoming ones. It can be shown that the mixture of incoming and reflected waves can lead to a resulting wave that is localized in space (for an elementary review see [16.5]. For more details see Chap. 2 of [3.34]). The constraint of low temperature is needed for this interference to occur, otherwise inelastic processes may violate the interference effect if the colliding waves do not have the same wavelength.

Note that, in general, to move in space from point A to B several paths can be followed by the electron in Fig. 16.3, and in quantum mechanics the total probability $W_{\mathrm{A}\to\mathrm{B}}$ for that movement to occur is given by $W_{\mathrm{A}\to\mathrm{B}} = |\sum_i \psi_i|^2 = \sum_i |\psi_i|^2 + \sum_{i\neq j} \psi_i\psi_j^*$, where ψ_i describes the wave function or amplitude associated with trajectory i (there are an infinite number of them, of course). This result tells us that the wave functions are added and, *then*, squared to construct the probability, unlike in classical physics. This leads to the possibility of interference between the many waves, i.e., between the many paths. The last term in the equation above shows that the total probability is the sum of the individual probabilities plus the contribution coming from interference, which is of quantum origin. Note that it may seem that the different lengths of the trajectories lead to substantial phase shifts in the wave functions, producing an overall cancellation of the last term (namely the sum over $i \neq j$ contains positive and negative contributions that could cancel each other). However, this does not take into account the effect of *self-intersecting* trajectories such as in Fig. 16.3 (lower panel). It can be shown that the two trajectories contained in Fig. 16.3 (lower panel), i.e., with and without the closed loop, are coherent or in phase, and as a

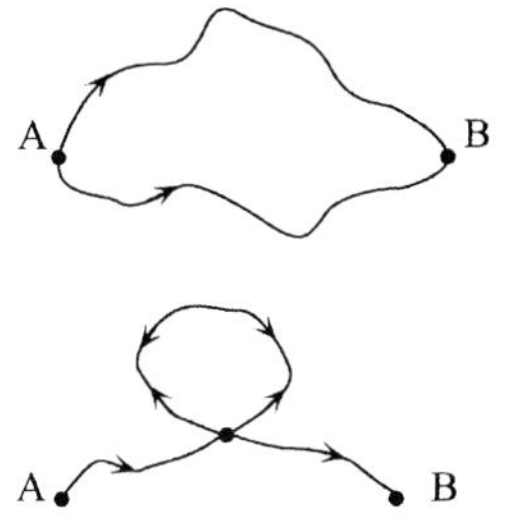

Fig. 16.3. *(Upper panel)* Two of the possible paths from positions A to B, to illustrate the discussion about interference in the text. *(Lower panel)* A self-intersecting path from A to B, which is important for Anderson localization effects

consequence the cancellation of the interference term is not complete. These self-intersecting trajectories are the ones responsible for the reduction in the conductivity of low-dimensional systems, and the eventual localization of the wave function. Of course, the present discussion is only very qualitative, and actual calculations and numerical simulations must be carried out to prove that total localization can be induced by the interference described above. After this elementary primer, the reader interested in the subject should consult more focused recent reviews, such as [16.6], to learn the fine details of this interesting phenomenon. For our purposes, the simple discussion in this section is sufficient to define the phenomenon of quantum percolation in the next section, and also to address issues related to Anderson-localization ideas for manganese oxides. Theories of manganites that rely on Anderson localization to produce the insulating state above the Curie temperature will be briefly described in Chap. 22. One of the main problems of these theories is that they do not reproduce the pseudogap effects found in photoemission experiments (see Chap. 18), since the density of states at the Fermi level is not expected to develop any singular structure due to the Anderson-localization process alone (see [3.34] and references therein).

Note that for a long time it has been widely assumed that localization should occur in any two-dimensional metal with an infinitesimal amount of disorder. However, recent experimental results have questioned this belief, and the subject has gathered new momentum. The current discussion is centered around the relevance of Coulombic interactions among electrons, and its competition with the interference caused by disorder that leads to localization [16.6].

16.4 Quantum Percolation

Let us consider once again a bond-percolation model, where the bonds can be metallic or insulating. However, in contrast with the previous calculations where the metallic and insulating character of a bond is associated with small and large values of a resistance at the bond, here the problem will be

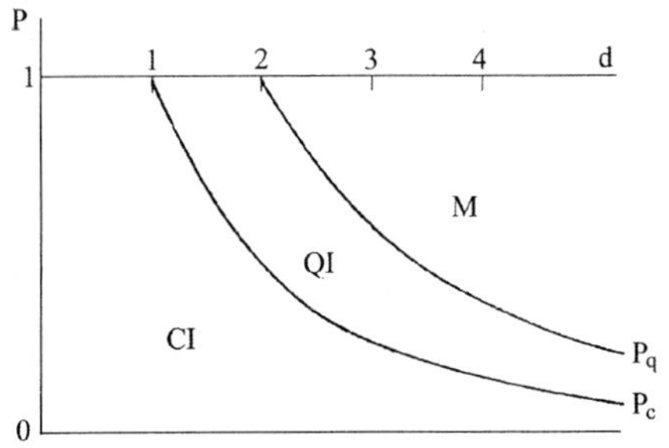

Fig. 16.4. Schematic representation of the dependence of the properties of a hypercubic lattice of randomly distributed wires, with its dimension and the probability P for the global number of those wires (from [16.7]). CI, QI, and M denote a classical insulator (due to the absence of percolation), quantum insulator (the metallic network is percolated but Anderson-localization effects render it still insulating), and a metal, respectively. The energy is assumed to be constant

treated quantum mechanically. The reader can consider this case as evolving smoothly from the previous classical analysis, as the size of the metallic and insulating clusters is reduced. Once that size is made smaller than the mean-free path, created itself, for instance, by collisions with impurities, then the propagation across the metallic regions is ballistic, namely, there are no collisions. Working on a square lattice, with electrons located at the sites of the lattice, this ballistic transport can be simply mimicked by a hopping term that destroys an electron at a given site and creates it in one of its nearest neighbors (hopping term). The "insulating link" can be mimicked in this context by a zero hopping amplitude. Then, the model to be studied with quantum effects incorporated is defined on a lattice, typically cubic or square, and its Hamiltonian has a standard hopping term with an amplitude to move from site i to j equal to t_{ij}. This hopping is either t or 0, following the same procedure that gave us conductance 1 or 0 for the classical case.

The Hamiltonian defined by this procedure cannot be diagonalized exactly analytically, but diagonalization routines are needed. The movement of electrons locally is ballistic when a metallic link is visited, but overall the insulating links are regions that the electrons collide with, inducing a net resistance. This effect can be caused by: (i) Classical geometric lack of percolation: if there are no clusters of metallic links joining one side of the sample with the other, then electrons simply can not propagate across the sample. (ii) Quantum effects due to Anderson localization. Even if the metallic fraction p is larger than the classical percolation threshold $p_c^{\text{classical}}$, the interference between reflected waves can lead to localization. Then, the true threshold for quantum percolation in a resistor network, as defined above, satisfies $p_c^{\text{quantum}} > p_c^{\text{classical}}$.

Figure 16.4 illustrates the expected behavior of percolation as the dimension of the lattice increases (from [16.7]). In dimension 1, there is no percolation for any $p < 1$. This obviously occurs for geometrical reasons, namely a single insulating link (having zero conductance in the classical description, or zero hopping in the quantum-mechanical model described above) leads to an overall insulating chain. The situation is more interesting in dimension 2.

Here, it is known that in the classical limit p_c is 0.5 for a square lattice, as discussed before. However, the discussion around Anderson localization shows that in dimension 2, the localization occurs for any amount of disorder. On the square lattice with hopping electrons, disorder in the links exists at any $p < 1$, thus at the quantum level there is no conductance in such a system. Finally in D=3, both classically and quantum mechanically, for p sufficiently large a percolation into a finite conductivity exists. Note that classically only the backbone of the infinite cluster matters for conduction, while in quantum mechanics it is the whole cluster. The dead ends provide reflected waves that contribute to the localization.

16.5 Appendix: Solution of the Random-Resistor-Network Equations

This brief Appendix contains some details at the solution of a random resistor network problem. It is adapted from results presented in the Ph.D. Thesis of M. Mayr (Florida State University, 2001).

Let us address the formal aspects of the solution of a random resistor network (RRN), namely, how to find its effective resistance or conductance G. The calculation of G for a RRN is based on a relaxation technique. From electrostatics one knows that the current is related to the voltage through $\boldsymbol{j}(\boldsymbol{r}) = \sigma(\boldsymbol{r})\boldsymbol{\nabla}V(\boldsymbol{r})$, where $V(\boldsymbol{r})$ and $\boldsymbol{j}(\boldsymbol{r}) = (\boldsymbol{j}_x, \boldsymbol{j}_y)$ (in two dimensions) denote the local voltage and current, respectively. In addition, the current must satisfy the zero-divergence condition $\nabla \cdot \boldsymbol{j}(\boldsymbol{r}) = 0$. In the case of a lattice with unit bond length $\Delta r = 1$, these equations can be discretized as follows

$$\begin{aligned} \nabla_x \boldsymbol{j}_x(\boldsymbol{r}) &= \boldsymbol{j}_x(\boldsymbol{r} + \boldsymbol{e}_x) - \boldsymbol{j}_x(\boldsymbol{r}) \\ &= \sigma^x(\boldsymbol{r} + \boldsymbol{e}_x)(V(\boldsymbol{r} + \boldsymbol{e}_x) - V(\boldsymbol{r})) - \sigma^x(\boldsymbol{r})(V(\boldsymbol{r}) - V(\boldsymbol{r} - \boldsymbol{e}_x)), \end{aligned}$$

where $\boldsymbol{j}_x(\boldsymbol{r}) = \sigma^x(\boldsymbol{r})[V(\boldsymbol{r}) - V(\boldsymbol{r} - \boldsymbol{e}_x)]$ was used, and a similar equation holds for the y-derivative. $\sigma^x(\boldsymbol{r} + \boldsymbol{e}_x)$ is the conductivity between sites $\boldsymbol{r}$ and $\boldsymbol{r} + \boldsymbol{e}_x(\boldsymbol{e}_x$ is the unit vector in the x-direction). From the previous equations and $\nabla_x \boldsymbol{j}_x(\boldsymbol{r}) + \nabla_y \boldsymbol{j}_y(\boldsymbol{r}) = 0$ $(d = 2)$ we immediately arrive at

$$V(\boldsymbol{r}) = \sum_{\boldsymbol{r}'} \sigma_{\boldsymbol{r}\boldsymbol{r}'} V(\boldsymbol{r}') / \sum_{\boldsymbol{r}'} \sigma_{\boldsymbol{r}\boldsymbol{r}'}, \tag{16.1}$$

where $\sigma_{\boldsymbol{r}\boldsymbol{r}'}$ is the conductivity between adjacent sites $\boldsymbol{r}, \boldsymbol{r}'$, i.e., $\boldsymbol{r}' = \boldsymbol{r} \pm \boldsymbol{e}_{x(y)}$, and, therefore, $\sum_{\boldsymbol{r}'} \sigma_{\boldsymbol{r}\boldsymbol{r}'} = \sigma^x(\boldsymbol{r}+\boldsymbol{e}_x)+\sigma^x(\boldsymbol{r})+\sigma^y(\boldsymbol{r}+\boldsymbol{e}_y)+\sigma^y(\boldsymbol{r})$. Depending on whether the two sites $\boldsymbol{r}, \boldsymbol{r}'$ are connected by a metallic or an insulating link, then $\sigma_{\boldsymbol{r}\boldsymbol{r}'} = 1$ or 0. The voltages $\{V^s(\boldsymbol{r})\}$ at two opposite surfaces of the system are kept fixed at $V^s(\boldsymbol{r}) = V_c \neq 0$ and 0, respectively, in order to generate a flow of charge. In the direction(s) perpendicular to the surfaces of constant voltage, cyclic boundary conditions are imposed by introducing an additional layer of bonds.

Starting from an initial configuration of voltages selected as $\{V^0(\boldsymbol{r})\} = V_c(1-a)/L$ (where a is the distance between point $\boldsymbol{r}$ and the boundary with $V = V_c$, and L is the linear size of the lattice), (16.1) is applied to obtain the next set of voltages $\{V^1(\boldsymbol{r})\}$ (in general $V^n(\boldsymbol{r})$ is determined according to the *updated* value of $V^n(\boldsymbol{r}-\boldsymbol{e}_x)$), and the whole procedure is repeated until $|V^n(\boldsymbol{r}) - V^{n+1}(\boldsymbol{r})| < \epsilon$ at each lattice site $\boldsymbol{r}$, with ϵ an arbitrary, small number. Typically less than 100 sweeps are necessary to reach convergency far away from the percolation threshold, but this number increases dramatically close to p_c, where up to 10^4 loops are required – a phenomenon known as "critical slowing down". To speed up the computation, it has also proved useful to apply an "overrelaxation" technique by choosing $V^{n+1}(\boldsymbol{r}) = (1+\alpha)\hat{V}^{n+1}(\boldsymbol{r}) - \alpha V^n(\boldsymbol{r})$, with $0 < \alpha < 1$ and $\hat{V}^{n+1}(\boldsymbol{r})$ the calculated value for the voltage at site $\boldsymbol{r}$ according to (16.1). Often, $\alpha = 0.8$ could be used without causing any instability. From the knowledge of the equilibrium voltage distribution $\{V(\boldsymbol{r})\}$, the currents at the surfaces of constant potential can be calculated (they must be equal on both sides; this fact may serve as a check on the correctness of the calculation), and the resistance as well as the conductivity follows from Ohm's law.

17. Competition of Phases as the Origin of the CMR

In previous chapters, it was argued that realistic models for manganites exhibit electronic phase separation. The two competing phases have different electronic density and, after the inclusion of long-range $1/r$ Coulomb interactions, nanometer clusters are expected to survive. This scale is compatible with a variety of results for manganites, reviewed in this book. However, other results suggest larger (micrometer) scale clusters. In this chapter, a possible mechanism to understand the origin of huge clusters is discussed. The discussion leads naturally to spin toy models, competition between ordered phases, and the influence of quenched disorder on uniform states, including a variety of concepts that are much used in condensed matter. For these reasons, this is an important chapter of this book. At the end of the chapter, a contribution by Y. Tokura is presented, detailing how quenched disorder is of fundamental importance to understand the behavior of CMR materials, in qualitative agreement with the theoretical description presented here. A second contribution by D. Khomskii explains the role of entropy in transitions from FM to CO states with increasing temperature.

17.1 Quenched Disorder Influence on First-Order Transitions and Mesoscopic Coexisting Clusters

The results of Uehara et al. [11.41], Fig. 11.13, and of Fäth et al., Fig. 11.15, strongly suggest the presence of large-scale submicrometer inhomogeneities in some manganites, with percolative and fractal-like properties. This scale is far larger than the expected nanometer-scale clusters that could be stabilized in regions of electronic phase separation, once the long-range Coulomb interaction is taken into account. Their formation requires another idea, some new mechanism that leads to huge coexisting clusters with the *same* electronic density (otherwise the energy penalization due to charge accumulation would be too large, preventing stability). This does not mean that nanocluster formation due to the electronic phase-separation mechanism (Chap. 6) is not operative in manganites. It is possible that both mechanism could be simultaneously at work. But, by what other procedure can we have submicrometer coexisting clusters?

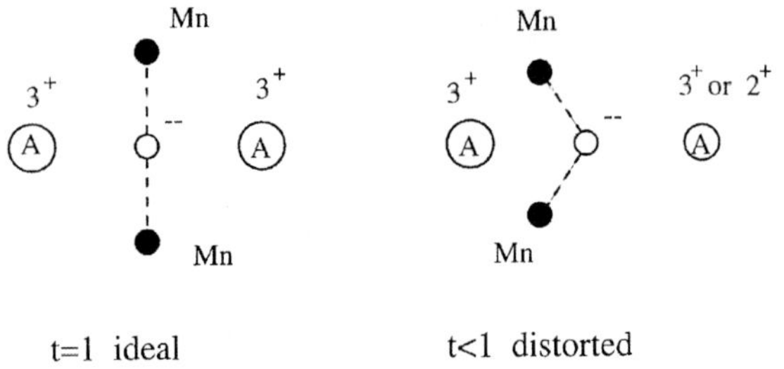

Fig. 17.1. Schematic representation of the dependence of the hopping Mn Mn on the size of the ion located at an A-site. Different sizes produce different angles in the Mn O Mn bridge, leading to different values for the hopping

Moreo et al. [6.41] recently presented a possible mechanism. This mechanism is built upon the following ideas: (1) In real systems two ordered phases are often in competition, with patterns of spontaneous symmetry breaking so different that the transition between them is of first order. This is actually the case observed in simulations involving the CE-state and the FM-state, within the context of the two-orbital model for manganites (see, e.g., Fig. 10.1). The metal and the insulator, or FM and AF states, are separated by first-order transitions in models widely considered as realistic for Mn oxides. (2) The second important ingredient is the presence of quenched disorder. Consider, for instance, the unavoidable disorder introduced by chemical doping in compounds such as $La_{1-x}Ca_xMnO_3$, where La is replaced by Ca. These ions are located at the A-sites of the perovskite structure, and they have different charge. This produces trapping centers for the e_g electrons moving in this environment. Even in compounds such as $(La_{1-y}Pr_y)_{1-x}Ca_xMnO_3$, the replacement of equal-valence La by Pr leads to disorder. This material was studied by Uehara et al. [11.41] in their discovery of submicrometer clusters, and here the different ionic sizes in the A-site are believed to alter locally the hopping matrix elements, as shown in Fig. 17.1. This is certainly another source of disorder. How does this intrinsic disorder affect the first-order metal insulator phase transition?

Moreo et al. [6.41] carried out a Monte Carlo study of the one- and two-orbital models, in the presence of disorder introduced in the form of a random component added to the uniform hopping and J_{AF} couplings. Different links are uncorrelated regarding this random component. Some of the results of Moreo et al. are shown in Fig. 17.2. For two orbitals (left) the density is fixed at 0.5 and the temperature is low. Without disorder, the transition FM AF is clearly first order, as shown in the upper part of the figure. The most interesting results are obtained with nonzero disorder, in the vicinity of the first-order transition, as in Fig. 17.2b. The presence of *cluster coexistence at a scale much larger than the lattice spacing* can be observed. In the limit when the disorder is very strong, these clusters should approximately be of the size of the lattice spacing but, in the weak disorder case, the clusters are much

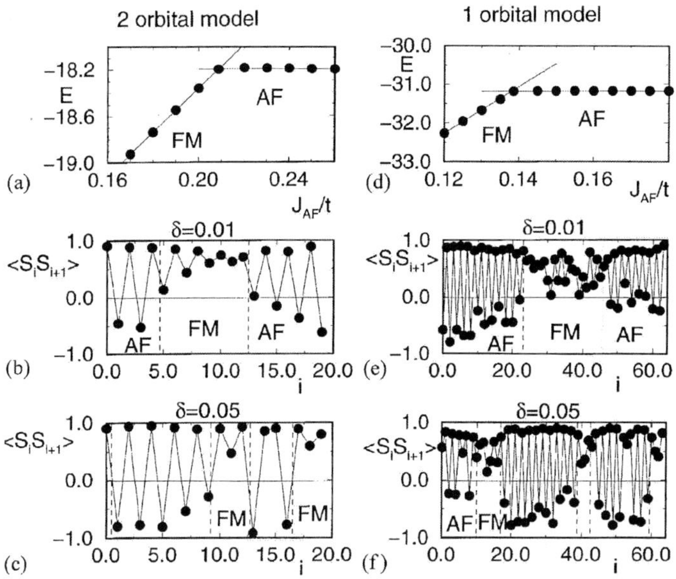

Fig. 17.2. Monte Carlo results for the two-orbital *(left)* and one-orbital *(right)* models for manganites, from Moreo et al. [6.41]. The dimension of the problem is one. (**a**) Energy vs. J_{AF} at density 0.5 and low temperature. A first-order transition separating the FM and AF phases can be observed. (**b–c**) Nearest-neighbor spin spin correlations at the J_{AF} of the first-order transition in (**a**), adding a weak random component δ to that coupling. Clusters are formed, as shown in the figure. (**d–f**) Results for the one-orbital model, analogous to those obtained for two orbitals. The density is 0.5 and the temperature is low. Large clusters can be observed

larger. In this case, creating several interfaces would cost too much energy, and the system settles to a compromise stabilizing just a few. Quite similar results were found in the case of the one-orbital model (parts d–f of Fig. 17.2), once again with disorder incorporated into the hoppings and J_{AF}.

Another related example, also from Moreo et al. and in Fig. 17.3, is the discontinuous transition in density vs. chemical potential μ observed in the simulations of Chap. 6. This is the behavior expected in the phase-separation regime, but this effect could be considered simply as a first-order transition varying μ. Here, the addition of disorder in the chemical potential creates large clusters, as shown in Fig. 17.3b, and now the density vs. μ no longer presents a discontinuity (part a of the figure), but a rapid crossover. Note also the very interesting result that appears in the density of states (part c of the figure) where a clear *pseudogap* is observed. Pseudogaps of this kind are reviewed extensively in Chap. 18. They seem to appear in inhomogeneous states, mixing metals and insulators. This pseudogap reduces the density of states at the Fermi energy and probably induces a large resistivity. Finally, in

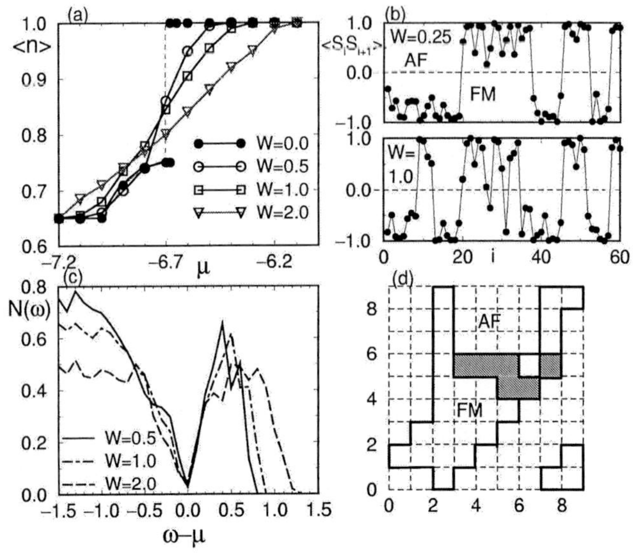

Fig. 17.3. Results of a Monte Carlo simulation of the one-orbital model with a random chemical potential, from Moreo et al. [6.41]. A chain is used as illustration. (**a**) Shows density $\langle n \rangle$ vs. chemical potential μ for various strengths of the disorder W. The discontinuous transition in the non disordered case, becomes continuous when disorder is added. (**b**) Nearest-neighbor spin spin correlations, showing the presence of clusters FM and AF. (**c**) Density of states at various strengths of the disorder, showing the existence of a pseudogap. (**d**) In the presence of small external magnetic fields, a percolative transition can be observed (details can be found in [6.41])

the presence of magnetic fields, a percolative behavior is observed, in agreement with experiments.

The results described thus far, with the generation of large clusters for weak disorder, do not appear to depend on the particular properties of the competing phases, or their microscopic origin. From this perspective, whether the Jahn–Teller coupling or the Coulombic interactions are the key ingredients to produce the physics of manganites is of relevance regarding the particular properties of each phase, and their pattern of symmetry breaking, but not regarding the issue of phase competition and concomitant inhomogeneities.

17.2 Random Field Ising and Heisenberg Models

The results outlined in the previous section are based on very general arguments. The generation of quite similar coexisting large clusters have been described before in the context of the random field Ising model (RFIM). This

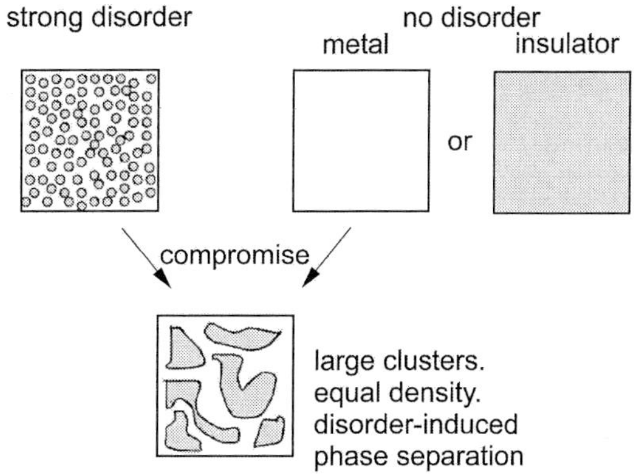

Fig. 17.4. Schematic representation of the competitions that arise when disorder is incorporated near a first-order transition. At zero disorder, no interfaces exist. At large disorder, small-size clusters are formed. At weak disorder, the compromise between the two tendencies leads to very large clusters

model, widely used in the context of disordered systems, has a Hamiltonian defined as $H = -J\sum_{\langle ij\rangle} S_i^z S_j^z - \sum_i h_i S_i^z$, where $S_i^z = +1$ or -1 is an Ising variable that represents two competing states near a first-order transition. Its study started with the pioneering work of Imry and Ma [17.1] (more recent literature can be found in [13.2] and [17.2]). The ferromagnetic-like interaction between the spin variables implies that the system wants to be uniformly all up or all down, and dislikes interfaces at low temperature. This is what is expected from a system at a first-order phase transition: by definition, it is either entirely of one phase or the other and walls between phases cost energy. The second term in the Hamiltonian is very important: it represents the influence of disorder, with a random field h_i taken from a distribution with zero mean value and width regulated by a constant h (this can be a box distribution, or Gaussian, or others). From the perspective of this term, the variables S_i^z arrange locally such that they point in the direction of the random field. In this respect the system wants to form small clusters, with a characteristic length scale given by the lattice spacing, since h_i is uncorrelated from site to site. From the competition between these two tendencies (uniform state favored by the first term, small clusters favored by the second term), a situation with clusters of variable size emerges. If h is small, the system wants to be as uniform as possible, but still have some clustering, as a compromise between the two tendencies. This leads to a situation where clusters of any size can be stabilized by varying h, namely, at large h the clusters are small, but as h is reduced compared to J, the clusters increase in size. For h/J small, the clusters can be truly huge. This very general situation is illustrated in Fig. 17.4.

The results corresponding to the RFIM are based on the competition between two tendencies: no interfaces favored by the first term, and small clusters favored by the second term. This competition depends on the dimension of the problem. Assuming a *sharp* interface between the two domains (in this case the up and down regions), then the penalization for having an interface in the form of a bubble centered at the origin of radius R grows like the surface, namely R^{d-1}, where d is the spatial dimension. On the other hand, the term induced by the disorder is more subtle: it is determined by the fluctuations of the disorder that has a standard deviation that grows as $R^{d/2}$. The derivation is simple: consider the field induced by the disorder term inside the bubble of radius R defined as $M_R = \sum_i h_i$. Its standard deviation is given by $\sigma_{M_R} = \sqrt{\langle \sum_{ij} h_i h_j \rangle} = hR^{d/2}$, if $\langle h_i h_j \rangle = h^2 \delta_{ij}$, which is the correct result for uncorrelated disorder. Then, in a cluster of radius R, the fluctuations in the effective field induced by disorder can be as large as $R^{d/2}$. For example, inside a disk with 100 spins in two dimensions, a distribution of random fields will tend to produce a similar number of ups and down, such that the average field is close to zero. However, fluctuations of order 10 are not uncommon, and this effect cannot be neglected. From this, it is clear that the *critical dimension* of the problem is $d = 2$, since at that dimension the two terms scale in the same way (i.e., $R^{d/2} = R^{d-1}$ if $d = 2$). This reasoning leads to the prediction that the two-dimensional Ising model generates an overall disorder state *immediately* upon switching on the strength of the disorder h, due to the formation of clusters of unlimited size with a given orientation of the spins, embedded in a region with the other orientation. Although the argument is rough, not including, for instance, the fractality of the usual clusters that may effectively change their dimension, and although numerical simulations are hard to carry out for very small h, this rationale is widely accepted. Some numerical results can be observed in Fig. 17.5 (the RFIM has also been used in the manganite context for other purposes [17.3]). It then seems that, in principle, the mechanism outlined here is promising for 2D manganites and cuprates, but not in 3D perovskites, and this is an important concern. If the disorder is strong enough, large clusters can be generated in 3D (see Fig. 17.5d), but for weak disorder the tendency is to have nearly uniform states. "Correlated" disorder, with a power-law behavior $\langle h_i h_j \rangle = h^2(1/|i-j|^\alpha)$, is currently being investigated as a way out of this problem [11.112].

However, the arguments leading to a critical dimension of two rely heavily on the behavior of the Ising variables of the RFIM. In particular, the transition between the two domains (represented by all spins up and all down) is assumed to be sharp (i.e. the wall between the two domains is of size one lattice spacing). Perhaps in real models this is too drastic, and a smoother transition from one phase to the other is found. Along this line of thinking, it is useful to recall that the random field *Heisenberg* model (RFHM) has a critical dimension d_c quite different from the d_c of the RFIM. This model is defined as $H = -J \sum_{\langle ij \rangle} \boldsymbol{S}_i \cdot \boldsymbol{S}_j - \sum_i \boldsymbol{h}_i \cdot \boldsymbol{S}_i$, where now the variable at each

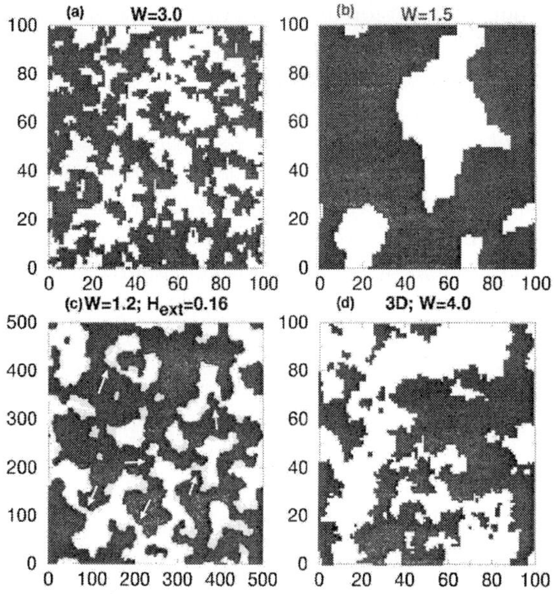

Fig. 17.5. Results of a Monte Carlo simulation of the random field Ising model, from Moreo et al. [6.41]. The *dark* (*white*) regions represent clusters with spins up (down). (**a**) corresponds to results using a 100×100 cluster, and the disorder W shown (for the definition of the disorder, temperature, and other details see the original reference). Note the fractal-like domains created. (**b**) Reducing the disorder strength, increases the cluster size. (**c**) Results on a 500×500 cluster for smaller disorder strengths are shown, also including an external field. In this case, gray regions can also be observed. They correspond to regions with spin down that flip to up upon the application of the field, illustrating the percolative nature of the transition. *Arrows* indicate regions where this field-induced switching leads to important changes in the connectivity of the clusters. (**d**) Results showing a slice of a 3D 100^3 lattice, showing coexisting clusters, at sufficiently large disorder

site is a classical vector instead of an Ising variable, and the random field is a vector as well, with the property $\langle \boldsymbol{h}_i \cdot \boldsymbol{h}_j \rangle = h^2 \delta_{ij}$.

It is instructive to address in detail this change of critical dimension. Consider as before a bubble of radius R. Both for the Ising and Heisenberg models, the standard deviation of the total random field inside the bubble scales as $R^{d/2}$. The difference between the two models is in the cost of creating a wall between the two domains. In the Ising case, the cost is proportional to $J \times R^{d-1}$ since at each link an energy proportional to J is paid, and that has to be multiplied by the surface. On the other hand, for the Heisenberg model, the energy per link scales with the inverse of the size of the wall l_{wall}, as shown in elementary textbooks [3.3]. A slow twist in the orientation of the classical spins, analogous to spin-wave excitations, costs very little energy. The wall-energy cost is then $(1/l_{\text{wall}}) \times R^{d-1}$. If we assume that l_{wall}

is comparable to R, then the penalization for the wall scales like R^{d-2}. When this last factor is compared with $R^{d/2}$ from the fluctuations of the total field, one finds that the critical dimension is now d_c=4, instead of two. For more details the reader can consult the RFHM literature [17.4].

This example shows how subtle is the estimation of critical dimensions. The use of the RFIM to explain coexisting huge clusters should be cautiously considered, since it appears to apply to 2D systems only. But could it be that deviations from this model give finite-size walls that can accommodate large coexisting clusters even in three dimensions, maybe with correlated disorder instead of uncorrelated? Certainly more work should be carried out to explain the micrometer-scale clusters found by Uehara et al. [11.41].

17.3 Effective Resistor Network Approximation

As explained in the first chapter, manganites are interesting for the rich phase diagram they show, and also due to the challenging intrinsic inhomogeneities they present even in good single crystals. However, at least historically, the main reason for the present widely renewed effort directed to Mn oxides is the unusual transport properties they present under the influence of external fields. Unfortunately, addressing this last issue is quite difficult for theorists. While the resistivity is among the first results that experimentalists usually report in their characterization of a newly synthesized material, the D.C. resistivity cannot be accurately computed using realistic models. The task is simply too complex, involving the zero-frequency limit of the optical conductivity given by current current correlations in space and time. In addition, a source of dissipation is needed to provide a width to the zero-frequency Drude δ-function if finite cluster systems are used. While approximations are available to handle these problems, they usually rely on homogeneous states, which is an incorrect assumption for manganites after an overwhelming amount of experimental data have revealed the presence of inhomogeneities in the temperature and density regions with the CMR effect. Complicating matters substantially, fully quantum problems need the use of the Landauer formalism where conductances are related to transmission probabilities, which are difficult to calculate for interacting electrons.

To address this complicated transport problem, Mayr et al. [17.5] recently used a simple percolative phenomenological approach, which captures at least the essence of the experimental results. These authors employ a coarse-grain approximation to the real system: they reason that real manganites are inhomogeneous at scales from nano- to micrometer (although the evidence for the latter emerges from a couple of experiments only). Since real systems are then inhomogeneous involving metallic and insulating competing phases, for the purposes of resistivity estimations they represent a Mn oxide as a random "resistor network" (see Chap. 16), namely, a randomly distributed array of resistances, arranged for simplicity on a cubic or square lattice. At the links of the lattice, a resistance with either metallic or insulating characteristics is located (note that each resistance represents a region of the sample large enough that Ohm's law is valid and transport is diffusive, not ballistic). The relative amounts of metal and insulator are regulated by the parameter p, which is the relative fraction of metallic resistances. Using simple iterative procedures the effective global resistance of the resistor network can be calculated. The regime of most interest is near percolation, which is around $p = 0.5$ in two dimensions and $p = 0.25$ in three dimensions, for perfectly insulating and metallic individual resistances (Chap. 16).

The simple picture that emerges from the studies of Mayr et al. [17.5] is depicted in Fig. 17.6a (see also related literature in [17.6]). Near percolation and at low temperatures, there are rather convoluted paths of metallic resistances that allow for transport of charge from one end to the other of the sample. In addition, at intermediate and high temperatures the insulating links of the resistor network actually do not have a large resistance (this is according to experiments, since the resistivity of metallic and insulating samples tends to merge near room temperature). For these reasons, the system appears to be effectively made up of two temperature-dependent resistances (Fig. 17.6b), representing the two dominant paths for charge transport. These are (1) a resistance $R_I(T)$ corresponding to transport across the insulator, growing fast as the temperature is reduced, and (2) a resistance $R_M^{\text{per}}(T)$ representing the percolative paths. This last effective resistance is metallic at low temperatures, and it has a residual large value at $T = 0$ due to the complex paths allowed in a percolative network. In addition, it grows fast as T increases since the percolative path is easily suppressed by temperature effects that may cut off the line of transport across the sample. From the combination of these two resistances, coupled in parallel, a resulting resistance vs. T curve similar to that shown in experiments emerges (see Fig. 17.6b, right): there is insulating behavior at high temperatures – where transport occurs mainly across the insulator – and "poor-metal" behavior at low temperatures near percolation, with a resulting inevitable peak in between.

Results of an actual resistor network calculation are shown in Fig. 17.6c. The individual resistances at the links of the square or cubic lattices employed were taken from Uehara et al. [11.41], who reported investigations on $(La_{1-y}Pr_y)_{1-x}Ca_xMnO_3$. At $y = 0.0$ and 0.42, the system at low temperature is metallic or insulating, respectively, and those resistances form the building blocks of the network. The nontrivial result of the analysis is the effective resistivity at $p \neq 0.0$ or 1.0, as shown in Fig. 17.6c. Compatible with the simple picture outlined before of competition between the resistance of the insulator and the percolative metallic path, the resulting resistances in the range from $p = 0.45$ to 0.60 present insulating-like behavior near room temperature, followed by a peak (which if using a linear scale in the vertical axis would be more prominent), and then metallic behavior at low temperature, with a large residual resistivity. These results are in qualitative agreement with

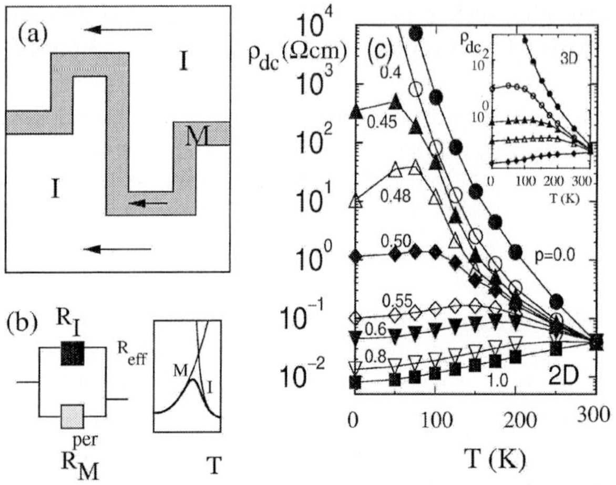

Fig. 17.6. (**a**) Schematic representation of a percolative metallic path across an insulating sample. (**b**) Two-resistance model for manganites. The parallel connection is shown on the *left*, involving the resistance of the percolative path and the insulator. On the *right* is the individual behavior with temperature and the resulting effective resistance. (**c**) Net resistivity obtained using a 100×100 resistance network vs. temperature, at the indicated metallic fractions p. The *inset* of (**c**) shows results for a 20^3 cluster at $p = 0.0$, 0.25, 0.30, 0.40, and 0.50, from the top. Both in 2D and 3D, the calculations are averages over 40 configurations. The $p = 1$ and $p = 0$ limits (purely metallic and purely insulating) are from [11.41]. From Mayr et al. [17.5]

a variety of experiments, as those reviewed in Chap. 3. In several cases the resistivity peak is not too prominent in real materials, and Fig. 17.6c is sufficient for a phenomenological description of the results. In addition, the shape of the resistivity curves in the insulating regimes suggests that they can be fit with exponentials, as in polaronic scenarios. However, it is clear that the phenomenological results of Fig. 17.6c are unrelated to polarons since they are produced by phase mixtures. Care must be taken when fits of resistivities with exponentials are presented as evidence of polaronic behavior.

In some manganites the resistivity peak is far more prominent than the one shown in Fig. 17.6c. Still, at a phenomenological level, the theoretical calculation can be improved by realizing that percolation can occur not only by varying chemical compositions at constant hole density (as for example replacing La by Pr), but also as a function of temperature involving metal, insulating, and probably paramagnetic clusters. In fact, the metallic resistivities in the network represent FM regions, which lose their spin polarization at T_C, and a reduction of the metallic fraction as the temperature increases appears inescapable. In Fig. 17.7a, results of a Monte Carlo simulation of the random field Ising model are shown at finite temperature. Paramagnetic regions appear between the ordered regions with spins up and down. These regions

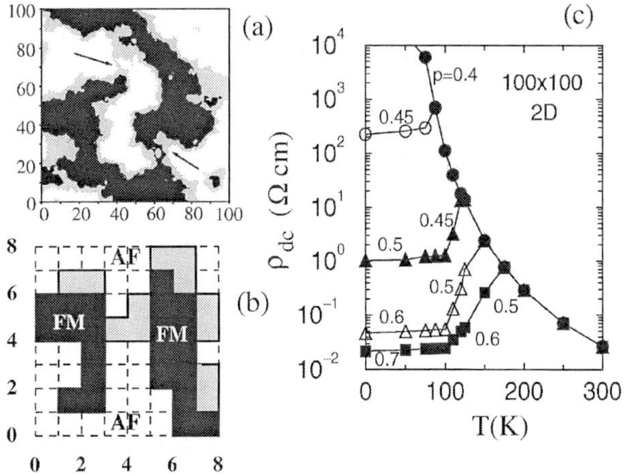

Fig. 17.7. Results from Mayr et al. [17.5] illustrating the phenomenological resistor network analysis of manganites, allowing for a temperature dependence of the metallic fraction to mimic percolation induced by temperature, at constant electronic density. (**a**) Results for the random field Ising model at finite temperature. *Black* and *white* represent spins mainly up and down. The gray region is paramagnetic and induced by temperature, with a small value of the order parameter. The *arrows* indicate places where black or white clusters are connected at zero temperature, but became disconnected at higher temperature due to the paramagnetic regions. (**b**) Results similar to (**a**) but showing FM, AF, and PM regions (gray), obtained for the one-orbital model in the regime of phase separation, with coexisting clusters stabilized by disorder in the chemical potential. (**c**) Net resistivity vs. temperature of a 100 × 100 random resistor network, with the metallic fraction p changing with temperature. Averages over 40 configurations are shown

can destroy narrow metallic paths between large clusters, reducing drastically the percolative conductance. Similar results are shown in Fig. 17.7b, where at low temperatures coexisting FM and AF regions in the phase regime of the one-orbital model are stabilized upon the introduction of disorder in the chemical potential. At finite temperature, paramagnetic regions destroy the connection between metallic islands. These effects can be phenomenologically mimicked in the resistor network by reducing with increasing temperature the metallic fraction p used initially at zero temperature. This reduction should be rapid near T_C, where the magnetization is known to change rapidly in real manganese oxides. Typical results of resistor calculations under these assumptions are shown in Fig. 17.7c (the individual link resistances are the same as for the previous figure). Now the peaks are clearly far more prominent than before, in agreement with experimental resistivity vs. temperature data for several manganite systems.

Mayr et al. also carried out calculations that incorporate quantum effects. One of them uses a 3D lattice model with nearest-neighbor hopping amplitudes randomly selected between large "metallic" and small "insulating" val-

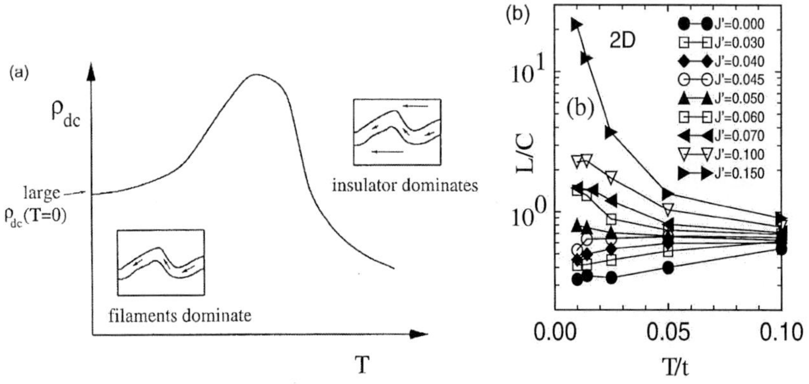

Fig. 17.8. (**a**) Schematic representation of the percolative view. For more details, see text (**b**) Inverse conductance of the half-doped one-orbital model for manganites, using an 8×8 cluster, $J_{\mathrm{H}} = \infty$, $t = 1$. J' is the Heisenberg coupling among localized spins. The disorder strength and other details can be found in the original reference Mayr et al. [17.5]. The Landauer formalism and setup of the conductance calculation were taken from Vergés, and Calderón et al. [17.7].

ues. The conductance was calculated using the Landauer formalism [17.7], which is based on the transmission across the sample, when the latter is assumed connected to ideal leads (see below). The results in [17.5] are in reasonable agreement with those of the random-resistor-network approximation. The conductance C for a realistic one-orbital model on small clusters was also calculated by Mayr et al. Finite disorder was incorporated in order to create coexisting cluster of metallic and insulating phases, in the vicinity of the first-order transition. Results are given in Fig. 17.8b. Varying the coupling J_{AF} (denoted by J' in the figure), we observe a transition from an insulator to a metal. However, note that the lattice size is too small to find truly percolative effects that would lead to a maximum in the inverse of the conductance at finite temperature. Plenty of work remains to be done in this context.

To mimic the influence of small magnetic fields we can also vary the metallic fraction p slightly. Near percolation, this produces large effects in the net resistivity. Another source of large MR effects, also described by Mayr et al. [17.5] involves the insulating regions. Even a small change in their conductivity from 0 to small – crudely representing the canting induced by a magnetic field in a zero-conductance antiferromagnetic state – can produce substantial changes in the overall resistance. This effect induces a large MR even in simulations of one-dimensional systems, as will be discussed further in Chap. 22.

Summarizing this section, Fig. 17.8a shows a representation of the main idea of Mayr et al. At low temperatures, transport is across percolative paths, while at high temperature the similarity in resistivities of metals and insulators allows for transport across the sample, leading to an insulating-like behavior near room temperature, as observed in experiments. Of course, this should be considered only as a starting point for a description of manganites. Substantially more work is needed to confirm this appealing albeit simple picture. Efforts in this direction are described below. Note also that other recent approaches [17.8] address the experimental resistivity vs. temperature curves with a mechanism exclusively based on polaron localization above the Curie temperature, without phase competition. This idea is fundamentally different from the phase-separation scenario. Since phase coexistence FM-CO is supported by recent neutron and X-ray scattering experiments (see for instance, Chap. 19), the effort in this chapter continues along the two-phase competition idea.

17.3.1 First-Order Percolation Transitions

Some of the percolative transitions found in real manganites also present first-order characteristics (note, for instance, the hysteresis loops of Fig. 11.12). Can a percolative-like transition be first order? This issue is currently under investigation by Burgy et al. [11.112]. Using realistic models, these authors observed that thin regions with couplings that would favor, e.g., antiferromagnetic correlations, change their properties when immersed in a ferromagnetic background (Fig. 17.9a and b), increasing substantially their conductivities. For this reason, it was argued that clusters that better represent manganite physics should not be as fractal-like as in standard percolation, but more rounded. Elasticity arguments lead to the same conclusion. Burgy et al. [11.112] used a set of algorithms (sketched in Fig. 17.9c, details can be found in [11.112]) where the metal or insulator character of a link depends on the properties of its neighbors. Different variations of these new rules indeed lead to clusters with far less fractal-like characteristics, as shown in d–f of Fig. 17.9. The latter, f, was in fact found to generate a first-order transition when the percolative critical density is reached. Details and previous references can be found in [11.112]. This is an interesting novel area of investigations.

17.3.2 Calculation of Conductances in Quantum Problems

The DC conductance is usually calculated using the Kubo formalism. The Monte Carlo procedure, extensively discussed in previous chapters, was applied to the one-orbital model to generate relevant configurations of the classical localized spins at fixed electronic density and temperature. It is in this background of localized spins that the Kubo formula is applied. The setup for the calculation involves connecting the cluster under investigation with two ideal leads. The current runs from one lead to the other and, in the case when Ohm's law is valid, the conductance C is related to the conductivity

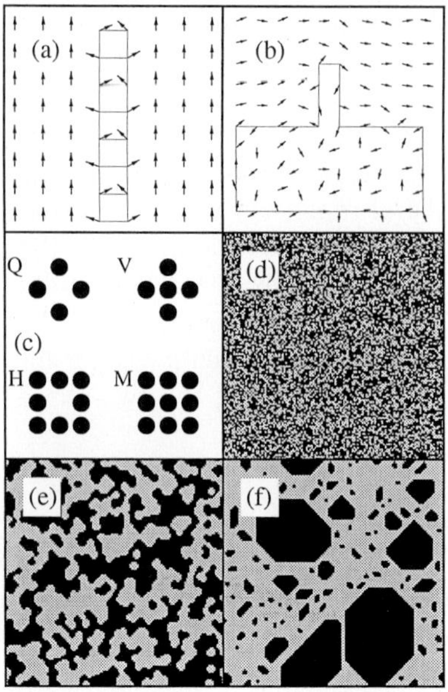

Fig. 17.9. (**a and b**) Typical Monte Carlo snapshots for the cases of a ladder, and a finger-like feature emerging out of a region with a "flux" phase, using the one-orbital model. The J_{AF} coupling favors an AF state in the ladder and a "flux" state in the finger and below, while it favors ferromagnetism elsewhere. However, the study of spin correlations shows that these properties are much affected by the FM environment, and their conductivity across is much larger than their values when in isolation. (**c**) Sketch of algorithms used by Burgy et al. [11.112]. The metal or insulator character of a link is decided based on the value of its neighbors. (**d–f**) Typical clusters obtained using (**d**) standard percolation and (**e and f**) the new rules sketched in (**c**), that mimic surface tension effects. The latter, (**f**), leads to a first-order transition at the critical density. Details can be found in [11.112]

σ through the formula $C = \sigma A/L$, where A is the cross section and L the length along which the current runs. For a cube of size N, $C = \sigma/N$. Of course, the system under study will not always show Ohmic behavior. In fact, there are two transport regimes: *diffusive and ballistic.* In the first case, diffusive, Ohm's law is satisfied and the carriers scatter often when crossing the sample. However, if little or no scattering occurs (as, for example, in the noninteracting limit of zero Hund and electron phonon coupling), the transport is ballistic (mean-free path larger than the size of the system). The conductivity σ is, then, independent of the size of the system and it can only be properly calculated in the diffusive regime.

The Kubo formalism evaluates the linear response of a system to an external perturbation. Linear means that the signal is proportional to the intensity of the external perturbation, which can be very small. The basic formulas to calculate the conductance are discussed in the literature, and they will not be repeated here. The formalism involves advanced and retarded Green's functions. The influence of the infinite leads to the left and right of the sample is incorporated through a self-energy, which can be evaluated exactly for ideal leads. The calculation of the Green's functions of the cluster is carried out numerically, using the Hamiltonian and the self-energies of the leads. More details can be found in the first study of manganite models using this formalism, reported by Calderón, Vergés, and Brey [17.7].

17.3.3 The Difficult Task of Calculating Resistivities

Here, it is interesting to make a brief detour into issues associated with the D.C. resistivity of materials. Resistivities are usually among the first properties that emerge from the characterization of a new compound. However, accurate calculations of resistivities are quite difficult from a theoretical perspective. Allen [17.9] has provided an excellent overview of conventional and unconventional transport properties of high-temperature superconductors, which can be used as a starting point to understand the subtleties involved in resistivity calculations of these and other materials as well. Many calculations start with the assumption of a "Fermi liquid", namely, a state made out of a dilute gas of quasiparticles (for a review see [17.10]). In this context Boltzmann transport equations can be used to estimate the resistivity. However, other effects can substantially affect the calculation such as localization à la Anderson (reviewed elsewhere in this book), where the resistivity grows as the disorder in the sample increases. Another interesting effect is "saturation" as found in some materials where the resistivity tends to a constant value as that disorder grows (see [17.11] for a brief recent review and references). The saturation effect is believed to occur when the distance between successive collisions of the carriers is reduced to such small values that the two collisions are not additive, and the Boltzmann equation is no longer applicable. In this context, the term "bad metal" is used to denote materials that fail to exhibit resistivity saturation in spite of the fact that the distance between collisions is rather small [17.12]. In a recent brief article, Allen [17.13] discusses the differences between saturation and "escalation" (increase of resistivity without saturation), concluding that both are mysterious and that a proper theory is still lacking for their understanding. These issues and the associated literature clearly show how difficult it is to calculate resistivities from first principles. In the area of manganites our effort is limited to either a phenomenological random-resistor-network approach, which does not start with a microscopic Hamiltonian but is based on a phenomenological approach, or quantum calculations of conductances for nondirectly interacting electrons, such as in the one- and two-orbital models. Extreme care should be taken when calculating resistivities. It is good advice for beginners to be highly suspicious of reports where this quantity is evaluated without a clear discussion of the assumptions made. Resistivities should be among the *last* quantities to be evaluated in reliable theoretical calculations, since their values rely heavily on the pattern of long-range or local spin, charge, and orbital order in a material.

17.4 Strain and Long-Range Coulombic Effects

Experimental investigations in thin films have shown that a competition between two phases, and their coexistence, can also be established through "strain" effects. This process leads to a final state similar to those outlined in the calculations described earlier in this chapter. The strain idea is that the substrate over which the manganite thin film grows has a lattice parameter that is different from those of bulk manganite ("lattice mismatch"). In cases discussed by Biswas et al. [17.14] and references therein, the lattice parameter of the substrate is smaller than that of the crystal of manganite at the same composition. In this case, "biaxial strain" is incorporated into the thin film. The lattice spacing in the direction perpendicular to the film can adjust, and it will expand if the biaxial strain induces a contraction in the plane (as compared with manganite bulk lattice spacings). Biswas et al. observed that biaxial strain leads to phase coexistence: the minimum energy is obtained with a distribution of lattice spacings, rather than a uniform one. The two phases observed experimentally for thin films of $La_{1-x}Ca_xMnO_3$ at $x = 0.30$ and low temperature correspond to the metallic FM usually stabilized in single crystals, together with the insulating CO/AF state that appears in $Pr_{1-x}Ca_xMnO_3$ at the same composition [17.15,17.16]. This type of result is reminiscent of those obtained for single crystals of $(La_{1-y}Pr_y)_{1-x}Ca_xMnO_3$ in [11.41] (the relevance of strain has also been discussed by Mathur and Littlewood [17.17] and Egami and Louca [17.18]). Given a chance, the system spontaneously generates a mixture of FM and AF regions, without involving the paramagnetic state. This is likely due to the close energy of these phases. Qualitatively, both strain effects in thin films and the quenched disorder effects discussed before lead to similar phase mixtures.

Another interesting effect that may contribute to the inhomogeneities in manganites was discussed by Schmalian and Wolynes [17.19] in the context of the cuprates. These authors observed that a system with short-range attraction and long-range Coulomb repulsion could undergo a self-generated glass transition, which is not induced by quenched disorder. An exponentially large number of metastable configurations was found. These effects may also contribute to the physics of manganites, where phase-separation tendencies can be frustrated by Coulombic effects (see also Kivelson et al. in [1.3]).

17.5 Competition Between Ordered States Leading to "Quantum-Critical"-Like Behavior

We have learned how qualitative phenomenological arguments lead to the resistor network explanation for the temperature dependence of the resistivity of manganites. First-principles calculations produced indeed coexisting metallic and insulating clusters when quenched disorder is introduced near a first-order transition at very low temperatures. In this section, the effect of a finite temperature is investigated in that same regime of phase coexistence. Besides being a natural extension of the previous results, the analysis of finite-temperature effects leads to qualitative understanding of experimentally intriguing phase diagrams of some manganites and cuprates, as those reproduced in the inset of Fig. 17.10. In the right inset, is the famous phase diagram of the one-layer high-temperature superconductor $La_{2-x}Sr_xCuO_4$. Between the antiferromagnetic and superconducting phases, a globally disordered regime is known to exist. This regime is widely considered to be a spin-glass, but its properties are not well understood. The left inset shows an early version of the phase diagram of the manganite $RE_{1-x}Sr_xMnO_3$ at $x = 0.45$ (RE stands for rare earth). Solutions involving rare-earth compounds result in modifications of the average radius of the A-site ion of the perovskite, believed to be equivalent to varying the average hopping amplitude Mn O Mn. A transition from the ferromagnetic phase to an antiferromagnetic charge-ordered regime is induced, with an interesting disordered regime in between where the transition temperatures drop substantially. What is the origin of such a phase diagram that resembles a quantum-critical behavior at zero temperature (namely, the presence of a continuous phase transition at $T = 0$, which must be induced by quantum effects since thermal effects are obviously absent)? This is analyzed below.

17.5.1 A Toy Model of Competing Phases

To study the competition of phases in the presence of disorder, one can use either the one- or two-orbital models. However, the expected percolative nature of the transition requires large clusters in numerical simulations, and that regime is hard to reach using realistic Hamiltonians. Fortunately, the general features of phase competition emerge from far simpler *toy models*, namely, models of spins with two competing phases. Burgy et al. [11.78] recently carried out an analysis of phase competition using a model where the spins are Ising variables, defined by $H = J_1 \sum_{ij} S_i^z S_j^z + J_2 \sum_{im} S_i^z S_m^z + J_4 \sum_{in} S_i^z S_n^z$. The first term links spins at a distance of one lattice spacing, the second at distance $\sqrt{2}$ (namely, along the diagonals of the plaquettes – elementary squares – of the lattice), and the third at distance $\sqrt{5}$ (two spacing in one direction, one in the other). The three couplings are positive. Why are so many couplings needed? The first one is just a scale that can be set to one. The second is necessary in order to generate a competing phase. If only one coupling were present, then just one low-temperature phase would exist. The last coupling J_4 is small in the study of Burgy et al. and set to $0.2J_1$ in order to generate a robust first-order transition between the competing phases. For the present model, these phases are an antiferromagnetic phase at small J_2/J_1, and a "collinear" state with rows or columns of spins up and down for large J_2/J_1. All this may sound complicated to the reader, but the expectation is that the results of the analysis presented by Burgy et al. do not depend on the details of the competing phases, and they could be valid for any pair of phases in competition, including those of real manganites.

The results of the Monte Carlo simulation of the toy model are shown in Fig. 17.10. In the absence of disorder ("clean" limit), the model has a first-order transition between the two ordered phases at a coupling $J_{2c} = 0.7$. When the disorder is increased, the first-order transitions turn into continuous transitions. The phase diagram far from J_{2c} is not much affected by disorder. However, important changes are observed near J_{2c}. With sufficiently strong disorder the phase diagram becomes similar to those in the insets of the figure, with substantial reductions of the critical temperatures to the point of inducing a region at zero temperature that does not possess long-range order. This regime could be labeled as "quantum critical", with reference to a transition that occurs at such a low temperature that quantum effects are more important than thermal effects [17.20]. These quantum-critical transitions are currently under much investigation in the contexts of materials other than manganites. However, note that this occurs here in a "classical" model (Ising variables do not have quantum fluctuations at zero temperature, unlike $S = 1/2$ Heisenberg spins). A model with both quantum and classical degrees of freedom, such as the one-orbital Hamiltonian, has also been numerically studied on small clusters, and the phase diagram is similar to that for the toy model (see Fig. 17.14d below).

The phase diagram in Fig. 17.10 was obtained by calculating the order parameter of each phase. In Fig. 17.11a and b, the antiferromagnetic order parameter is presented for several lattice sizes in two dimensions. At $J_2 = 0.65$, a convergence to a finite value is observed below a critical temperature, which is a canonical behavior. However, at $J_2 = 0.69$, the order parameter tends to cancel as the lattice increases, in the entire range of temperatures shown.

Representative spin configurations are shown in Fig. 17.11c. This information is very important since it provides an intuitive picture in real space of the dominant states in the competition of phases once quenched disorder is included. In panel (1) of Fig. 17.11c, the dominance of the white color indicates a fairly standard paramagnetic state (white is used for the case where spins spend an approximately equal amount of time in states up and down, leading to a zero mean value). In panel (2) a more complicated, and far more interesting state is formed, with substantial white, but also darker tones. The reader

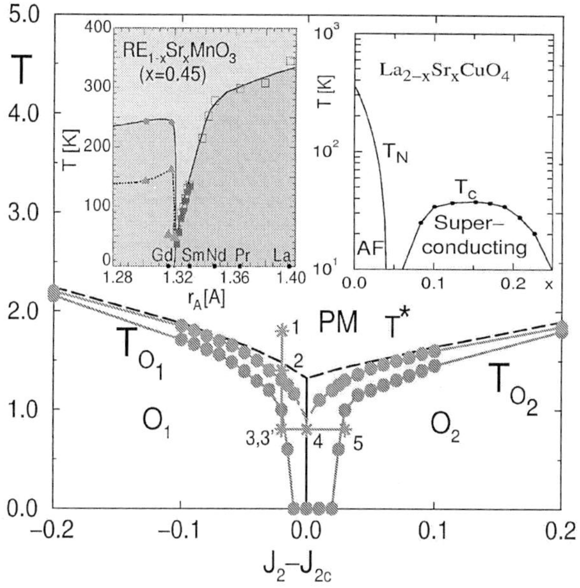

Fig. 17.10. Phase diagram of the "toy model" of Ising variables with J_1–J_2–J_4 couplings in two dimensions ($J_4 = 0.2\,J_1$, and $J_1 = 1$ providing the energy scale). This phase diagram is believed to represent the behavior of any pair of competing phases, when disorder is introduced into the coupling J_2 that allows the system to change from phase 1 to phase 2, which in this case are antiferromagnetic and collinear, respectively. Details of the definition of the disorder, lattices and techniques used, can be found in the original reference Burgy et al. [11.78]. T_{O_1} and T_{O_2} are the true ordering temperatures, while T^* is the clean-limit ordering temperature, which survives as a rapid "crossover" for cluster formation when disorder is introduced. The *insets* are phase diagrams of Mn oxides on the *left* (private communication from Y. Tokura and Y. Tomioka), and the single layer cuprate on the *right*

is advised to consult the original publication where the figure is presented in color. The dark gray and black regions of the figure denote the same antiferromagnetic state, but with order parameters of *different signs*: if the staggered order parameter is defined as positive for $\ldots\uparrow\downarrow\uparrow\downarrow\ldots$, then it is negative for $\ldots\downarrow\uparrow\downarrow\uparrow\ldots$. Their relative amount is approximately the same, according to the simulation and, as a consequence, there is no global order parameter due to a nearly exact cancellation. But, locally order can be found. There are also small regions of lighter gray, which correspond to locations where collinear order(the state competing with antiferromagnetism) is stabilized due to the influence of the disorder. This state (2) is the most interesting: if the state stabilized at small J_2 is assumed to have metallic characteristics, and the stable state at large J_2 is made insulating, for a given species of spins traveling the sample (say, spin up), both the insulating and metallic regions with order parameter down cannot be crossed, and these obstacles will substantially in-

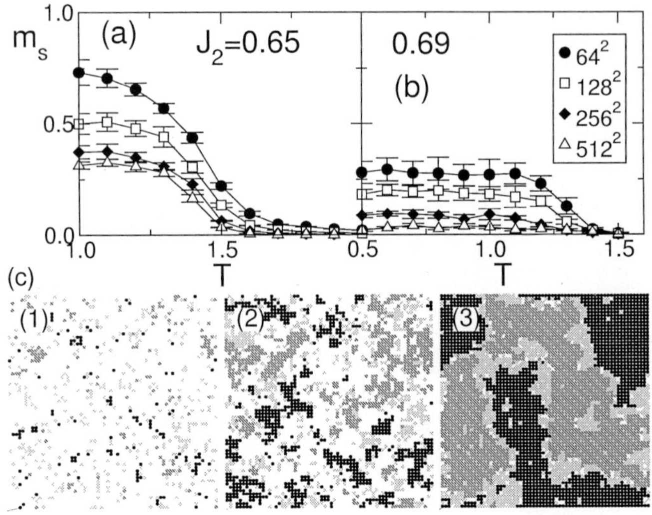

Fig. 17.11. (**a and b**) AF order parameter vs. temperature for the toy model defined in the text at fixed disorder strength and for two couplings J_2. Results for several lattice sizes are shown. In (**a**) the order parameter converges to a finite number as the lattice grows, indicating long-range order. In (**b**) the order parameter is suppressed when the size of the cluster increases, indicating a globally disordered regime (but the system still has local order). (**c**) Typical spin configurations representative of dominant states, generated by the MC simulation. Shown are averages over nearly 100 MC sweeps. The coupling and temperature for the three cases (1–2–3) are the same as in Fig. 17.10. The conventions used are the following: the *darkest regions* correspond to the AF phase with positive order parameter; the *next dark tone* is AF with negative order parameter; the *light gray* is the competing collinear phase; while *white* corresponds to a paramagnetic region. The original colors can be found in Burgy et al. [11.78]

crease the resistivity (see below). State (2) appears in the temperature regime above the true ordering temperature T_{O_1}, but below T^* (here defined as the crossover temperature remnant of the clean limit original ordering temperature, below which local order is favored). This scale plays an important role in the discussion below. Finally, in (3) the low-temperature state is reached. Here, one of the two possibilities for the antiferromagnetic state (i.e., order parameter positive or negative) wins over the other by a small amount leading to global order. There are virtually no paramagnetic regions. Note that the collinear phase is neatly stabilized between the two antiferromagnetic orders: *domain walls can be stabilized by inducing a competing phase inside.* This is an interesting mechanism that deserves further investigations, and it is illustrated in Fig. 17.12. It is remarkable that Palanisami et al. [17.21] arrived at similar conclusions through the study of resistance fluctuations (see also [17.22]).

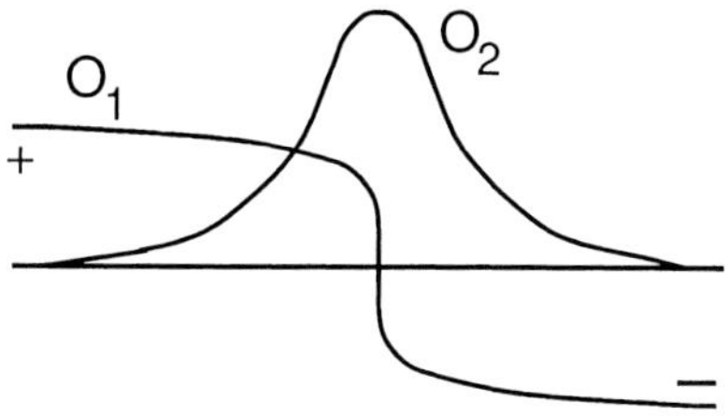

Fig. 17.12. Schematic representation of the stabilization of the competing order O_2 (for example, the CO phase of manganites) in domain walls of the order parameter O_1 (for example, the FM phase of manganites)

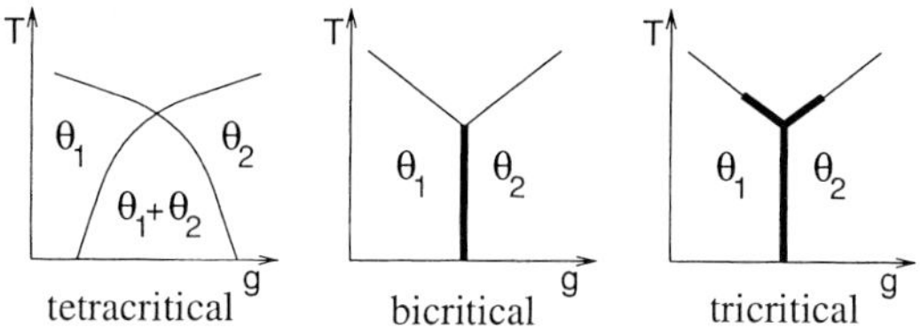

Fig. 17.13. Schematic possible phase diagrams when two distinct ordered phases compete. g is a generic parameter that allows a transition from one phase to the other. The high-T phase is disordered. *Thick* (*thin*) lines denote first- (second-) order transitions

The phase diagram in Fig. 17.10 and the MC snapshots of Fig. 17.11 are a natural result of the competition of very different phases, once a source of disorder is explicitly incorporated. The results are sufficiently simple that the phase diagram can be easily accepted to be general enough to apply to *any pair of competing ordered states in the presence of any source of uncorrelated disorder*. This is certainly a conjecture, but a reasonable one. The critical exponents associated with the transitions may well be different, as well as the critical dimensions, but the overall topology of the phase diagram is believed to remain independent of details.

Note: Multicritical Points. The competition between two distinct types of ordering leads to multicritical points, as discussed in the 1970s by Kosterlitz et al. [17.23] and others. Figure 17.13 shows the three types of phase diagrams expected for phase competition, from general considerations. They are : (1) The diagram with a "tetracritical" point. Here the two orders coexist at low temperatures. (2) Phase diagram with a "bicritical" point, where a line of first-order transitions separates the two phases. When the line splits with increasing temperature the transitions become of second order. (3) Phase diagram with "tricritical" points, where the first-order nature of the transition persists into the order disorder line. Figure 17.10 corresponds to a tricritical case in the clean limit.

17.5.2 Estimating Resistances in Mixed-Phase Manganites

The general form of the phase diagram of competing phases with quenched disorder led Burgy et al. [11.78] to make predictions about the behavior of competing ferromagnetic metallic and antiferromagnetic insulating phases, as occurs in manganites. A realistic model for the manganites would be difficult to simulate in sufficiently large lattices for percolative properties to show, but one can conjecture that it is sufficient to consider the phase diagram Fig. 17.10 and simply "translate" the phases studied there into the manganite language. A possible translation identifies the O_1 phase with ferromagnetism, and then the associated clusters in Fig. 17.11 have positive or negative magnetization. The phase O_2 corresponds to the insulator, for which the orientation of the order parameter is irrelevant for transport. In this context, panel (2) of Fig. 17.11c, the relevant state in the regime $T_{O_1} < T < T^*$, can be thought as consisting of preformed FM clusters with uncorrelated moment orientations, leading to a zero global magnetization. The "walls" that stabilize the many FM clusters are made out of the insulating phase. Figure 17.14a shows the idealized state proposed for manganites, in the CMR regime.

The proposed state Fig. 17.14a can be tested by calculating its resistivity vs. temperature, and its behavior in the presence of magnetic fields. For this purpose, a resistor-network calculation can be set up. The individual resistances of the network emerge from the "dictionary" described in the previous paragraph. The regions with *positive* order parameter O_1 are associated with ferromagnetic regions with up polarization. As a consequence, this regime has good conductivity for the spin-up carriers, and nearly zero conductivity for spin-down. Reciprocally, the negative O_1 clusters have good conductivity for spin-down only. The insulating regions of phase O_2 being insulating, have bad conductivity for both spin-up and -down. The white regions, correspond to paramagnetic clusters, and have an intermediate conductivity. The actual values of the conductivities used can be found in Burgy et al. [11.78]. The qualitative behavior is not expected to depend crucially on details.

After carrying out the proposed translation from the MC results to a resistance network, the corresponding Kirchoff equations are solved iteratively to obtain the effective resistance (for details, see Chap. 16). Results are shown in Fig. 17.14b. In agreement with the qualitative description outlined before, the state in the intermediate temperature regime between T_{O_1} and T^*, with the complicated cluster coexistence, has the largest resistance. The other two states – the paramagnet at high T and the percolated at low T – have far smaller values. In addition, the figure shows that the effects are largest when the coupling J_2 gets close to the original critical coupling. This is reasonable since this is the regime where coexisting clusters are found.

17.5.3 Huge MR in the Presence of Small Magnetic Fields

The characteristics of the proposed CMR state in Fig. 17.14a suggest that strong effects could occur upon the application of small magnetic fields. The reason is that there are "preformed" FM clusters, each with a large magnetic moment. When these large moments are coupled to a relatively small field,

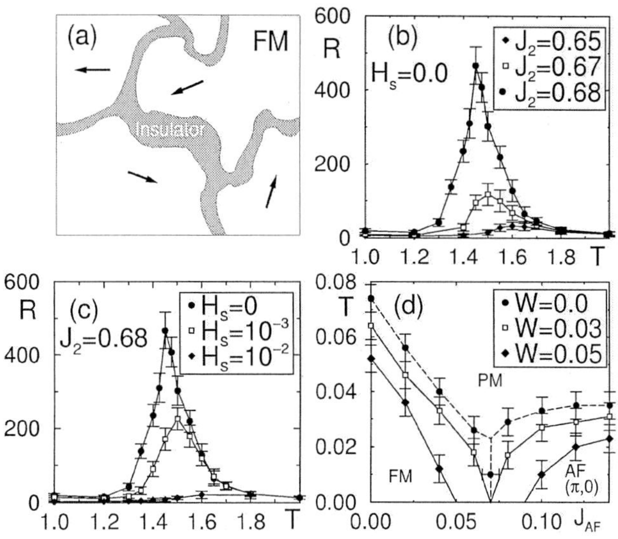

Fig. 17.14. (**a**) Sketch of the proposed state for Mn oxides in the CMR regime, after Burgy et al. [11.78]. (**b**) Resistance at several values of J_2, obtained from the MC configurations of the toy model J_1–J_2–J_4 by a suitable translation to manganite language of the many phases, supplemented by mild assumptions on the individual resistances that form the resistor network. Note the large value of the resistance in the intermediate-temperature region with preformed clusters, together with global cancellation of the order parameters. Note also the increase of resistance as the critical region at $J_2 = 0.70$ is reached. (**c**) Resistance obtained by a procedure similar to that in (**b**), but including an external field. There is a strong dependence of the resistance on external field, leading to a huge MR ratio, comparable to experiments. (**d**) Phase diagram analog of Fig. 17.10, but for the one-orbital model for manganites, at density $x = 0.5$, infinite Hund coupling and with the hopping as energy scale. Disorder is added to J_{AF}. With increasing disorder, the results are very similar to those found in the toy model J_1–J_2–J_4 using far larger clusters

the net effect can be substantial. Rotation of these preformed moments can rapidly lead to a robust magnetization and finite conductivity. In the toy model analyzed thus far, the equivalent of the effect of magnetic fields in manganites can be studied by adding external fields that favor phase O_1. The clusters of phase O_1 will tend to align their order parameter in the direction of the external field. In [11.78], it was shown that indeed a small field of just 1% in the natural units of the problem (J_1 in the toy model), leads to a robust dominance of only one sign of the O_1 order parameter. This is particularly important in the regime between T_{O_1} and T^* where the resistivity is very high since electrons of, say, spin up had difficulty propagating through the insulating regions or through the clusters with the wrong orientation of the moment. If all clusters orient their moments in the field direction, one obstacle is removed. In addition, the insulating walls melt when the FM regions on

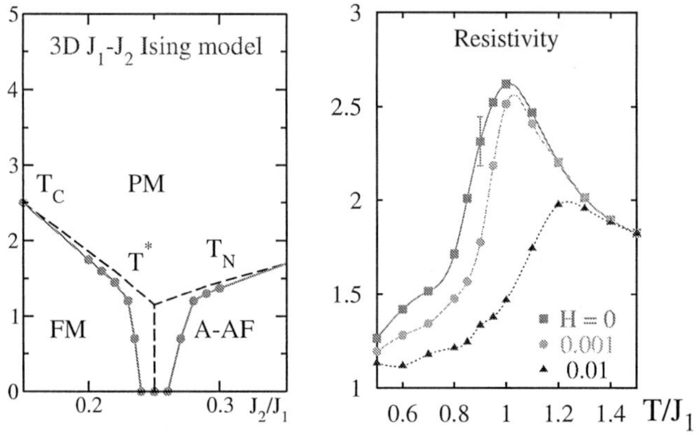

Fig. 17.15. *(Left)* Phase diagram of the J_1–J_2 toy model using a three-dimensional lattice. The couplings were selected such that ferromagnetic and collinear (or A-AF) phases are in competition. The T^* scale is indicated. Results are quite similar to those in two dimensions. *(Right)* Resistivity vs. temperature, at the external fields indicated. A large magnetoresistance is observed. The calculation was done using similar resistor-network rules as in two dimensions. Details, such as couplings used and strength of disorder, can be found in [17.24]

both sides have the same moment orientation. All this suggests that the resistivity will substantially change upon the application of magnetic fields.

Results for the magnetic-field dependence of the resistivity are shown in Fig. 17.14c. The calculation is carried out as in panel b of the same figure, but including the external field. Panel c clearly shows a *dramatic magnetoresistance effect.* For a field as small as $0.001 J_1$, the resistance decreases by about 50%. When the field is $0.01 J_1$, the entire peak is virtually suppressed. Figure 17.14c contains one of the most important results of the book. The associated MR ratios can be as large as those found in experiments.

Results similar to those described thus far for 2D systems have also been obtained by Burgy et al. [17.24] in simulations of toy models in 3D, following a similar procedure. The phase diagram is in Fig. 17.15(left), which should be contrasted against Fig. 17.10. Clearly, there are no qualitative differences between two and three dimensions, although the magnitud of the effect changes with the dimension and this is currently under investigation. Figure 17.15(right) contains the results for the resistivity vs. temperature of the three-dimensional model. The curves closely resemble experimental results. As in two dimensions, small fields produce substantial effects on the transport properties.

17.5.4 Prediction of a New High-Temperature Scale T^*

The phase diagram shown in Fig. 17.10 and the associated relevant configurations, Fig. 17.11c, suggest that, in addition to the true ordering temperatures

T_{O_1} and T_{O_2}, there should be a higher-temperature scale T^*, where the clusters start forming locally. In this region is where the resistivity grows rapidly with decreasing temperature, as found in the resistor-network studies. The new scale T^* does not depend much on the coupling J_2. The experimental information currently available about T^* will be discussed in Chap. 19. There are experimental probes that suggest the existence of a new characteristic temperature at approximately 300 to 400 K, related to cluster formation.

The temperature scale T^* can also be associated with a "Griffiths temperature" [17.25], which is the temperature where a clean system would have its transition in the absence of disorder. Below this temperature, disorder leads to cluster formation. What is remarkable in the previous discussion is that the Griffiths phase and the singularities appear abnormally enhanced when two phases compete, leading to a strong response to external fields, as remarked by Burgy et al. [11.78]. The analogy with Griffiths phases was also observed by Salamon et al. [17.26]. Combined effects of intrinsic randomness, doping, tendency to phase separate, and self-trapping of polarons appear to contribute to its formation.

17.6 Experimental Phase Diagrams in the FM CO Region of Competition

The theoretical studies that lead to Figs. 17.10 and 17.15 predict two kinds of phase diagrams in the region where the FM and CO states compete. For the case of "weak disorder", a bicritical or tricritical behavior should be observed. For "strong disorder" the first-order transition of the clean limit should entirely disappear, and be replaced by a clustered state with no long-range order (presumably with glassy properties). It is interesting to search for manganite compounds exhibiting these extreme behaviors.

Bicritical behavior. A material that presents bicritical behavior in the FM-CO region of competition is $Pr_{1-x}(Ca_{1-y}Sr_y)_xMnO_3$ at $x = 0.45$. The phase diagram was recently reported by Tomioka and Tokura [17.27]. An extended version including other materials is reproduced in Fig. 17.16.

Glassy behavior. A material with strong A-site cation size mismatch (i.e., with strong disorder) is $(La,Y)_{1-x}Ca_xMnO_3$, according to studies by Sundaresan et al. [17.28]. In the other extreme, $Pr_{1-x}Ca_xMnO_3$ is expected to have negligible A-site ionic size mismatch since both the Pr and Ca ions have approximately the same size of 1.18 Å. The La and Y ions in the compound studied in [17.28] have sizes of 1.216 Å and 1.075 Å, respectively. Spin-glass behavior has been observed in this compound [17.28]. A related material is $(La,Tb)_{1-x}Ca_xMnO_3$, which also has strong disorder. The phase diagram at $x = 1/3$ was already discussed in Chap. 11, but it is reproduced in Fig. 17.17 again to help the reader. The shape of the phase diagram clearly resembles the theoretical predictions in Fig. 17.10 for the case of strong disorder, with

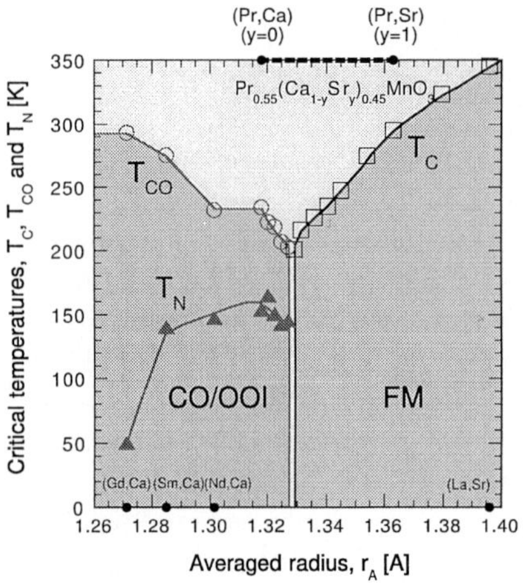

Fig. 17.16. Extended phase diagram at x=0.45 over a wide range of the A-site average ionic radius, as an example of bicritical behavior. The results shown correspond mainly to $Pr_{0.55}(Ca_{1-y}Sr_y)_{0.45}MnO_3$, but results from several other x =0.45 compounds are also indicated. From [17.27]

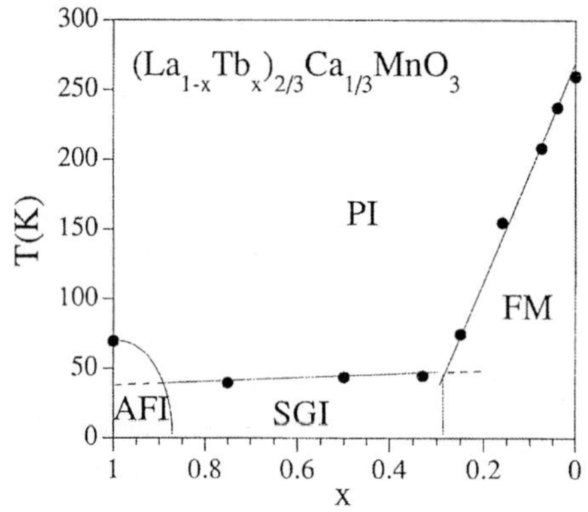

Fig. 17.17. Magnetic and electrical phase diagram of the series $(La_{1-x}Tb_x)_{2/3}Ca_{1/3}MnO_3$, from [11.36]. SGI is a spin-glass insulator phase. This figure was already shown in Fig. 11.11, but it is reproduced here again to facilitate the reading of this chapter

an intermediate cluster-glass region between the two competing phases. Other theoretical studies have obtained phase diagrams that also resemble Fig. 17.17 even without using disorder, in regimes where several phases compete [6.48].

Intermediate Regime with Quantum-Critical-like Behavior. If the theoretical discussion is correct, there must be materials with a phase diagram intermediate between those shown in Figs. 17.16 and 17.17. In such an intermediate regime, the phase diagram may appear to contain a quantum-critical point, in the sense that the ordering temperatures collapse to nearly zero temperature at one particular value of the mean A-site ionic radius $\langle r_A \rangle$. The left inset of Fig. 17.10 agrees with this picture, although more experimental work is needed to confirm the results. Related with this issue, an interesting phase diagram was reported by Kuwahara and Tokura [17.30] including results for several materials with x = 0.5. This phase diagram is in Fig. 17.18. Clearly, the suppression of ordering temperatures at the tolerance

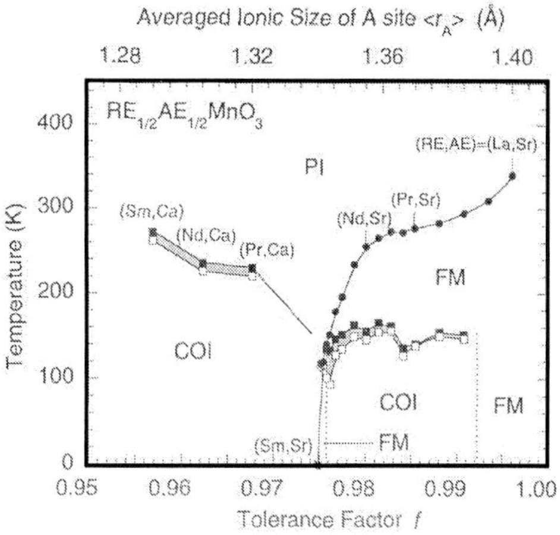

Fig. 17.18. Phase diagram of manganites with $x = 0.5$, from [17.30]. Results for many compounds are presented. The notation is standard. From [17.30]

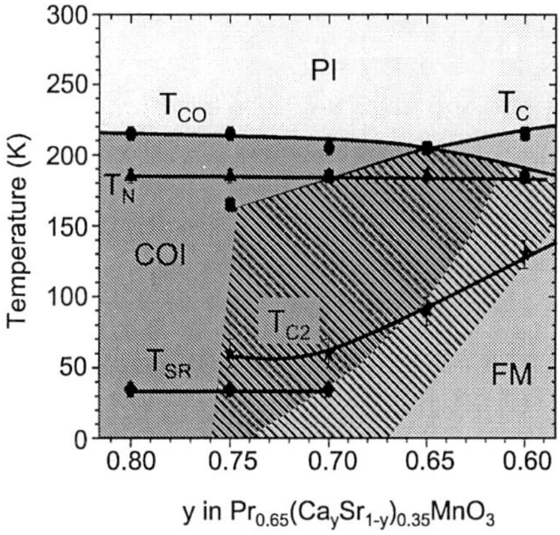

Fig. 17.19. Schematic phase diagram of $Pr_{0.65}(Ca_ySr_{1-y})_{0.35}MnO_3$. T_{CO}, T_N, and T_C are the onset temperatures of charge-ordering, AF ordering, and FM ordering, respectively. A sudden increase in the FM phase fraction and electrical conductivity occurs at T_{C2} and a spin reorientation in the AF phase occurs at T_{SR}. The *hatched area* indicates phase coexistence. From Blake et al. [17.34]

factor 0.975 is stronger than in the bicritical case, Fig. 17.16. The Curie and CO critical temperatures meet at $\sim 140\,K$.

Recent studies of polycrystalline samples of $Pr_{0.65}Ca_{0.35-x}Sr_xMnO_3$ with several values of x, by Niebieskikwiat et al. [17.31], reported thermoelectric power, electrical resistivity, and DC magnetization results that lead to a phase diagram with competing CO and FM states. The shape of the diagram has strong similarities with the results of theoretical simulations reported in this chapter, including quantum-critical-like behavior. Similar results were also presented in [17.27]. However, more work is needed to fully confirm the theoretical predictions discussed here. Other theoretical studies are based on the notion of multicriticality [17.33], where disorder does not play an important role. Clarifying the origin of the quantum-critical-like behavior of certain manganites is among the most interesting subareas of research in the Mn-oxide context at present.

Very recently, $Pr_{0.65}(Ca_ySr_{1-y})_{0.35}MnO_3$ has also been investigated by Blake et al. [17.34] using neutron powder diffraction. In the region between $y = 0.6$ and 0.75, CO/AF insulating and FM metallic states coexist at a *mesoscopic* scale. In this study, no quantum-critical point was observed from the CO-FM competition. The phase diagram is shown in Fig. 17.19. Clearly, more work should be devoted to this crucial CO-FM competition in manganites, and to the search for quantum-critical-like phase diagrams, intermediate between Figs. 17.16 and 17.17.

17.7 Control of Quenched Disorder in Manganites *by Y. Tokura*

The close interplay between spin, charge, orbital, and lattice degrees of freedom is the source of the complex behavior found in manganites. The competing interactions/orders inherent in the manganites, such as double-exchange ferromagnetism vs. superexchange antiferromagnetism and charge-orbital order vs. metallic states, tend to produce the multicritical state where external stimuli occasionally cause a dramatic phase conversion. The CMR is, of course, one such example. As explained in this book, quenched disorder in a critical region may produce the phenomenon of phase separation into the competing two ordered phases on various time scales and length scales from nano- to micrometer. The most typical example is the relaxor-like effect induced by Cr-doping on the Mn sites in the charge/orbital-ordered state, which gives rise to ferromagnetic metallic clusters coexisting with the charge/orbital-ordered state, as described in Chap. 11 of this book.

Such a remarkable effect of the random potential arising from chemical disorder is seen not only in the case of impurity doping on the Mn sites (perovskite B-site) but also in the conventional random alloying on the perovskite A-site with trivalent rare-earth (Ln^{3+}) and divalent alkaline-earth (Ae^{2+}) ions. The latter effect is sometimes discussed in terms of the difference (standard deviation) in ionic radii of the constituent Ln and Ae components. In this chapter, the profound effect that quenched disorder has on CMR-related phenomena has been discussed at length, including the generation of quantum-critical-like features. Here, we discuss the case where the *least* chemical disorder has been obtained in the perovskite manganite, which can in turn help visualize more clearly the dramatic effect of quenched disorder in the critical region.

The strategy adopted by Akohoshi and coworkers at the Correlated Electron Research Center (CERC), AIST, is to investigate the perovskite A-site ordered structure for the half-doped ($x = 0.5$) manganites, $Ln_{1/2}Ba_{1/2}MnO_3$, in comparison with the A-site disordered analogs. As first demonstrated by Ueda and coworkers at ISSP, University of Tokyo, a carefully processed sample of $Ln_{1/2}Ba_{1/2}MnO_3$ shows nearly perfect ordering of the A-site, namely,

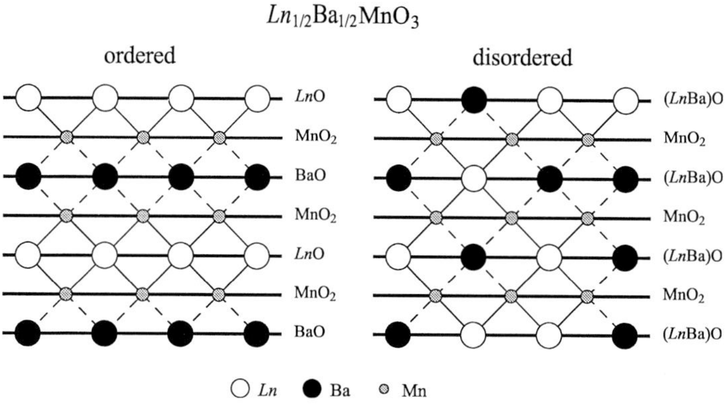

Fig. 17.20. Schematic structure of $Ln_{1/2}Ba_{1/2}MnO_3$ viewed along the c-axis for the A-site ordered (*left*) and disordered (*right*) perovskites with half-doping

the atomic LnO and BaO sheets alternately stack along the c-axis with intervening MnO_2 sheets (left panel of Fig. 17.20). This ordering is possibly due to the large difference in the ionic radii of Ln^{3+} and Ba^{2+} and is reminiscent of the structure of the well-known 123-type high-T_c superconductors $LnBa_2Cu_3O_7$. As a result, the individual MnO_2 sheets in this tetragonal form do not suffer from randomness, which would otherwise arise from the A-site Coulomb potential or local strain. By contrast, the melt-quenched sample with the same A-site composition can show the complete solid-solution range of the Ln and Ba ions on the A-sites, with the simple cubic structure as the average structure (right panel of Fig. 17.20).

Thus, it is interesting to compare the electronic states of the two compounds, with the perfect order and disorder of the A-sites and an identical composition. In Fig. 17.21, we show the electronic phase diagram for $Ln_{1/2}Ba_{1/2}MnO_3$ with A-site order (solid line) and disorder (broken line), plotted as a function of the Ln ionic size. For the ordered perovskite, the charge/orbital ordered state and the ferromagnetic metallic state compete with each other in a bicritical manner: The charge/orbital ordering temperature T_{CO} is as high as 500 K for the ordered form of $Y_{1/2}Ba_{1/2}MnO_3$ and a small-Ln compound, but it decreases steeply with the Ln ionic size. On the other hand, the ferromagnetic transition temperature T_C decreases with decreasing Ln ionic size. Finally, both transition temperatures meet at around Ln = Nd, forming a prototypical bicritical point.

The A-site randomness as realized in the solid solutions was found to result in a dramatic modification of the electronic phase diagram, as indicated by broken lines in Fig. 17.21. The long-range order in the charge and orbital sectors, with a relatively high transition temperature T_{CO}, totally collapses in the disordered system and remains dynamical down to the spin (and perhaps orbital) glass temperature of about 50 K, as shown in the lower-left panel of Fig. 17.21. The glass state persists up to near the original bicritical

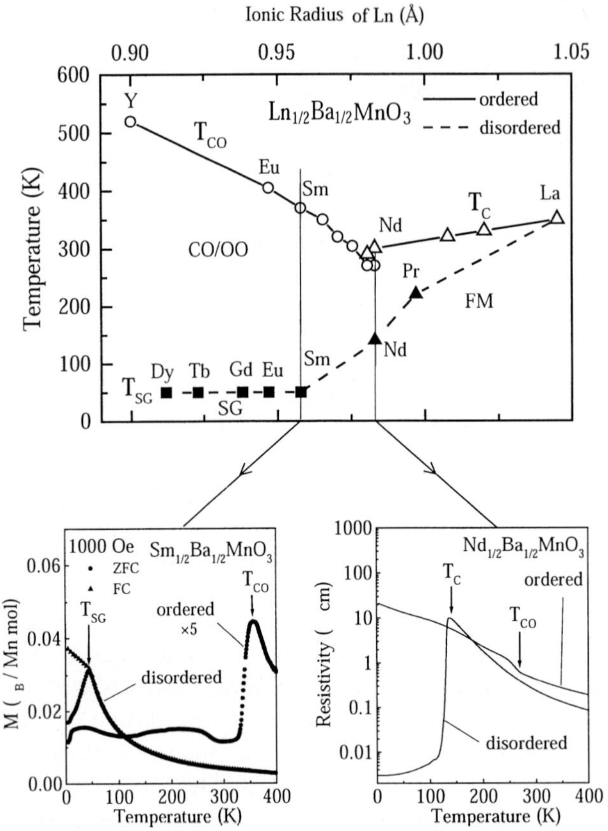

Fig. 17.21. Electronic phase diagrams for the A-site ordered (*open symbols*) and disordered (*closed symbols*) perovskites with half-doping, $Ln_{1/2}Ba_{1/2}MnO_3$, as a function of the ionic radius of Ln. CO/OO, FM, and SG stand for the charge/orbital ordered, ferromagnetic metallic, and spin-glass state, respectively. T_{CO}, T_C, and T_{SG} represent the transition temperatures. *Lower-left panel*: Change in the temperature dependence of the magnetization for the A-site ordered and disordered perovskites of $Sm_{1/2}Ba_{1/2}MnO_3$ (Ln = Sm). FC and ZFC represent the field-cooling and zero-field-cooling process, respectively. *Lower-right panel*: Change in the temperature dependence of the resistivity for the A-site ordered and disordered perovskites of $Nd_{1/2}Ba_{1/2}MnO_3$ (Ln = Nd). The A-site disordered sample shows typical CMR behavior around T_C

point, Ln = Nd. The quenched disorder on the perovskite A site produces another interesting feature near that original bicritical point. The ferromagnetic transition temperature T_C is critically suppressed on decreasing the Ln ionic radius, as compared with the A-site ordered case. For Ln=Nd near the original bicritical point, T_C is suppressed the most. The charge-orbital correlation as evidenced by X-ray diffuse scattering, grows with lowering temperature, but suddenly disappears at T_C, in agreement with the sharp and gigantic drop of resistivity at T_C (lower-right panel of Fig. 17.21). It is well known that the CMR is maximized in materials with the most reduced T_C, as is the case of the A-site disordered $Nd_{0.5}Ba_{0.5}MnO_3$.

Thus, quenched disorder not only produces the glassy state and the microscopically phase-separated region but also enhances the fluctuations of the competing orders, i.e., the charge/orbital order vs. the metallic ferromagnetism, especially near the original bicritical point [17.35]. Such large quantum-critical-like fluctuations induced by the random potential are certainly one of the most essential ingredients of the CMR physics.

17.8 More on the Influence of Disorder on T_C

In this chapter, it has been extensively argued that phase competition FM-CO is very important to understand manganites. The relevance of disorder was also noted, in order to understand the drastic reduction of the critical temperature in the region of phase competition of some of these compounds. Other independent investigations also arrived at the conclusion that disorder in manganites is important. For example, Rodriguez-Martinez and Attfield [17.36] studied eight samples at fixed doping $x = 0.3$ and A-cation radius $\langle r_A \rangle = 1.23$ Å, using a variety of trivalent and divalent ions such that the *variance* σ^2 of the A-cation radius changed from sample to sample. They found that T_C depends strongly on σ^2, and that T_C was the highest when the disorder was the smallest. Studies by Sundaresan et al. [17.28] also lead to similar conclusions. This is in agreement with Figs. 17.10 and 17.15, showing that at a fixed J_2 the critical temperature decreases with increasing disorder. Figure 17.22 shows the critical temperature vs σ^2, with a linear behavior at small σ^2. The extrapolated critical temperature in the ultra-clean limit is $T_{\mathrm{extrapolated}} = 400$ K. For smaller values of $\langle r_A \rangle$, extrapolated temperatures as high as 530 K were found. According to the theoretical studies of Figs. 17.10 and 17.15, $T_{\mathrm{extrapolated}}$ should correspond to T^*, also described there as the clean-limit critical temperature. It is remarkable that the 400 K scale mentioned here is quite similar to the T^* described in previous sections and in Chap. 19, for the same $x = 0.3$ family of compounds.

The importance of disorder has also been emphasized by Raveau et al. [17.29]. Figures 16 and 17 of [17.29] are very interesting. They show that the critical temperatures (Néel or Curie) remain robust for materials with small mismatch disorder, whereas they are much reduced if the disorder is substantial. The reduction of the Curie temperature by alloy disorder was also discussed in [17.32]. Raveau et al. [17.29] write "... all attempts to study $\langle r_A \rangle$ effects on the properties of manganites without controlling σ^2 (disorder) are doomed to failure". Their results and conclusions are in good agreement with the theory contained in Fig. 17.10.

In summary, in this chapter the relevance of the competition between FM and CO states has been discussed. This competition leads to nontrivial

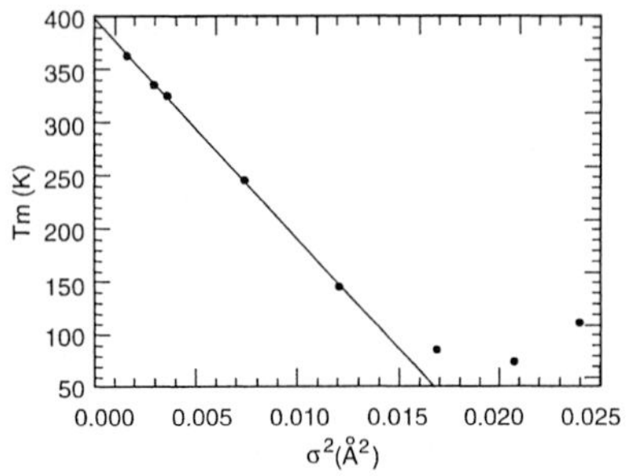

Fig. 17.22. Critical temperatures (T_{m}) as a function of the A-cation size variance σ^2, for eight samples with $x = 0.3$ and $\langle r_{\mathrm{A}} \rangle = 1.23$ Å. From Rodriguez-Martinez and Attfield [17.36]

behavior as parameters are changed in experiments or simulations, allowing for a transition from one phase to the other. The influence of disorder is substantial, leading to different phase diagrams depending on its strength. A new temperature scale T^* appears, signaling the formation of clusters upon cooling. T^* is a remnant of the clean-limit critical temperature. CMR is believed to be caused by the rapid alignment of preformed FM clusters when magnetic fields are turned on. Charge-ordered clusters also influence CMR, through their melting with magnetic fields. Experimental results appear to be in close agreement with theory.

17.9 Transition From Charge-Ordered to Ferromagnetic Metallic States and Giant Isotope Effect *by D.I. Khomskii*

As discussed in several parts of this book, one of the key properties of manganites is the competition between the charge-ordered (CO) insulating and the ferromagnetic metallic (FM) states. In some cases, one of these states can be present only in the form of short-range correlations (see, e.g., Chap. 19). However, in many situations there exists long-range order of one kind or the other, and transitions between different states, say between CO and FM ones, can be caused by changes of temperature, magnetic field, or pressure. Many interesting phenomena exist in such situations, and some of the related problems are still not completely clear.

Typical phase diagrams for the system $Pr_{1-x}Ca_xMnO_3$ are shown in Fig. 3.10. Usually, one can suppress the antiferromagnetic CO state by applying a magnetic field, which causes a first-order phase transition to the FM state, accompanied by large hysteresis. An unexpected feature was discovered in [3.18]: whereas the slope $\frac{\mathrm{d}T_{\mathrm{c}}}{\mathrm{d}H}$ of the CO FM transition line is negative for the half-doped commensurate case $x = 0.5$ (if the CO state exists, it survives down to $T = 0$), $\frac{\mathrm{d}T_{\mathrm{c}}}{\mathrm{d}H}$ becomes positive for $x < 0.5$, i.e., there exists a re-entrant transition from CO to FM states with decreasing temperature.

This behavior seems counterintuitive at first glance: one is used to thinking of a CO-phase as an electronic crystal, and the FM state as a liquid. In normal cases crystals melt with increasing temperature and not otherwise. Here, however, the opposite situation exists for $x < 0.5$. How is this possible?

One reason may be that for $x < 0.5$ the CO-state is actually not perfectly ordered even at $T = 0$. Experimentally, it is established for $Pr_{1-x}Ca_xMnO_3$ ($0.3 < x \leq 0.5$) that there exists a superstructure with wavevector $q = (\frac{1}{2}, \frac{1}{2}, 0)$ (checkerboard CO in a basal plane) [2.8]. But, while this is fine for $x = 0.5$, such superstructure at $x < 0.5$ implies that there should exist a certain disorder in the system. One should locate extra electrons somewhere, and this must be done randomly, so as not to change the total periodicity. There are two options: either these extra electrons (extra Mn^{3+} ions) are randomly distributed on the background of the perfect checkerboard CO, or there exist inhomogeneous states, with the electron-rich regions forming random clusters. Of course, both are nonequilibrium states and they do not correspond to an absolute minimum of the free energy; but, as follows from the existing data, such states actually appear in experiments.

Based on these ideas, it is possible to qualitatively understand the main features of the phase diagrams mentioned above, in particular those with the re-entrant CO FM behavior [6.12] (for another study in this context, see [17.37]). One should remember that the FM state is not an ordinary liquid, but a *Fermi liquid*, which has a unique ground state with zero entropy at $T = 0$. Its specific heat and entropy behave at small temperatures as γT, so that the total free energy is $F_{\mathrm{FM}} = E_{\mathrm{FM}} - \gamma T^2$. At the same time, as we argued above, for $x \neq 0.5$ the "checkerboard" CO-state should have a frozen-in disorder (random extra Mn^{3+} – ions), which gives a finite entropy S_0. In effect, the free energy of such a CO-state is $F_{\mathrm{CO}} = E_{\mathrm{CO}} - TS_0$. We have a situation where $E_{\mathrm{FM}} < E_{\mathrm{CO}}$, such that the FM state is the ground state at $T = 0$. However, due to a faster decrease with temperature, the curve of the F_{CO} may cross that of the FM state at some temperature. Thus, a first-order transition from the FM to the "partially disordered" CO state occurs with increasing temperature. This sequence of phases is, thus, determined by the extra entropy of the partially disordered CO phase for $x < 0.5$. This simple scenario allows us to explain the qualitative features of the phase diagrams in the case of competing CO and FM phases, although more robust theoretical justification and calculations are still lacking.

A convenient system to study the competition between the CO and FM states is the mixed compound $(La_{1-y}Pr_y)_{1-x}Ca_xMnO_3$ (LPCMO). For $x = 0.3$ [6.12, 17.38, 17.39], or $x = 0.375$ [11.41, 11.43] one can transform from the FM state ($y = 0$) to the CO state ($y = 1$) by changing the Pr/La ratio y. The CO FM transition now occurs at a certain y_c and at zero magnetic field, in contrast to the case of $Pr_{1-x}Ca_xMnO_3$ [3.18]. It turns out that this transition also has a re-entrant character, with a large hysteresis region and with all the features of phase separation, such as percolative behavior,

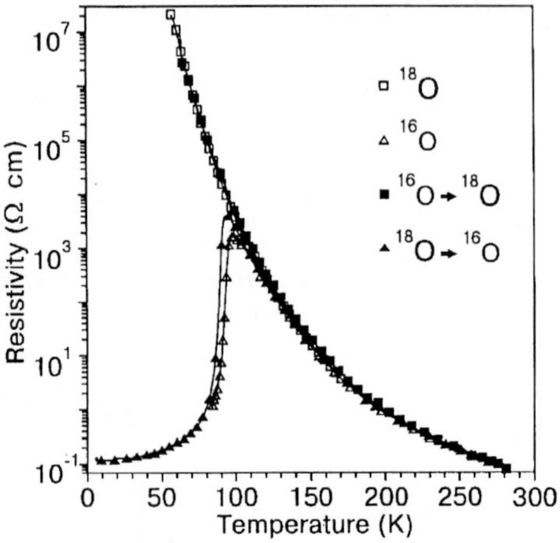

Fig. 17.23. Metal insulator transition induced by $^{16}O \rightarrow {}^{18}O$ substitution in $La_{0.175}Pr_{0.525}Ca_{0.3}MnO_3$ [17.38]. The reverse exchange proves that this process is reversible

etc. [11.41, 11.43, 17.40]. In particular, the different coexisting phases can be visualized explicitly using electron microscopy [11.41]. In addition, strongly nonlinear I–V characteristics exist [17.40], as well as giant $1/f$ noise [11.98], strange "shifted hysteresis" in the M(H) dependence [13.4], etc.

The properties of the system in such a regime are very sensitive even to very small perturbations. One of the most spectacular effects observed here is the metal–insulator transition induced by the isotope substitution $^{16}O \rightarrow {}^{18}O$ [17.38, 17.39, 17.41], see Fig. 17.23. For a properly chosen composition the ground state of the ^{16}O sample is metallic, but it transforms to a CO insulator with increasing T. At the same time, the ^{18}O sample remains an antiferromagnetic CO insulator down to the lowest temperatures. This interpretation is confirmed by neutron scattering data [17.42], in which the change of ordering following the isotope substitution was clearly seen. One can suppress the insulating phase of the ^{18}O sample by a magnetic field and transform it into a FM state. The isotope shift of T_C in this case is huge: 65 K in the field of 2 T [17.38] – much larger than the shift of 20 K observed in another system [3.12].

The most probable explanation of this giant isotope effect is the dependence on the isotope composition of the effective bandwidth, e.g., due to polaronic effects. The simplest model to describe the competition between CO and FM states is the model of a "Wigner crystal on a lattice". For spinless fermions, it has the form

$$H = -\sum_{ij} t_{ij} c_i^\dagger c_j + V \sum_{\langle ij \rangle} n_i n_j , \qquad (17.1)$$

where the first term describes the hopping of electrons from site to site, and the second term is the repulsion of electrons, e.g., at neighboring sites. Ignoring nesting (e.g., by including further-neighbor hopping), one can easily show [17.39, 17.43] that for $x = 0.5$ the CO-state ("Wigner crystal") exists if the hopping integral t is small enough, $t < V$. However, if t and the corresponding bandwidth $W = 2zt$ increase, there is a "cold-melting" transition to a metallic state.

Apparently, the effective hopping and the resulting bandwidth decrease for heavier ions (^{18}O). Then, the ^{18}O sample may be a CO insulator, whereas the ^{16}O samples are still metallic. The detailed shape of the resulting phase diagram would then be determined by the factors described above. In LPCMO it is re-entrant, with large hysteresis and with all the rich properties resulting from this. We see that isotope substitution is yet another tool to influence the delicate balance between different states. The giant isotope effect proves that the electron lattice interaction is very important in the physics of manganites.

18. Pseudogaps and Photoemission Experiments

The study of the density of states (DOS) of a material is important to understand its properties, particularly regarding its metallic or insulating character. If a gap exists at the Fermi energy E_F, then the system is insulating. However, if at that Fermi energy the DOS is nonzero and featureless, the system should be metallic (unless Anderson localization is at work). In many transition-metal oxides, such as the-high temperature superconductors and the manganites, evidence is accumulating that a "pseudogap" exists at E_F. A pseudogap is an intermediate case between a fully established gap, and a standard metallic case with no gap. The DOS of a pseudogapped system has a substantially depleted DOS at E_F, but its actual value does not reach exactly zero. The explanation of pseudogap behavior has been at the forefront of investigations in cuprates [18.1], and a similar concept is rapidly reaching similar relevance in manganites. In fact, a pseudogap has already been observed in theoretical investigations of models for Mn oxides, as well as in photoemission experiments in bilayer manganites. Both issues are reviewed in this chapter. Clearly, this is a fairly novel subarea of investigations in manganites, and the research is at its early stages. Considerable work should be devoted to the study of this phenomenon in Mn oxides and related compounds.

18.1 Brief Introduction to Photoemission Experiments

Let us start this chapter with a brief review of the photoemission techniques, which are much used to investigate the DOS of a system. The term "photoemission spectroscopy" refers to several experimental techniques that are based on the emission of electrons from a material after stimulation by photons. This is basically the photoemission effect that we learn in elementary classes of modern physics. In photoemission (PES) experiments a monochromatic (i.e., one dominant wavelength) photon beam obtained from a discharge lamp or a monochromator of a synchrotron source is used to strike the surface of the material under study, in ultrahigh vacuum. Photons can be absorbed by the electrons in that material, and those electrons are excited to a higher energy level. If these energies are higher than the work function ϕ of the sample (typically a few eVs), then the electrons are ejected from the system. Since the energy of the incoming photons is known, the energy distribution of the excited ejected electrons (which can be measured with an energy analyzer) will provide information on the energy distribution those electrons had inside the material before the photon electron collision. This is the information searched for with the PES method. Reviews on the subject have been prepared by Shen and Dessau [18.2], more recently by Damascelli et al. [18.3], and others. The reader should consult those reviews for details, and extra references. The goal of this section is to provide just a brief description of photoemission experiments, and the text is actually inspired by the reviews mentioned before [18.2, 18.3].

If electrons inside the metal are assumed to be weakly interacting, then the measured energy distribution will directly provide the density of occupied states of the material. However, for interacting electrons the situation is more complicated and the interpretation of the PES results is not straightforward. For example, after absorbing the photon the soon-to-be ejected electron may collide with other electrons on its way to the surface of the sample. These collisions, if they occur, will reduce the energy of the electron. Then, typically the outcome of the experiment contains results from primary electrons, those that did not collide with others on their way out, and secondary electrons, which collided and lost energy. By this procedure, sharp features in the density of states caused by primary electrons are combined with a background caused by the secondary electrons. This effect causes the well-known *surface sensitivity* of PES experiments: the primary electrons detected usually come from the first layers near the surface, since electrons deeper in the material cannot make it to the surface without strong collisions. In fact, surface preparation is a critical issue for the success of these experiments. A typical question every time a new PES result is reported is " How good is the sample's surface?" In discussions with PES experts the issue of surface quality often arises, and sometimes dominates the discussion. In typical experiments carried out with incoming energies in the range 20–200 eV, the electrons that escaped from the solid were originally located within a thin region next to the surface of width about 10 Å. This is just one or two layers for typical transition-metal oxides as studied in this book. Very often, PES experimentalists studying a particular family of compounds use just one or two "good" materials that present clean surfaces. A typical example is Bi2212 in the family of high-temperature superconductors.

Figure 18.1 summarizes the processes involved in a typical PES experiment. Electrons are originally in their valence band, below the chemical potential, or even deeper in energy, in core levels of the ions under investigation. When the photon provides energy $h\nu$, the electron can be ejected, but before exiting the material it has to reach the surface. Then, it flies to the energy detector, which registers its presence. Accumulating events, a histogram of electrons vs. energy is constructed from the experiment, the histogram that is associated with the original density of states of electrons inside the

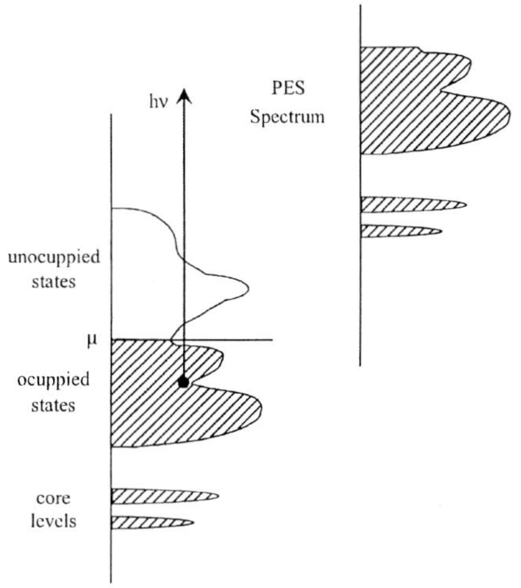

Fig. 18.1. Schematic representation of the processes of relevance in a PES experiment. On the *left*, the actual density of states of a material is shown, with occupied and empty states, and core levels. The energy $h\nu$ provided by the photon allows the reconstruction of the occupied portion of the spectrum, on the *right*

solid, up to smearing effects due to secondary electron collisions, and electron correlations as described below.

Photoemission results in the literature can be angle integrated or angle resolved. In the first case, the simplest, the electrons are collected in a large solid angle. Only their energy is analyzed but not the angle at which they were ejected from the solid. On the other hand, in the angle-resolved mode shown schematically in Fig. 18.2a, the angle at which the electrons are ejected is registered as part of the outcome of the experiment. Typically, the momentum of the incoming photon, at low energies, is negligible (the typical wavelengths associated with photons of energies between 20 and 100 eVs are much larger than the size of atoms or interatomic distances in crystals). For this reason the transitions of the electrons, from the initial to the final state, are assumed to occur at constant momentum. They are the so-called direct transitions. This information allows the PES method to provide the structure of bands of the solid under investigation, see Fig. 18.2b. Note in this figure that the width of the peaks becomes smaller as the Fermi energy is reached. In addition, correlated electron systems typically present many more peaks than shown in the simple case of nearly free electrons sketched in Fig. 18.2b.

Suppose a detector at an angle θ with respect to the vertical axis observes an electron with energy $E_{\rm kin}$. How do we relate this information with the energy that the electron had inside the material and with its momentum? From simple conservation of energy $h\nu - \phi - E_{\rm B} = E_{\rm kin}$, namely, the energy of the incoming photon minus the work function ϕ minus the actual energy of the electron inside the metal, the so-called binding energy $E_{\rm B}$, must match the kinetic energy observed by the detector. A second equation tell us that the momentum of the electron parallel to the surface must satisfy $p_{||} = \sqrt{2mE_{\rm kin}}\sin\theta$, where the square root is the modulus of the momentum

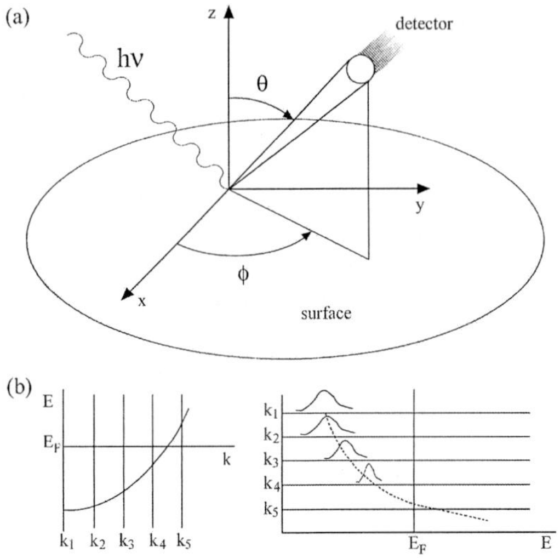

Fig. 18.2. (**a**) Geometrical setup for angle-resolved PES experiments. (**b**) Idealized outcome of the experiment described in (**a**) and text, for the case of nearly free electrons, moving in a band. The case of correlated electrons is much more complicated, sometimes with several broad peaks instead of just one

in terms of the energy, and θ is the polar angle. In this last equation the conservation of momentum along the parallel direction has been used, and also that the momentum of the incoming photon for the energies usually used in PES experiments can be neglected, as already explained. Thus, the $p_{||}$s of the electron inside the material before being hit by the photon, and outside after the photon absorption, are actually the same. However, note that regarding the direction perpendicular to the plane there is no reason to conserve momentum since there is *no* translation invariance in the system (momentum conservation occurs when the system is invariant under translations in real space). Then, $p_{\perp}$ cannot be determined in an angle-resolved PES experiment. However, this uncertainty in the momentum perpendicular to the surface of the sample is not so relevant in quasi-two-dimensional systems, such as high-temperature superconductors and single- or double-layered manganites. In these materials the electrons naturally move within 2D planes and the dispersion in the direction perpendicular is very small. Then, 2D systems are ideal for photoemission experiments.

There are several other techniques that are related to PES, which are not reviewed here. Among them are the "inverse" PES (IPES) and BIS (bremstrahlung isochromat spectroscopy), where the process of photoemission is reversed: electron beams are aimed at the sample, and photons emitted are analyzed. By this procedure information about the *empty* states is obtained (while the photoemission method described above provides information about the occupied states). IPES and BIS unfortunately have less resolution, and for this reason they have not reached the same level of accuracy that PES has reached to study transition-metal oxides in recent years. Another technique in this context which is used frequently is XAS (X-ray absorption spectroscopy) where electrons deep in the core levels (as in Fig. 18.1) are promoted to empty states, providing in the process information about those empty states, as in IPES and BIS.

On the theory front, the previous simple PES description suggests that the plain density of states is the natural outcome of the experiments, if the complications of secondary electrons and others are neglected. However, the density of states is truly obtained in PES only for materials that are very accurately described by free-electron models, such as some standard metals. In other situations, notably electrons in manganites or cuprates, electrons are strongly interacting and the formulas to be used to theoretically predict the outcome of a PES experiment do not just involve the density of states. Without derivation (for references see [1.20, 18.2, 18.3]), here it is just stated that the outcome of an angle-resolved photoemission experiment is proportional to the *spectral weight function* $A(\boldsymbol{k}, \omega)$, which can be formally written at zero temperature as

$$A(\boldsymbol{k}, \omega) = \sum_n |\langle \phi_n^{N-1} | c_{\boldsymbol{k},\sigma} | \phi_0 \rangle|^2 \delta(\omega - (E_n^{N-1} - E_0^N)) , \tag{18.1}$$

where the state $|\phi_0\rangle$ is the ground state of the N-electron system with energy E_0^N, $|\phi_n^{N-1}\rangle$ is the n-th state of the $N-1$ electron subspace with energy E_n^{N-1}, and $c_{\boldsymbol{k},\sigma}$ is the operator that destroys an electron with momentum $\boldsymbol{k}$ and z-projection of the spin σ. The sum is over all the states of the $N-1$ electron subspace. The δ functions enforce the conservation of energy between the incoming photon and the electron. Note that $\hbar$ is set to one above. To relate (18.1), which is the formula usually calculated by many-body theorists (see for example [1.20]), with experiments, it is necessary to work in the *sudden approximation*. In this approximation, the electron when hit by the photon is instantaneously ejected from the sample, without allowing for relaxation of the environment. Mathematically, this property is shown in the use of the operator $c_{\boldsymbol{k},\sigma}$ in the expression for the spectral weight, since this operator simply eliminates an electron with quantum numbers $(\boldsymbol{k}, \sigma)$ from the ground-state wave function of N electrons. Finally, note that the comparison theory experiment must also account for temperature effects through a Fermi function modulation of the intensity, and a one-electron matrix element corresponding to the dipole interaction that is responsible for the photon electron interaction (for details see [18.3]). Certainly the interpretation of PES results is nontrivial.

The most clear difference between the case of nearly free electrons and strongly correlated electrons is that the spectral weight function in the former consists of just a few sharp peaks, while in the latter there is more structure, with satellite peaks and, in general, a nontrivial redistribution of weight. In fact, for nearly free electrons, the operation of acting with $c_{\boldsymbol{k},\sigma}$

over $|\phi_0^N\rangle$ produces an eigenstate of the $N-1$ electron subspace, and most of the delta functions (but one) in the spectral weight carry negligible weight. On the other hand, in correlated electrons $c_{\boldsymbol{k},\sigma}|\phi_0^N\rangle$ is usually far from being an eigenstate of the problem. For instance, often the lowest state of holes in antiferromagnetic backgrounds contains a distortion of the spin order around the hole (magnetic polaron) [1.20]. This modification of the immediate hole surroundings can also manifest itself in the lattice degrees of freedom (lattice polaron), as happens in manganites. This "cloud" around the hole is not contained in the expression $c_{\boldsymbol{k},\sigma}|\phi_0^N\rangle$, and as a consequence the overlap of this state with the true lowest energy state of the $N-1$ electron subspace can be small. This overlap, usually denoted by $Z_{\boldsymbol{k}}$, decreases in value as the size of the distortion grows. The more $Z_{\boldsymbol{k}}$ deviates from one, the stronger is the deviation from a canonical behavior where hole *quasiparticles* are formed in the system. As a consequence, weight will appear in other delta functions since the ω-integrated spectral weight is one due to standard sum-rules [1.20,18.2,18.3], leading to overall fairly complicated spectral weight functions. The reader can find many results of this variety for the case of high-temperature superconductors in the review [1.20]. In the context of manganites, PES experiments and calculations are reviewed below in this chapter.

18.2 Pseudogap in Manganites: Experimental Investigations

The first set of high energy resolution angle-resolved photoemission (ARPES) measurements in the context of manganites was reported by Dessau et al. [18.4], using the bilayer compound $La_{2-2x}Sr_{1+2x}Mn_2O_7$ with $x = 0.4$. This material was investigated because it cleaves easily, as occurs for the Bi-family Bi2212 compounds of the high-temperature superconductors that are also bilayered. The energy E vs. momentum k relation of the hole excitations was investigated in the manganite study mentioned above. Overall, the conclusion of Dessau et al. [18.4] is that the low-temperature ferromagnetic state of the bilayered manganite is very different from a prototypical metal, indicating new and anomalous physics. Some results are reproduced in Fig. 18.3 with the photoemission intensity along several directions near the Fermi energy, measured at 10 K in the ferromagnetic regime. One of the most surprising results is that the peaks, presumably indicative of a quasiparticle state, never reach the Fermi energy. In addition, the widths of the ARPES features are anomalously broad, and do not sharpen as they approach the Fermi momentum. This seems to indicate that the peaks do not correspond to quasiparticles in the usual sense. Correlated to this result, only a small spectral weight is found at the Fermi energy, although the measurements were carried out in a ferromagnetic state. This is consistent with the high resistivity of this compound in the regime investigated (see Chap. 3), at the border between what can be

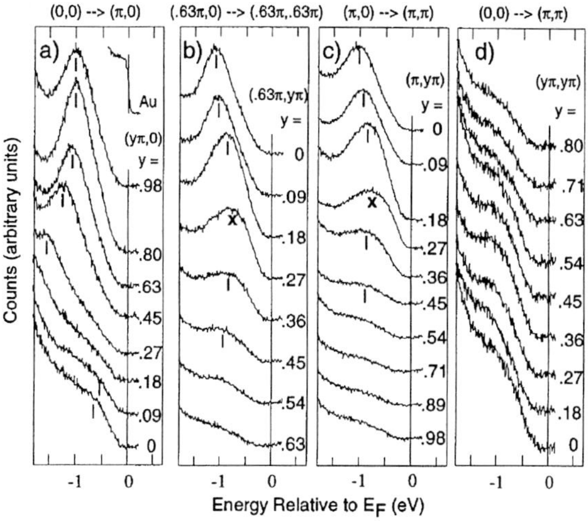

Fig. 18.3. (a-d) Low-temperature (10 K) ARPES spectra of the $x = 0.4$ bilayered manganite, which has a Curie temperature of 126 K, along several directions as indicated at the top of each panel. Note the broad character of the peaks, and lack of dispersion across the Fermi energy. From [18.4]

considered a metal or an insulator. Dessau et al. [18.4] properly describe the absence of spectral weight at the Fermi energy as a "pseudogap". They found that the pseudogap is present both at low temperatures in the ferromagnet, as well as at higher temperatures above T_C in a magnetically long-range disorder state. In fact in the latter the pseudogap is larger than in the former. More recent investigations by Saitoh et al. [18.5] provide further support to the existence of a pseudogap in manganites. Fujimori et al. [18.6] have also discussed these issues. Scanning tunneling microscopy experiments by Biswas et al. [11.46] have also revealed features compatible with a gap above T_C. A pseudogap has also been found in bilayer manganites using Raman scattering spectroscopy [18.7]. All these results clearly show that *Anderson localization is not at work in manganites*, since such localization does not open a gap in the DOS, contrary to the experiments described here. The mobility of the carriers is not the important issue, rather the number of states available at the Fermi energy.

Dessau et al. searched for possible origins of the pseudogap. The experimental results provide a pseudogap that involves the entire Fermi surface to a similar degree. Regarding a gap opened as in the Hubbard model (Mott-Hubbard gap), this is expected to be present at particular densities such as $\langle n \rangle = 1.0$, but not at the intermediate density investigated in the bilayered manganite discussed here. The same applies with splitting due to the Jahn Teller effect. The double-exchange ideas do not predict a pseudogap in the ferromagnetic state, on the contrary that state is supposed to be a good metal.

Anderson localization does not lead to a pseudogap, as already explained. Discarding all these standard alternatives, Dessau et al. [18.4], studying a model of electrons in a bath of optical phonons, proposed that the ARPES features should actually correspond to many peaks, each corresponding to an electron and a number of excited phonons. Other theoretical explanations based on highly inhomogeneous states are described in the next section.

More recent photoemission investigations by Chuang et al. [18.8] (see also the perspective on the subject written by Keimer [18.9]) has provided an even more complex picture of the 40%-doped bilayered manganite. The pseudogap was found to remove 90% of the spectral weight at the Fermi energy, confirming the results of previous investigations [18.4], but now with greatly improved energy and angular resolution spectral weight was observed (with dramatically dropping intensity) all the way to the Fermi energy. Even more fascinating, the Fermi surface topology of the faint ARPES quasiparticle feature that disperses toward the Fermi energy has straight parallel segments indicative of quasi-1D behavior, which in addition can lead to nesting effects. Chuang et al. believe that these combined features are indicative of *charge-density-wave* distortions, at least at short distances. The effect occurs both above and below the Curie temperature, and a discussion of similarities with results for the cuprates can be found in [18.8]. Chuang et al. concluded that nanoscale fluctuating charge/orbital modulations are present in the system. They write that nanoscale phase separation having been shown to be critical for cuprates and manganites appears to be a "*hallmark of modern condensed-matter physics*". The author strongly agrees with this statement.

18.3 Pseudogap in Manganites: Theoretical Investigations

Calculations of the DOS in models for manganites have been carried out by Moreo et al. [18.10] using the one- and two-orbital models, with and without quenched disorder, with the help of Monte Carlo techniques. Some results are shown in Fig. 18.4. Part (a) contains the DOS of the one-orbital model at the parameters indicated in the caption, and electronic densities close to 1. The regime investigated is the one where a large compressibility was reported in Chap. 6, and upon further cooling the regime of phase separation is reached. A clear precursor of phase separation at finite temperature consists of a dynamical mixture of clusters of the two phases that are in competition, in this case a hole-undoped antiferromagnetic and a hole-doped ferromagnet. The word dynamical is used here because the clusters observed in the simulations evolve with Monte Carlo time, changing shapes, and locations. The minimum in DOS at the chemical potential, which becomes E_{F} as the temperature is reduced, is indicative of pseudogap behavior, and it is natural to associate such DOS features with a mixture of a gapped insulating state and a gapless metallic state.

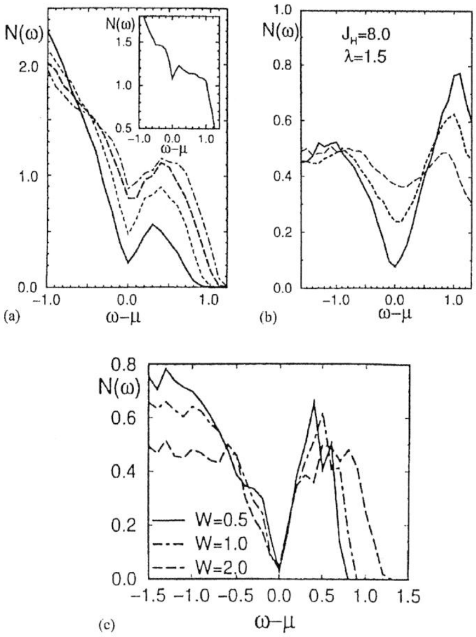

Fig. 18.4. (**a**) DOS of the one-orbital model using a 10×10 cluster at infinite Hund coupling, and temperature $T = 1/30$ in units of the hopping. The four lines from the top correspond to densities of mobile carriers equal to 0.90, 0.92, 0.94, and 0.97. The *inset* has results at density 0.86. (**b**) DOS for the two-orbital model using a 20-site chain, electronic density 0.7, Hund coupling 8, and $\lambda = 1.5$. The three lines at the chemical potential from the top correspond to temperatures $T = 1/5$, $1/10$, and $1/20$. Both (**a**) and (**b**) are reproduced from Moreo et al. [18.10], where more details can be found. (**c**) DOS in the presence of disorder, from [6.41]. W is the strength of disorder, and shown are results for the one-orbital model on a chain of 20 sites, with temperature $T = 1/75$ (hopping units), Hund coupling 8, and density 0.87. This corresponds to a regime of phase separation for zero disorder. The disorder stabilizes the system, and creates a pseudogap. For more details see [6.41]

The phenomenon appears in other dimensions and models as well. In Fig. 18.4b results for two orbitals are shown, using a robust electron phonon coupling and this time varying temperatures instead of densities. The pseudogap is clearly observed. Even with disorder added to the one-orbital model, as in the case of Fig. 18.4c, a similar feature exists in the density of states. Extensive theoretical studies show that pseudogaps appear when mixed-phase insulating/metal states dominate the physics of the problem [18.10]. The existence of a pseudogap is very likely in the regime between the Curie temperature and T^* (see Chap. 17), which is also a regime of mixed-phase characteristics. Investigations of the spin-fermion model for cuprates (see Chap. 6) have also identified a pseudogap for the case of dynamical stripes in cuprates. A robust gap appears in striped states in the manganite context as well, see Chap. 10. In Fig. 18.5 the spectral function obtained numerically is shown. The pseudogap features can be identified. Note the large width of the quasi-

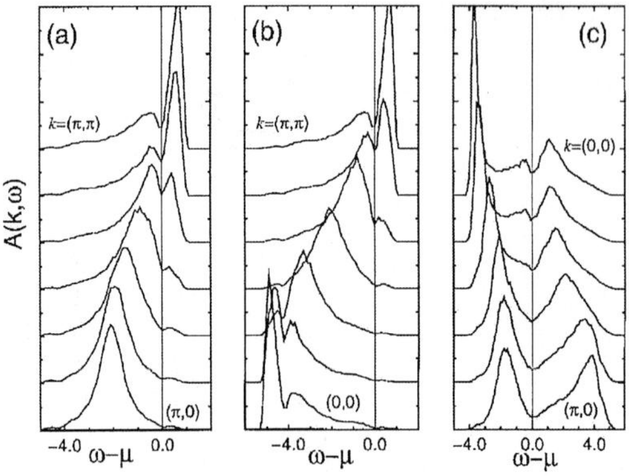

Fig. 18.5. Spectral function $A(\boldsymbol{k}, \omega)$ at $J_{\mathrm{H}} = \infty$, from Moreo et al. [18.10]. (**a**) corresponds to results for a 2D one-orbital model at low temperature and hole density 8%, along the $(\pi, 0)$ to (π, π) direction. (**b**) Same as (**a**), but along (0,0) to (π, π). (**c**) are results for a 2D two-orbital model at low temperature, with robust electron phonon coupling and density of 30% holes, along (0,0) to $(\pi, 0)$. Note in the latter the presence of a pseudogap, which also appears in (**a**) and (**b**) although not as prominently. Also note the large widths of the peaks, in agreement with experiments

particle peaks, which are caused by the nonuniform background of classical phonons and localized spins.

A possible intuitive explanation of the pseudogap behavior was provided by Moreo et al. [18.10] for the case without explicit disorder. Consider part (a) of Fig. 18.6, where a rough sketch of a mixed AF/FM state is shown. From the discussions in previous chapters it is clear that electrons can move easily in the ferromagnetic regions (which carry a mobile electron density less than 1), while they cannot travel large distances in the antiferromagnetic state. The hole density is shown in part (b), indicating that holes indeed prefer to be located in the FM regions. This same physics can be obtained using an effective potential with finite wells, as in part (c). Every time a hole is added to the system a new well is created. The overlap of the states of the well create a narrow band for holes, which lies entirely below the chemical potential, as shown in part (d). The DOS is not "rigid" but it changes with density, i.e., new states are created (new wells) with the addition of holes. Figure 18.6d shows the formation of a pseudogap feature. This reasoning should be valid for more complicated models as well, and even including disorder.

Other calculations, such as those reported by Perroni et al. [6.16] with variational techniques, have also observed the presence of a pseudogap in mixed-phase regimes. A pseudogap was also recently reported by Craco et al. [18.11] using dynamical mean-field theory. The pseudogap feature seems to be quite robust, and of relevance for manganites as described before when the experiments in this context were analyzed. It is conceivable that the origin of the pseudogap in real experiments is the mixed-phase character of

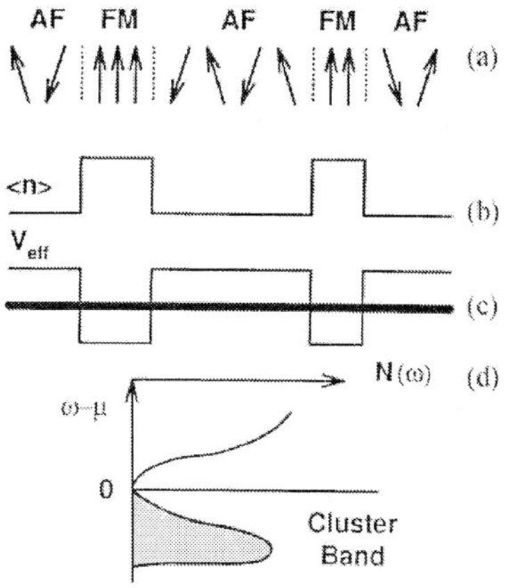

Fig. 18.6. Schematic explanation of the pseudogap found in computer simulations, from Moreo et al. [18.10]. (**a**) shows a mixed-phase AF/FM state. (**b**) is the associated density of hole carriers (they prefer the FM regions). (**c**) is an effective potential acting on the holes, which favors the FM portions of the sample. The *thick line* is a crude representation of a band produced by the small overlap between states in different clusters. (**d**) The resulting DOS. The states in gray represent the holes in the effective wells

the manganite state, as conjectured by Moreo et al. [18.10]. In this context a related important question is: can a homogeneous state induce a pseudogap in the density of states? It is easy to induce a gapped DOS or a metallic featureless DOS, but it appears to be more complicated to have a pseudogap.

18.4 Properties of One Hole

In this section devoted to photoemission, it is also important to discuss the properties of just one hole in the system. In the context of high-temperature superconductors, calculations of the properties of one hole were very important to understand how this carrier is "dressed" by spin excitations [1.20]. A large body of literature concluded that the bandwidth of the dressed hole was severely reduced from a scale of order "t" in the noninteracting limit, to a scale "J" when the hole is immersed in an antiferromagnetic background governed by the Heisenberg model with exchange J. The predictions of theorists were confirmed by experiments, providing strong support to the use of t–J-like models to describe the cuprates [18.3].

In the area of manganites, similar calculations would indicate how the hole is dressed by the spin, orbital, and lattice degrees of freedom, since manganites are more complicated than cuprates regarding the number of active modes. Calculations in this context have been produced almost simultaneously by several groups. For example, Perebeinos and Allen [18.12] studied the two-orbital model with cooperative phonons, and predicted the presence of multiple Franck–Condon broadening due to the phonon dressing of the hole. These effects could be observed in photoemission experiments. Van den Brink et al. [18.13] used the orbital t–J model (see Chap. 9) to also address the behavior of one hole dressed by orbital excitations (i.e., "orbital polaron"). (See brief discussion on polarons in Chap. 6). A large mass was

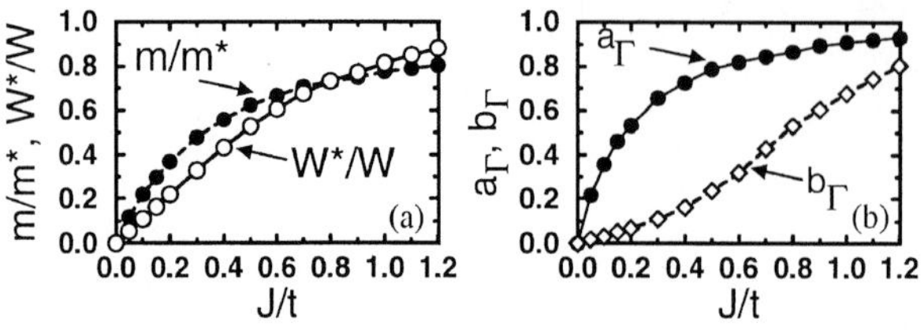

Fig. 18.7. Quasiparticle properties as a function of J/t. (**a**) Inverse effective mass m/m^* (*full circles*) and the hole bandwidth W^*/W (*empty circles*). Clearly, the effective mass grows dramatically as J/t is reduced to zero. (**b**) Weights of the high-energy quasiparticle (*filled*) and low-energy quasiparticle (*open*) at the Γ point. At small J/t the weights are negligible. The present results are from van den Brink et al. [18.13]. Very similar results have been found in models for high-temperature superconductors [1.20]

observed as well as a small quasiparticle weight (for definitions of these concepts, see review [1.20]), as shown in Fig. 18.7 below. Yin et al. [18.14] also studied the properties of one hole including Jahn–Teller distortions in the 2D orbital t–J model, using the Born approximation and exact diagonalization as the many-body techniques. The quasiparticle bandwidth was found to be small, scaling as the coupling J (which is the orbital nearest-neighbor coupling, since the background is assumed to be ferromagnetic), similarly as found in the context of the high-T_c superconductors where J governs the strength of the antiferromagnetic spin background [1.20]. It is clear that a hole carrier is much affected by the low-energy excitations of the background in which it is moving, and angle-resolved experiments in the undoped limit $LaMnO_3$ would be important to confirm the theoretical calculations. Similar experiments in the area of cuprates reported an excellent agreement between theory and experiments, and a similar agreement could be unveiled in manganites.

Summarizing, similarly as occurs in cuprates, the photoemission studies of bilayer manganites have produced very important results that help in clarifying the physics behind these complex materials. A pseudogap has been clearly identified. Similar conclusions were found in theoretical studies in mixed-phase states. The pseudogap regime of manganites promises to be as important as the analogous regime in cuprates (see also Chap. 22). The study of a small number of holes has also led to nontrivial predictions that can be tested in experiments. Further photoemission experiments in this context are needed to analyze the evolution of the density of states and quasiparticle properties with densities and temperature.

19. Charge-Ordered Nanoclusters above T_C: the Smoking Gun of Phase Separation?

This chapter describes interesting experimental results that support the theory of phase separation for manganites outlined in this book. The information presented here was gathered using a variety of techniques, convincingly showing the existence of charge-ordered nanoclusters above the Curie temperature of CMR ferromagnetic manganites. In addition, it has been shown in many experimental measurements that the "strength" of these clusters correlates with the resistivity, both as a function of temperature at zero magnetic field and as a function of magnetic field at constant temperature. These experimental investigations have also shown that free polarons exist in the paramagnetic insulating (PI) state, but their presence does not seem correlated with the magnetotransport CMR phenomenon. The key feature of the PI state that induces the well-known MR behavior of its resistivity appears to be the charge-ordered nanoclusters described here. This result, if confirmed, would provide key support to the theory of manganites that is based on mixed-phase tendencies, and would rule out theories based on a gas of uncorrelated small polarons in the PI state. Work in the near future will hopefully further clarify these issues. It is recommended that the results of this chapter should be read in combination with those of some previous chapters, such as Chap. 11, where experimental evidence for inhomogeneities was presented.

19.1 Introduction and Relevance of Results Discussed in this Chapter

It is intuitively expected that individual polarons will generate "diffuse" scattering, known in this context as Huang scattering. For a perfect crystal, it is well known that only Bragg peaks should be observed in neutron or X-ray experiments. Due to the presence of lattice-polaronic distortions, locally there should be small deviations from the perfect crystal structure. These deviations induce a finite neutron or X-ray intensity in the vicinity of the Bragg peak (i.e., not at the peak, but close to it). For Jahn–Teller polarons this extra intensity has a particular "butterfly" shape in momentum space. Its presence in manganite experiments, as shown below, implies the existence of JT polarons in the paramagnetic insulating state of Mn oxides above T_C. However, in addition to this "diffuse polaron scattering," *satellite peaks* were also found in the vicinity of the Bragg peaks. These extra peaks are well defined, and in some cases located at positions corresponding to well-known charge arrangements, such as the CE-state. As a consequence, the PI state should be visualized as composed of (i) independent polarons contributing to the diffusive scattering and (ii) so-called "correlated polarons" that are expected to form small charge-order clusters at temperatures between T_C and some new scale T^* where they form upon cooling. The temperature-dependent intensity of the peak associated with (ii) was shown to scale well with the temperature-dependent resistivity. As a consequence, it is natural to believe that the charge-ordered clusters are responsible for that resistivity, as well as its anomalous behavior as a function of magnetic field. These results are in agreement with the theoretical ideas described in Chap. 17 where a T^* scale was predicted and precisely assigned to the coexistence of clusters of two phases, insulating and metallic.

The reader should note how remarkable it is that at relatively high temperatures there are small islands of a phase, such as the CE, that is not stable at low temperatures. In materials of the LCMO family this CE phase is stable at $T = 0$ only upon further increasing the hole density beyond $x = 0.5$. It seems obvious to this author that the CE islands cannot be called "polarons", as they are sometimes referred to in the literature, since a polaron is one charge surrounded by a lattice and/or spin distortion. In each of the CE nanoregions more than one carrier may be present. Alternatively, these CE regions are sometimes attributed to the presence of "correlated polarons", but even this language is not sufficiently descriptive, since the observed internal structure in those clusters corresponds to a specific phase involving charge ordering. This CE phase is clearly competing with the FM metallic phase, which eventually wins as the system is cooled down. The CMR phenomenon seems to be caused by phase competition, not by standard polaronic behavior. Unfortunately, the experimental literature often refers to any kind of inhomogeneity as a polaron, but it would be better to use a more proper language. "CE-clusters" or "charge-ordered clusters" are a possibility, bringing the mixed-phase nature of the insulating state to the forefront, likely a key issue behind manganite physics. Nevertheless, in this chapter the term correlated polarons will still be used as an alternative to clusters, simply to adjust to the language widely used in current experimental literature.

19.2 Diffuse Scattering and Butterfly-Shaped Patterns

19.2.1 Bilayers

X-ray and neutron scattering studies of the bilayer manganite $La_{1.2}Sr_{1.8}Mn_2O_7$ by Vasiliu-Doloc et al. [15.11] have revealed the presence of diffuse scattering

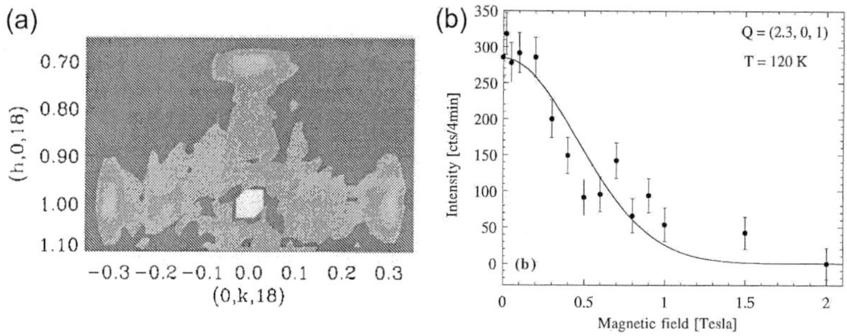

Fig. 19.1. (**a**) X-ray intensity at $T = 125\,\mathrm{K}$ for the bilayer $La_{1.2}Sr_{1.8}Mn_2O_7$ (with $T_C = 112\,\mathrm{K}$), from Vasiliu-Doloc et al. [15.11]. The butterfly-shaped pattern – if (**a**) is rotated by 45 degrees – is observed together with incommensurate peaks at $(\pm 0.3, 0, 18)$ and $(0, \pm 0.3, 18)$ that correspond to charge correlations. (**b**) Intensity of a peak associated with charge correlations at 120 K, as a function of magnetic fields. The rapid decrease in the signal shows its relevance for the CMR phenomenon. From Vasiliu-Doloc et al. [15.11]

associated with lattice distortions around localized charges (polarons) above the ferromagnetic transition temperature T_C (see also Argyriou et al. [15.12] and Campbell et al. [15.13]). The evidence came in the form of a butterfly-shaped diffuse intensity pattern compatible with JT distortions. Moreover, short-range correlations between those polarons were revealed in this study as well (see Fig. 19.1a) since the butterfly patterns in addition carry small peaks at incommensurate locations, which were interpreted as being caused by polaron correlations. The size of the correlated regions was found to be about six lattice spacings in the planes. Perpendicular to the bilayers, only the two planes of each bilayer appear correlated. Vasiliu-Doloc et al. remarked that finding short-range charge ordering in an optimally doped CMR ferromagnet was an important novel feature, a conclusion that this author certainly agrees with. In addition, it was observed that the intensity of the peak related to charge ordering collapsed in the presence of a magnetic field, similarly as the resistivity does (see Fig. 19.1b). Clearly, the correlated regions are important for the CMR phenomenon.

19.2.2 Perovskites

Conclusions very similar to those described in the previous section were independently reached by Shimomura et al. [15.14] using X-rays to study single crystals of $(Nd_{0.125}Sm_{0.875})_{0.52}Sr_{0.48}MnO_3$. Typical results are shown in Fig. 19.2a, where the temperature dependence of the X-ray scattering intensity is presented in the vicinity of a Bragg peak at momentum (0,10,0). The tails away from the Bragg peak are due to diffuse scattering. The results in the figure show that the diffuse intensity increases as T_C is approached from room temperature, and then falls sharply below T_C. Shimomura et al. also

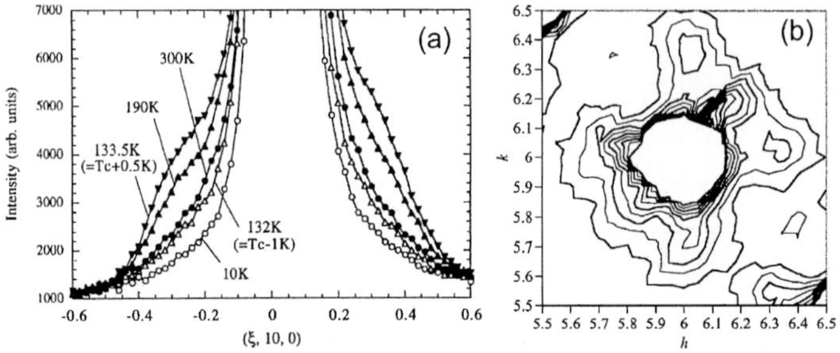

Fig. 19.2. X-ray scattering results for $(Nd_{0.125}Sm_{0.875})_{0.52}Sr_{0.48}MnO_3$, from Shimomura et al. [15.14]. (**a**) Temperature dependence of the scattering intensity in the vicinity of one Bragg point. The signal shown is caused by diffuse scattering due to JT distortions. Note the hump at $\xi \sim -0.3$, which is due to correlated polarons (clusters with charge ordering). (**b**) Butterfly-shaped pattern of diffuse scattering around a Bragg peak. Note the presence of small peaks at $(6 \pm 0.3, 6, 0)$ and $(6, 6 \pm 0.3, 0)$, which cannot be attributed to individual JT polarons. Details of how these scattering intensities were obtained from the actual data are in [15.14]

calculated the dependence of the diffuse scattering intensities with momentum. The patterns obtained are butterfly-shaped, as were those reported also by Vasiliu-Doloc et al., and they are believed to be caused by the contractions and elongations in the Mn O distance caused by the Jahn Teller effect at a Mn^{3+} site. The actual butterfly-shaped pattern is in Fig. 19.2b. While the shape agrees with simple expectations based on the distortion caused by one JT polaron, there are small peaks at $(6 \pm 0.3, 6, 0)$ and $(6, 6 \pm 0.3, 0)$ which cannot be explained by a picture of noninteracting polarons. Those peaks can also be seen in part (a) of the figure as humps, and they were tentatively attributed to the correlations between polarons.

Further X-ray scattering work by Shimomura et al. [19.1], this time on $Pr_{1-x}Ca_xMnO_3$ with $x = 0.35$, 0.4, and 0.5, further clarified the previous investigations. This material does not have a low-temperature metallic FM ground state, as do the others reported in this section. However, the systematic behavior of the diffuse scattering above the charge-ordering temperature T_{CO} is *qualitatively similar* to that in $(Nd_{0.125}Sm_{0.875})_{0.52}Sr_{0.48}MnO_3$, with satellites developing in the diffuse-scattering region as the system is cooled.

Thus, it appears that manganites with a ferromagnetic ground state, nevertheless develop charge-ordering correlations above T_C, a quite remarkable result. Of course, considerable evidence reviewed already in Chap. 11 has shown that manganites with a large MR have ferromagnetic clusters above T_C as well. As a consequence, the coexistence of both FM and CO regions appears to be present in that regime of temperatures, in agreement with theoretical expectations (Chap. 17).

19.3 CE-type Clusters above T_C

19.3.1 Neutron Scattering Applied to LCMO $x = 0.3$

More recently, two independent experimental studies of $La_{1-x}Ca_xMnO_3$ using neutron scattering have provided important information regarding the existence of charge correlations above T_C. These studies analyzed in detail the relevance of the small peaks illustrated in Fig. 19.2b; in particular, the studies not only showed that the small peaks correspond to correlated polarons (or clusters), but that they actually specify the pattern of charge order, and they have a temperature dependence related to that of the resistivity.

Dai et al. [19.2] studied $x = 0.15$, 0.20, and 0.30 LCMO, observing neutron diffuse scattering associated with polarons and their short-range polaron polaron correlations. The latter can also be considered as short-range charge ordering. These authors observed that the populations of uncorrelated (or free, or individual) polarons and the so-called correlated polarons are intimately related to the transport properties of the materials analyzed. Figure 19.3 contains the key information of the effort of Dai et al. On the left, it shows the well-known behavior of the resistivity with temperature. On the right, the intensities of two peaks are presented: the open circles correspond to a momentum where there is diffuse scattering but no polaron order. This peak has a weak temperature dependence. On the other hand, the black circles correspond to a momentum associated with charge order (i.e., polaron correlations). This peak has a temperature dependence similar to that of the resistivity. The correlation length of the polaron correlation increases from ~ 13 Å at room temperature to ~ 28 Å just above T_C. Dai et al. write "*the resistivity of these manganites is directly imaged by the population of the correlated polarons*". This is a clear and important statement that assigns to the charge-order regions above T_C the responsibility for the curious behavior of manganites. This result is in agreement with the theoretical studies of Chap. 17 and other chapters, where a T^* for cluster formation was introduced and the CMR phenomenon was quantitatively explained by a mixed-phase state. Theories involving just individual polarons are not sufficient to understand manganites.

In agreement with the previous results, studies of $x = 0.30$ LCMO by Adams et al. [19.3] also reported the presence of individual lattice polarons above T_C, together with short-range polaron correlations. They also found a remarkable similarity between the temperature dependence of those short-range correlations, the resistivity, and the anomalous quasielastic $\omega = 0$ peak above T_C found in previous neutron scattering studies, which were already reviewed in Chap. 11. Regarding the signal corresponding to individual lattice polarons, results corresponding to the intensity of the diffuse component are shown in Fig. 19.4a. Clearly, below T_C the signal is much reduced (although it is still nonzero about 30 K below T_C), while above the Curie temperature it is nearly constant up to room temperature. It is likely that this signal will not

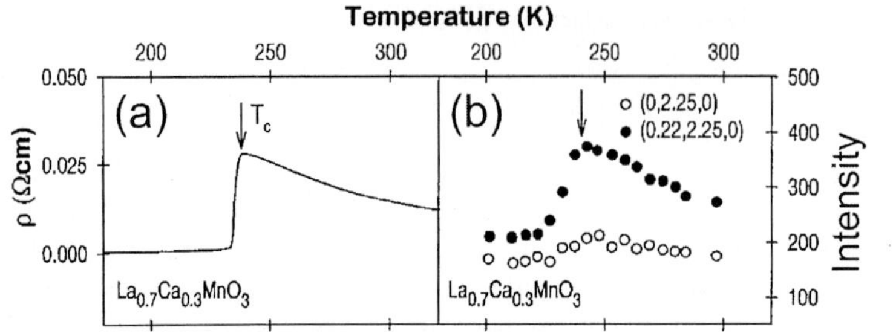

Fig. 19.3. (**a**) Temperature dependence of the resistivity corresponding to LCMO at $x = 0.30$. (**b**) Intensity of the neutron diffuse scattering at the momentum corresponding to a charge ordering modulation (*black circles*), and away from them (*white circles*). The plots show that the charge-ordering modulation temperature behavior is similar to the resistivity, while the *white circles*, corresponding to uncorrelated or independent polarons, have a near independence with temperature which does not explain the transport behavior. From Dai et al. [19.2]

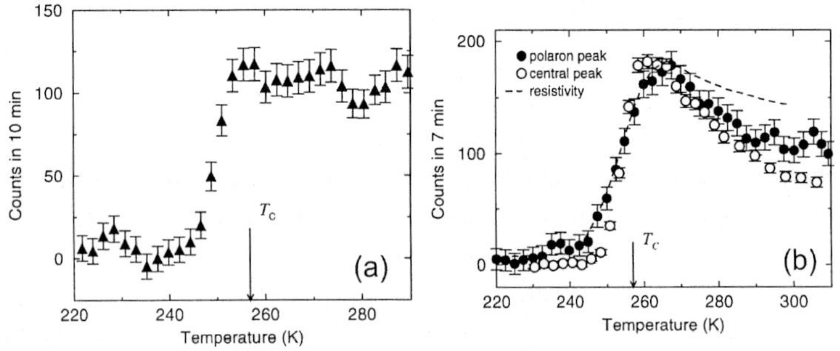

Fig. 19.4. (**a**) Temperature dependence of the diffuse polaron scattering (uncorrelated polarons). Note that above T_C the results are nearly temperature independent in the range investigated. (**b**) Temperature dependence of the intensity of the "polaron peak" – corresponding to charge ordering – compared to the central peak intensity (Chap. 11) discovered by Lynn et al. [11.12], as well as the resistivity. The data have been scaled so the peak heights match. The similarity of the data indicates a common origin. From Adams et al. [19.3], where more details can be found

disappear until much higher temperatures are reached. On the other hand, the signal intensity corresponding to a momentum where there is a satellite peak has a different temperature behavior, shown in Fig. 19.4b: while close to T_C part (b) is similar to (a), clearly the correlated polaron or charge-order cluster peak intensity decreases with increasing temperature and it may vanish roughly around 350–400 K. Figure 19.4b is remarkable and quite convincing: the correlated polaron peak, the resistivity, as well as the anomalous quasielastic peak intensity of Fig. 11.5, are clearly correlated, "*indicating that they have a common origin*", as remarked in [19.3].

The results become even more interesting when the momentum of the satellite peak is analyzed: it is compatible with the CE-state that is known to be a strong competitor of the ferromagnetic metallic state, both in the theoretical studies of Chap. 10 and the many experiments described in Chap. 11. The fact that this state is usually associated with half-doping is not a problem, since a CE-like state could be stabilized at smaller hole densities either through defects or increasing the amount of charge at the 4+ sites.

The experimental results of Adams et al. are crucial for the development of a theory of manganites. They tell us that (a) theories based on small individual polarons, such as those of [5.3, 5.4], may not be sufficient to explain manganites since their behavior is not correlated with the resistivity. Clearly, polarons can and do exist above T_C but if they were forming a gas, the resistivity would be nearly temperature independent in a wide range above T_C following the intensity of the peak in Fig. 19.4a. In addition, (b) the results of Adams et al. tell us that *CE-like clusters exist in a range of temperatures between T_C and a new scale T^*, dominating the behavior of the resistivity in that range.* The size of the cluster is about 10 Å. A paramagnetic insulating state containing clusters of a phase (CE) that can be made stable by varying the chemical composition (e.g., replacing La by Pr, as in $(La_{1-y}Pr_y)_{1-x}Ca_xMnO_3$) is conceptually totally different from a state of free polarons.

19.3.2 Size and Internal Structure of Charge-Ordered Clusters

Additional information about the presence of charge-ordered clusters above the Curie temperature of manganites was reported by Kiryukhin et al. [19.4]. In this study, X-ray scattering results for NSMO and $La_{1-x}(Ca,Sr)_xMnO_3$ were obtained. The results confirm the presence of scattering caused by independent polarons as well as scattering caused by correlated regions, as shown in Fig. 19.5a. In addition, Kiryukhin et al. provided the temperature dependence of the correlation length ξ (i.e., size of the charge-ordered clusters) from the width of the satellite peak, and the modulation of charge *inside* the charge-ordered cluster, from the position of the peak in momentum space. The results lead to the surprising conclusion that ξ is approximately constant, equal to 2–3 lattice spacings, and is independent of both temperature *and* hole density (studies of temperature and doping dependence appear also in [19.3] and [15.11], and they reached similar conclusions). This result is unexpected since, at least naively, one would have anticipated some dependence with the hole density of the optimal size of the correlated regions. On the other hand, the periodicity of the charge inside the correlated regions was found to vary with doping concentration x following an approximately linear relation, as shown in Fig. 19.5b. These simple relations provide important constraints on possible models of the structure of the correlated domains.

In addition, Kiryuhkin *et al.* showed that the size $\sim (2\xi)^3$ of the correlated regions can be fairly large, involving about one hundred unit cells. They also

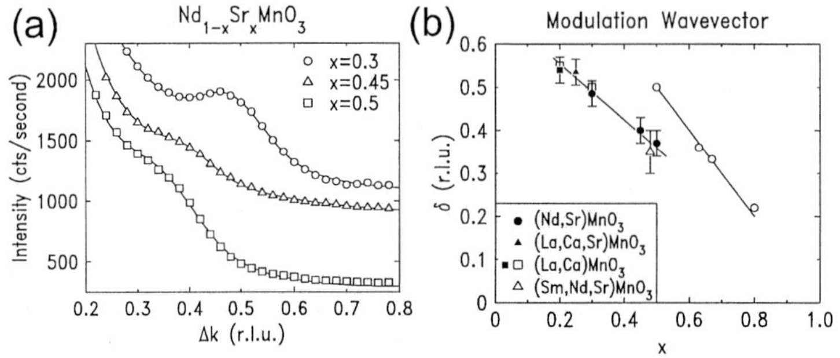

Fig. 19.5. (**a**) Intensity of X-ray scattering for $Nd_{1-x}Sr_xMnO_3$ along a particular direction in momentum space in the vicinity of a Bragg peak, showing the presence of diffuse scattering and a satellite peak characteristic of charge-ordered clusters. The temperatures for each density are close to T_C. This result shows that NSMO presents a behavior quite similar to other compounds described in this section. (**b**) Period of the lattice modulation inside the correlated regions (i.e., inside the region with charge ordering) as a function of doping x, for the materials shown, revealing a universal behavior. The *open circles* on the *right* are results for densities larger than 0.5. Details can be found in Kiryukhin et al. [19.4]

concluded that, at least at $x = 0.30$, the correlated regions are small domains with CE-type order, namely, checkerboard charge ordering and a characteristic orbital ordering. It is clear that robust islands of the CE-phase are present above the Curie temperature in many manganites. Once again, it is worth remarking that these results appear to rule out theories of the insulating state above T_C based on small polarons, and they highlight the key role of phase competition. Kiryukhin et al. [19.4] also made the important observation, contained in Fig. 19.5b that the order inside the correlated regions follows a behavior similar to that observed before at hole densities *higher* than 0.50. This leads to the possibility of electronic phase separation in the sense of a true separation between charge-rich and charge-poor regions, on the nanometer scale, as discussed in theoretical studies of manganite models.

In agreement with the previous results, recent X-ray scattering studies of $Pr_{1-x}Ca_xMnO_3$ and $La_{1-x}Ca_xMnO_3$ with doping $x \sim 0.3$ by Nelson et al. [19.5] and Hill et al. [19.6] revealed weak charge and orbital correlations well above the phase transition, which leads to a charge-ordered state in the former and a FM metallic state in the latter. The correlation length appears to be just 1 or 2 lattice constants. A T^* scale was reported around 400 K [19.5], in excellent agreement with results of other techniques, as discussed later in this chapter. A sketch of the small cluster that the authors of [19.5, 19.6] use to explain their results is in Fig. 19.6. It consists of a small segment of a zigzag chain, similar to those that appear in the CE-state, with a FM character. These results are compatible with all the other X-ray and neutron scattering results presented in this section, including the lack of a temperature

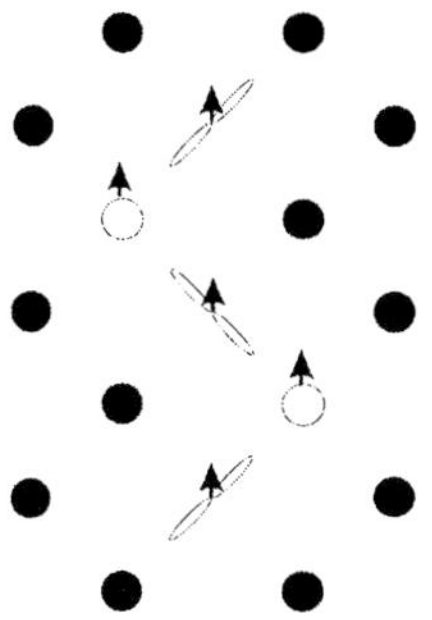

Fig. 19.6. Schematic diagram of the small CE-type cluster proposed by Nelson et al. [19.5] and Hill et al. [19.6] to explain their X-ray scattering data at high temperature, using $Pr_{1-x}Ca_xMnO_3$ and $La_{1-x}Ca_xMnO_3$ with $x = 0.3$

dependence of the size of the CE-clusters up to temperatures ~ 400 K where they disappear.

19.3.3 Behavior of Nanoclusters with Magnetic Fields

The potential relevance of the charge-ordered nanoclusters discussed in this section for the phenomenon of colossal magnetoresistance was addressed in an X-ray scattering study of $Nd_{0.7}Sr_{0.3}MnO_3$ by Koo et al. [19.7]. Their main result in this context is in Fig. 19.7, where are shown (a) the resistivity vs. magnetic field, with the rapid decrease characteristic of CMR, (b) the intensity of the satellite peak that corresponds to the charge-ordered correlated regions (i.e. the CE-type domains) vs. magnetic field, showing a behavior quite similar to the resistivity (for similar conclusions see Fig. 19.1b), and (c) the size of those correlated regions that remain constant with magnetic fields, as previously observed as temperatures and densities are varied. Note that the nanoclusters do not disappear at high fields (part (b) of figure), showing that the conducting state could be inhomogeneous. In addition, in contrast to the intensity of the peak associated with nanoclusters, Koo et al. observed that the diffuse intensity at other momenta show only small changes with magnetic fields (although this analysis is not conclusive yet, and it is being revisited at the time of printing). This other peak intensity is attributed to single polarons, and as a consequence it appears that for the CMR phenomenon the nanoclusters are more important than the isolated JT polarons, although it is likely that both peak intensities (correlated and uncorrelated polarons) must be somewhat affected by magnetic fields. Clearly more of these investigations are needed to confirm the dominance of correlated polarons (or charge-order clusters) to understand CMR. These are key experiments to distinguish among phase-separation vs. small-polaron theories of manganites.

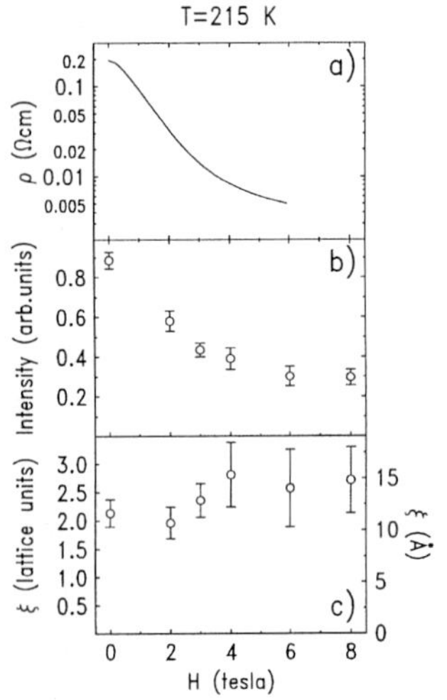

Fig. 19.7. Results for $Nd_{0.7}Sr_{0.3}MnO_3$ just above T_C, from Koo et al. [19.7], showing the relevance of the charge-ordered nanoclusters in the CMR phenomenon. (**a**) Resistivity vs. magnetic field. (**b**) Intensity of the X-ray scattering peak related to the charge-ordered nanoclusters. (**c**) Correlation length (size) of those ordered regions

19.3.4 Electron Diffraction for LCMO $x = 0.3$

In agreement with the results of X-ray and neutron scattering, recent electron diffraction studies of $La_{1-x}Ca_xMnO_3$ with $x = 1/3$ by Zuo and Tao [19.8] have revealed temperature-dependent diffraction spots, which are explained as originating in nanometer-sized domains with charge ordering. The average domain is ~ 3.6 nm in diameter, and the charge arrangement resembles results at $x = 1/2$. Then, the PM to FM phase transition is clearly accompanied by the formation and melting of random charge-ordered droplets. This important conclusion is in agreement with results elsewhere in this section. Observing that the use of different experimental techniques leads to the same result is very encouraging.

19.3.5 Neutron Diffraction and Reverse Monte Carlo

Recent neutron powder diffraction experiments on LCMO $x = 0.30$ over a broad temperature range by García-Hernández et al. [19.9] have been analyzed using reverse Monte Carlo simulation. This approach allows the calculation of both the atomic pair-distribution functions as well as *the spin correlation functions.* The results show that the phase separation observed above T_C in this compound involves both ferro and antiferromagnetic domains, since the distribution of spin correlations corresponding to nearest-neighbors and next-nearest-neighbors Mn ions can be both positive and negative. This suggests that the CE-type clusters reported in the X-ray and neutron diffuse scattering studies could contain antiferromagnetically aligned spins, as is indeed present in the long-range CE structure.

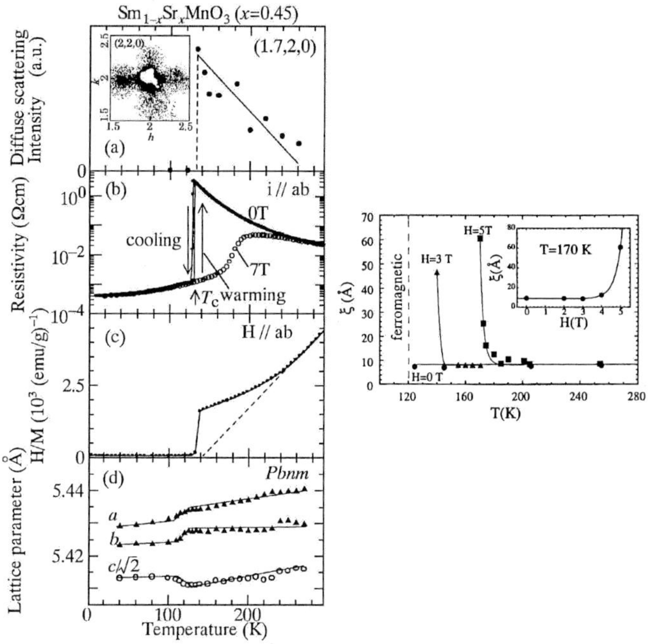

Fig. 19.8. *(Left)* Results from Saitoh et al. [19.10] corresponding to $Sm_{1-x}Sr_xMnO_3$ ($x = 0.45$). (**a**) Intensity of an X-ray diffuse scattering peak associated with charge ordering (the *inset* is a typical butterfly-shaped plot). The result in (**a**) is qualitatively similar to the resistivity shown in (**b**). The starting of the nonzero intensity in (**a**) coincides with the deviation in the inverse magnetization (**c**) from the Curie-law behavior. The lattice parameters in (**d**) also show some changes at the T^* where the intensity in (**a**) starts, coming from high temperature. *(Right)* Results from De Teresa et al. [19.12] regarding the existence of small ferromagnetic regions in the state above T_C of $Sm_{1-x}Sr_xMnO_3$ ($x = 0.45$). Shown is the magnetic correlation length ξ as a function of temperature, parametric with magnetic fields. Note that this correlation increases with magnetic fields, contrary to the size of the coexisting CE-type islands in the same material, which decrease with magnetic fields

19.3.6 The Interesting Case of $Sm_{1-x}Sr_xMnO_3$

Recent studies of $Sm_{1-x}Sr_xMnO_3$ with $x = 0.45$ by Saitoh et al. [19.10] have shown the existence of a state above T_C with properties quite similar to those described before in this section. The evidence is in Fig. 19.8 (left) where the intensity of an X-ray diffuse scattering peak related to charge ordering is shown, together with resistivity, inverse magnetization, and lattice parameters. The phenomenology is totally analogous to that of the rest of the manganites presented above, and small charge/orbital-ordered clusters are expected to be present in the paramagnetic insulating state. Note that the broad character of the peaks associated with the clusters obviously also indicate the absence of long-range order in the state above T_C. Raman scattering results [19.10] also support the presence of dynamical charge/orbital ordering in that temperature range. It is also interesting to note the discontinuous first-order character of the metal insulator transition (MIT) in the absence of a magnetic field. Recent theoretical investigations unveiled percolative processes with first-order characteristics [11.112]. Other low- or intermediate-bandwidth manganites present an abrupt MIT transition, but not as abrupt as in this case. Martin et al. [11.113] have also studied this compound showing that the Sm and Sr ions involved have a large mismatch. Also, Saitoh et al. [19.11] have shown that the optical conductivity in the regime above T_C has characteristics similar to other manganites in a charge/orbital-ordered state.

While in $Sm_{1-x}Sr_xMnO_3$ ($x = 0.45$) the presence of charge/orbital-ordered clusters above T_C appears clear, the coexistence of those clusters with ferromagnetic regions has been under discussion. However, recent crucial results by De Teresa et al. [19.12] have convincingly shown that there are ferromagnetic clusters *different from the CE-type clusters* in the regime above T_C. Using high-resolution neutron diffraction and small-angle scattering, these authors found small ferromagnetic clusters having a size of about one nanometer. The size of these clusters *increases* with the magnetic field, contrary to the behavior of the CE-type clusters (see Fig. 19.8 (right)). As a consequence, the ferromagnetic clusters observed by De Teresa et al. [19.12] are not the CE-like regions detected with X-ray diffuse scattering. This is a very important result, showing the coexistence in the paramagnetic state of both FM- and CE-type islands.

19.4 A New Temperature Scale T^* above T_C

Thermal Expansion. By now, a variety of experimental techniques have suggested the presence of a new temperature scale T^* above T_C, where CE-type nanoclusters start forming upon cooling from high temperatures. This new scale appears to be near or slightly above room temperature, according to the X-ray, neutron scattering, and other studies described in this section. This conclusion is in good agreement with the experiments of Ibarra et al. reported in Chap. 11. For instance, in Fig. 11.10 the linear thermal expansion of $La_{0.60}Y_{0.07}Ca_{0.33}MnO_3$ clearly revealed a new scale T_p where deviations from conventional behavior were detected. That temperature was identified with individual polarons in the original studies of Ibarra et al., but in view of the evidence reported in this chapter it may be better to assign it to the scale where CE-clusters start forming. The scale T_p was also studied by Ibarra et al. in the case of highly hole-doped LCMO, and its value in that case is also about 400 K [19.13]. In [11.35] results for PCMO were reported as well, including a discussion on T_p, which occurs above the CO ordering temperature.

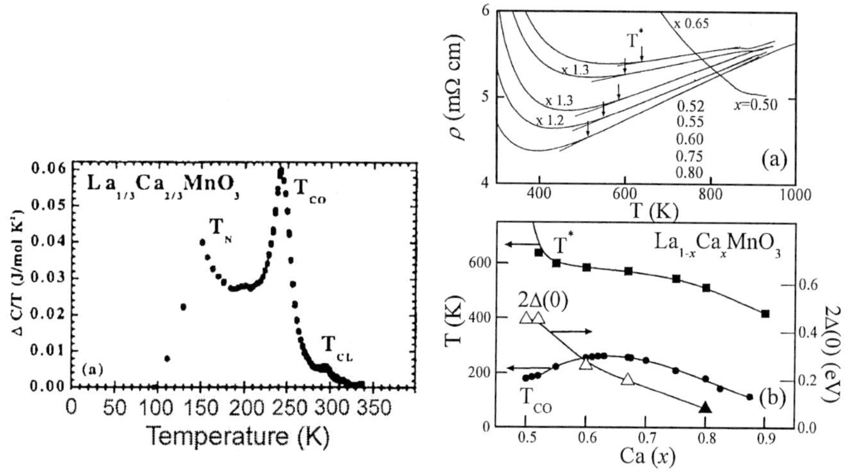

Fig. 19.9. *(Left)* Temperature dependence of the electronic and magnetic specific heat of $La_{1/3}Ca_{2/3}MnO_3$, from Martinez et al. [19.17]. Note that together with peaks at the expected Néel and charge-ordering temperatures, there is an extra peak at "T_{CL}", that the authors of [19.17] associated with charge localization. *(Right)* (**a**) High-temperature resistivity data for $La_{1-x}Ca_xMnO_3$, from Kim et al. [19.18]. T^* is defined as the temperature where the metallic resistivity deviates from linearity. (**b**) Phase diagram showing T^*, T_{CO}, and the gap $2\Delta(T=0)$

Lattice Parameters. The studies of the temperature dependence of Mn O bonds by Argyriou et al. [15.12] and Mitchell et al. [19.14] have also revealed a characteristic temperature T^* around room temperature, that appears correlated with short-range charge ordering since their associated intensity in X-ray experiments disappears at approximately the same temperature.

Specific Heat. It is also remarkable that specific heat experiments by Gordon et al. [19.15] revealed a peak at 260 K in the $x = 0.4$ bilayer compound, well above the actual ordering temperature. Perhaps this is another "smoking gun" for T^*. In agreement with this result, Mira et al. [19.16] also found indications of a T^* in the heat capacity of $La_{2/3}(Ca_{1-x}Sr_x)_{1/3}MnO_3$. Moreover, specific-heat measurements by Martinez et al. [19.17] for $La_{1/3}Ca_{2/3}MnO_3$ revealed a "charge localization" CL temperature above the charge-ordering phase transition. This CL temperature is, once again, about 300 K, in agreement with previous discussions. The result is given in Fig. 19.9 (left). It is clear that thermodynamic measurements are potentially very important for revealing the complex nanoscale structure of the paramagnetic insulating state of manganites. Systematic studies in other compounds should be pursued to address the phase diagrams predicted theoretically in Chap. 17.

Optical Conductivity. Very recently, optical spectroscopy was also used to investigate the properties of $La_{1-x}Ca_xMnO_3$ with x between 0.48 and 0.67. This investigation by Kim et al. [19.18] showed the presence of an optical

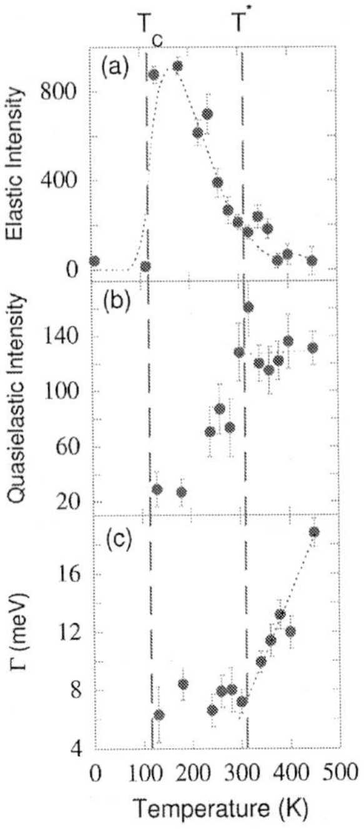

Fig. 19.10. Temperature dependence of the elastic (**a**) and quasielastic intensity (**b**) at the momentum corresponding to the CE clusters, from [19.20]. Results were obtained on a bilayer manganite at $x = 0.38$. From (**a**), it is clear that in the range between T_C and T^* the intensity is large. There the CE clusters are frozen since the elastic peak is considered. In part (**b**), the fact that the quasielastic signal (finite energy in the inelastic neutron data) grows beyond T^* indicates that the clusters become dynamic. In (**c**), the width of the quasielastic contribution is shown

pseudogap above the charge-ordering temperature T_{CO}. Anomalies in the resistivity were also detected in this range, at a temperature T^* between 500 and 600 K. The experimental analysis is carried out quite similarly to that in the cuprates, materials in which a pseudogap temperature was also detected many years ago. The results of Kim et al. are shown in Fig. 19.9 (right). Other studies of Mn oxides at very high temperatures have also been reported [19.19].

T^* as a Glass Transition. Very recent neutron scattering measurements by Argyriou et al. [19.20] on a bilayer manganite at $x = 0.38$ have been explained as being caused by short-range CE-like polaron correlations that *freeze* upon cooling to a temperature $T^* \approx 310$ K, causing a transition from a polaron liquid to a polaron glass. Figure 19.10 shows the experimental data upon which these conclusions are based. The results suggest that the insulating state above $T_C = 114$ K is caused by orbital frustration and/or disorder, since both inhibit the formation of long-range orbital and charge ordering. The glass transition identified by Argyriou et al. seems to be related to the T^* predicted by theoretical studies (see Chap. 17), after introducing quenched disorder in a system with two competing order parameters. Indications of T^* in bilayers, obtained using spectroscopic ellipsometry, have also been recently reported by Kunze et al. [19.21].

Summarizing, in this chapter a robust experimental case has been made for (i) the existence of charge-ordered CE-type clusters above T_C, (ii) their correlation with the resistivity and CMR effect, and (iii) the existence of a new temperature scale T^*. All these results are in agreement with predictions from the theory of phase separation for manganites (Chap. 17), and appear to rule out other theories based exclusively on weakly interacting polarons. However, more work should be devoted to these issues to totally confirm the picture discussed here.

20. Other Compounds with Large MR and/or Competing FM AF Phases

In this chapter, several materials with a behavior similar to that of the manganites are described. The analogies with Mn oxides are based on the presence of a large magnetoresistance, a competition between FM and AF states, or both. The quite remarkable particular case of Eu-based semiconductors, with a CMR effect, is not only addressed by the author below, but also in an invited contribution by S.L. Cooper and collaborators. In addition, in the same contribution a review of inelastic light scattering techniques is provided. This chapter is conceptually very important: it shows that the complex and interesting behavior of electrons in Mn oxides described in this book can be extended to a variety of other materials as well. This reveals a *universal* behavior, which was unexpected until very recently: inhomogeneities dominate the physics of a large number of materials. The fine details appear irrelevant, and the key issue emerging is phase competition.

20.1 Eu-based Compounds

Let us start with the well-known Eu-based materials with a large MR effect. For a long time, these materials were believed to be simply described by the physics of "spin polarons" above the ordering temperature. However, in view of the results for manganites, perhaps the word "polaron" should be replaced by "cluster" in this context as well. This is not a mere language change, but it sheds light on the presence of a variety of length scales and carrier numbers in the spin-polarized regions. Polarons have a fixed size and one carrier, at least in the usual sense of the word.

We can begin with **EuB_6**, a ferromagnet with Curie temperature $\sim 12\,K$ involving the ordering of localized Eu^{2+} spins. NMR investigations at low temperatures revealed the coexistence of two magnetically similar, but electronically different, phases [20.1]. Recent studies by Süllow et al. [20.2] unveiled a large MR effect, as shown in Fig. 20.1a. Clearly, the resistivity curves have similarities with results for manganese oxides, although EuB_6 is not expected to present large Jahn Teller effects: it appears that a large electron-phonon coupling is not a fundamental ingredient of large MR materials.

Note that the resistivity of a material such as **Eu-rich EuO** (see [20.3]) also has a behavior very similar to that of manganites, namely, insulating behavior at high temperatures, and an abrupt crossover to metallic behavior at low temperatures with a conductivity change of 13 orders of magnitude. Once again, these results are compatible with our expectations (Chap. 17) that, in order to generate CMR, the actual fine properties of the competing FM-metal and insulator phases involved is not as important as the competition itself. Süllow et al. already realized that it is *not* the suppression of spin scattering but the increase in size of the "polaronic" regions that matters the more for the MR effect of this material, similarly as in the manganites. A related picture, as well as giant MR, was also reported in $\mathbf{Eu_{1-x}Gd_xSe}$ [20.4]. **EuSe** by itself already has a substantial MR, as shown in Fig. 20.1b, also clearly similar to results for Mn oxides.

When boron, B, is partially substituted by carbon, C, forming $\mathbf{EuB_{6-x}C_x}$, the material undergoes a transformation from a $x = 0$ ferromagnet, to an antiferromagnet at $x \sim 0.2$ [20.6], with an intermediate composition where the material is inhomogeneous (see question mark in Fig. 20.1c). This regime could have interesting transport properties, and it resembles similar results found for bilayer ruthenates, described below. Recently, $\mathbf{Eu_{1-x}La_xB_6}$ was studied by Snow et al. [20.8] with inelastic light scattering techniques. This compound presents a phase diagram similar to that of $EuB_{6-x}C_x$, with competing FM and AF regions. The existence of FM clusters in the paramagnetic regime above the Curie temperature was also observed by Snow et al. [20.8], a result compatible with the prediction of the existence of a T^* for cluster formation (Chap. 17). More details can be found in the contribution by S.L. Cooper in this chapter. Finally, note that large MR effects also appear in $\mathbf{Eu_{1-x}Ca_xB_6}$ [20.9], of the family of hexaborides such as EuB_6, which have attracted considerable attention recently.

Overall, the similarities between Eu-based compounds and manganites is quite strong. While for many years the results for those Eu-based materials have been explained within the picture of ferromagnetic polarons [20.10], it seems that a more complex state may be at work. Instead of ferro-polarons, namely one carrier with a cloud of polarized spins, a distribution of ferromagnetic cluster sizes, with a concomitant distribution of carrier numbers inside those clusters, may be more representative. The word polaron loses meaning in such an intricate landscape, and mixed-phase or phase-separation are more appropriate terms considering that each cluster may contain several carriers. More experimental and theoretical work should be devoted to these interesting materials, to clarify their behavior with doping or the application of a magnetic field. In fact, very recent work by Rho et al. [20.11] on $Eu_{1-x}Gd_xO$ clearly shows that these compounds behave quite similarly to the manganites, including the existence of a T^* temperature for ferromagnetic cluster formation, and the polarization of those preformed FM clusters upon the application of small magnetic external fields.

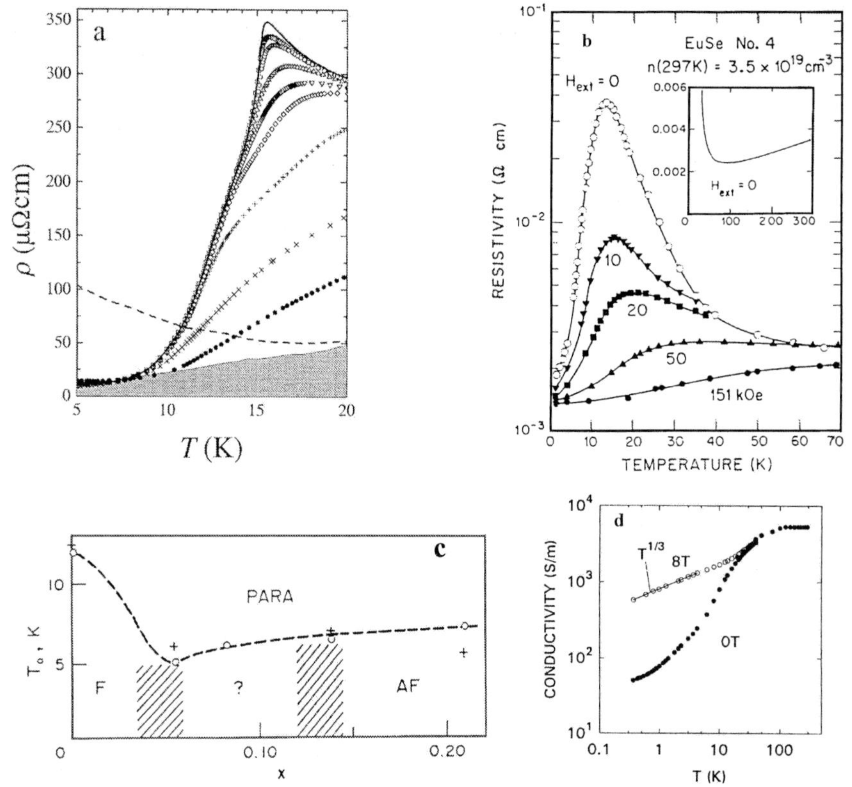

Fig. 20.1. (**a**) Temperature dependence of the resistivity of EuB_6 in zero field (*solid line*) and in fields B of value 0.021 T, 0.029 T, 0.05 T, 0.076 T, 0.1 T, 0.2 T, 0.5 T, 1 T (from top down) and 5 T (*dashed line*). The *gray region* is an upper limit for phononic contributions. Details can be found in [20.2]. (**b**) Resistivity vs. temperature of EuSe (from Shapira et al. [20.5]), for the values of magnetic fields shown. The inset contains the zero-field resistivity vs. temperature, at high temperatures. (**c**) Magnetic-ordering temperatures of $EuB_{6-x}C_x$, from [20.6]. (**d**) Conductivity vs. temperature, with and without fields (indicated), for the diluted magnetic semiconductors described by Ohno et al. [20.7]

20.2 Diluted Magnetic Semiconductors

Diluted magnetic semiconductors (DMS), such as **p-type (In,Mn)As**, are materials with localized spins and mobile carriers in interaction. They also have properties explained by the existence of clusters of size as large as 100 Å, each carrying a substantial number of carriers [20.7]. As remarked before, the term polaron loses its meaning in this context, since a polaron is generally associated with just one carrier. These compounds also exhibit a large MR effect (see Fig. 20.1d). Important MR effects have also been observed in amorphous Si doped with magnetic rare-earth ions, near the metal insula-

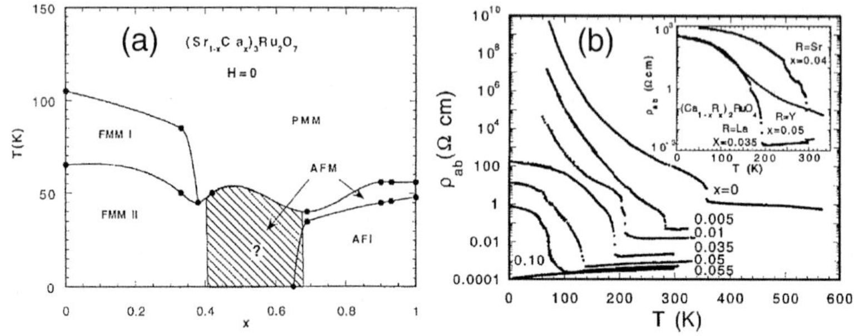

Fig. 20.2. (**a**) Magnetic phase diagram showing multiple phases in $(Sr_{1-x}Ca_x)_3Ru_2O_7$, a bilayer ruthenate compound where FM and AF states have been stabilized (see Cao et al. [20.14]). (**b**) Resistivity of the basal plane of the single-layer ruthenate $(Ca_{1-x}La_x)_2RuO_4$, from Cao et al. [20.15], varying the La concentration. The *inset* shows results using Sr and Y, in addition to La

tor transition [20.26]. This is a new class of dilute magnetic semiconductors. DMS compounds have potential importance in the recently developed area of spintronics, and the reader can easily find more information about these materials in specialized journals. Due to lack of space, elaborating more on these promising compounds is not possible here. Some extra information, including the presence of a T^* in this context, is given in Chap. 22.

20.3 Ruthenates

Ruthenates are materials that are also receiving considerable attention currently [20.12]. The Sr-based single-layer ruthenate presents a superconducting phase at low temperatures, and other interesting properties that have captured the attention of condensed-matter experts. But it is only recently that robust analogies between manganites and ruthenates are starting to be unveiled. For example, when Sr is replaced by Ca in the chemical composition, an antiferromagnetic Mott insulator is stabilized that competes with the metal [20.13]. Bilayer ruthenates also have an interesting phase diagram, with a competition between ferromagnetic and antiferromagnetic states [20.14]. Although the ruthenate literature is too vast to be properly reviewed here, for our purposes it is sufficient to remark that the bilayer $\mathbf{(Sr_{1-x}Ca_x)_3Ru_2O_7}$ has an intermediate region at $x = 0.5$ where the properties are not well-established and appear to arise from a mixture FM AF, somewhat similar to the competition in some manganites (see Fig. 20.2a). In addition, in the single-layer material $\mathbf{(Ca_{1-x}La_x)_2RuO_4}$ a dramatic decrease in the resistivitiy is observed upon La-doping the Mott insulator at $x = 0$, eventually reaching a metallic state (see Fig. 20.2b). The resistivity vs. temperature curves are similar in shape to those found in manganites.

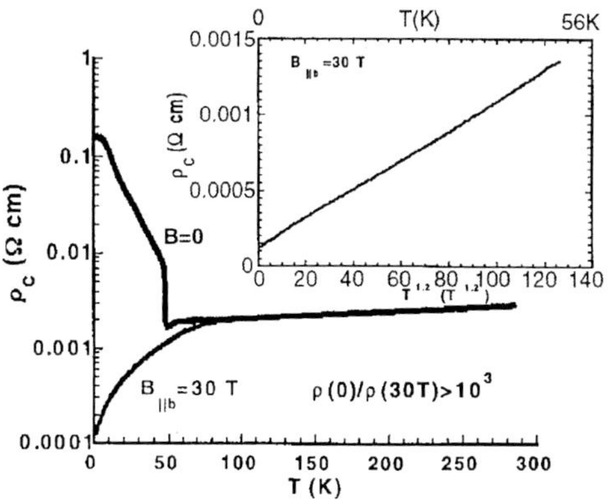

Fig. 20.3. Resistivity of $Ca_3Ru_2O_7$ in the direction perpendicular to the bilayers, from Cao et al. [20.18]. The resistivity at low temperature changes by three orders of magnitude upon the introduction of a field of 30 T along the in-plane axis *b*. The inset shows a nearly linear behavior of the resistivity vs. temperature along the *c*-axis, unexpected in a fully spin polarized metal

It is clear that both types of materials – manganites and ruthenates – present metallic and insulating states that can compete in some regions of parameter space, as shown before. In addition, ruthenates may present orbital ordering as well (Hotta and Dagotto [20.16], is a reference where related literature can be found). Moreover, Nakatsuji et al. [20.17] reported a *cluster glass phase* induced by the competition between a nearly FM state and an orbitally ordered nearly AF state, in the single-layer Ca-Sr compound. In addition, *tunneling colossal magnetoresistance*, due in part to the magnetic valve-like effect, has been recently discovered, with a drop in the resistivity of the bilayer $\mathbf{Ca_3Ru_2O_7}$ by three orders of magnitude, according to recent investigations by Cao et al. [20.18] (see Fig. 20.3). Even in nearly fully spin polarized states the resistivity between bilayers does not follow a Fermi-liquid behavior, suggesting the presence of other relevant degrees of freedom [20.18], such as orbitals [20.16]. It is clear that not only manganites but ruthenates as well, and several other compounds, share a similar phenomenology: competition of states, relevance of many degrees of freedom combined (spin, charge, orbital, lattice), large magnetoresistance effects, glassy characteristics, and deviations from standard metallic behavior.

20.4 More Materials with Large MR or AF FM Competition

Several other materials present the phenomenon of magnetoresistance, although not with such huge values as in the manganites. Nevertheless, some of them have substantial potential technological application since their Curie temperatures are high. Among them, a promising compound is the double perovskite $\mathbf{Sr_2FeMoO_6}$ studied by Kobayashi et al. [3.25], which has $T_C = 415\,K$. Resistivity vs. temperature curves at several external magnetic fields are shown in Fig. 20.4a. Although the MR ratios as usually defined are not spectacular, the MR is not negligible at room temperature. The analysis

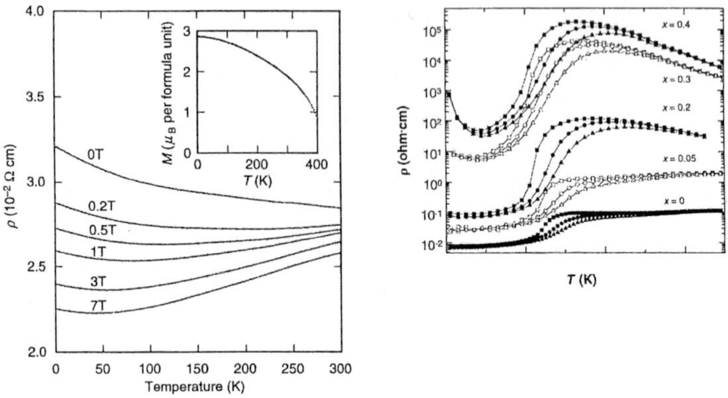

Fig. 20.4. (**a**) Temperature dependence of the resistivity vs. temperature of polycrystaline samples of Sr_2FeMoO_6, at the magnetic fields indicated. From Kobayashi et al. [3.25]. (**b**) Resistivity vs. temperature of $Tl_{2-x}Sc_xMn_2O_7$, reproduced from Ramirez and Subramanian [20.21]. The *upper*, *middle*, and *lower curves* for each x correspond to applied fields $H = 0$, 3, and 6 T, respectively

of Kobayashi et al. [3.25] suggest that the effect is caused by grain-boundary effects, rather than being intrinsic to the material. Low-field MR in FeMo double perovskites was also studied by García-Hernández et al. [20.19], finding a correlation of the effect with cationic disorder. This observation has similarities with ideas in the context of manganites described in Chap. 17. More recently, Mira et al. [20.20] found that MnAs has a CMR at 313 K. Analogics with manganites were also discussed by these authors.

The pyrochlore compound $\mathbf{Tl_{2-x}Sc_xMn_2O_7}$ [20.21] also has a substantial MR effect. Its resistivity curves are shown in Fig. 20.4b, with and without fields. The overall shape of the curves suggest analogies with the manganites. Room-temperature magnetoresistance and cluster glass behavior was found in $\mathbf{Tl_{2-x}Bi_xMn_2O_7}$ by Alonso et al. [20.22]. Large MR effects have also been found in a Co-based compound $\mathbf{La_{1-x}Sr_xCoO_3}$ by Briceno et al. [20.23], replacing Mn by cobalt, Co. The existence of inhomogeneities in these compounds was reported by Caciuffo et al. [20.24]. Clearly, Co oxides and Mn oxides have several common features, including a large MR effect. This by itself also suggests that neighboring oxides in the periodic table, such as Ni oxides, will likely have inhomogeneities as well. A metal insulator transition in $\mathbf{La_{1-x}Eu_xNiO_3}$ is characterized by a resistivity vs. temperature curve with similarities to those found in manganites [20.25]. Negative MR effects have been observed in several other materials not listed here, including Ge doped with impurities. $\mathbf{LaSb_2}$ also has a large MR [20.38]. For recent references, the reader can consult [20.37].

A compound that presents a competition between antiferromagnetic and ferromagnetic states is $\mathbf{SrFe_{1-y}Co_yO_3}$, with a transition among them at

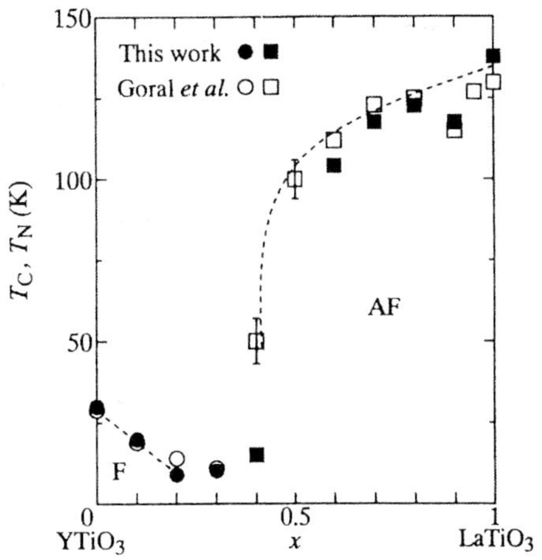

Fig. 20.5. Magnetic phase diagram of $La_xY_{1-x}TiO_3$, from [20.30]. The notation is standard. Note the presence of an intermediate doping regime where both Néel and Curie temperatures have been depleted, as in the discussion of Chap. 17. "This work" in the caption refers to [20.30], where information about the other reference, Goral et al., can be found

$y \sim 0.15$. The resistivity vs. temperature results obtained by Kawasaki et al. [20.27] are similar to those of manganites. Moreover, fixing y to zero and studying $\mathbf{Ca_{1-x}Sr_xFeO_3}$ instead, the resistivity curves obtained by Takeda et al. [20.28] have strong similarities with those of Mn oxides, and it is natural to conjecture that the transition from the $x = 0$ insulator to the $x = 1$ metal must be percolative through a mixed-phase state. Another compound interpolating between a FM and an AF state is $\mathbf{Se_{1-x}Te_xCuO_3}$ [20.29]. In the intermediate-x samples spin-glass behavior was found. **Ti oxides** appear to have a competition between FM and AF as well, as shown in Fig. 20.5, from Okimoto et al. [20.30]. Both critical temperatures are clearly suppressed as the transition from AF to FM is reached near $x = 0.30$. This phase diagram is quite similar to the theoretical predictions of Chap. 17, where "quantum-critical" behavior was induced by the influence of disorder on first-order transitions. Percolative properties of the related compound $\mathbf{La_{1-x}Sr_xTiO_3}$ were also discussed by Hays et al. [20.31].

Other compounds with properties related to those described in this book are $\mathbf{NiS_{2-x}Se_x}$ [20.32], with a resistivity that is insulating at high temperatures, turning metallic at low temperatures. Also, using NMR techniques, Gavilano et al. [20.36] found a two-component spectrum in the heavy-electron material $\mathbf{CeAl_3}$ indicative of a spatially inhomogeneous system. The list of compounds with inhomogeneities and large MR is truly huge.

Another material with a competition between FM and AF phases is the pseudobinary $\mathbf{Ce(Fe_{0.96}Al_{0.04})_2}$, recently reported by Singh et al. [20.33] and Manekar et al. [20.34]. This material has hysteresis effects and phase coexistence. An AF–FM competition is also present in the Kondo system $\mathbf{CeNi_{1-x}Cu_x}$ [20.35]. Recent experiments have identified a spin-glass phase at temperatures higher than the Curie temperature. Indications of phase separation have been found also in the spin-chain compound $\mathbf{Sr_3CuIrO_6}$ [13.37], rationalized as being caused by defects that render the system inhomogeneous, as in the discussion of Chap. 17.

It is also interesting to remark that the coexistence of two phases has also been observed at the metal insulator transition in a *two-dimensional hole gas* using GaAs/AlGaAs samples [20.39]. For a long time it was assumed that two-dimensional systems were always insulating due to Anderson localization (see Chap. 16), but recent experiments have shown this assumption to be false [16.6, 20.40].

The list of compounds with large MR and/or FM AF competition is very extensive, and only a small fraction has been described here. It is natural to expect that all these compounds share a common phenomenology. If this universality assumption is correct, then the conclusions of this book regarding inhomogeneities in manganites are far more general than anticipated, and it becomes a new paradigm for theoretical studies in condensed-matter physics.

20.5 Illuminating Magnetic Cluster Formation with Inelastic Light Scattering

by S.L. Cooper, H. Rho, and C.S. Snow

20.5.1 Introduction

Strongly correlated systems are characterized not only by diverse phase diagrams, but also by the dramatic phenomena they exhibit, including "colossal magnetoresistance" [2.15], electronic phase separation [1.10, 11.78, 20.41], magnetic polaron formation [20.8, 20.42], and charge ordering [3.8]. These extraordinary properties result from a strong competition among various important interactions that occur in strongly coupled systems, i.e., electron electron, spin-exchange, short-range Coulomb, and electron lattice, whose intricate interplay as a function of composition, temperature, pressure, magnetic field, etc., spawns the exotic properties and myriad phases observed in these systems.

While contributing greatly to the rich physical properties of strongly coupled systems, the importance of spin, charge, lattice, and orbital degrees of freedom in these materials presents a significant experimental challenge. In particular, conventional spectroscopic probes of the dynamical properties in strongly coupled systems generally offer only a small "window" on the vast panorama of physics important in these systems: for example, inelastic neutron scattering provides a means of studying spin and lattice dynamics in strongly coupled systems, but offers no information regarding the important charge and orbital dynamics; and infrared reflectivity typically conveys important details regarding only the charge and lattice dynamics.

An important technique that has emerged over the past few years as a powerful method for probing most of the relevant degrees of freedom in strongly coupled systems is inelastic light (Raman) scattering. The great versatility of this technique can be seen by first noting that inelastic light scattering results from fluctuations in the macroscopic polarization $\boldsymbol{P}(\boldsymbol{r}, t)$ induced in the material by an incident electric field $\boldsymbol{E}_{\mathrm{I}}(\boldsymbol{r}, t)$ [20.43],

$$\boldsymbol{P}(\boldsymbol{r}, t) = \chi \boldsymbol{E}_{\mathrm{I}}(\boldsymbol{r}, t) , \tag{20.1}$$

where χ is a second-rank tensor describing the linear electric susceptibility of the material. Elastic contributions to the light-scattering response arise from the induced polarization associated with the static susceptibility, χ_o. On the other hand, *inelastic contributions* to the light scattering response arise due to excited states of the material that *dynamically modulate* its polarization, specifically, by modulating the wave functions and energy levels in the material. Consequently, since most excitations influence the polarization – for example via the electron lattice interaction in the case of lattice vibrations (phonons), and via the spin-orbit interaction in the case of spin-excitations – the energy, symmetry, and lifetime of many of the important excitations involved in "complex" phase transitions of strongly coupled systems can be simultaneously investigated using inelastic light scattering [20.8, 20.42, 20.44, 20.45]. In addition to providing a unique means by which many of the important excitations of strongly coupled systems can be explored through "complex" phase transitions, inelastic light scattering affords a number of key benefits in comparison to spectroscopies like infrared reflectivity and neutron scattering: (1) By obtaining the spectral response in different scattering geometries, light scattering provides important symmetry information about excitations studied, and in particular affords a means by which the anisotropic carrier dynamics of a system can be investigated. (2) Inelastic light scattering is particularly sensitive to the charge response in low carrier density systems, where charge-density fluctuations are not substantially screened by the Coulomb interaction – this is of particular utility in systems having mixed electronic phases [20.8, 20.46]. (3) Light scattering can be used to study very small samples and microstructures (< 80 micrometers). (4) As light-scattering studies typically employ visible lasers as excitation sources, these techniques lend themselves to spectroscopic studies of complex phase transitions under "extreme" conditions: e.g., in high-magnetic fields of small-bore magnets that are often difficult to access using other spectroscopic techniques; under the high pressures provided by diamond anvil cells and other pressure-transmission materials that are transparent to visible light, but not to much of the infrared spectrum; and with fast time resolution, which is readily obtainable using pico- and femto-second pulsed lasers and pump-probe techniques.

The inelastic light scattering response includes two contributions, (a) the Stokes component, in which the incident light field $\boldsymbol{E}_{\mathrm{I}}(\boldsymbol{r}, t)$ creates an excitation of energy ω and wavevector $\boldsymbol{q}$, causing a reduction in the energy of the scattered field $\boldsymbol{E}_{\mathrm{S}}(\boldsymbol{r}, t)$, $\omega_{\mathrm{S}} = \omega_{\mathrm{I}} - \omega$; and (b) the anti-Stokes component, in which the scattered field gains energy from thermally excited excitations in the sample, $\omega_S = \omega_I + \omega$. In the semiclassical description of the Stokes inelastic light scattering process, excited states in the material are represented by a dynamical variable, $\boldsymbol{\xi}(\boldsymbol{r}, t)$, which represents space- and time-dependent deviations of, for example, the atomic position, charge density, spin density, etc., from their equilibrium values. The resulting induced polarization in the presence of excited states in the material is obtained by expanding the dynamical susceptibility χ in powers of $\boldsymbol{\xi}$ around the static susceptibility, χ_o. This gives rise to a second-order contribution to the induced polarization,

$$P_S^j(\boldsymbol{q}_S, \omega_S) = \sum_j \sum_k \frac{\partial \chi^{ij}}{\partial \xi^k_{q,\omega}} \xi^k(\boldsymbol{q}, \omega) E_I^j(\boldsymbol{q}_I, \omega_I) , \tag{20.2}$$

where $\boldsymbol{P}_S(\boldsymbol{q}_S, \omega_S)$, $\boldsymbol{E}_I(\boldsymbol{q}_I, \omega_I)$, and $\boldsymbol{\xi}(\boldsymbol{q}, \omega)$ are the space and time Fourier transforms of the polarization, incident electric field, and dynamical variable, respectively, $\boldsymbol{q}_S = \boldsymbol{q}_I - \boldsymbol{q}$ is the wavevector of the scattered photon inside the scattering medium, and $\boldsymbol{q}$ and $\boldsymbol{q}_I$ are the wavevector of the excitation and incident photon inside the scattering medium. For example, in the case of magnetic excitations, which will be of particular interest to us in our discussion of CMR-type systems below, the second-order contribution to the induced polarization is given by [20.43],

$$\boldsymbol{P}_S^{(2)} = iG\boldsymbol{M} \times \boldsymbol{E}_I , \tag{20.3}$$

where $\boldsymbol{M}$ is the local magnetization of the material, and $iG = \partial \chi^{ij}(\boldsymbol{M})/\partial M^k$ is related to the Faraday coefficient, and depends in general upon the strength of the spin-orbit coupling in the material.

From (20.2), one can write the photon differential cross section per unit scattered frequency range, $\mathrm{d}^2\sigma/\mathrm{d}\Omega\mathrm{d}\omega_S$, which reflects the fraction of incident photons inelastically scattered in the energy interval $\mathrm{d}\omega_S$ per unit solid angle $\mathrm{d}\Omega$ [20.43]

$$\frac{\mathrm{d}^2\sigma}{\mathrm{d}\Omega \mathrm{d}\omega_S} = \frac{\omega_I \omega_S^3 V^2 n_S \left\langle \epsilon_S \cdot \boldsymbol{P}_S^*(\boldsymbol{q}_S\, \epsilon_S \cdot \boldsymbol{P}_S(\boldsymbol{q}_S))\right\rangle_{\omega_S}}{c^4 n_I |\boldsymbol{E}_I|^2} , \tag{20.4}$$

where n_I and n_S are the indices of refraction in the incident and scattered media, respectively, V is the scattering volume, $|\boldsymbol{E}_I|$ is the magnitude of the incident electric field, and $\boldsymbol{\epsilon}_S$ is the unit polarization vector for the scattered electric field. By combining (20.2) and (20.4), one obtains the following form for the differential light scattering cross section:

$$\frac{\mathrm{d}^2\sigma}{\mathrm{d}\Omega \mathrm{d}\omega_S} = \frac{\omega_I \omega_S^3 V^2 n_S}{c^4 n_I |\boldsymbol{E}_I|^2} \left| \epsilon_S^i \chi'^{ij} \epsilon_I^j \right|^2 \langle \xi(\boldsymbol{q}) \xi^*(\boldsymbol{q}) \rangle_\omega , \tag{20.5}$$

where $\chi'^{ij} = \partial \chi^{ij}/\partial \xi$ is the susceptibility derivative. There are several essential features to note in (20.5). First, (20.5) indicates that the differential cross section for photons scattering from a medium with wavevector transfer $\boldsymbol{q}$ and energy transfer $\hbar\omega$ directly measures a single Fourier component of the correlation function of the dynamical variable $\xi(\boldsymbol{q})$; this dynamical variable can be associated with, for example, the charge density, spin density, lattice displacement, or other excitation "displacement" that modulates the induced polarization. Second, implicit in (20.5) is a selection rule on the light scattering cross section that is imposed by the susceptibility tensor χ'^{ij}. Specifically, the selection rule on χ'^{ij} is imposed by the requirement that the relationship between the Stokes polarization, excitation amplitude, and incident electric field in (20.2) be covariant under all the spatial transformations of the scattering medium. Since $\boldsymbol{P}_{\mathrm{S}}$ and $\boldsymbol{E}_{\mathrm{I}}$ are both polar vectors, this is tantamount to stating that an excitation is Raman-active only if it has a symmetry contained in the decomposition of $\Gamma^*_{\mathrm{PV}} \otimes \Gamma_{\mathrm{PV}}$, where Γ_{PV} is the irreducible representation of the material's point group that transforms like a polar vector. The $|\epsilon^i_{\mathrm{S}} \chi'^{ij} \epsilon^j_{\mathrm{I}}|$ term in (20.5) also illustrates an extremely powerful feature of inelastic light scattering: one can experimentally couple to different components of the susceptibility tensor, and thereby identify the symmetry of a particular excitation, by obtaining the light-scattering cross section for different incident and scattered light polarizations, $\boldsymbol{\epsilon}_{\mathrm{I}}$ and $\boldsymbol{\epsilon}_{\mathrm{S}}$.

By way of comparison, in the quantum-mechanical derivation of the inelastic light-scattering cross section, one considers the two-photon process in which a photon changes its frequency ω, wavevector $\boldsymbol{q}$, and polarization $\boldsymbol{\epsilon}$ via scattering from the values $(\omega_{\mathrm{I}}, \boldsymbol{q}_{\mathrm{I}}, \boldsymbol{\epsilon}_{\mathrm{I}})$ to the values $(\omega_{\mathrm{S}}, \boldsymbol{q}_{\mathrm{S}}, \boldsymbol{\epsilon}_{\mathrm{S}})$, while the scattering material experiences a transition from initial state $|i>$ to final state $|f>$; the resulting light-scattering cross section is obtained by considering the $\boldsymbol{p} \cdot \boldsymbol{A}$ term in the interaction Hamiltonian to second order in perturbation theory, and the $\boldsymbol{A}^2$ term in the interaction Hamiltonian to first order in perturbation theory, where $\boldsymbol{p}$ is the electron momentum and $\boldsymbol{A}$ is the vector potential associated with the incident radiation field. The resulting scattering cross section has a transition susceptibility tensor χ that connects initial and final states, $|i>$ and $|f>$, of the scattering medium, and is the quantum-mechanical analog of the classical susceptibility in (20.2). Inelastic light scattering from phonon or spin excitations is typically treated by considering, respectively, the electron phonon and spin-orbit interaction as perturbations, and the associated light scattering cross sections are calculated using third- or higher-order perturbation theory.

A final note is in order concerning the time and spatial scales associated with the inelastic light scattering process. First, since the frequency of incident and scattered light is, under normal circumstances, substantially larger than the excitation frequency, i.e., ω_{I}, $\omega_{\mathrm{S}} \gg \omega$, the time dependence of the excitation is small compared to that of the probe, and the excitation can generally be thought to modify the susceptibility χ via a *static* "displacement." Second, concerning the spatial scale measured using inelastic light scattering, it is important to first recognize the limitations on inelastic light scattering due to the kinematical constraint in the Stokes scattering process,

$$\boldsymbol{q} = \boldsymbol{q}_{\mathrm{I}} - \boldsymbol{q}_{\mathrm{S}} \,. \qquad (20.6)$$

In typical experimental situations, one has $\omega \ll \omega_I$ and $|\boldsymbol{q}_I| \approx |\boldsymbol{q}_S|$, indicating that the wavevector transfer in (20.6) can to a good approximation be written as $|\boldsymbol{q}| = 2|\boldsymbol{q}_I|\sin(\theta/2)$, where θ is the angle between the incident and scattered wavevector directions. As inelastic light scattering experiments are typically performed with incident frequencies in the visible frequency range, $\omega_I \approx 10^{14}$ Hz, light-scattering measurements can generally couple to excitations with wavevectors no larger than $|\boldsymbol{q}| \sim 10^{-3}\,\text{Å}^{-1}$, which is several orders of magnitude smaller than the Brillouin zone boundary, $|\boldsymbol{q}_{BZ}| \approx 1\,\text{Å}^{-1}$. Consequently, one strong limitation of inelastic light scattering compared with neutron scattering, is that light scattering can generally probe excitations only very near the Brillouin zone center, $|\boldsymbol{q}| \approx 0$, and therefore one cannot in general obtain enough q-dependence for the light-scattering response function to deduce useful information regarding an excitation's spatial extent.

20.5.2 Application to Eu-based Semiconductors

To illustrate the efficacy of inelastic light scattering as a probe of strongly correlated systems, we focus on recent inelastic light scattering studies of spin carrier interactions and magnetic cluster formation in Eu-based ferromagnetic semiconductors such as EuO and EuB_6. These Eu-based compounds are of interest because they share many of the properties of "colossal-magnetoresistance"-type systems, such as the Mn- and Co-based perovskites ("manganites" and "cobaltites"), the telluride systems $Ge_{1-x}Mn_xTe$ and $Eu_{1-x}Gd_xTe$, and the Mn pyrochlores (see Fig. 20.6): [20.47] they exhibit paramagnetic semiconductor-to-ferromagnetic metal transitions near T_C ($\sim$ 69 K in EuO [20.41, 20.48]; $\sim$ 12 K in EuB_6 [20.49]), which are accompanied by large negative magnetoresistances and magnetic cluster ("polaron") formation. On the other hand, these Eu-based systems have few of the structural complexities of other CMR-type systems, such as strong electron lattice coupling effects due to Jahn–Teller and breathing-mode distortions, enabling one to study in particularly simple systems a range of phenomena pertinent to a wide class of CMR-type systems: metallic ferromagnetism, inhomogeneous electronic and magnetic phases [1.10, 11.78, 20.41]; the effects of temperature and magnetic field on magnetic cluster formation, and the competition between ferromagnetic and antiferromagnetic correlations.

As shown in Fig. 20.7 for the case of EuO, the inelastic light scattering response in the high-temperature paramagnetic phase ($T > T_C$) of EuO and EuB_6 is primarily characterized by an increase in scattering at low frequencies, which is associated with electronic light scattering from the charge carriers. To understand the origin of this scattering response, consider first that the inelastic light scattering cross section from conduction electrons is described by [20.8],

$$\frac{\mathrm{d}^2\sigma}{\mathrm{d}\Omega\mathrm{d}\omega_S} = -\frac{r_o^2}{\pi}\frac{\omega_S}{\omega_I}\frac{n_S}{n_I}\,[n(\omega)+1]\,\mathrm{Im}\chi_{\rho\rho}(\boldsymbol{q},\omega)\,, \tag{20.7}$$

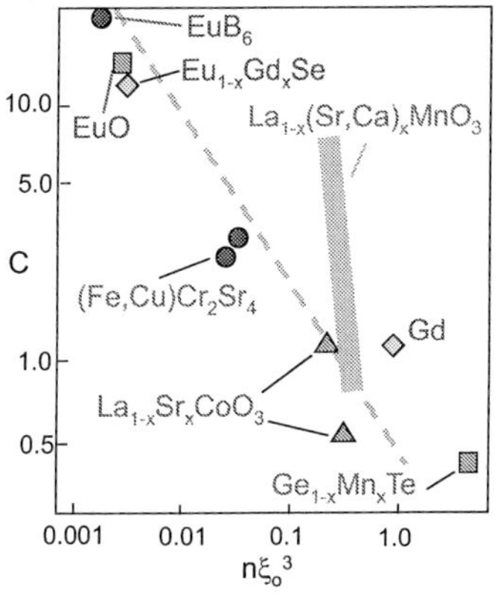

Fig. 20.6. Magnetoresistance (C) vs. the carrier density scaled to the magnetic lattice spacing (n ξ_o^3) for a number of CMR-type systems [20.47], including estimated values for the materials under discussion here, EuB_6 and EuO. *Dashed line* represents $n^{2/3}$ trend predicted for Born scattering by spin fluctuations. Adapted from Majumdar and Littlewood in [20.47]

where $n(\omega) = [\exp(\hbar\omega/k_{\mathrm{B}}T) - 1]^{-1}$ is the Bose thermal factor, $r_o = e^2/mc^2$ is the Thomson radius of the electron, and $\mathrm{Im}\chi_{\rho\rho}(\boldsymbol{q}, \omega)$ is the imaginary part of the Raman susceptibility tensor associated with fluctuations in an effective charge density, $\rho_{\boldsymbol{q}}$, which under typical circumstances can be related to the inverse effective mass tensor $\boldsymbol{\mu}^{-1}$. Consequently, it is appropriate to think of the light scattering cross section from conduction electrons as being associated with fluctuations in the inverse effective mass tensor [20.50]. In isotropic systems with high carrier densities, Coulomb screening effects push the spectral weight associated with charge-density fluctuations up to the plasma frequency, and hence very little low-frequency electronic light-scattering intensity is expected for such systems. However, in anisotropic systems, one also expects (mass) density fluctuations that do not involve a net change in the charge density, and consequently these contributions are not screened and can have appreciable light-scattering intensities. One also expects a measurable electronic light-scattering intensity in low carrier density systems, making light scattering particularly suited to probing low carrier density phases of electronically inhomogeneous systems.

Notably, in the presence of strong electronic scattering ($k_{\mathrm{F}}l < 1$), the response function for intraband electronic light scattering is best described by a "collision-dominated" (diffusive) form [20.8, 20.42, 20.51, 20.52],

$$\mathrm{Im}\chi_{\rho\rho}(\omega) = \frac{|\gamma_L|^2 \omega \Gamma_L}{\omega^2 + \Gamma_L^2}, \tag{20.8}$$

where γ_L is the Raman scattering vertex associated with scattering geometry L, which is proportional to the inverse effective mass tensor $\boldsymbol{\mu}^{-1}$, and Γ_L is the carrier scattering rate associated with scattering geometry L. This response function is associated with light scattering from mobile electrons, and hence is the analog of the Drude optical response due to free carriers. Thus, electronic light scattering can provide a means by which the electronic scattering rate Γ can be optically probed under a variety of conditions.

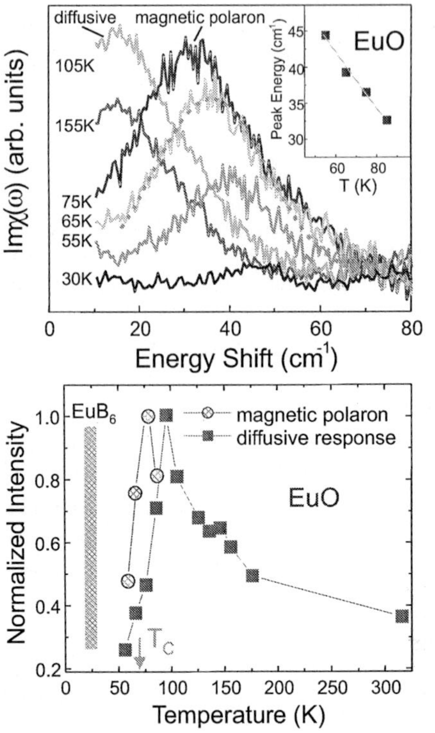

Fig. 20.7. (*Top*) $H = 0$ Raman scattering spectrum of EuO for various temperatures, illustrating both "collision-dominated" low-frequency electronic scattering for ($T > 105\,\mathrm{K}$), and spin-flip (SF) scattering due to the formation of magnetic polarons ($T^* < 75\,\mathrm{K}$). *Dashed line* is a fit to the Gaussian response function in (20.9). (*inset*) Peak frequency of SF peak as a function of temperature. (*Bottom*) Normalized intensity of the "collision-dominated" electronic scattering (*squares*) and SF Raman scattering (*circles*) as a function of temperature, illustrating the "two-phase" coexistence regime near T_C. The *solid vertical bar* illustrates the approximate temperature range over which magnetic polarons are stable in EuB_6 [20.42]

This "collision-dominated" response function describes quite well the low frequency electronic scattering response in the high-temperature paramagnetic phase ($T \gg T_C$) of both EuB_6 and EuO ($T > 105\,\mathrm{K}$ in Fig. 20.7 (top)), indicating that the charge dynamics in this phase regime is characterized by strong electronic scattering. Typically, the scattering rate Γ is dominated by electronic scattering from static impurities [20.45, 20.46]. However, in CMR-type systems such as EuB_6 [20.42] and EuO [20.41, 20.48], the carrier scattering rate is dominated by spin-fluctuation scattering over a substantial temperature range near T_C; indeed, this fact is reflected in the light scattering spectra of EuB_6 and EuO by the fact that the "collision-dominated" electronic scattering intensity increases as the temperature approaches T_C in the paramagnetic phase (see Fig. 20.7 (top)), and the scattering rate in this temperature regime scales according to $\Gamma \propto T\chi$, where χ is the magnetic susceptibility. This result not only confirms that the carrier dynamics is dominated by spin-fluctuation scattering over a substantial temperature range above T_C in these systems, but it demonstrates that electronic light scattering can be used as a means of optically detecting magnetotransport and spin dynamics in magnetic systems. This can be fruitful, for example when contactless methods for measuring magnetotransport would be beneficial. It is also useful for studying secondary, low carrier density phases below the percolation transitions of electronically inhomogeneous systems, as these are phases to which conventional magnetotransport measurements are not sensitive [20.46].

20.5.3 Cluster Formation Temperature T^* and Percolation

The temperature range over which strong "collision-dominated" electronic light scattering is dominated by strong spin-fluctuation scattering corresponds well to the temperature regime in which the resistivity exhibits "activated" behavior in both EuB_6 [20.49] and EuO [20.41,20.48]; this confirms that at least part of this behavior can be associated with critical scattering due to spin fluctuations near T_C. However, that not all of the "activated" resistivity behavior is associated with critical scattering in these systems is betrayed by interesting changes in the Raman scattering response in EuO and EuB_6 as $T \to T_C$: as shown in Fig. 20.7 (top) for EuO, below a temperature $T^*(> T_C)$ [$T^* \sim 75$ K in EuO; $T^* \sim 30$ K in EuB_6], the collision-dominated electronic scattering response gradually diminishes, and is replaced by an inelastic peak that is well described by a Gaussian response function (dashed line) [20.8]

$$\mathrm{Im}\chi_{\rho\rho}(\omega) \propto \sqrt{\frac{2}{\pi}}\frac{1}{W}\exp\left[-2\left(\frac{\omega-\omega_o}{W}\right)^2\right], \tag{20.9}$$

where ω_o is the peak frequency of the inelastic response, and W is the Gaussian linewidth. Notably, this inelastic response develops in the $\boldsymbol{E}_\mathrm{I} \perp \boldsymbol{E}_\mathrm{S}$ geometry, but not in the $\boldsymbol{E}_\mathrm{I} \parallel \boldsymbol{E}_\mathrm{S}$ scattering geometry, and hence has the transformation properties of the totally antisymmetric Raman susceptibility tensor. This distinctive symmetry is a clear signature of light scattering from magnetic excitations (this is related to the fact that magnetic excitations modulate the induced polarization via the $\boldsymbol{M} \times \boldsymbol{E}_\mathrm{I}$ term in (20.3) [20.53]). More specifically, the distinctive Gaussian scattering response and $\boldsymbol{E}_\mathrm{I} \perp \boldsymbol{E}_\mathrm{S}$ symmetry associated with the inelastic peak in Fig. 20.7 (top) has been identified in many studies of both dilute [20.54–20.56] and "dense" [20.8, 20.42] magnetic systems as $H = 0$ spin-flip (SF) Raman scattering associated with the development of magnetic polarons. Consequently, the development of $H = 0$ SF Raman scattering in both the EuO and EuB_6 systems betrays a distinct magnetic "cluster" or "polaron" formation temperature $T^*(> T_C)$ below which local FM clusters nucleate prior to the development of the FM ground state. More importantly, the observation of this response provides us with a unique opportunity to obtain substantial information about the evolution of magnetic clusters through temperature- and field-driven metal insulator transitions in CMR-type systems. In particular, the peak frequency ω_o in the Gaussian profile in Fig. 20.7 (top) represents the spin-splitting of the localized charge by the d–f exchange field from the Eu^{2+} moments in the cluster – consequently, this energy provides a measure of the exchange energy of the clusters as a function of temperature, magnetic field, etc. Moreover, the Gaussian linewidth W reflects a distribution in the number of magnetic moments contributing to the magnetization in different magnetic clusters.

A careful analysis of the results in Fig. 20.7 reveals several important features of magnetic-cluster development in Eu-based CMR-type systems. First, the $H = 0$ spin-flip Raman energies are much larger than typical Zeeman energies, reflecting the large value of the d–f exchange-induced magnetization in these systems. Second, the $H = 0$ SF Raman energy ($\hbar\omega_o$), which is associated with an effective exchange field within the polaron, $H_{\text{ex}} = \hbar\omega_o/g\mu_{\text{B}}$ (where g is the Lande g factor, and μ_{B} is the Bohr magneton), increases systematically with decreasing temperature (inset, Fig. 20.7 (top)), indicative of a systematic increase in the magnetic cluster size with decreasing temperature below T^*. This evolution of magnetic cluster size is perhaps the most direct evidence that the paramagnetic semiconductor to ferromagnetic metal transitions in these systems may in fact be driven by a "percolation" transition of the magnetic polarons (see schematic in Fig. 20.8). Indeed, as a function of decreasing temperature below T_{C}, the $H = 0$ SF response gradually decreases in intensity as the system transitions into the FM metal phase, reflecting the dissolution of the magnetic clusters as carriers trapped in the polarons are ionized in the increasingly spin-polarized environment of the FM phase. Interestingly, this nucleation and evolution of magnetic polarons below T^* provides a natural explanation for the large negative magnetoresistivities observed near T_{C} in these systems, since field-induced growth and alignment of magnetic clusters increases the hopping of carriers between sites. In this situation one expects, for sufficiently high magnetic fields, a field-induced insulator metal transition driven by the percolation of magnetic polarons; indeed, such transitions are clearly observed in these systems, as shown in Fig. 20.8 (top) for EuB_6 [20.8, 20.42].

It is also interesting to note from Fig. 20.7 (bottom) that careful fits to the spectra show that the temperature regime below the "cluster" formation temperature T^* is characterized by the coexistence of $H = 0$ SF (circles) and collision-dominated electronic scattering (squares) responses, providing strong evidence that this temperature range is a two-phase regime, where FM clusters coexist with remnants of the PM phase. This demonstrates clearly that intrinsic electronic inhomogeneity and magnetic-cluster formation are *not* restricted to perovskite-related oxides [1.10, 11.78], but indeed are present even in structurally simple binary systems that exhibit a similar strong competition between carrier kinetic and spin-exchange energies.

A comparison of the temperature and field dependence of magnetic-cluster formation in various Eu-based systems illustrates the important role that disorder plays in stabilizing magnetic-cluster formation in these systems. In particular, the series of compounds EuB_6 ($T_{\text{C}} \sim 12$ K), EuO ($T_{\text{C}} \sim 69$ K), $Eu_{0.994}Gd_{0.006}O$ ($T_{\text{C}} \sim 70$ K), and $Eu_{0.965}Gd_{0.035}O$ ($T_{\text{C}} \sim 115$ K) exhibit a systematic increase in disorder – as inferred from the resistivities of these

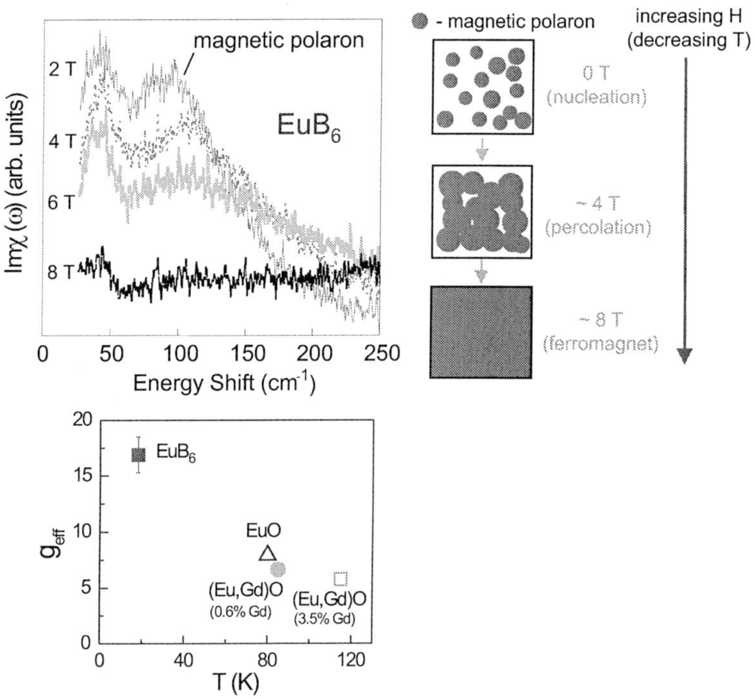

Fig. 20.8. (*Top*) $T = 25\,\mathrm{K}$ Raman scattering spectrum of EuB_6 for various magnetic field strengths, illustrating the increasing SF peak energy ω_o with increasing magnetic field (reflecting growth of magnetic polaron cluster sizes), and the broadening of the SF peak linewidth with increasing magnetic field (reflecting increasing distribution of cluster sizes), as the system transits through a field-induced metal insulator transition. Picture to *right* schematically depicts the nucleation and percolation of magnetic clusters that is believed to occur as a function of increasing field (or decreasing temperature) between the paramagnetic semiconductor and ferromagnetic metal phases. (*Bottom*) Plot of $g_{\mathrm{eff}}(= \hbar\Delta\omega_o/\mu_{\mathrm{B}}\Delta H)$ vs. T_{C} for various Eu-based CMR-type systems, illustrating the systematic decrease in g_{eff} with both increasing T_{C} and increasing disorder

materials – and an associated increase in both the "cluster" formation temperature (T^*) and the temperature range over which magnetic clusters are found to be stable. Furthermore, our results show that the effective g-values ($g_{\mathrm{eff}} = \hbar\Delta\omega_o/\mu_{\mathrm{B}}\Delta H$) of this series of compounds, inferred from the field-dependence of the SF Raman energies in the linear "high-field" regime ($H > 1T$), systematically decrease as a function of increasing T_{C} (see Fig. 20.8 (bottom)). This indicates that the growth rate of the clusters with applied magnetic field diminishes with increasing amounts of disorder in these materials. All of these results suggest a picture in which disorder at once stabilizes the formation of polarons, presumably through the Coulomb interaction, but also restricts the field-induced evolution of the magnetic polarons. These results are in good agreement with the theoretical predictions discussed in Chap. 17.

To summarize, inelastic light scattering studies of Eu-based ferromagnetic semiconductors afford a unique and powerful means of exploring the evolution of spin-carrier interactions and magnetic cluster formation expected in a broad class of doped magnetic systems [1.10, 11.78, 20.41]. These studies reveal that magnetic clusters evolve below T^* ($> T_C$) in various low carrier density ferromagnetic semiconductor systems, and that these clusters coexist with remnants of the PM phase throughout a narrow range of temperatures near T_C. This behavior in turn provides strong evidence that magnetic-cluster formation plays a key role in "colossal magnetoresistance" behavior and the temperature- and field-dependent metal insulator transitions observed in these systems. These studies further demonstrate the important role that disorder plays in stabilizing magnetic clusters in low carrier density magnetic systems, presumably by favoring – along with the magnetic exchange energy – the tendency for carrier localization.

We acknowledge support of this work by the Department of Energy (DEFG02-96ER45439) and the National Science Foundation (DMR97-00716).

21. Brief Introduction to Giant Magnetoresistance (GMR)

Although this book is mainly devoted to the understanding of CMR materials, it is also important to include a chapter on the famous "giant" – as opposed to colossal – MR devices. These devices have already found their way into real technological applications, after a very rapid transition from their discovery, as part of basic research investigations, to commercialization in the form of *read* heads for magnetic hard-disk drives. The information in this chapter will allow the reader to judge what properties of the CMR materials must be improved in order to reach similar success. From the beginning, it will become clear that a key difference between CMR and giant MR (GMR) is that the former deals with intrinsic effects of a material, and concomitant magnetic fields that are typically large compared with those needed for real applications. On the other hand, the GMR devices are made out of individual components that by themselves are not so remarkable, but that in combination produce a large MR effect. This occurs at room temperature – thus no expensive and bulky equipment to refrigerate your computer is needed – and at small fields such as those produced by bits in magnetic disks, as discussed below. On the other hand, CMR occurs mainly at low temperature and upon the application of magnetic fields that are small in natural electronic units, but still large compared with those needed in computers. As a consequence, it is clear that a potential CMR commercialization relies on:

1. increasing the Curie temperatures beyond room temperature,
2. achieving CMR at those temperatures at fields about 100 times smaller than one Tesla, the current scale of interest for computer applications.

It is conceivable that further research on manganites may allow us to reach these goals in the future. The rest of this chapter will focus on GMR materials, where applications have already found their way into real computers. CMR experts have plenty to learn from the GMR success story!

21.1 Giant Magnetoresistance

Recent advances in electronics have been possible due to the exploitation in some devices of the spin-up or -down properties of the carriers, rather than relying on electrons or holes as in semiconductor electronics. This new area

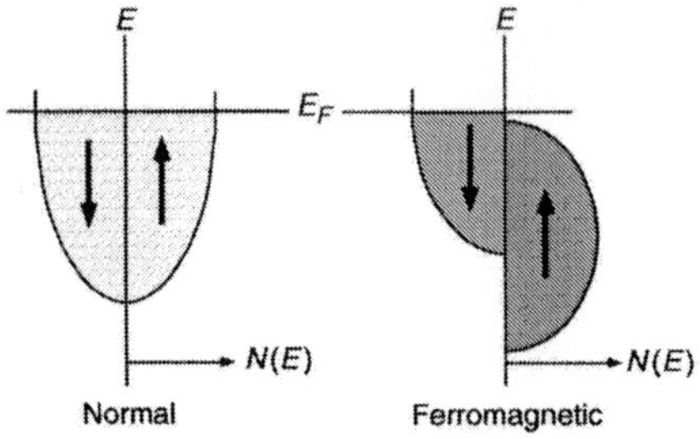

Fig. 21.1. Schematic representation of the density of states available for electrons in a normal non magnetic metal (*left*) and in a ferromagnetic metal (*right*). E is the energy, E_F the Fermi energy, and $N(E)$ the density of states. Reproduced from [21.2]

of investigations and applications receives the name of *magnetoelectronics* or more colloquially *spintronics*. This is defined as the branch of electronics that utilizes the spin degree of freedom of the electron together with its charge, to store, process, and transmit information. Sometimes the subject is also covered under the area of *spin-polarized transport*. In particular, in recent years there has been considerable interest in the study of artificially made nanoscale devices with interesting physical properties that take advantage of the spin of the electrons. Among these devices, much work centered on multilayer systems, consisting of thin magnetic and non magnetic layers, grown one over the other. The magnetic layers are made out of ferromagnetic materials, such as Fe or Co, while the non magnetic ones consist of Cu or Cr. These devices present a large MR effect, much larger than the MR of the individual magnetic layers that form the device. This phenomenon is the giant MR effect, which is common to many metallic multilayers. The distance between ferromagnetic layers (i.e., the thickness of the non magnetic layers) allows for a coupling of consecutive magnetizations, which can be ferromagnetic or antiferromagnetic, allowing for the creation of complicated magnetic structures. Many applications of these GMR devices have been proposed, and several have already found their way into real computers. Here, a brief review of the GMR effect will be provided, with emphasis on its origin in simple cases, and on some of its applications. The literature on the subject is vast and the reader can easily find extra information if needed. Introductory review-like articles, upon which this chapter is based, can be found in [21.1–21.3]. A simple cartoon-like representation of the basic notions related with spin-polarized transport is given in Fig. 21.1. There, the density of states of a non magnetic normal metal, such as copper, contains an equal number of spins up and down available at the Fermi level. The current induced by small electric fields in copper is, thus, not spin polarized. However, if we consider a ferromagnetic metal, the density of states of spins up and down are shifted one from the other by the so-called *exchange splitting*, which lowers the total energy of the system. This occurs also in the case of large and intermediate bandwidth manganites, both in models and real compounds. In the example of Fig. 21.1(right), the net moment points up, since there are more states filled with spin up than down. No current can flow with

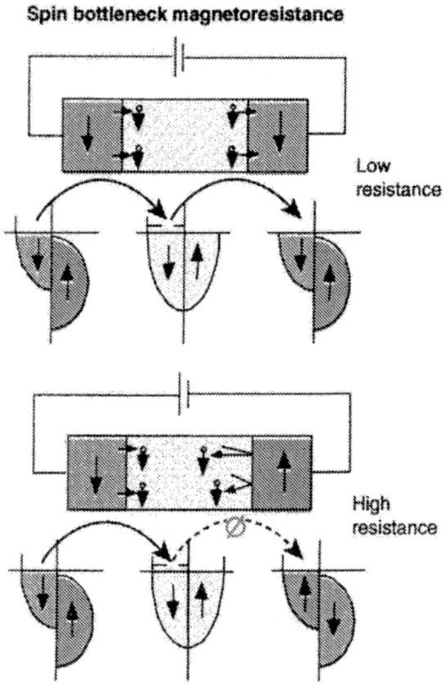

Fig. 21.2. Schematic representation of a spin-valve effect, as described in the text. *(Top)* is made out of two ferromagnetic regions with aligned spins. In this case, low resistance is achieved. *(Bottom)* consists of antiparallel ferromagnetic moments, and in this case the device has a high resistance. Reproduced from [21.2]

spin up since the Fermi level is located above the up-spin band. However, current can flow within the down-spin band. This current is certainly spin polarized. The main lesson to be learned from such a simple idea is that in a ferromagnetic metal the carriers will have a net spin. Note that Fig. 21.1 is certainly an idealization. In practice, the fraction P characterizing the degree of polarization, defined as the difference in currents with spin up and down normalized to the total current, is only as large as 44% for Fe, and 11% for Ni. Note that these "half-metallic" systems have been discussed already in Chap. 3, where more references can be found.

The main idea behind the GMR effect can be illustrated with the simple device sketched in Fig. 21.2, reproduced from the review article of Prinz [21.2]. Electrons in a circuit are assumed to travel from a ferromagnetic metallic region, to a normal metal that has no net magnetization, to another ferromagnetic metallic region. When the two macroscopic moments of the ferromagnetic regions are parallel to each other, then the current is maximized, as is easily understood from the density-of-states considerations also shown in the figure. The active spins (those that can move, in this case spin down) are those with spin pointing opposite to the overall magnetization. On the other hand, if the two magnetizations are antiferromagnetically aligned as in the lower portion of Fig. 21.2, then the resistance must be very high. The spins that can easily move from the first magnetic component to the non magnetic region, cannot continue moving through the oppositely aligned ferromagnetic domain, since they do not have states available at the Fermi level for such movement. Each layer acts as a *spin-filter* since ideally only one spin species is allowed to go through. Clearly if one were able to rotate the magnetization

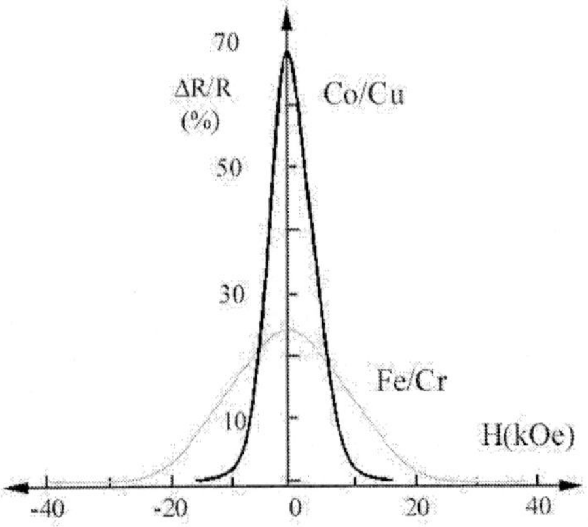

Fig. 21.3. Change ΔR in the room-temperature resistance of Fe/Cr and Co/Cu multilayers, as a function of external magnetic field. R in the vertical axis (denominator) refers to the saturated resistance at high fields, which is the reference resistance in ΔR. (From [21.1])

of one magnetic region with respect to the other, from parallel to antiparallel, a considerable change in the resistance will occur, thus inducing a large magnetoresistance, the basis for the GMR effect.

In practice, GMR systems are made out of *thin films* suitable for technological applications in tiny devices, and the current runs along the thin films, rather than perpendicular to them. Devices as shown in Figs. 21.2 with only two magnetic layers are called *spin-valves* or *magnetic valves*, and they are constructed such that one of the magnetic layers has a moment that is hard in the sense that small fields do not affect it, while the other layer has a soft moment, namely small fields can change its direction. Then, those small fields can change the configuration from parallel to antiparallel, thus drastically modifying the resistance of the system.

In Fig. 21.3, examples of GMR measurements are given for the cases of Co/Cu and Fe/Cr multilayers. $\Delta R/R(\infty)$ multiplied by 100 appears in the vertical axis. $R(\infty)$ is the resistance at large fields, $\Delta R = R(H) - R(\infty)$, and $R(H)$ is the resistance at field H. From the figure, the effect is largest at small fields, approximately in the range from 0 to 10 kOe, which is 1 T. Other GMR devices can achieve equally large MR effects at much smaller magnetic fields (see [21.1]). The temperature at which the measurements were made is room temperature. The measured current is *in the plane* of the layers, and the magnetic field is applied perpendicular to the current but also in the plane of the layers. Figure 21.3 shows that the GMR effect at room temperature can be as large as 70%. At lower temperature it is even higher. As discussed already, it has been found experimentally that the change in resistance is related to the relative change in the orientation of the magnetic moments between adjacent magnetic layers (as illustrated in Fig. 21.2). The effect is largest when these layers are coupled such that at $H = 0$, those moments are antiparallel. As H increases, it is natural that the moments tend to align and in such a configuration the resistance is minimum. If the moments are already aligned at $H = 0$, then the MR is negligible. It is important also to have thin

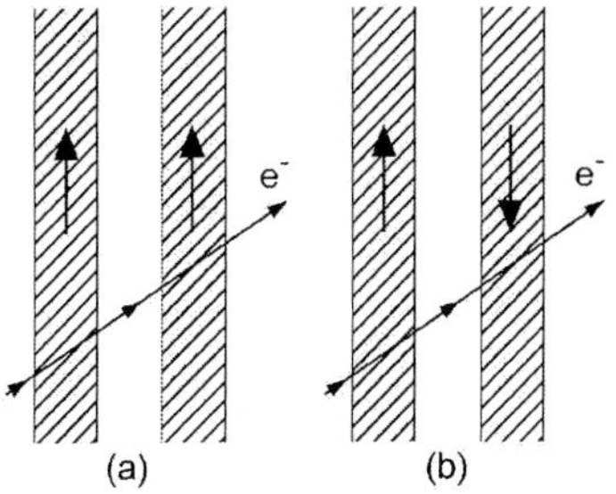

Fig. 21.4. Schematic representation of transport across magnetic multilayers. (**a**) corresponds to ferromagnetically aligned layers. The *solid tilted line* represents an electron traveling across these layers. (**b**) Same as (**a**) but for antiferromagnetically aligned layers

layers of the order of just a few nanometers – for thick layers, the coupling among the moments is negligible, thus the orientation of the moments is random, and the effect is not important. Actually, oscillations in the sign of the coupling between moments have been observed (*oscillatory interlayer exchange coupling*) as the distance between magnetic layers changes. The period of oscillation is about one nanometer. Theoretical work has shown that the GMR is the result of electrons propagating from one magnetic layer to another.

In the case of currents running along thin films, the explanation of the GMR effect is more complicated than when the current is perpendicular to those films. Consider first the *ferro* configuration, as in Fig. 21.2 (upper portion). We are interested in estimating its resistance, and compare that with the resistance of the *antiferro* configuration. If they are different, then the device has a magnetoresistance: flipping the magnetization of one of the layers using an external magnetic field, then the resistance of the system changes. Following the work of Parkin [21.1], the effective resistance $R^C_{\rm eff}$ ($C = F$ or AF) is assumed to be made out of the individual resistances of the spin up R^C_+ and down R^C_- components, in a *parallel* connection since they are considered to be independent of one another. This assumption holds for both the ferro and antiferro configurations, thus, in general

$$R^C_{\rm eff} = \frac{R^C_+ R^C_-}{R^C_+ + R^C_-} \,. \tag{21.1}$$

Suppose now that the mean-free path for spin-scattering events affecting electrons in each of the layers of the device is *larger* than the thickness of each layer (nonspin-dependent scattering, where the spin is unchanged in the event, is much more frequent than spin scattering). In such a case, electrons can easily follow paths such as sketched by solid lines in Fig. 21.4, namely they can travel the full extent of the device along the short direction with little change in spin (and not necessarily in a straight line as suggested for simplicity in Fig. 21.4 since the spin-independent scattering can be substantial). Actual experiments have shown that a spin-polarized current injected from a ferromagnetic metal to a non magnetic metal can travel close to 0.1 mm at low temperature without changing its spin polarization, so the assumption of having large mean-free paths for spin scattering is justified. In such a regime, the electron with up spin (Fig. 21.4a if the FM configuration is considered) moves across regions with (i) resistance $\rho_+ N$, where ρ_+ is the resistivity in the magnetic layer for spins with the proper orientation for good conduction and N is the thickness of the layer, and (ii) resistance $\rho_s M$, where ρ_s is the resistivity of the non magnetic layer and M is its thickness. The path shown in Fig. 21.4 suggests that these resistances must be considered as being connected in series, since the electron experiences them sequentially one after the other. For this reason, $R_+^F = (\rho_+ N + \rho_s M)/2$. A similar reasoning applies to the down-spin electron, simply replacing ρ_+ by ρ_-, the latter being the (expected to be high) resistivity in a magnetic layer of electrons with the wrong spin for conduction. Defining $\alpha = \rho_+/\rho_s$ and $\beta = \rho_-/\rho_s$, it can then be shown that

$$R_{\text{eff}}^F = \frac{\rho_s M}{2} \frac{(\alpha + \frac{N}{M})(\beta + \frac{N}{M})}{(\alpha + \beta + 2\frac{N}{M})} . \tag{21.2}$$

Let us now repeat the calculation for the AF configuration shown in Fig. 21.4b. In this case, an electron in its typical path (solid line) encounters sequentially two magnetic layers, one with the right spin and another with the wrong spin, separated by non magnetic layers. As a consequence,

$$R_+^{AF} = \frac{2\rho_s M + \rho_+ N + \rho_- N}{4} . \tag{21.3}$$

In the AF configuration, it is clear that $R_+^{AF} = R_-^{AF}$ since both types of spins move through the same environment, with one magnetic layer pointing in each direction. As a consequence, R_{eff}^{AF} is simply $R_+^{AF}/2$. Defining a measure of the magnetoresistance effect through $\Delta R/R = (R_{\text{eff}}^{AF} - R_{\text{eff}}^F)/R_{\text{eff}}^F$, it can be shown that

$$\frac{\Delta R}{R} = \frac{(\alpha - \beta)^2}{4(\alpha + \frac{N}{M})(\beta + \frac{N}{M})} , \tag{21.4}$$

which is the main result of this analysis. The equations show that it clearly makes a difference to consider the FM or AF configurations, and certainly a nonzero MR effect can be obtained with the multilayer device shown in Fig. 21.4. The net effect depends on the ratio N/M and the values of α and β. For instance, if $\alpha/\beta \approx 10$, and $N/M = 1$ for simplicity, then $\Delta R/R$ can easily be about 100%. Since the larger the ratio α/β the larger the effect, it is important to find conducting materials that are as close as possible to being 100% polarized, namely with truly only one active spin component in the system, the so-called *half-metals* (see Chap. 3). These materials are actively pursued for potential applications in magnetoelectronics. In the case of manganites, note that it is still controversial how large is the degree of polarization of Sr- or Ca-doped $LaMnO_3$, even at the hole density with the highest Curie temperature.

To qualitatively illustrate the main point of the previous mathematical discussion, consider the limits of very small α (spins moving with little dissipation through a magnetized layer when they have the proper orientation), and very large β (spins having a large resistivity when in a magnetic layer with the wrong orientation). In such a case, $R^F_{\text{eff}} \approx \rho_s N/2$, which is only due to the easily conductive spin species, and all its resistance comes from crossing the non magnetic layer. On the other hand, $R^{AF}_{\text{eff}} \approx \beta M \rho_s/8$ which grows as β, since in the AF configuration for both orientations of the spin it is unavoidable to travel across regions where spin and magnetizations are in the wrong relative orientation. Both species contribute in the same amount to the resistance in this case. Then, clearly R^{AF}_{eff} can be much larger than R^F_{eff}, where in the former one spin species effectively short circuits across the magnetic layers. This is the key effect behind GMR devices. Note that if the mean-free path for spin scattering is smaller than the size of the layers, then $R^C_{\pm}$ does not arise from the connection in series of individual resistivities, but from its *parallel* connection. In this case it can be easily shown that there is no MR effect, namely the FM and AF configurations give the same resistance.

As already remarked, the GMR effect lies in the existence of ferromagnetic regions in the device, where the relative orientation of their moments can be changed with small fields or other variables. Thus, there is no need to have multilayers but granular alloys can be equally suitable for GMR (see [21.1]). This observation suggests a potentially interesting relation between GMR and CMR, since the latter is believed to be caused by mixed-phase tendencies that manifest in a coexistence of metallic and insulating clusters in the regions of interest, leading to a large MR (see Chap. 17). This analogy should be more carefully investigated.

21.2 Applications

GMR devices have found their way into real computers and have contributed considerably to recent advances in the increase of memory bit densities. Here this application is analyzed qualitatively. More speculative applications (such as nonvolatile memories) are also discussed. Readers interested in CMR materials may receive motivation or inspiration from this subsection to imagine useful devices based on manganese oxides!

21.2.1 Magnetic Recording

GMR devices are already in read heads of magnetic disk recorders. A key measure for disk magnetic recording is the *areal density*, the number of bits that can be stored per unit area. The larger this number the better, for the insatiable need for more and more available memory in computers. The state-of-the-art for bit density within a track of recording medium was

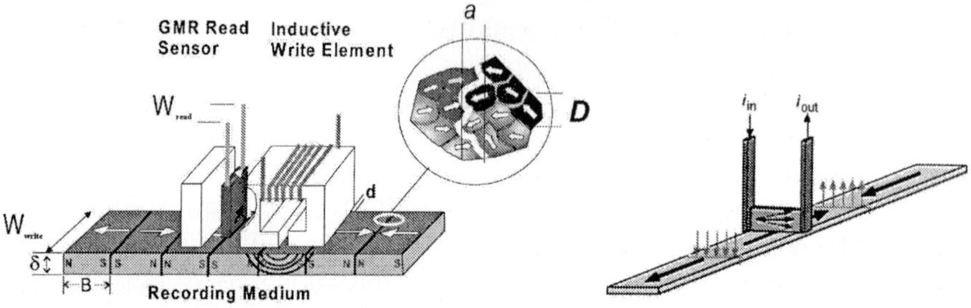

Fig. 21.5. *(Left)* Sketch of a basic recording process, reproduced from [21.5]. An "inductive write element" is used to record horizontally aligned magnetic regions. The "GMR read sensor" detects the fields these magnetic regions produce. The recording medium is granular as shown. *(Right)* Schematic representation of a GMR read head passing over recording media. The *long arrows* are the magnetization of the bits. The *short ones* are the domain walls between bits (see text). The GMR device senses these domain-wall fields since one of its metallic layers can easily adjust its orientation. This produces changes in the GMR device resistance, which is measured with a current as indicated. (From [21.2])

129 000 bits/inch in 1995 (see [21.3]) and this number grows rapidly. Each bit is a magnetized region of size less than a μm long, along the track of the magnetic disk. Predicted bit sizes in the near future are of the order of 100 nm (ultrahigh-density nanowire arrays recently developed can increase the areal density by another large factor [21.4]). As the density of bits grows, the bit size naturally decreases and the magnetic fields associated to each one, that must be read by a head, are also reduced. Then, very sensitive magnetic-field sensors are needed as the areal density grows, and the GMR devices efficiently play such a role.

A schematic illustration of the recording process is shown in Fig. 21.5 (left) [21.5]. The recording medium is typically a granular CoCr-based magnetic alloy composed of 100–500 weakly coupled ferromagnetic grains per bit cell. Each grain is about 10 to 15 nm in diameter. The read head must quickly sense magnetic information about the individual *bits* that form the magnetic disk. These bits are basically small magnetic domains, with magnetization pointing in either of the two directions along a track, as sketched in Fig. 21.5 (right). If the magnetization points in one direction or the another, we either have a 0 or a 1 stored in the bit. In the region between two bits, a very small domain wall of size 100 Å to 1000 Å exists, producing a field that extends sufficiently out of the track that can be detected by the reading device.

Depending on whether this domain-wall field points up or down out of the track, the magnetic field of the bit region that is being read can be determined indirectly (see Fig. 21.5, right). The GMR device needs to be able to determine the direction of the field caused by the domain wall. The GMR reader in the figure has a hard magnetic layer that points out of the plane defined by the recording media, while the soft one either points parallel or antiparallel to the hard one depending on the direction of the field of the domain wall. The GMR configurations are then FM or AF, as discussed before, depending on whether the bits are 0 or 1, and the resistance of the GMR component, as sensed by a current, changes accordingly. The high sensitivity of GMR devices to small fields is very important here. In typical GMR readers, changes of 1% per Gauss are achieved. This also illustrates why colossal magnetoresistant materials are potentially important, since the changes of their resistances are much larger than for GMR devices upon the application of high fields. However, a large CMR still does not occur either at room temperature or in the range of the small fields produced by the domain walls separating bits in the recording medium of Fig. 21.5 (right). Further work is needed to achieve commercial applications of manganites.

21.2.2 Nonvolatile Memories

A potential area of application of GMR devices is in *nonvolatile* memories, those that do not vanish when the power is switched off to the computer. Certainly magnetic disks and tapes satisfy this requirement. However, the currently widely used core memory in computers (random access memory or RAM) is volatile. It would be desirable to have memory as fast as the currently available RAM, but with nonvolatile properties. For this purpose devices exploiting the GMR effect have been proposed. An example from the review [21.2], is shown in Fig. 21.6. Along a line usually referred to as the *sense* line, small spin-valve structures are arranged in series. Each spin-valve can be in two states, moments parallel or antiparallel, and in this respect it itself behaves like a bit 0 or 1. As in the usual spin-valves previously discussed, one of the magnetic layers is soft and the orientation of its moment can be easily changed with small external fields, while the other one is hard.

To manipulate the GMR bits, wires run above and below the sense lines forming an xy grid pattern. When a current runs through one of those wires it produces a magnetic field that affects the GMR spin-valve. The intensity of the pulse of current is selected such that it produces a field that is just *half* of what is needed to rotate the moment of the soft component of the spin-valve or GMR bit. To produce the full rotation, another pulse must be run along the wire running perpendicular to the sense line. In this way, through pulses in two perpendicular wires, only one bit is modified. Using this xy grid, any element (bit) of an array can be selectively addressed, either to store information or to read its value. The exact form in which this simple idea is implemented depends on a wide variety of details, and more information can be found in [21.2], and references therein.

21.2.3 Magnetic Tunnel Junctions

Another device with potential for applications is the *magnetic tunnel junction* (MTJ) sketched in Fig. 21.7. As in any tunneling device, there are two

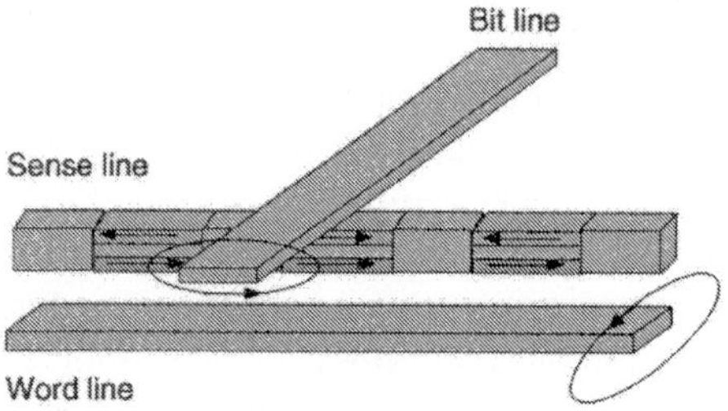

Fig. 21.6. Schematic representation of RAM constructed out of GMR elements connected in series. Wires above and below, in an xy pattern, are used for reading or writing (see text). From [21.2]

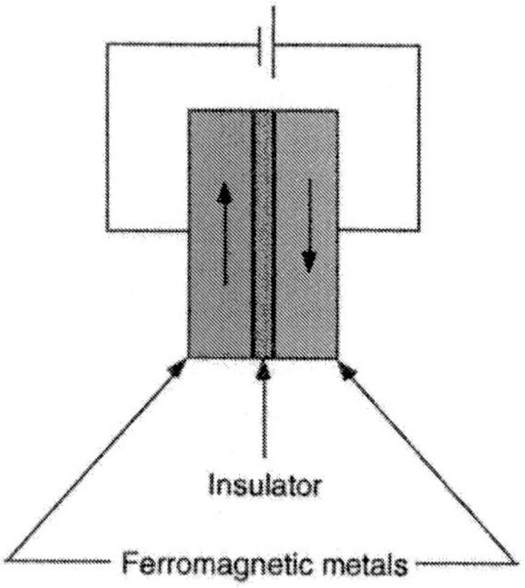

Fig. 21.7. Schematic representation of a magnetic tunnel junction formed by a thin insulating barrier separating two ferromagnetic metallic films. As in the spin-valve case (Fig. 21.2), current flowing through the junction will find high resistance for antiparallel domains and low resistance for parallel domains. From [21.2]

conducting layers separated by a very thin insulating layer. Electrons can quantum-mechanically tunnel from one side to the other upon the application of voltage. Suppose now that the two layers are ferromagnetic. Considerations similar to those described before in the spin-valve case will apply here. If the moments of the layers point in the same direction, tunneling will occur. However, in the antiparallel configuration for the moments of the layers, this is no longer true: the tunneling barrier is *spin dependent*. As in spin-valve devices, one layer is magnetization hard and the other soft. This type of MTJ can find applications in a variety of contexts, as discussed in [21.2].

In tunneling magnetoresistance devices it is possible to use manganites as ferromagnetic layers, as recently briefly reviewed by Kimura and Tokura [21.6]. Resistance changes of a factor 2 were obtained in tunneling trilayers, with manganites as metallic layers, at fields as small as 200 Gauss, although still at very low temperature 4.2 K [21.7]. In spite of this last limitation, the example shows that large MR responses can be achieved at low fields using manganites in layered arrangements similar to those used in GMR devices. This is a potentially fruitful approach to manganite applications in the near future, namely, exploiting the ferromagnetic features of these compounds rather than their CMR intrinsic properties, and using devices geometrically similar to those employed in the GMR context. Devices for potential applications using thin films of manganites have been recently reviewed [21.8] (see also [21.9] and [21.10,21.11]). More recently, low-field (500 Oe) MR has been observed in LCMO/PCMO superlattices [21.12], involving a metal and an insulator, as occurs in the phase-separation scenario described in Chap. 17. There seems to be plenty of room for progress in this area of research!

Let us end this chapter with other useful definitions in the area of magnetoresistive sensors that may be of interest to the readers:

(1) *Anisotropic* MR (AMR) refers to the changes in resistance when the internal magnetization of a sample is rotated with respect to the direction of the current (typically from parallel to perpendicular). A common material for AMR in device applications is permalloy ($Ni_{0.80}Fe_{0.20}$). However, GMR devices are several times more sensitive than permalloy.

(2) Among the simplest magnetoresistors are those based on the *Lorentz* force. Consider some metallic material with carriers that move in the presence of a voltage, with some resistance, in a thin slab geometry. If a magnetic field is applied perpendicular to the slab, the Lorentz force deflects the carriers and the length of their paths increases, increasing the resistance (positive MR).

(3) Another important concept in this context is *exchange bias*, a phenomenon that occurs at the interface between antiferro- and ferromagnetic materials in the form of an exchange anisotropy. Nogués and Schuller [21.13] have presented a very complete review on exchange bias, where the reader can find further information about this topic.

In summary, giant MR is a great example of basic science applied to devices of much relevance in technological applications. The key concept is that electrons can propagate in ferromagnetic regions with the same orientation of the spins, but they cannot propagate when the moment is in the opposite direction. A very similar phenomenon is believed to occur in manganites (Chap. 17) at a microscopic level, with phase coexistence leading to ferromagnetic clusters with random orientation of their moments. In this respect, GMR and CMR appear to rely on similar physics, and this observation may provide guidance to design useful devices involving manganites.

22. Discussion and Open Questions

In this chapter, we present a list of "things we know" and "things we do not know" about manganites and related compounds. The chapter is informally organized to keep the discussion fluid. When results are mentioned that have already been considered in previous chapters, the associated references and details are omitted to avoid repetition. The reader can form a better opinion about the statements in this chapter by reading the appropriate chapters in the present book. Some of the comments and ideas are somewhat sketchy, but this is natural when addressing issues that are still not understood. Also, some items reflect the personal opinion of the author and may be debatable. An important portion of the chapter discusses results for other families of compounds that exhibit similarities with manganites. It is conceivable that the knowledge accumulated in Mn oxides may be applicable to the famous high-T_c cuprates, as well as other materials discussed here and in Chap. 20. Coexisting nanoclusters appear to be a common phenomenon of a variety of compounds, and "colossal" effects of one form or another could be expected. There is plenty of work ahead, and surprises waiting to be unveiled.

22.1 Facts About Manganites Believed to be Understood

After the huge effort carried out in recent years in the study of manganites, involving both experimental and theoretical investigation, much progress has been achieved. This has been described in the bulk of this book. Here, only a brief summary of established results is provided (in the next section, there will be more detail in the discussion when issues that remain to be explained are addressed):

1. At hole density $x = 0$, as in $LaMnO_3$, the ground state is an A-type antiferromagnet with staggered orbital order. Several other Mn oxides have orbital order as well. The orbital degree of freedom plays a role as important as the spin, charge, and lattice degrees of freedom.

2. At $x = 1$, fully hole-doped, the spin arrangement is of G-type.

3. There are many manganites with a ferromagnetic ground state at intermediate densities. Zener ferromagnetism, widely known as "double exchange", appears to capture the essence of the tendency to ferromagnetism.

This FM metallic state is a poor metal, and it is only one of the several possible states of manganites. Some authors believe that orbital correlations may play a role in this ferromagnet. Thus, there is still some discussion about the fine details of this state, but the essential features of this phase appear to be well understood.

4. At other densities such as $x = 0.5$, charge/orbital/spin-ordered states can be stabilized, leading to an asymmetric phase diagram with respect to half-doping. The existence of these states has been established in simulations as well as in mean-field studies. The CO states strongly compete with the ferromagnetic state, both in real experiments and theoretical studies. This is a key observation leading to potential explanations of the CMR, since this phenomenon occurs when two phases – metal and insulator – are in competition. It happens at the boundaries between FM metal and paramagnetic insulator or FM metal and CO/OO/AF states.

5. The presence of phase-separation tendencies in theoretical studies has been unveiled and confirmed. This can be caused by at least two tendencies: electronic phase separation where competing states have different densities, leading to nanocluster coexistence, or disorder-driven phase separation with competing states of equal density that can lead to larger clusters.

6. The existence of intrinsic inhomogeneities in single crystals has been established experimentally. This can occur within the ordered phases, or above the ordering temperatures. In fact, charge-ordered nanoclusters above T_C have been found in many low- or intermediate-bandwidth manganites. In the interesting CMR regime this behavior is correlated with the resistivity. The nano- or mesoclusters are believed to be crucial for the occurrence of CMR.

7. Percolative tendencies in manganites have been unveiled experimentally and in theoretical studies. But sharp first-order metal insulator transitions have also been reported, again both in theory and experiments. In some compounds, a mixture of these behaviors – percolation and first-order features – have been found.

22.2 Why Theories Alternative to Phase Separation May Not Work

The enormous experimental effort on Mn oxides has already provided sufficient results to address whether some of the proposed theories are realistic. It appears that some of the early proposals for theories of CMR manganites have already been put to rest, and that the leading effort centers around inhomogeneities. The detail is the following [22.1]:

1. In theories sometimes referred to as "double exchange", electron hopping above T_C is simply described by a renormalized hopping "$t\langle\cos\theta/2\rangle$". These theories are based on the movement of electrons in a disordered spin-localized background (see Fig. 22.1a), without invoking other phases. However, this is not sufficient to produce an insulating state. These ideas may be

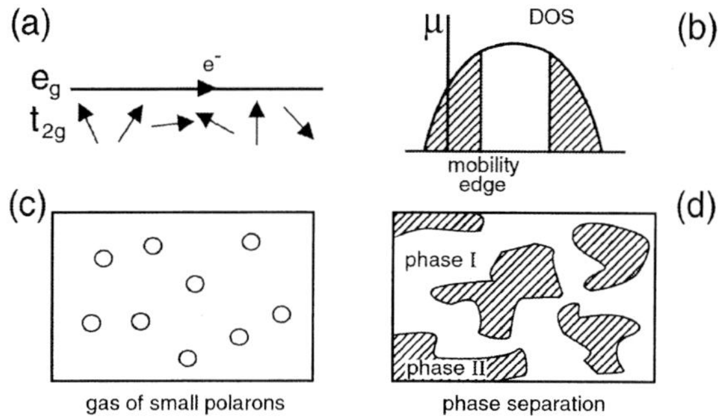

Fig. 22.1. Schematic representation of theories for manganites. (**a**) is a simple double exchange. (**b**) relies on Anderson localization. (**c**) is based on a gas of polarons above T_C. In (**d**), a phase-separated state above the ordering temperatures is sketched. Details can be found in the text

suitable for large-bandwidth manganites, but they do not explain the CMR phenomenon.

2 Some theories rely on Anderson localization to explain the insulating state above T_C [22.2]. However, the amount of disorder needed to achieve localization at the densities of relevance is very large (at least in a simple three-dimensional Anderson model). Recent studies by Slevin et al. [22.3] locate the critical value at $W_c \sim 16$ (in units of the hopping), assuming a uniform box distribution of random on site energies $[-W/2, +W/2]$ in the Anderson model, and with the Fermi energy at the band center. Perhaps this large disorder strength crudely mimics the influence of phonons, correlations, nanoclusters, strain, and quenched disorder present in the real materials. But, even in this case, it is difficult to explain the density-of-states pseudogap found in photoemission experiments (Chap. 18). Anderson localization does not produce such a pseudogap (Fig. 22.1b). In addition, experimentally it is clear that the CMR originates in the competition between phases, typically FM metallic and AF/CO insulating. This fundamental effect is not included in simple Anderson-localization scenarios, where ordering and phase competition are absent. For these reasons, Anderson localization does not seem to be the best approach to explain manganite physics.

3. Some theories are based on a picture in which the paramagnetic insulating state is made out of a gas of small and heavy polarons (Fig. 22.1c). These theories do not explain either recent experiments reporting CE charge ordering above T_C correlated to the resistivity (Chap. 19), or the many indications of inhomogeneities. A polaron gas may be a good description at much higher temperatures, well above room temperature, but such a state does not appear realistic in the important region for CMR. The charge-ordered clusters found experimentally above T_C have properties corresponding to phases that are stable at low temperatures, sometimes in other regimes of densities such as the CE-phase. These complex clusters cannot be considered to be mere polarons.

Theories based on microscopic phase separation (Fig. 22.1d) provide a more realistic approach to manganites since they are compatible with dozens of experiments. However, as shown below, there is plenty of room for improvement as well.

22.3 Open Issues in Mn Oxides

22.3.1 Potentially Important Experiments

• **The existence of the predicted new temperature scale T^* should be further investigated.** (i) Thermal expansion, magnetic susceptibility, X-ray, neutron scattering, and other techniques have already provided results supporting the existence of a new scale T^* (see Chap. 19). This scale should manifest itself even in the DC resistivity, as it does in the high-temperature superconductors at the analog "pseudogap temperature". Preliminary results in Chap. 19 support this notion. (ii) In addition, the specific heat should systematically show the existence of structure at T^* due to the development of short-range order (at T^* an "almost" true phase transition may occur). (iii) The dependence of T^* with doping and tolerance factors should be analyzed systematically. Theoretical studies in Chap. 17 suggest that the tolerance factor may not change T^* substantially.

• **Quantum-critical-like features in the metal insulator competition (Chap. 17) should be studied experimentally.** This issue is related to the previous one. In the quantum-critical-like region, the CMR effect is enhanced and, therefore, it is crucial to investigate the nature of the paramagnetic insulating state above T_C.

• **X-ray and neutron scattering studies of $(La_{1-y}Pr_y)_{1-x}Ca_xMnO_3$ are needed to analyze the evolution of charge-ordered nanoclusters.** As described in Chap. 19, there is considerable evidence that LCMO $x = 0.3$ at temperatures above T_C presents charge ordered nanoclusters, responsible for the behavior of the resistivity. It is important to track the intensity and location in momentum space of the peaks associated with charge ordering as La is replaced by Pr. This replacement enhances charge-ordering tendencies. It is also important that the replacement of Ca by Sr decreases that tendency. Is there a smooth evolution from LCMO to the regime found in LPCMO with phase separation at low temperature (Chap. 11) and to the "double-exchange" homogeneous regime of LSMO? Is the crude picture of the state between T_C and T^* shown in Fig. 22.2 qualitatively correct?

• **The evidence for charge-ordered nanoclusters above T_C should be further confirmed.** It is important that a variety of techniques reach the same conclusions regarding the presence of nanoclusters above T_C (Chap. 19).

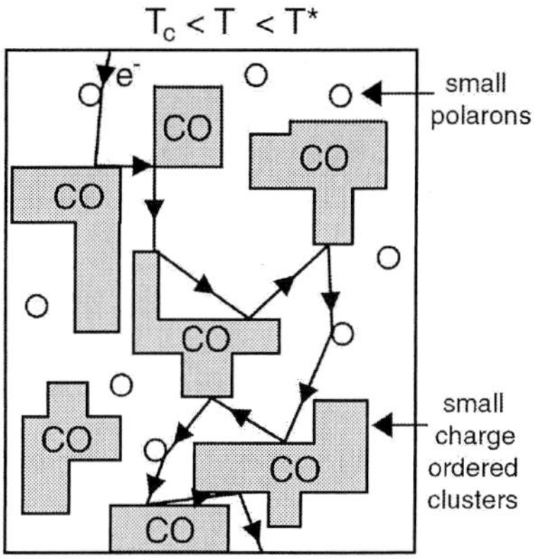

Fig. 22.2. Schematic representing the proposed state in the regime between T_C and T^*. Small polarons are represented by *circles*, and electrons traveling across the sample (*thick line*) may not scatter much from them. The *gray areas* are the CO clusters where scattering is more severe. The FM regions, expected to be present in this state (Chap. 17) are not shown

For instance, it would be important to determine the role played by nanoclusters on optical conductivity results, which thus far has been described mainly as consisting of "polarons". How do the nanoclusters manifest themselves in the optical spectra? Recent results at $x > 0.5$ (Chap. 19) contribute to this issue.

• **Are the mesoscopic clusters found in $(La_{1-y}Pr_y)_{1-x}Ca_xMnO_3$ using electron diffraction representative of the bulk? Are there other compounds with the same behavior?** As reviewed in this book, the evidence for nanocluster formation is very strong in manganites. However, the evidence for larger structures of the mesoscopic kind is based on a smaller number of experiments: electron diffraction and STM techniques. Are these results affected by the surface of the sample or are they representative of the bulk? To what extent should one consider two kinds of phase separation, i.e., nanoscale and microscale? Can one evolve smoothly into the other as the Curie temperature decreases?

• **What is the nature of the ferromagnetic insulating phases?** Is this phase truly qualitatively different from the ferromagnetic metallic phase, or are they very similar microscopically? In other words, is there a spontaneously broken symmetry in the ferro insulator transition? Does charge ordering and/or orbital ordering exist there? In Chap. 10, theoretical studies have shown the presence of many spin ferromagnetic phases, with or without charge and orbital order. There is no reason why these phases could not be stabilized in experiments. In fact, a new phase of manganites has recently been unveiled at $x = 0$ (Chap. 9).

• **Atomic-resolution STM experiments should be performed, at many hole densities.** The very recent atomic-resolution STM results for BiCaMnO (Chap. 11) are very important for the clarification of the nature of manganites, and for the explicit visual confirmation of phase-separation ideas. Extending experiments of this variety to other hole densities, particularly

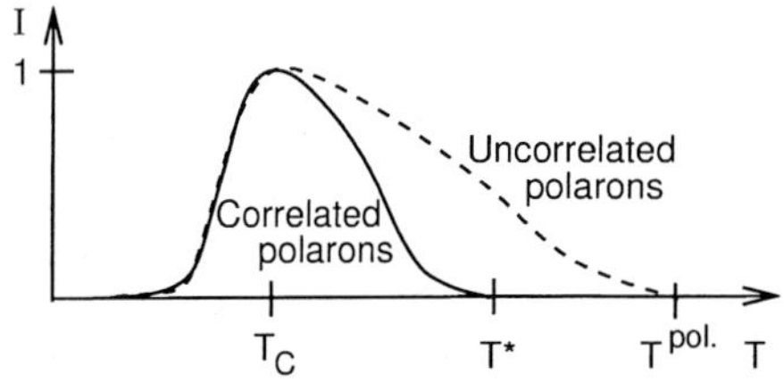

Fig. 22.3. Schematic representation of an idealized neutron scattering experiment. Intensities I in the vertical axis correspond to momenta associated with (1) correlated polarons (i.e. CE clusters), as in Chap. 19, and (2) uncorrelated polarons. Two proposed temperature scales, T^* and $T^{\rm pol}$, are shown. The results are normalized to the same intensity at $T_{\rm C}$

those where the system becomes ferromagnetic metallic at low temperatures, is of much importance.

• **The regime of temperatures well above room temperature must be carefully explored, even beyond T^*.** At $T > T^*$, the CO clusters are no longer formed but a gas of polarons may still be present. Perhaps this leads to another temperature scale, denoted by $T^{\rm pol}$ in Fig. 22.3, where individual polarons start forming. Is the system metallic upon further increasing the temperature?

• **Is the glassy state in some manganites of the same variety as in spin-glasses, or does it belong to a new class of "phase-separated glasses"?** The discussion in this context is in Chap. 13. The issue is subtle since, as of today, proper definitions of a glass and associated transitions are still under much discussion. Can glasses be separated into different classes? Are the glasses in manganites a new class? What is the actual nature of the "cluster glasses" frequently mentioned in manganites?

• **Should nanocluster phase-separated states be considered as a new state of matter?** This issue is obviously very important. In order to arrive at an answer, the origin of the nanocluster formation must be clarified. Is it purely electronic, disorder driven, strain driven? If the inevitable disorder related to tolerance factors could be tuned, what happens with the nanoclusters? If experiments such as those of Schön et al. [22.4] (for results in this direction see [22.5]) could be carried out for manganites, with doping caused by a gate device rather than chemically, will we get different results?

• **What is the nature of the charge-ordered states at $x < 0.5$, such as those in PCMO?** It is accepted that these states "resemble" the CE-state in their charge distribution, but what is the arrangement of spin and orbitals? Is the excess of electronic charge distributed randomly in the $x = 0.5$ CE structure or is it uniformly distributed, for instance in the Mn^{4+} sites? Theoretical studies are difficult at these densities.

• **Is there any compound with a truly spin-canted homogeneous ground state?** Theoretical studies have not been able to find spin-canted homogeneous states in reasonable models for manganites. Are there experiments suggesting otherwise? Thus far, experimental evidence for canting can be alternatively explained through inhomogeneities in the ground state. A counterexample are perhaps bilayers in the direction perpendicular to the planes, but this may be a different kind of state, unrelated to the original proposal that postulated a homogeneous spin-canted state interpolating between FM and AF.

• **Are the nanoclusters found in manganites and cuprates (see below) characteristic of other materials?** The answer seems to be positive. Chapter 20 lists materials that behave similar to Mn and Cu oxides. Nickelates are another family of compounds that have stripes and charge ordering competing with antiferromagnetism. Below, other materials with similar characteristics are reviewed. These results are more than accidents. They suggest that many compounds are intrinsically inhomogeneous. Theories based on homogeneous states may be unrealistic.

• **Are Eu-based semiconductors truly described by ferromagnetic polarons as believed until recently, or is a nanocluster picture more appropriate?** The more recent studies of Chap. 20 using Raman scattering suggest the existence of close analogies between Eu semiconductors and manganites. Perhaps phase separation dominates in Eu compounds as well.

22.3.2 Unsolved Theoretical Issues

• **The study of models for manganites is far from over.** Although much progress has been made, as reviewed in previous chapters, many important issues are still unexplored or under discussion. For example, the phase diagram in *three dimensions* of the two-orbital model may contain many surprises. It is already known that the 1D model with non cooperative phonons has a rich phase diagram, with a variety of competing phases. Studies in two dimensions including cooperative effects have also revealed the presence of stripes (see Chap. 10) at densities $x = 1/3$ and $1/4$, and probably others. Thus, one can easily imagine a surprisingly rich phase diagram for bilayers or 3D systems. Perhaps the dominant phases will still be the A-type AF at $x = 0$, ferro metal at $x \sim 0.3$, CE-type at $x = 0.5$, C-type at $x \sim 0.75$, and G-type at $x = 1.0$. However, the details remain unclear.

• **The theory of phase separation should be made more quantitative.** For example, the temperature dependence of the DC resistivity should be predicted within the percolative scenario. This is a complicated task due to the difficulty in handling inhomogeneities.

• **Is the "small" J_{AF} truly an important coupling for manganites?** The Heisenberg coupling between localized spins appears to play a key role in the stabilization of the A-type AF state at $x = 0$ (this coupling selects whether the system is in a FM, A-, C- or G-type state), in the stabilization of the correct spin arrangement for the CE-phase (if J_{AF} is too small, ferromagnetism wins, if too large G-type antiferromagnetism wins), and in the charge stacking of the CE-phase (see Chap. 10). This key role appears in Jahn–Teller and Coulombic theories as well. However, are there other mechanisms for stabilization of A-, CE- and charge-stacking states? The importance of J_{AF} may be magnified by the competition between many phases in manganites, with several nearly degenerate states.

• **Why is the CE-state so sensitive to Cr doping?** Experimentally it is not expected that a robust charge-ordered state could be destabilized by a relatively very small percentage of impurities. Can this be reproduced in MC simulations?

• **For the explanation of CMR, is there a fundamental difference between JT- and Coulomb-based theories?** Thus far, for CMR phenomena the origin (JT vs. Coulomb) of the competing phases does not appear to be important, but the competition itself is. Is this correct?

• **In Chap. 17, phenomenological models were used for CMR. Can a large MR effect be obtained with more realistic models?** Of course the calculations would be very complicated in this context, if unbiased robust many-body techniques are used.

• **Are there models with spin-canted homogeneous ground states?** Thus far, when models for manganites were seriously studied with unbiased techniques, no homogeneous spin-canted states have been identified. Perhaps other models exist?

• **Can a model develop a charge-ordered AF phase at intermediate temperature, while having a ferromagnetic metallic phase at low temperatures?** This is quite difficult and perhaps can only occur if the CO phase has an associated high entropy (see contribution of D. Khomskii in Chap. 17).

22.3.3 Are There Two Types of CMR?

The manganite $(Nd_{1-y}Sm_y)_{1/2}Sr_{1/2}MnO_3$ investigated by Tokura et al. [22.6] has a curious behavior, shown in Fig. 22.4. This compound presents *two* types of CMR phenomena: (1) At temperatures in the vicinity of 250 K ($y = 0$) and 150 K ($y = 0.75$) a somewhat "standard" CMR is observed. Here, standard refers to the CMR behavior described in the introductory chapters of this book. However, (2) at lower temperatures where the system is insulating, a huge MR effect is observed as well. Results for two values of "y" are shown in the figure. Is this indicative of two independent mechanisms for CMR?

Related with this issue is the recent work of Fernandez-Baca et al. [22.7] studying $Pr_{0.70}Ca_{0.30}MnO_3$, where they reported the discontinuous character of the insulator-to-metal transition induced by an external field. The authors argued that the CMR phenomenon is more complex than a simple percolation of FM clusters. Their results are in qualitative agreement with those of Fig. 22.4 where, at low temperatures, a fairly abrupt transition is observed. In fact, it is known that even in percolative processes such as those in LPCMO (Fig. 11.12), hysteresis has been found in the resistivity. This suggests that a *mixture* of percolation with first-order features could be at work in manganites [11.112].

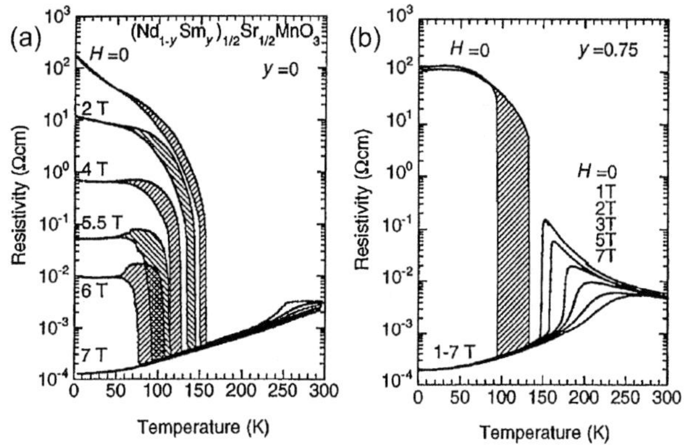

Fig. 22.4. Temperature dependence of resistivity under various magnetic fields for $(Nd_{1-y}Sm_y)_{1/2}Sr_{1/2}MnO_3$ with $y = 0$ (**a**) and 0.75 (**b**). The *hatched area* represents thermal hysteresis. Results from Tokura et al. [22.6]

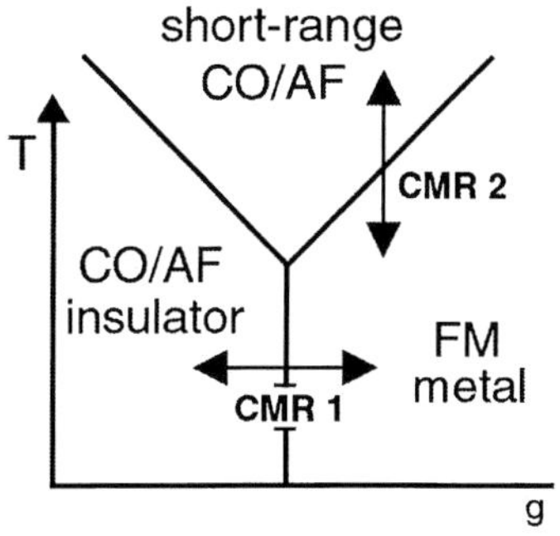

Fig. 22.5. Schematic representation of the phase diagram in the presence of competing metal and insulator (see Chap. 17). g is a generic variable needed to transfer the system from one phase to the other. CMR1 and CMR2 are the two MR transitions described in the text

Can theory explain the presence of two types of CMR transitions? The answer is tentatively yes, but more work is needed to confirm this. The main reason is contained in the phase diagram of the spin toy model (Fig. 17.10) used to address the competition of two phases, as schematically reproduced in Fig. 22.5. There, two CMRs are shown. One corresponds to the "standard" transition from a clustered short-range ordered phase to the FM metallic phase with decreasing temperature (CMR2). At low temperatures, and if the disorder is not too strong, the transition between the competing phases remains of first order. As a consequence, a field can cause a first-order transition between the two phases (CMR1 in the figure).

22.3.4 Other Very General Open Questions

• **Can quasi-one-dimensional manganites be prepared?** In the area of cuprates it was quite instructive to analyze copper oxides with chain or ladder structures. These quasi-1D systems can usually be studied fairly accurately by theorists, and concrete predictions can be made. Regarding CMR,

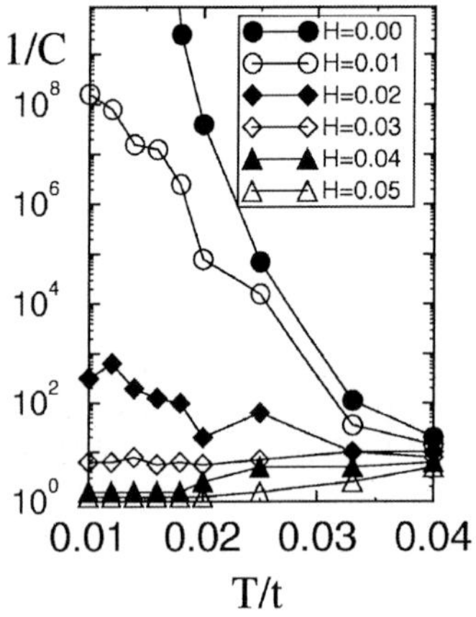

Fig. 22.6. Inverse conductivity of the half-doped one-orbital model on a 64-site chain at many values of the magnetic field H (in units of the hopping) as indicated. Note the large changes in the conductivity with increasing fields. Other couplings and details can be found in [17.5]

there are already calculations (see Fig. 22.6) that show a very large MR effect in 1D models. In 1D, a single perfectly antiferromagnetic link is sufficient to block the movement of charge (its effective hopping is zero), creating a huge resistance. Small fields can slightly bend those AF-oriented spins, allowing for charge movement and decreasing the resistivity by several orders of magnitude.

• **Should one totally exclude superconductivity in manganites?** Naively, it appears that the presence of superconductivity in transition-metal oxides of the $n = 3$ shell should not be necessarily restricted to cuprates. In the Cu-oxide context, superconductivity appears when the insulator is rendered unstable by hole doping. In the high-doping side of the superconducting region, a Fermi liquid exists even at low temperature. Is there a manganite compound, likely a large-bandwidth one, that does not order at low temperature and maintains a metallic character? Searching for metal insulator transitions in those manganites may lead to surprises. If it is confirmed that, even in these circumstances, no superconductivity is found, then the $S = 1/2$ spins of cuprates (as opposed to the higher spins of other materials) is likely to play a key role in the process.

• **Technical applications of manganites remain a possibility.** There are two areas of technologically motivated investigations. One is based on materials with high T_C at or above room temperature, which may be useful as plain nearly half-metals (see Chap. 3) in the construction of multilayers spin-valve-like devices (see Chap. 21). Another area is the investigations of thin films, exploiting CMR as an intrinsic property of these materials. Special treatments of those films may still lead to a large MR at high temperatures.

22.4 Is the Tendency to Nanocluster Formation Present in Other Materials?

We end this book with a description of materials that present properties similar to those of the Mn oxides, regarding inhomogeneities, phase competition, and the occurrence of T^* (this section adds to a related discussion already presented in Chap. 20). It is interesting to speculate that all of these compounds share a similar phenomenology. Then, by investigating one particular system progress would be made in understanding the others as well. We start with the famous high-T_c compounds, continue with diluted semiconductors (of much interest these days, and already partially addressed in Chap. 20), and finish with organic and heavy-fermion materials.

22.4.1 Phase Separation in Cuprates

Considerable work has been recently devoted to the study of inhomogeneous states in cuprates. In this context the issue of stripes as a form of inhomogeneity was raised several years ago, as described in the introductory chapters [1.3]. However, recent results obtained with STM techniques have revealed inhomogeneous states of a more complex nature. They appear in the surface of one of the most studied superconductors Bi2212 (but they may be representative of the bulk), at low temperatures in the superconducting phase, both in the underdoped and optimally doped regions. Figure 22.7 shows the spatial distribution of *d*-wave superconducting gaps. The inhomogeneities are notorious. Indications of similar inhomogeneities have been observed using a variety of other techniques as well (see for instance [1.3, 22.9], and references therein). The list of relevant experiments is simply too large to be reproduced here. Currently, this is a much-debated area of research in high critical temperature superconductors.

On the other hand, recent NMR work on YBCO by Bobroff et al. [22.11] has not reported inhomogeneities, suggesting that phase separation may not be a generic feature of the cuprates. Similar conclusions were reached by Yeh et al. [22.12] through the analysis of quasiparticle tunneling spectra. More work is clearly needed to clarify the role of inhomogeneities in the cuprate compounds.

22.4.2 Colossal Effects in Cuprates?

The results in Chap. 17 related to general aspects of the competition between two ordered phases in the presence of quenched disorder (see for instance the phase diagram in Fig. 17.10) can be applied to the cuprates as well. In fact, the phase diagram of the single-layer high-T_c compound $La_{2-x}Sr_xCuO_4$ contains a regime widely known as the "spin-glass" region. In view of the results in Chap. 17, it is conceivable that this spin-glass region could have emerged,

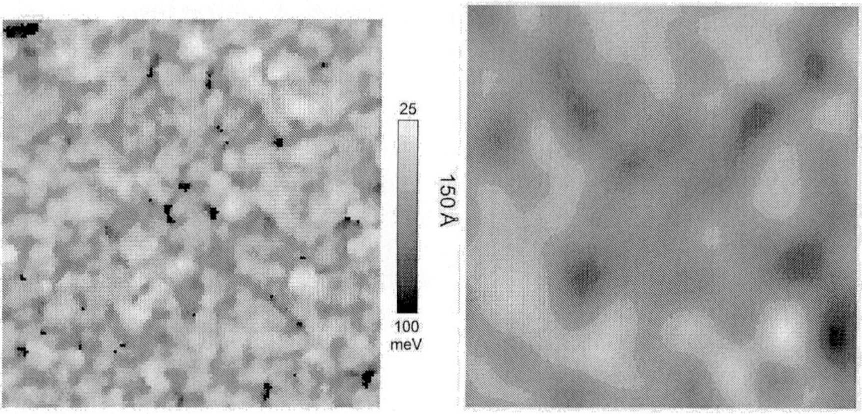

Fig. 22.7. Spatial variation of the superconducting energy gap of Bi2212 as reported by Pan et al. [22.8], using STM techniques. Shown on the *left* is a 600×600 Å area. On the *right* a 150×150 Å subset is enlarged. The gap scales are also shown

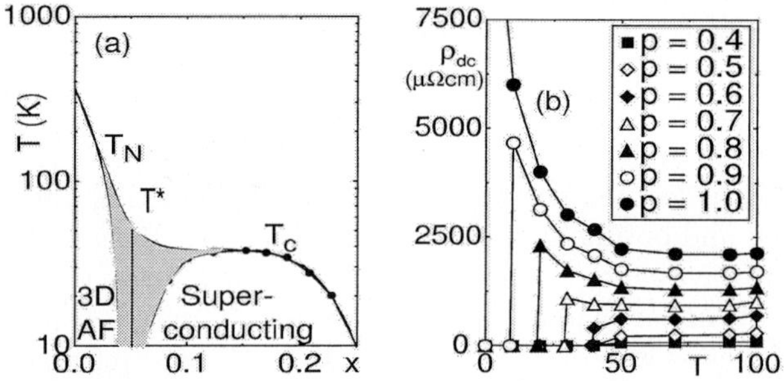

Fig. 22.8. (**a**) Phase diagram of the 214 high-temperature superconductor, as conjectured by Burgy et al. [11.78]. A possible first-order line separating the clean-limit superconducting and antiferromagnetic insulating regimes in the low hole density region is shown. With disorder, the *gray region* is generated, exhibiting coexisting clusters. As in Chap. 17, T^* is the remnant of the clean-limit ordering temperature. (**b**) Resistivity of the state conjectured in (**a**) using a crude resistor-network approximation. More details can be found in the original reference [11.78]. p is insulating fraction at 100 K. These fractions are also temperature dependent

due to the influence of disorder, in place of the original first-order phase transition separating the doped antiferromagnetic state (probably with stripes) and the superconducting state. The resulting phase diagram would be as in Fig. 22.8a. The gray region exhibits a possible coexistence of small locally-ordered clusters, globally forming a disordered phase. If this conjecture is correct, then small superconducting islands should exist in the spin-glass regime, an effect that may have already been observed [22.10].

If preformed superconducting regions are indeed present in underdoped cuprates, then the "alignment" of their order parameter should be possible upon the influence of small external perturbations. This is similar to the development of preformed ferromagnetic clusters in manganites, which align their moments when a relatively small external magnetic field is applied (see Chap. 17). Could it be that "colossal" effects occur in cuprates and other materials as well? This is an intriguing possibility raised by Burgy et al. [11.78]. There are already experiments by Decca et al. [22.13] that have reported an anomalously large "proximity effect" in underdoped $YBa_2Cu_3O_{6+\delta}$. Decca et al. referred to the effect as "colossal proximity effect". Perhaps colossal-type effects are more frequent than previously expected, and they are prominent in manganites simply because one of the competing phases can be favored by an easily generated uniform magnetic field. If a simple external perturbation would favor charge-ordered, antiferromagnetic, or superconducting states, large effects could be observed in the region of competition with other phases. If this scenario is correct, several problems should be percolative, and preliminary studies by Burgy et al. [11.78] suggest that a resistor-network mixing superconducting and insulating islands (similar to the calculation leading to Fig. 17.14b–c in the manganite context) could roughly mimic results for cuprates (see Fig. 22.8b). Certainly, more work is needed to test these challenging ideas.

The results for manganites and cuprates also have similarities with the physics of superconductivity in granular and highly disordered metals [22.14]. In this context, a critical temperature is obtained when the phases of the order parameters in different grains locks. This temperature is much smaller than the critical temperature of the homogeneous system. In manganites, the Curie or charge-ordering temperature may also be much higher in a clean system than in the real system, as discussed in this and other chapters.

22.4.3 Relaxor Ferroelectrics and T^*

The so-called *relaxor ferroelectrics* are an interesting family of compounds, with ferroelectric properties and diffuse phase transitions [11.85]. The analogies with manganites are notorious. Representative cubic materials, such as $PbMg_{1/3}Nb_{2/3}O_3$, exhibit a glass-like phase transition at a freezing temperature $T_g \sim 230\,K$. Besides the usual features of a glassy transition, a broad frequency-dependent peak in the dielectric constant has been found. The transition does not involve long-range ferroelectric order, providing a difference with most manganites that tend to have some form of long-range order at low temperature (although there are several with glassy behavior even in that temperature regime). However, a remarkable similarity is the appearance of another temperature scale, the so-called Burns temperature [11.86], which is $\sim 650\,K$ in $PbMg_{1/3}Nb_{2/3}O_3$. Below this temperature, polar *nanoregions* are formed. The Burns temperature appears to be the analog of the T^* temperature of Burgy et al. [11.78], discussed in Chap. 17. These analogies between ferroelectrics and Mn oxides should be explored further. For exam-

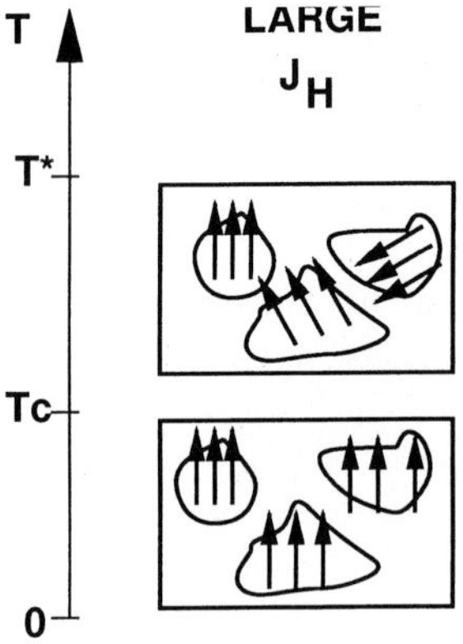

Fig. 22.9. Schematic representation of the clustered state proposed for diluted magnetic semiconductors in [22.17] and [22.18]. At T^*, uncorrelated ferromagnetic clusters are formed. At T_C, they orient their moments in the same direction

ple, is there some sort of "colossal" effect in the relaxor ferroelectrics? Are there two or more phases in competition?

22.4.4 Diluted Magnetic Semiconductors and T^*

Diluted magnetic semiconductors (DMS) based on III-V compounds are recently attracting considerable attention due to their combination of magnetic and semiconducting properties, which may lead to spintronic applications [22.15]. $Ga_{1-x}Mn_xAs$ is the most studied of these compounds with a maximum Curie temperature $T_C \approx 110\,K$ at low doping x, and with a carrier concentration $p = 0.1x$ [22.15]. The ferromagnetism is carrier induced, with holes introduced by doping mediating the interaction between $S = 5/2\,Mn^{2+}$-spins. Models similar to those proposed for manganites, with coupled spins and carriers, have been used for these compounds [22.16]. Recently, the presence of a T^* has also been proposed by Mayr et al. [22.17] and Alvarez et al. [22.18], using Monte Carlo methods quite similar to those discussed in Chap. 7. The idea is that in the random distribution of Mn spins, some of them form clusters (i.e., they lie close to each other). These clusters can magnetically order by the standard Zener mechanism at some temperature, but different clusters may not correlate with each other until a much lower temperature is reached. This situation is illustrated in Fig. 22.9. Naively, we expect that above the true T_C, the clustered state will lead to an insulating behavior in the resistivity, as happens in Mn oxides. This, together with the Zener character of the ferromagnetism, unveils unexpected similarities between DMS and Mn oxides.

22.4.5 Organic Superconductors, Heavy Fermions, and T^*

There are other families of materials that also present a competition between superconductivity and antiferromagnetism, as the cuprates do. One of them is the group of organic superconductors [22.19]. The field of organic

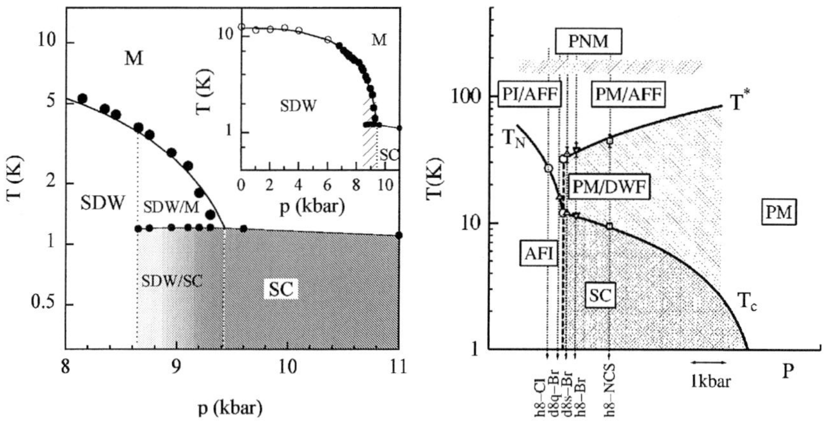

Fig. 22.10. *(Left)* Phase diagram of $(TMTST)_2PF_6$ [22.20]. SDW, M, and SC indicate spin-density-wave, metallic, and superconducting regimes. The gradient in shading indicates the amount of SC phase. *(Right)* Phase diagram of κ-$(ET)_2X$ [22.22]. PNM means paramagnetic non metallic, AFF a metallic phase with large AF fluctuations, and DWF is a density-wave with fluctuations. Note the first-order SC AF transition (*dashed lines*), as well as the presence of a T^* scale. Other details can be found in [22.22]

superconductivity will not be reviewed here, but some recent references will be provided to guide the reader into this interesting area of research. In Fig. 22.10 (left), the phase diagram [22.20] of a much-studied material known as $(TMTST)_2PF_6$ is shown. In a narrow region of pressures, a mixture of SDW (spin-density-wave) and SC (superconductivity) is observed at low temperatures. This region may result from coexisting domains of both phases, as in the FM CO competition in manganites. The two competing phases have the same electronic density and the domains can be large. The layered organic superconductor κ-$(ET)_2Cu[N(CN)_2]Cl$ also has a phase diagram with coexisting SC and AF [22.21]. In addition, other materials of the same family present a *first-order* SC-AF transition at low temperature, according to the phase diagram recently reported in [22.22]. This result is reproduced in Fig. 22.10 (right), where a characteristic scale T^* produced by charge fluctuations is also shown. The similarities with results for manganites are strong: first-order transitions or phase coexistence, and the existence of a T^*. It remains to be investigated as to whether the similarities are accidental or reveal the same phenomenology observed in competing phases of manganites.

Another family of materials with competing SC and AF are the heavy fermions. In Fig. 22.11, a schematic phase diagram of Ce-based heavy fermions is shown [22.23]. Note the presence of a pseudogap (PG) region, with non-Fermi-liquid (NFL) behavior at higher temperatures. The PG phase is reminiscent of the region between ordering temperatures and T^*, discussed in this chapter and Chap. 17, and it may arise from phase competition. In addition, clusters in U- and Ce-alloys were discussed in [22.24], and spin-glass behav-

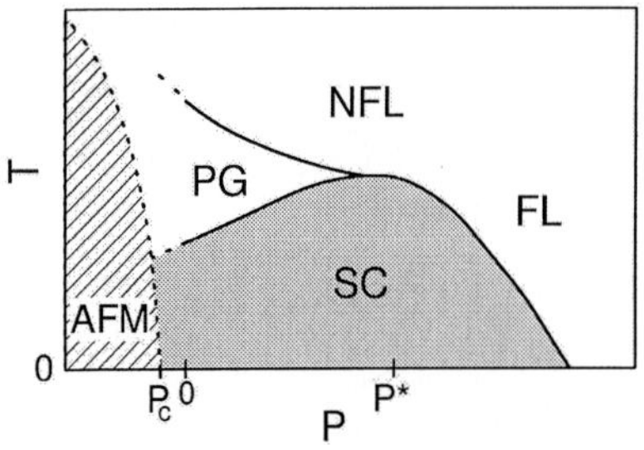

Fig. 22.11. Schematic temperature-pressure phase diagram of some Ce-based heavy fermions, from [22.23]. The notation is standard

ior in $CeNi_{1-x}Cu_x$ was reported in [22.25]. Recently, Takimoto et al. [22.26] found interesting formal analogies between models for manganites and for f-electron systems that deserve further studies.

The analogies between organic, cuprate, and heavy-fermion superconductors are strong. In fact, the SC AF phase competition with the presence of a T^*, pseudogaps, phase coexistence, and first-order transitions, is formally similar to analogous phenomena unveiled in the FM CO competition for manganites. The self-organization of clustered structures in the ground state appears to be a characteristic of many interesting materials, and work in this promising area of investigations is just starting.

References

Chapter 1

1.1 J. Bardeen, L.N. Cooper, and J.R. Schrieffer: Phys. Rev. **108**, 1175 (1957)
1.2 J.G. Bednorz and K.A. Müller: Z. Phys. B **64**, 189 (1986); Rev. Mod. Phys. **60**, 585 (1988)
1.3 J.M. Tranquada et al.: Nature **375**, 561 (1995); V. Emery, S. Kivelson, and H. Lin: Phys. Rev. Lett. **64**, 475 (1990); S.A. Kivelson: Physica C **209**, 597 (1993); S.A. Kivelson and V.J. Emery: Synth. Met. **65**, 249 (1994); U. Löw et al.: Phys. Rev. Lett. **72**, 1918 (1994); S.A. Kivelson and V.J. Emery: Synth. Met. **80**, 151 (1996); J. Zaanen: J. Phys. Chem. Solids **59**, 1769 (1998); D. Poilblanc and T.M. Rice: Phys. Rev. B **39**, 9749 (1989); C. Castellani, C. Di Castro, and M. Grilli: Phys. Rev. Lett. **75**, 4650 (1995); A.H. Castro Neto and D. Hone: Phys. Rev. Lett. **76**, 2165 (1996); S.R. White and D.J. Scalapino: Phys. Rev. Lett. **80**, 1272 (1998); S.A. Kivelson, E. Fradkin, and V.J. Emery: Nature **393**, 550 (1998); C.S. Hellberg and E. Manousakis: Phys. Rev. Lett. **83**, 132 (1999); C. Buhler, S. Yunoki, and A. Moreo: Phys. Rev. Lett. **84**, 2690 (2000); A.L. Chernyshev, A.H. Castro Neto, and A.R. Bishop: Phys. Rev. Lett. **84**, 4922 (2000); G. Martins et al.: Phys. Rev. Lett. **84**, 5844 (2000); E. Dagotto et al.: Phys. Rev. B **49**, 3548 (1994); and references therein. See also L.P. Gorḱov and A. Sokol: JETP Lett. **46**, 420 (1987). The list is just partial, there are several hundreds of papers on the subject
1.4 Y. Tokura and N. Nagaosa: Science **288**, 462 (2000); and references therein
1.5 K.I. Kugel and D.I. Khomskii: Sov. Phys. Usp. **25**, 231 (1982); and references therein. See also D.I. Khomskii and G.A. Sawatzky: Solid State Commun. **102**, 87 (1997)
1.6 R.B. Laughlin and D. Pines: PNAS, vol. **97**, page 28 (Jan. 4, 2001); and references therein
1.7 T. Vicsek, Nature **418**, 131 (2002).
1.8 J. Langer: Phys. Today, July 1999, p. 11
1.9 E. Dagotto, T. Hotta, and A. Moreo: Phys. Rep. **344**, 1 (2001)
1.10 A. Moreo, S. Yunoki and E. Dagotto: Science **283**, 2034 (1999)
1.11 A. Ramirez: J. Phys. Condens. Matter **9**, 8171 (1997)
1.12 J. Coey, M. Viret, and S. von Molnar: Adv. Phys. **48**, 167 (1999)
1.13 Y. Tokura and Y. Tomioka: J. Magn. Magn. Mater. **200**, 1 (1999)
1.14 M.R. Ibarra and J.M. De Teresa: Mater. Sci. Forum. **302–303**, 125 (1999); Trans Tech Publications, Switzerland
1.15 C. Rao et al.: J. Phys.: Condens. Matter **12**, R83 (2000)
1.16 A. Moreo: J. Electron Spectrosc. Relat. Phenom. **117–118**, 251 (2001)
1.17 M. Ziese, cond-mat/0111263
1.18 M.B. Salamon and M. Jaime: Rev. Mod. Phys., July 2001
1.19 M. Imada, A. Fujimori, and Y. Tokura: Rev. Mod. Phys. **70**, 1039 (1998)

1.20 E. Dagotto: Rev. Mod. Phys. **66**, 763 (1994); and references therein
1.21 C.N.R. Rao and B. Raveau (Eds.) *Colossal Magnetoresistance, Charge Ordering, and Related Properties of Manganese Oxides* (World Scientific, Singapore 1998)
1.22 T.A. Kaplan and S.D. Mahanti (Eds.) *Physics of Manganites* (Kluwer Academic/Plenum Publishers, New York 1998)
1.23 Y. Tokura (Ed.) *Colossal Magnetoresistive Oxides* (Gordon and Breach, London, 2000)
1.24 N. Tsuda, K. Nasu, A. Fujimori, and K. Siratori: *Electronic Conduction in Oxides*, 2nd edn. (Springer, Berlin Heidelberg New York 2000)

Chapter 2

2.1 G.H. Jonker and J.H. van Santen: Physica (Utrecht) **16**, 337 (1950)
2.2 J.H. van Santen and G.H. Jonker: Physica (Utrecht) **16**, 599 (1950)
2.3 G.H. Jonker: Physica 22, 707 (1956). See also G.H. Jonker and J.H. Van Santen: Physica **19**, 120 (1953)
2.4 J. Volger: Physica **20**, 49 (1954)
2.5 E.O. Wollan and W.C. Koehler: Phys. Rev. **100**, 545 (1955)
2.6 Z. Jirák, S. Vratislav, and J. Zajícek: Phys. Stat. Sol. (a) **52**, K39 (1979)
2.7 E. Pollert, S. Krupicka, and E. Kuzmicova: J. Phys. Chem. Solids **43**, 1137 (1982)
2.8 Z. Jirák, S. Krupicka, Z. Simsa, M. Dlouha, and S. Vratislav: J. Magn. Magn. Mater. **53**, 153 (1985). See also Z. Jirák et al.: J. Magn. Magn. Mater. **15**, 519 (1980)
2.9 C.W. Searle and S.T. Wang: Can. J. Phys. **48**, 2023 (1970); and references therein
2.10 G. Matsumoto: J. Phys. Soc. Jpn. **29**, 606 (1970). See also G. Matsumoto: J. Phys. Soc. Jpn. **29**, 615 (1970)
2.11 R.M. Kusters et al.: Physica (Amsterdam) **155B**, 362 (1989)
2.12 R. von Helmolt et al.: Phys. Rev. Lett. **71**, 2331 (1993)
2.13 K. Chahara et al.: Appl. Phys. Lett. **63**, 1990 (1993)
2.14 H. Ju et al.: Appl. Phys. Lett. **65**, 2108 (1994)
2.15 S. Jin, T.H. Tiefel, M. McCormack, R.A. Fastnacht, R. Ramesh, and L.H. Chen: Science **264**, 413 (1994)
2.16 G.C. Xiong, Q. Li, H.L. Ju, S.M. Bhagat, S.E. Lofland, R.L. Greene and T. Venkatesan: Appl. Phys. Lett. **67**, 3031 (1995)
2.17 For a brief review see Y. Tokura, Y. Tomioka, H. Kuwahara, A. Asamitsu, Y. Moritomo, and M. Kasai: J. Appl. Phys. **79**, 5288 (1996). Other important references describing the first-order nature of the CO-FM transition are Y. Tomioka, A. Asamitsu, Y. Moritomo, H. Kuwahara, and Y. Tokura: Phys. Rev. Lett. **74**, 5108 (1995); H. Kuwahara, Y. Tomioka, A. Asamitsu, Y. Moritomo, and Y. Tokura: Science **270**, 961 (1995)
2.18 C. Zener: Phys. Rev. **81**, 440 (1951). See also C. Zener: Phys. Rev. **83**, 299 (1951)
2.19 C. Zener: Phys. Rev. **82**, 403 (1951)
2.20 P.W. Anderson and H. Hasegawa: Phys. Rev. **100**, 675 (1955)
2.21 K. Kubo and N. Ohata: J. Phys. Soc. Jpn. **33**, 21 (1972). See also K. Kubo: J. Phys. Soc. Jpn. **51**, 782 (1982)
2.22 M. Cieplak: Phys. Rev. B **18**, 3470 (1973)
2.23 P.-G. de Gennes: Phys. Rev. **118**, 141 (1960)

2.24 J.B. Goodenough: Phys. Rev. **100**, 564 (1955). See also John B. Goodenough: *Magnetism and the Chemical Bond* (Interscience, New York 1963)
2.25 W.A. Harrison: *Electronic Structure and the Properties of Solids* (Dover Books on Physics and Chemistry, 1989)
2.26 For more information on magnetic fields see, for instance, *Physical Phenomena at High Magnetic Fields*, edited by E. Manousakis, P. Schlottmann, P. Kumar, K. Bedell, and F.M. Mueller (Addison-Wesley 1992) More information about strong fields can be found in the introductory review articles by G. Boebinger, Phys. Today, June 1996, p. 36, and by G. Boebinger, A. Passner, and J. Bevk: Sci. Am., June 1995, p. 58

Chapter 3

3.1 P. Schiffer, A.P. Ramirez, W. Bao, and S.-W. Cheong: Phys. Rev. Lett. **75**, 3336 (1995)
3.2 P.G. Radaelli, D.E. Cox, M. Marezio, S-W. Cheong, P.E. Schiffer, and A.P. Ramirez: Phys. Rev. Lett. **75**, 4488 (1995)
3.3 N.W. Ashcroft and N.D. Mermin: *Solid State Physics* (Saunders College, 1976)
3.4 S.-W. Cheong and H.Y. Hwang: *Ferromagnetism vs Charge/Orbital Ordering in Mixed-Valent Manganites*, in *Colossal Magnetoresistance Oxides*, edited by Y. Tokura (Gordon & Breach, Monographs in Condensed Matter Science, London 1999)
3.5 Y-K. Yoo, F. Duewer, H. Yang, D. Yi, J-W. Li, and X-D. Xiang: Nature **406**, 704 (2000). See also R. W. Cahn: Nature **410**, 643 (2001) and references therein
3.6 Y. Nagaoka: Phys. Rev. **147**, 392 (1966)
3.7 R. Mahendiran, S. Tiwari, A. Raychaudhuri, T. Ramakrishnan, R. Mahesh, N. Rangavittal, and C.N.R. Rao: Phys. Rev. B **53**, 3348 (1996); H. Chiba, M. Kikuchi, K. Kusaba, Y. Muroba, and Y. Syono: Solid State Commun. **99**, 499 (1996); A. Maignan, C. Martin, F. Damay, and B. Raveau: Chem. Mater. **10**, 950 (1998); C. Martin, A. Maignan, M. Hervieu, and B. Raveau: J. Solid State Chem. **134**, 198 (1997)
3.8 S. Mori, C.H. Chen, and S.-W. Cheong: Nature **392**, 473 (1998)
3.9 P.G. Radaelli, D.E. Cox, L. Capogna, S.-W. Cheong, and M. Marezio: Phys. Rev. B **59**, 14440 (1999)
3.10 J.J. Neumeier, M.F. Hundley, J.D. Thompson, and R.H. Heffner: Phys. Rev. B **52**, R7006 (1995)
3.11 H.Y. Hwang, T.T.M. Palstra, S-W. Cheong, and B. Batlogg: Phys. Rev. B **52**, 15046 (1995)
3.12 G-M. Zhao, K. Conder, H. Keller, and K.A. Müller: Nature **381**, 676 (1996). See also G-M. Zhao et al.: cond-mat/9912355, cond-mat/0008029, and references therein. For work in cuprates, see G-M. Zhao, H. Keller, and K. Conder: J. Phys. Condens. Matter **13**, R569 (2001) and references therein. See also C.A. Perroni, G. De Filippis, V. Cataudella, G. Iadonisi, V. Marigliano Ramaglia, and F. Ventriglia: cond-mat/0203121
3.13 I. Isaac and J.P. Franck: Phys. Rev. B **57**, R5602 (1998); J.P. Franck et al.: Phys. Rev. B **58**, 5189 (1998); J.P. Franck et al.: J. Supercond. **12**, 263 (1999); J.E. Gordon et al.: submitted to PRL
3.14 Y. Tomioka and Y. Tokura: *Metal-Insulator Phenomena Relevant to Charge/Orbital-Ordering in Perovskite-Type Manganese Oxides*, 1999 preprint

3.15 T. Asaka, S. Yamada, S. Tsutsumi, C. Tsuruta, K. Kimoto, T. Arima, and Y. Matsui: Phys. Rev. Lett. **88**, 097201 (2002)
3.16 H.Y. Hwang, T.T.M. Palstra, S-W. Cheong, and B. Batlogg: Phys. Rev. B **52**, 15046 (1995)
3.17 Y. Moritomo, H. Kuwahara, Y. Tomioka, and Y. Tokura: Phys. Rev. B **55**, 7549 (1997)
3.18 Y. Tomioka, A. Asamitsu, H. Kuwahara, Y. Moritomo, and Y. Tokura: Phys. Rev. B **53**, R1689 (1996)
3.19 A. Urushibara et al.: Phys. Rev. B**51**, 14103 (1995). See also Tokura et al.: J. Phys. Soc. Jpn. **63**, 3931 (1994)
3.20 H. Fujishiro, M. Ikebe, and Y. Konno: J. Phys. Soc. Jpn. **67**, 1799 (1998)
3.21 Y. Moritomo et al.: Phys. Rev. B **58**, 5544 (1998)
3.22 I.E. Dzyaloshinskii: J. Phys. Chem. Solids **4**, 241 (1958); T. Moriya:, Phys. Rev. **120**, 91 (1960). For more recent literature see D. Coffey, K. Bedell, and S. Trugman: Phys. Rev. B **42**, 6509 (1990); D. Coffey, T.M. Rice, and F.C. Zhang: Phys. Rev. B **44**, 10112 (1991). For recent work on manganites addressing the Dzyaloshinksy Moriya interaction see J. Deisenhofer et al., cond-mat/0108515.
3.23 W.E. Pickett and J.S. Moodera: Phys. Today, May 2001, p. 39
3.24 R.A. de Groot et al.: Phys. Rev. Lett. **50**, 2024 (1983)
3.25 K.-I. Kobayashi et al.: Nature **395**, 677 (1998)
3.26 J.-H. Park et al.: Nature **392**, 794 (1998). See also J.-H. Park et al.: Phys. Rev. Lett. **76**, 4215 (1996)
3.27 Y. Tokura: *Fundamental Features of Colossal Magnetoresistive Manganese Oxides*, in *Colossal Magnetoresistance Oxides* edited by Y. Tokura (Gordon & Breach, Monographs in Condensed Matter Science, London 1999)
3.28 R. Kajimoto et al.: Phys. Rev. B **60**, 9506 (1999)
3.29 H. Kawano et al.: cond-mat/9808286
3.30 H. Kawano, R. Kajimoto, H. Yoshizawa, Y. Tomioka, H. Kuwahara, and Y. Tokura: Phys. Rev. Lett. **78**, 4253 (1997)
3.31 R. Kajimoto, H. Yoshizawa, Y. Tomioka, and Y. Tokura: cond-mat/0110170, preprint
3.32 T. Hotta, Y. Takada, H. Koizumi, and E. Dagotto: Phys. Rev. Lett. **84**, 2477 (2000)
3.33 Y. Moritomo, A. Asamitsu, H. Kuwahara, and Y. Tokura: Nature **480**, 141 (1996)
3.34 Y. Imry: *Introduction to Mesoscopic Physics* (Oxford University Press, 1997)
3.35 N.F. Mott: *Metal Insulator Transitions* (Taylor and Francis, London 1990)
3.36 T. Kimura, Y. Tomioka, H. Kuwahara, A. Asamitsu, M. Tamura, and Y. Tokura: Science **274**, 1698 (1996)
3.37 D.N. Argyriou, J.F. Mitchell, P.G. Radaelli, H.N. Bordallo, D.E. Cox, M. Medarde, and J.D. Jorgensen: Phys. Rev. B **59**, 8695 (1999)
3.38 T. Kimura, R. Kumai, Y. Tokura, J. Q. Li, and Y. Matsui: Phys. Rev. B **58**, 11081 (1998). See also J. Q. Li, Y. Matsui, T. Kimura, and Y. Tokura: Phys. Rev. B **57**, R3205 (1998). For studies of the MR effect at high pressure and $x = 0.3$, see T. Kimura, A. Asamitsu, Y. Tomioka, and Y. Tokura: Phys. Rev. Lett. **79**, 3720 (1997)
3.39 C.D. Ling, J.E. Millburn, J.F. Mitchell, D.N. Argyriou, J. Linton, and H.N. Bordallo: Phys. Rev. B **62**, 15096 (2000)
3.40 D.N. Argyriou, J.F. Mitchell, J.B. Goodenough, O. Chmaissem, S. Short, and J.D. Jorgensen: Phys. Rev. Lett. **78**, 1568 (1997). See also R. Osborn, S. Rosenkranz, D.N. Argyriou, L. Vasiliu-Doloc, J.W. Lynn, S.K. Sinha, J.F. Mitchell, K.E. Gray, and S.D. Bader: Phys. Rev. Lett. **81**, 3964 (1998)

3.41 K. Hirota, Y. Moritomo, H. Fujioka, M. Kubota, H. Yoshizawa, and Y. Endoh: J. Phys. Soc. Jpn. **67**, 3380 (1998); M. Kubota, H. Fujioka, K. Hirota, K. Ohoyama, Y. Moritomo, H. Yoshizawa, and Y. Endoh: J. Phys. Soc. Jpn. **69**, 1606 (2000). See also comment by P.D. Battle, M.J. Rosseinsky, and P.G. Radaelli: J. Phys. Soc. Jpn. **68**, 1462 (1999), and reply by K. Hirota et al.: J. Phys. Soc. Jpn. **68**, 1463 (1999)

3.42 M. Tokunaga, Y. Tokunaga, M. Yasugaki, and T. Tamegai: (Physica B, to be published) found FM AF phase separation at $x = 0.45$ in bilayer manganites and at $x = 0.30$ in $La_{1-x}Ca_xMnO_3$, with clusters as large as μm in size

3.43 S.H. Chun, Y. Lyanda-Geller, M.B. Salamon, R. Suryanarayanan, G. Dhalenne, and A. Revcolevschi: preprint, cond-mat/0007249

3.44 Y. Moritomo, Y. Tomioka, A. Asamitsu, Y. Tokura, and Y. Matsui: Phys. Rev. B **51**, 3297 (1995). References to previous literature using ceramics can be found in this paper

3.45 W. Bao, C.H. Chen, S.A. Carter, and S-W. Cheong: Solid State Commun. **98**, 55 (1996)

3.46 B.J. Sternlieb, J.P. Hill, U.C. Wildgruber, G.M. Luke, B. Nachumi, Y. Moritomo, and Y. Tokura: Phys. Rev. Lett. **76**, 2169 (1996)

3.47 Y. Murakami, H. Kawada, H. Kawata, M. Tanaka, T. Arima, Y. Moritomo, and Y. Tokura: Phys. Rev. Lett. **80**, 1932 (1998)

3.48 M. Tokunaga, N. Miura, Y. Moritomo, and Y. Tokura: Phys. Rev. B **59**, 11151 (1999)

Chapter 4

4.1 A. Fetter and J.D. Walecka: *Quantum Theory of Many-Particle Systems* (McGraw-Hill, 1971)

4.2 J. Kanamori: Prog. Theor. Phys. **30**, 275 (1963)

4.3 See, for instance, Chap. 4 of J.S. Griffiths: *The Theory of Transition-Metal Ions* (Cambridge University Press, 1964)

4.4 S. Sugano, Y. Tanabe, and H. Kamimura: *Multiplets of Transition-Metal Ions in Crystals* (Academic Press, 1970). See also J. Otsuka: J. Phys. Soc. Jpn. **21**, 596 (1966), and C. Castellani, C.R. Natoli, and J. Ranninger: Phys. Rev. B **18**, 4945 (1978)

4.5 H. Tang, M. Plihal, and D.L. Mills: J. Magn. Magn. Mater. **187**, 23 (1998)

4.6 M.T. Hutchings: *Solid State Physics*, vol. 16, edited by F. Seitz and D. Turnbull (Academic Press, New York 1964) p. 227

4.7 E. Merzbacher: *Quantum Mechanics*, 3rd edn. (John Wiley & Sons, 1998)

4.8 In the case of 3*d* ions with nondegenerate ground states (there is always some crystal-field perturbation that breaks all degeneracies), it can be shown that the orbital angular momentum is *quenched*, meaning that its expectation value cancels

4.9 K. Yosida: *Theory of Magnetism* (Springer, Berlin Heidelberg New York 1998)

4.10 A. Abragam and B. Bleaney: *Electron Paramagnetic Resonance of Transition Ions* (Dover Publications, New York 1970)

4.11 R. Orbach and H.J. Stapleton: *Electron Spin-Lattice Relaxation*, Chap. 2 of *Electron Paramagnetic Resonance*, edited by S. Geschwind (Plenum Press, New York 1972)

4.12 See p. 624 of Abragam and Bleaney [4.10]

4.13 See C. Cohen-Tannoudji, B. Diu, and F. Laloe: *Quantum Mechanics* (Wiley-Interscience, New York 1977) p. 1047

4.14 H.A. Buckmaster: Can. J. Phys. **40**, 1670 (1962)
4.15 D. Smith and J.H.M. Thornley: Proc. Phys. Soc. **89**, 170 (1966)
4.16 K.W.H. Stevens: Proc. Phys. Soc. **65**, 209 (1952)
4.17 J. Kanamori: J. Appl. Phys. Suppl. **31**, 14S (1960)

Chapter 5

5.1 E. Müller-Hartmann and E. Dagotto: Phys. Rev. B **54**, R6819 (1996)
5.2 A.J. Millis, P.B. Littlewood, and B.I. Shraiman: Phys. Rev. Lett. **74**, 5144 (1995)
5.3 A.J. Millis, B.I. Shraiman, and R. Mueller: Phys. Rev. Lett. **77**, 175 (1996). See also A.J. Millis, B.I. Shraiman, and R. Mueller: Phys. Rev. B **54**, 5389 (1996); and A.J. Millis: Nature **392**, 147 (1998)
5.4 A.J. Millis et al.: Phys. Rev. B **54**, 5405 (1996)
5.5 H. Röder, J. Zang, and A.R. Bishop: Phys. Rev. Lett. **76**, 1356 (1996). See also J. Zang, A.R. Bishop, and H. Röder: Phys. Rev. B **53**, R8840 (1996)
5.6 J.C. Slater and G.F. Koster: Phys. Rev. **94**, 1498 (1954)
5.7 See for instance G. Martins et al.: Phys. Rev. Lett. **84**, 5844 (2000), and references therein
5.8 H. Shiba, R. Shiina, and A. Takahashi: J. Phys. Soc. Jpn. **66** 941 (1997)
5.9 J.-H. Park, C.T. Chen, S.-W. Cheong, W. Bao, G. Meigs, V. Chakarian, and Y.U. Idzerda: Phys. Rev. Lett. **76**, 4215 (1996)
5.10 J.E. Medvedeva, V.I. Anisimov, O.N. Mryasov, and A.J. Freeman: cond-mat/0111038. See also J.E. Medvedeva et al.: cond-mat/0104316, and references therein
5.11 For reviews on density functional theory, see: W. Kohn: Rev. Mod. Phys. **71**, S59 (1999); J. Bernholc: Phys. Today, Sept. 1999, p. 30; M.L. Cohen: Physics Today, July 1979, p. 40; N. Argaman and G. Makov: cond-mat/9806013
5.12 G.M. Zhao: Phys. Rev. B **62**, 11639 (2000)
5.13 D.S. Dessau and Z.-X. Shen: *Colossal Magnetoresistance Oxides*, in *Monographs in Condensed Matter Science*, edited by Y. Tokura (Gordon & Breach, London, 2000)
5.14 Y. Okimoto, T. Katsufuji, T. Ishikawa, A. Urushibara, T. Arima, and Y. Tokura: Phys. Rev. Lett. **75**, 109 (1995)
5.15 M. Quijada, J. Cerne, J.R. Simpson, H.D. Drew, K.H. Ahn, A.J. Millis, R. Shreekala, R. Ramesh, M. Rajeswari, and T. Venkatesan: Phys. Rev. B**58**, 16093 (1998)
5.16 A. Machida, Y. Moritomo, and A. Nakamura: Phys. Rev. B **58**, R4281 (1998)
5.17 S. Satpathy, Z.S. Popovic, and F.R. Vukajlovic: Phys. Rev. Lett. **76**, 960 (1996). See also W.E. Pickett and D.J. Singh: Phys. Rev. B **53**, 1146 (1996)
5.18 K. Yoshida: *Theory of Magnetism* (Springer, Berlin Heidelberg New York 1998)
5.19 A. Bocquet, T. Mizokawa, T. Saitoh, H. Namatame, and A. Fujimori: Phys. Rev. B **46**, 3771 (1992); T. Arima, Y. Tokura, and J.B. Torrance: Phys. Rev. B **48**, 17006 (1993); T. Saitoh, A. Bocquet, T. Mizokawa, H. Namatame, A. Fujimori, M. Abbate, Y. Takeda, and M. Takano: Phys. Rev. B **51**, 13942 (1995)
5.20 V. Perebeinos and P.B. Allen: Phys. Rev. Lett. **85**, 5178 (2000); P.B. Allen and V. Perebeinos: Phys. Rev. Lett. **83**, 4828 (1999)
5.21 T.G. Perring, G. Aeppli, Y. Moritomo, and Y. Tokura: Phys. Rev. Lett. **78**, 3197 (1997)

Chapter 6

6.1 S. Yunoki, J. Hu, A.L. Malvezzi, A. Moreo, N. Furukawa, and E. Dagotto: Phys. Rev. Lett. **80**, 845 (1998)
6.2 E. Dagotto, S. Yunoki, A.L. Malvezzi, A. Moreo, J. Hu, S. Capponi, D. Poilblanc, and N. Furukawa: Phys. Rev. B **58**, 6414 (1998)
6.3 D. Vollhardt et al.: cond-mat/9701150; T. Okabe: cond-mat/9707032; K. Held and D. Vollhardt: Eur. Phys. J. B **5**, 473 (1998); J. Wahle et al.: Phys. Rev. B **58**, 12749 (1998); and references therein
6.4 J.L. Alonso, L.A. Fernández, F. Guinea, V. Laliena, and V. Martín-Mayor: Nucl. Phys. B **596**, 587 (2001)
6.5 Y. Motome and N. Furukawa: cond-mat/0007407. See also N. Furukawa, Y. Motome, and H. Nakata: cond-mat/0103316
6.6 H. Yi, N.H. Hur, and J. Yu: Phys. Rev. B **61**, 9501 (2000)
6.7 H. Röder, R.R.P. Singh, and J. Zang: Phys. Rev. B **56**, 5084 (1997)
6.8 M.J. Calderón and L. Brey: Phys. Rev. B **58**, 3286 (1998)
6.9 Y. Motome and N. Furukawa: cond-mat/0007408. See also Y. Motome and N. Furukawa: cond-mat/0103230
6.10 N. Furukawa: J. Phys. Soc. Jpn. **64**, 2754 (1995)
6.11 M. Yu. Kagan, D.I. Khomskii and M.V. Mostovoy: Eur. Phys. J B **12**, 217 (1999). See also M.Y. Kagan, K.I. Kugel, and D.I. Khomskii: JETP **93**, 415 (2001) (cond-mat/0001245)
6.12 D.I. Khomskii: Physica B **280**, 325 (2000)
6.13 D. Arovas and F. Guinea: Phys. Rev. B **58**, 9150 (1998). See also D. Arovas, G. Gómez-Santos, and F. Guinea: Phys. Rev. B **59**, 13569 (1999), and F. Guinea, G. Gómez-Santos, and D. Arovas: Phys. Rev. B **62**, 391 (2000)
6.14 K. Nagai, T. Momoi, and K. Kubo: preprint, cond-mat/9911091
6.15 Using symmetry arguments F. Zhong and Z.D. Wang: Phys. Rev. B **60**, 11883 (1999) also arrived at the conclusion that PS is present in manganites. M. Guerrero and R.M. Noack: cond-mat/0004265 found PS in the periodic Anderson model as well. P. Schlottmann: Phys. Rev. B **59**, 11484 (1999) also found PS in a simple alloy-analogy model. S-Q. Shen and Z.D. Wang: Phys. Rev. B **58**, R8877 (1998) found PS in studies of the one-orbital model. See also R.Y. Gu et al.: cond-mat/9905152. B.M. Letfulov and J.K. Freericks: (cond-mat/0203262) report phase separation in the Falicov–Kimball model with phonons
6.16 C. Perroni, G. De Filippis, V. Cataudella and G. Iadonisi: cond-mat/0106588, preprint 2001
6.17 B.M. Letfulov: Eur. Phys. J. B **14**, 19 (2000)
6.18 V. Cataudella, G. De Filippis, and G. Iadonisi: cond-mat/0011156
6.19 D.I. Golosov: cond-mat/0110322; to appear in J. Appl. Phys. (2002). See also D.I. Golosov, cond-mat/0206257.
6.20 Y. Motome and M. Imada: ISSN 0082-4798, preprint
6.21 K.A. Müller, and G. Benedek (Eds): *Proc. of the Conference Phase Separation in Cuprate Superconductors* (World Scientific, Singapore 1993)
6.22 P.B. Visscher: Phys. Rev. B **10**, 943 (1974)
6.23 T. Hotta, A.L. Malvezzi, and E. Dagotto: Phys. Rev. B **62**, 9432 (2000)
6.24 E.L. Nagaev: JETP Lett. **6**, 18 (1967); E.L. Nagaev: Sov. Phys. Lett. **27**, 122 (1968); E.L. Nagaev: JETP Lett. **16**, 394 (1972)
6.25 E.L. Nagaev: Phys. Stat. Sol. (b) **186**, 9 (1994); E.L. Nagaev: Phys.-Usp. **38**, 497 (1995)
6.26 E.L. Nagaev: Phys.-Usp. **39**, 781 (1996); E.L. Nagaev: Phys. Rev. B **58**, 2415 (1998)

6.27 A. Mauger and D.L. Mills: Phys. Rev. Lett. **53**, 1594 (1984)
6.28 M.P. Marder: *Condensed Matter Physics* (Wiley & Sons, 2000)
6.29 Studies using the dynamical mean-field approximation ($D = \infty$) applied to the double-exchange model are in good agreement with the phase diagram found numerically in the opposite limit of low dimensions. Not only AF, FM, and phase separation but also spin-incommensurate regimes clearly appear in the $T = 0$ phase diagram. See A. Chattopadhyay, A.J. Millis, and S. Das Sarma: cond-mat/0004151. See also Fig. 3 of A. Chattopadhyay, A.J. Millis, and S. Das Sarma: Phys. Rev. B **61**, 10738 (2000)
6.30 S.R. White: Phys. Rev. Lett. **69**, 2863 (1992). For a review see K. Hallberg: cond-mat/9910082; and references therein
6.31 A. Weisse, J. Loos, and H. Fehske: cond-mat/0101234 and cond-mat/0101235, preprints
6.32 E. Dagotto and T.M. Rice: Science **271**, 618 (1996)
6.33 J. Riera, K. Hallberg, and E. Dagotto: Phys. Rev. Lett. **79**, 713 (1997)
6.34 L.-C. Ku, S.A. Trugman, and J. Bonca: cond-mat/0109282; and references therein
6.35 C. Gazza, G. Martins, J. Riera, and E. Dagotto: Phys. Rev. B **59**, R709 (1999)
6.36 A.L. Malvezzi, S. Yunoki, and E. Dagotto: Phys. Rev. B **59**, 7033 (1999)
6.37 J. Lorenzana, C. Castellani, C. Di Castro: cond-mat/0010092, and references therein
6.38 E.L. Nagaev: Phys. Stat. Sol. (b) **186**, 9 (1994); Phys.-Usp. **38**, 497 (1995). For more recent references, see E.L. Nagaev: cond-mat/0012321
6.39 C.S. Hellberg: cond-mat/0010429. See also Q. Yuan and P. Thalmeier: Phys. Rev. Lett. **83**, 3502 (1999); R. Pietig, R. Bulla, and S. Blawid: Phys. Rev. Lett. **82**, 4046 (1999)
6.40 S. Yunoki and A. Moreo, Phys. Rev. B **58**, 6403 (1998)
6.41 A. Moreo, M. Mayr, A. Feiguin, S. Yunoki, and E. Dagotto: Phys. Rev. Lett. **84**, 5568 (2000)
6.42 S. Yunoki, T. Hotta, and E. Dagotto: Phys. Rev. Lett. **84**, 3714 (2000)
6.43 D.J. Garcia, K. Hallberg, C.D. Batista, M. Avignon, and B. Alascio: Phys. Rev. Lett. **85**, 3720 (2000)
6.44 For other studies on the relevance of $J_{\rm AF}$, see H. Yi and J. Yu: Phys. Rev. B **58**, 11123 (1998); H. Yi and S-I. Lee: Phys. Rev. B **60**, 6250 (1999); H. Yi, J. Yu, and S-I. Lee: cond-mat/9910152. See also D.I. Golosov, M.R. Norman, and K. Levin: Phys. Rev. B **58**, 8617 (1998)
6.45 H. Aliaga, B. Normand, K. Hallberg, M. Avignon, and B. Alascio: Phys. Rev. B **64**, 024422 (2001)
6.46 M. Yamanaka, W. Koshibae and S. Maekawa: Phys. Rev. Lett. **81**, 5604 (1998)
6.47 D.F. Agterberg and S. Yunoki: Phys. Rev. B **62**, 13816 (2000)
6.48 J.L. Alonso, J.A. Capitán, L.A. Fernández, F. Guinea, and V. Martín-Mayor: Phys. Rev. B **64**, 054408 (2001)
6.49 J.L. Alonso, L.A. Fernández, F. Guinea, V. Laliena, and V. Martín-Mayor: Phys. Rev. B **63**, 064416 (2001)
6.50 J.L. Alonso, L.A. Fernández, F. Guinea, V. Laliena, and V. Martín-Mayor: Phys. Rev. B **63**, 054411 (2001)
6.51 For other studies on the relevance of the complex phase of the hopping matrix element, see V.Z. Cerovski, S.D. Mahanti, T.A. Kaplan, and A. Taraphder: Phys. Rev. B **59**, 13977 (1999)
6.52 G.-M. Zhao: Phys. Rev. B **62**, 11639 (2000); M. Vogt, C. Santos, and W. Nolting: Phys. Stat. Sol. (b) **223**, 679 (2001); D. Meyer, C. Santos, and W. Nolting: J. Phys.: Condens. Matter **13**, 2531 (2001); W. Nolting, G.G. Reddy,

A. Ramakanth and D. Meyer: Phys. Rev. B **64**, 155109 (2001); D.M. Edwards and A.C.M. Green: cond-mat/0109266
6.53 H. Tsunetsugu, M. Sigrist and K. Ueda: Rev. Mod. Phys. **69**, 809 (1997)
6.54 J. Kondo: Prog. Theor. Phys. **32**, 37 (1964)
6.55 L. Kouwenhoven and L. Glazman: Phys. World, January 2001, p. 33
6.56 J.R. Schrieffer: J. Low Temp. Phys. **99**, 397 (1995); B.L. Altshuler et al.: Phys. Rev. B **52**, 5563 (1995); A. Chubukov: Phys. Rev. B **52**, R3840 (1995); C.-X. Chen et al.: Phys. Rev. B **43**, 3771 (1991)
6.57 C. Buhler et al.: Phys. Rev. Lett. **84**, 2690 (2000)
6.58 C. Buhler et al.: Phys. Rev. B **62**, R3620 (2000)
6.59 M. Moraghebi et al.: Phys. Rev. B **63**, 214513 (2001)
6.60 M. Moraghebi, S. Yunoki, and A. Moreo: Phys. Rev. Lett. **88**, 187001 (2002)
6.61 M. Moraghebi, S. Yunoki, and A. Moreo: cond-mat/0205201

Chapter 7

7.1 K. Binder, and D.W. Heerman: *Monte Carlo Simulation in Statistical Physics* (Springer, Berlin Heidelberg New York 1992)
7.2 K. Binder (Ed.): *The Monte Carlo Method in Condensed Matter Physics*. Topics Appl. Phys. Vol. 71 (Springer, Berlin Heidelberg New York 1992)
7.3 J.J. Binney, N.J. Dowrick, A.J. Fisher, and M.E.J. Newman: *The Theory of Critical Phenomena, An Introduction to the Renormalization Group* (Oxford Science Pub., Oxford 1993)
7.4 J.W. Negele and H. Orland: *Quantum Many-Particle Systems* (Addison-Wesley, 1988)
7.5 E. Manousakis: Rev. Mod. Phys. **63**, 1 (1991)
7.6 N. Metropolis, A. Rosenbluth, M. Rosenbluth, and A. Teller: J. Chem. Phys. **21**, 1087 (1953)
7.7 *Numerical Recipes in Fortran 90: The Art of Parallel Scientific Computing* (Fortran Numerical Recipes, Vol. 2) W.H. Press et al. (Eds.)
7.8 R.H. Swendsen, and J.-S. Wang: Phys. Rev. Lett. **58**, 86 (1987)
7.9 U. Wolff: Phys. Rev. Lett. **62**, 361 (1989)
7.10 L. Onsager: Phys. Rev. **65**, 117 (1944); B. Kauffman and L. Onsager: Phys. Rev. **76**, 1244 (1949)
7.11 D.M. Ceperley: Rev. Mod. Phys. **67**, 279 (1995); N. Trivedi and D.M, Ceperley: Phys. Rev. B **41**, 4552 (1990)
7.12 W. von der Linden: *A Quantum Monte Carlo Approach to Many-Body Physics*, Phys. Rep. **230**, 53 (1992)
7.13 R. Blankenbecler, D.J. Scalapino, and R.L. Sugar: Phys. Rev. D **24**, 2278 (1981); J.E. Gubernatis, D.J. Scalapino, R.L. Sugar, and W.D. Toussaint: Phys. Rev. B **32**, 103 (1985)
7.14 S.R. White, D.J. Scalapino, R.L. Sugar, E.Y. Loh, J.E. Gubernatis, and R.T. Scalettar: Phys. Rev. B **40**, 506 (1989)
7.15 M. Suzuki: Prog. Theor. Phys. **56**, 1454 (1976)
7.16 H. Trotter: Proc. Am. Math. Soc. **10**, 545 (1959)
7.17 R.J. Glauber: J. Math. Phys. **4**, 294 (1963)
7.18 N. Shibata, K. Ueda: Phys. Rev. B **51**, 6 (1995)
7.19 J. Zang, H. Röder, A.R. Bishop, and S.A. Trugman: J. Phys. Condens. Matter. **9**, L157 (1997)
7.20 E. Dagotto, S. Yunoki, A.L. Malvezzi, A. Moreo, J. Hu: Phys. Rev. B **58**, 6414 (1998)

7.21 J. Riera, K. Hallberg, E. Dagotto: Phys. Rev. Lett. **79**, 713 (1997)
7.22 P. Gambardella et al.: Nature **416**, 301 (2002)
7.23 A. Auerbach: *Interacting Electrons and Quantum Magnetism* (Springer, Berlin Heidelberg New York 1994)
7.24 N.D. Mermin and H. Wagner: Phys. Rev. Lett. **17**, 1133 (1966)
7.25 E. Manousakis: Rev. Mod. Phys. **63**, 1 (1991)
7.26 I. Affleck, T. Kennedy, E.H. Lieb and H. Tasaki: Commun. Math. Phys. **115**, 477 (1988)
7.27 J. Frolich, R. Israel, E.H. Lieb and B. Simon: Commun. Math. Phys. **62**, 1 (1978)
7.28 H. Nakano and M. Takahashi: Phys. Rev. B **52**, 6606 (1995)
7.29 B.S. Shastry: Phys. Rev. Lett. **60**, 639 (1988)
7.30 F.D.M. Haldane: Phys. Rev. Lett. **60**, 635 (1988)
7.31 F.J. Dyson: Commun. Math. Phys. **21**, 269 (1971)
7.32 B. Simon: J. Stat. Phys. **26**, 307 (1981)
7.33 R.B. Griffiths: Phys. Rev. **136A**, 327 (1964)
7.34 L. Onsager: Phys. Rev. **65**, 117 (1944)
7.35 D. Knuth, *The Art of Computer Programming* (Addison-Wesley, 1973)

Chapter 8

8.1 J.R. Schrieffer, X.-G. Wen, and S.-C. Zhang: Phys. Rev. B **39**, 11663 (1989); and references therein
8.2 H. Koizumi, T. Hotta, Y. Takada: Phys. Rev. Lett. **80**, 4518 (1998); H. Koizumi, T. Hotta, Y. Takada: Phys. Rev. Lett. **81**, 3803 (1998). See also T. Hotta, Y. Takada, and H. Koizumi: Int. J. Mod. Phys. B **12**, 3437 (1998)
8.3 T. Hotta, A. Malvezzi, and E. Dagotto: Phys. Rev. B **62**, 9432 (2000)

Chapter 9

9.1 M. Fabrizio, M. Altarelli, and M. Benfatto: Phys. Rev. Lett. **80**, 3400 (1998); C.W.M. Castleton and M. Altarelli: Phys. Rev. B **62**, 1033 (2000)
9.2 L. Paolosini et al.: Phys. Rev. Lett. **82**, 4719 (1999)
9.3 K. Held and D. Vollhardt: Phys. Rev. Lett. **84**, 5168 (2000)
9.4 S. Yunoki, A. Moreo, and E. Dagotto: Phys. Rev. Lett. **81**, 5612 (1998)
9.5 More exotic ferromagnetic states with complex orbital ordering have also been considered [see J. van den Brink and D. Khomskii: Phys. Rev. B **63**, 140416(R) (2001); R. Maezono and N. Nagaosa: Phys. Rev. B **62**, 11576 (2000)]
9.6 K.I. Kugel, and D. Khomskii: Sov. Phys.-JETP **37**, 725 (1974); and references therein
9.7 T. Hotta, S. Yunoki, M. Mayr, and E. Dagotto: Phys. Rev. B **60**, R15009 (1999)
9.8 M. Capone, D. Feinberg, and M. Grilli: Euro. Phys. J. B**17**, 103 (2000)
9.9 S. Fratini, D. Feinberg, and M. Grilli: cond-mat/0011419 (preprint 2000)
9.10 W. Koshibae, Y. Kawamura, S. Ishihara, S. Okamoto, J. Inoue, and S. Maekawa: J. Phys. Soc. Jpn. **66**, 957 (1997)
9.11 L.F. Feiner and A.M. Oleś: Phys. Rev. B **59**, 3295 (1999)
9.12 I. Solovyev, N. Hamada, and K. Terakura: Phys. Rev. Lett. **76**, 4825 (1996)
9.13 S. Ishihara, M. Yamanaka, and N. Nagaosa: Phys. Rev. B **56**, 686 (1997)

9.14 S. Ishihara, J. Inoue, and S. Maekawa: Phys. Rev. B **55**, 8280 (1997)
9.15 J.J. Betouras and S. Fujimoto: Phys. Rev. B **59**, 529 (1999)
9.16 H. Yi and S.-I. Lee: Phys. Rev. B **60**, 6250 (1999)
9.17 R. Maezono, S. Ishihara, and N. Nagaosa: Phys. Rev. B **58**, 11583 (1998)
9.18 T. Mizokawa and A. Fujimori: Phys. Rev. B **51**, R12880 (1995); ibid. **54**, 5368 (1996)
9.19 J. van den Brink, P. Horsch, F. Mack, and A.M. Oleś: Phys. Rev. B **59**, 6795 (1999)
9.20 A. Muñoz et al.: Inorg. Chem. **40**, 1020 (2001)
9.21 T. Kimura et al.: preprint, 2002
9.22 T. Hotta, M. Moraghebi, A. Feiguin. A. Moreo, S. Yunoki, and E. Dagotto, preprint, 2002
9.23 J.-S. Zhou, H.Q. Yin, and J.B. Goodenough: Phys. Rev. B **63**, 184423 (2001)
9.24 J.M. De Teresa et al.: in preparation
9.25 J.L. Alonso, L.A. Fernandez, F. Guinea, V. Laliena, and V. Martin-Mayor: cond-mat/0111244
9.26 Y. Motome and N. Furukawa: cond-mat/0203213
9.27 W. Koller, A. Prüll, H. G. Evertz, and W. von der Linden, cond-mat/0206049 also found phase separation in the two-orbitals model.
9.28 S. Okamoto, S. Ishihara, and S. Maekawa: Phys. Rev. B **61**, 451 (2000)
9.29 M. Guerrero, and R.M. Noack: preprint 2000, cond-mat/0004265
9.30 G. Varelogiannis: Phys. Rev. Lett. **85**, 4172 (2000)
9.31 G. Jackeli and N.B. Perkins: cond-mat/0110185
9.32 P. Horsch, J. Jaklic, and F. Mack: Phys. Rev. B **59**, 6217 (1999)
9.33 A.M. Oleś, M. Cuoco, and N.B. Perkins, in: *Lectures on the Physics of Highly Correlated Electron Systems IV*, edited by F. Mancini, AIP Conference Proceedings Vol. 527 (AIP, New York 2000) pp. 226–380. See also [9.11] and references therein
9.34 K. Hirota, N. Kaneko, A. Nishizawa, and Y. Endoh: J. Phys. Soc. Jpn. **65**, 3736 (1996)
9.35 See for instance: G. Khaliullin and V. Oudovenko: Phys. Rev. B **56**, R14243 (1997); J. van den Brink, W. Stekelenburg, D. Khomskii, G. Sawatzky, and K. Kugel: Phys. Rev. B **58**, 10276 (1998); V. Perebeinos and P.B. Allen: Phys. Stat. Sol. B **215**, 607 (1999)
9.36 E. Saitoh et al.: Nature **410**, 180 (2001). See also P.B. Allen and V. Perebeinos: Nature **410**, 155 (2001). For a mixture of electron electron and lattice interactions applied to orbital excitations, see J. van den Brink: cond-mat/0107572
9.37 Y. Murakami et al.: Phys. Rev. Lett. **80**, 1932 (1998)
9.38 Y. Wakabayashi et al.: J. Phys. Soc. Jpn. **70**, 1194 (2001)
9.39 Y. Murakami et al.: Phys. Rev. Lett. **81**, 582 (1998)
9.40 Y. Endo et al.: Phys. Rev. Lett. **82**, 4328 (1999)
9.41 P. Wochner et al.: (unpublished)
9.42 K. Nakamura et al.: Phys. Rev. B **60**, 2425 (1999)
9.43 Y. Tanaka et al.: J. Phys.: Condens. Matter **11**, L505 (1999)
9.44 K. Hirota et al.: Phys. Rev. Lett. **84**, 2706 (2000)
9.45 M. v. Zimmermann et al.: Phys. Rev. Lett. **83**, 4872 (1999)
9.46 M. v. Zimmermann et al.: cond-mat/0007231, Phys. Rev. B (in press)
9.47 J. Garcia et al.: Phys. Rev. Lett. **85**, 578 (2000)
9.48 Y. Wakabayashi et al.: J. Phys. Soc. Jpn. **69**, 2731 (2000)
9.49 M. Noguchi et al.: Phys. Rev. B **62**, R9271 (2000)
9.50 B. Keimer et al.: Phys. Rev. Lett. **85**, 3946 (2000)
9.51 H. Nakao et al.: Phys. Rev. Lett. **85**, 4349 (2000)
9.52 H. Nakao et al.: (unpublished)

9.53 H. Nakao et al.: J. Phys. Soc. Jpn. **70**, 1857 (2001)
9.54 D.F. McMorrow et al.: Phys. Rev. Lett. **87**, 057201 (2001)
9.55 I.S. Elfimov, V.I. Anisimov, and G.A. Sawatzky: Phys. Rev. Lett. **82**, 4264 (1999)
9.56 S. Ishihara and S. Maekawa: Phys. Rev. B **58**, 13442 (1998)
9.57 S. Ishihara et al.: Physica C **263**, 130 (1996), Phys. Rev. B **55**, 8280 (1997)
9.58 Y. Murakami et al.: Phys. Rev. Lett. **80**, 1932 (1998)
9.59 Y. Murakami et al.: Phys. Rev. Lett. **81**, 582 (1998)
9.60 K. Nakamura et al.: Phys. Rev. B **60**, 2425 (1999)
9.61 Y. Endoh et al.: Phys. Rev. Lett. **82**, 4328 (2000)
9.62 M. von Zimmermann et al.: Phys. Rev. Lett. **83**, 4872 (2000)
9.63 S. Ishihara and S. Maekawa: Phys. Rev. Lett. **80**, 3799 (1998)
9.64 S. Ishihara and S. Maekawa: Phys. Rev. B **58**, 13442 (1998)
9.65 S. Ishihara and S. Maekawa: Phys. Rev. B **62**, 5690 (2000)
9.66 I.S. Elfimov et al.: Phys. Rev. Lett. **82**, 4264 (1999)
9.67 M. Benfatto et al.: Phys. Rev. Lett. **83**, 636 (1999)
9.68 M. Takahashi et al.: J. Phys. Soc. Jpn. **68**, 2530 (1999)
9.69 H. Nakao et al.: (unpublished)
9.70 S. Ishihara and S. Maekawa: Phys. Rev. B **62**, 2338 (2000)
9.71 S. Ishihara and S. Maekawa: Phys. Rev. B **62**, R9252 (2000)
9.72 H. Kondo et al.: Phys. Rev. B **64**, 014414 (2001)
9.73 T. Inami et al.: (unpublished) cond-mat/0109509

Chapter 10

10.1 R.Y. Gu and C.S. Ting: cond-mat/0204095
10.2 T. Mizokawa and A. Fujimori: Phys. Rev. B **56**, R493 (1997)
10.3 J.D. Lee and B.I. Min: Phys. Rev. B **55**, R14713 (1997)
10.4 G. Jackeli, N.B. Perkins, and N.M. Plakida: Phys. Rev. B **62**, 372 (2000)
10.5 V.I. Anisimov et al.: Phys. Rev. B **55**, 15494 (1997)
10.6 T. Mizokawa, D.I. Khomskii, and G.A. Sawatzky: Phys. Rev. B **61**, R3776 (2000)
10.7 M. Korotin et al.: cond-mat/9912456
10.8 T. Hotta and E. Dagotto: Phys. Rev. B **61**, R11879 (2000)
10.9 T. Hotta, A. Feiguin, and E. Dagotto: Phys. Rev. Lett. **86**, 4922 (2001)
10.10 For a review on $La_{2-x}Sr_xCuO_4$, see M.A. Kastner, R.J. Birgeneau, G. Shirane, and Y. Endoh: Rev. Mod. Phys. **70**, 897 (1998)
10.11 P. Dai, H.A. Mook, and F. Dogan: Phys. Rev. Lett. **80**, 1738 (1998)
10.12 I. Solovyev and K. Terakura: Phys. Rev. Lett. **83**, 2825 (1999)
10.13 J. van den Brink, G. Khaliullin, and D. Khomskii: Phys. Rev. Lett. **83**, 5118 (1999). See also the comment and reply, S.-Q. Shen: Phys. Rev. Lett. **86**, 5842 (2001); J. van den Brink, G. Khaliullin, and D. Khomskii: Phys. Rev. Lett. **86**, 5843 (2001). See also S.Q. Shen and Z.D. Wang: Phys. Rev. B **59**, 14484 (1999)
10.14 J. Garcia, M. Concepcion Sanchez, J. Blasco, G. Subías, and M. Grazia Proietti: J. Phys. Condens. Matter. **13**, 3243 (2001). See also J. Garcia, M. Concepcion Sanchez, G. Subías, and J. Blasco: J. Phys.: Condens. Matter. **13**, 3229 (2001); P. Mahadevan, K. Terakura, and D.D. Sarma: Phys. Rev. Lett. **87**, 66404 (2001) The absence of charge ordering in the Verwey transition of related compounds was discussed in J. Garcia, G. Subías, M.G. Proietti, H. Renevier, Y. Joly, J.L. Hodeau, J. Blasco, M.C. Sanchez and J.F. Berar:

Phys. Rev. Lett. 85, 578 (2000). See also J. Garcia et al: Phys. Rev. B **63**, 054110 (2001), and G. Subías et al.: J. Phys.: Condens. Matter **14**, 5017 (2002)
10.15 A. Daoud-Aladine, J. Rodriguez-Carvajal, L. Pinsard-Gaudard, M.T. Fernandez-Diaz, and A. Revcolevschi: Phys. Rev. Lett. **89**, 097205 (2002) (see also Publisher's Note, Phys. Rev. Lett. **89**, 129902 (2002)).
10.16 D.I. Khomskii and K.I. Kugel: cond-mat/0112340

Chapter 11

11.1 E.O. Wollan and C.G. Shull: Phys. Rev. **73**, 830 (1948)
11.2 O. Halpern and M.H. Johnson: Phys. Rev. **55**, 895 (1939)
11.3 C.G. Shull, E.O. Wollan and W.A. Strauser: Phys. Rev. **81**, 483 (1951)
11.4 L. Néel: Ann. Phys. **3**, 137 (1948)
11.5 C.G. Shull, W.A. Strauser and E.O. Wollan: Phys. Rev. **83**, 333 (1951)
11.6 E.O. Wollan and W.C. Koehler: Phys. Rev. **100**, 545 (1955)
11.7 S.W. Lovesey: *Theory of Neutron Scattering from Condensed Matter* (Oxford University Press, Oxford 1984)
11.8 G.L. Squires: *Introduction to the Theory of Thermal Neutron Scattering* (Cambridge, 1978)
11.9 D.L. Price and K. Sköld, in *Methods of Experimental Physics*, vol. 23 Neutron Scattering, Part A, Chap. 1. K. Sköld and D.L. Price, (Eds.) (Academic Press, 1986)
11.10 J. Fernandez-Baca et al.: Phys. Rev. Lett. **80**, 4012 (1998)
11.11 P. Dai, J.A. Fernandez-Baca, E.W. Plummer, Y. Tomioka and Y. Tokura: Phys. Rev. B **64**, 224429 (2001)
11.12 J.W. Lynn et al.: Phys. Rev. Lett. **76**, 4046 (1996). See also J.W. Lynn et al.: J. Appl. Phys. **81**, 5488 (1997)
11.13 This peak is unrelated to the static central peak associated with a soft phonon/structural distortion observed in $SrTiO_3$ (see, for instance, R. Wang et al.: Phys. Rev. Lett. **80**, 2370 (1998), and references therein). The effect in the latter is extrinsic and sample dependent, while in manganites it is intrinsic. The author thanks J. Lynn and J.A. Fernandez-Baca for conversations on these issues
11.14 A.M. Oleś and L.F. Feiner: Phys. Rev. B **65**, 052414 (2002)
11.15 B.I. Halperin and P.C. Hohenberg: Phys. Rev. **188**, 898 (1969), and references therein
11.16 R.S. Fishman: Phys. Rev. B **62**, R3600 (2000), and references therein. See also A.L. Chernyshev and R.S. Fishman, cond-mat/0207305.
11.17 T.G. Perring, G. Aeppli, S.M. Hayden, S.A. Carter, J.P. Remeika, and S.-W. Cheong: Phys. Rev. Lett. **77**, 711 (1996)
11.18 N. Furukawa: J. Phys. Soc. Jpn. **65**, 1174 (1996). See also D.I. Golosov: cond-mat/9909213, and [11.14]
11.19 L. Vasiliu-Doloc, J.W. Lynn, A.H. Moudden, A.M. de Leon-Guevara, and A. Revcolevschi: Phys. Rev. B **58**, 14913 (1998)
11.20 M. Martin, G. Shirane, Y. Endoh, K. Hirota, Y. Moritomo, and Y. Tokura: Phys. Rev. B **53**, 14285 (1996)
11.21 H.Y. Hwang, P. Dai, S.-W. Cheong, G. Aeppli, D.A. Tennant, and H.A. Mook: Phys. Rev. Lett. **80**, 1316 (1998)
11.22 P. Dai, H.Y. Hwang, J. Zhang, J.A. Fernandez-Baca, S.-W. Cheong, C. Kloc, Y. Tomioka, and Y. Tokura: Phys. Rev. B **61**, 9553 (2000)

11.23 N. Furukawa: J. Phys. Soc. Jpn. **68**, 2522 (1999); and cond-mat/9907362
11.24 G. Khaliullin and R. Kilian: Phys. Rev. B **61**, 3494 (2000)
11.25 D.I. Golosov: Phys. Rev. Lett. **84**, 3974 (2000); R. Maezono and N. Nagaosa: Phys. Rev. B **61**, 1189 (2000)
11.26 T.A. Kaplan and S.D. Mahanti, and Y.-S. Su: Phys. Rev. Lett. **86**, 3634 (2001). See also T.A. Kaplan and S.D. Mahanti: J. Phys. Condens. Matter **9**, L291 (1997)
11.27 F. Mancini, N.B. Perkins, and N.M. Plakida: cond-mat/0011464
11.28 D.M. Edwards: cond-mat/0201558. See also M. Hohenadler and D.M. Edwards: cond-mat/0111175
11.29 R. Kajimoto, H. Yoshizawa, H. Kawano-Furukawa, H. Kuwahara, Y. Tomioka, and Y. Tokura: J. Magn. Magn. Mater. **226**, 892 (2001)
11.30 N. Shannon and A.V. Chubukov: cond-mat/0204049
11.31 H. Fujioka, M. Kubota, K. Hirota, H. Yoshizawa, Y. Moritomo, and Y. Endoh: J. Phys. Chem. Solids **60**, 1165 (1999); K. Hirota, S. Ishihara, H. Fujioka, M. Kubota, H. Yoshizawa, Y. Moritomo, Y. Endoh, and S. Maekawa: cond-mat/0104535. See also T. Chatterji, L.P. Regnault, P. Thalmeier, R. Suryanarayanan, G. Dhalenne, and A. Revcolevschi: Phys. Rev. B **60**, R6965 (1999)
11.32 H. Yoshizawa, H. Kawano, J.A. Fernandez-Baca, H. Kuwahara, and Y. Tokura: Phys. Rev. B **58**, R571 (1998)
11.33 M.R. Ibarra et al.: Phys. Rev. Lett. **75**, 3541 (1995)
11.34 M.R. Ibarra et al.: Phys. Rev. B **57**, 7446 (1998)
11.35 J.M. De Teresa et al.: Phys. Rev. B **54**, R12689 (1996)
11.36 J.M. De Teresa et al.: Phys. Rev. B **56**, 3317 (1997)
11.37 See Fig. 8 of J. Blasco et al.: J. Phys. Condens. Matter **8**, 7427 (1996)
11.38 C. Kapusta, P.C. Riedi, W. Kocemba, G. Tomka, M.R. Ibarra, J.M. De Teresa, M. Viret, and J.M.D. Coey: J. Phys. Condens. Matter **11**, 4079 (1999)
11.39 J.M. De Teresa et al.: Nature **386**, 256 (1997)
11.40 C.A. Ramos et al.: contribution to the ICM 2000 conference, Recife, Brazil
11.41 M. Uehara, S. Mori, C.H. Chen, and S.-W. Cheong: Nature **399**, 560 (1999)
11.42 C.H. Chen, S. Mori, and S.-W. Cheong: Phys. Rev. Lett. **83**, 4792 (1999). See also C.H. Chen and S.W. Cheong: Phys. Rev. Lett. **76**, 4042 (1996); P.G. Radaelli, D.E. Cox, M. Marezio, and S.-W. Cheong: Phys. Rev. B **55**, 3015 (1997)
11.43 V. Kiryukhin et al.: Phys. Rev. B **63**, 024420 (2001). Results for $Nd_{1-x}Sr_xMnO_3$ at $x = 0.5$ can be found in V. Kiryukhin et al.: Phys. Rev. B **63**, 144406 (2001)
11.44 K.H. Kim, M. Uehara, C. Hess, P.A. Sharma, and S.-W. Cheong: Phys. Rev. Lett. **84**, 2961 (2000)
11.45 M. Fäth, S. Freisem, A.A. Menovsky, Y. Tomioka, J. Aarts, and J.A. Mydosh: Science **285**, 1540 (1999)
11.46 A. Biswas, S. Elizabeth, A.K. Raychaudhuri, and H.L. Bhat: Phys. Rev. B **59**, 5368 (1999)
11.47 V.N. Smolyaninova et al.: cond-mat/9903238
11.48 M. Bibes, L. Balcells, S. Valencia, J. Fontcuberta, M. Wojcik, E. Jedryka, and S. Nadolski: Phys. Rev. Lett. **87**, 067210 (2001)
11.49 C. Renner, G. Aeppli, B.-G. Kim, Y.-A. Soh, and S.-W. Cheong: Nature **416**, 518 (2002)
11.50 S.J. Billinge et al.: cond-mat/9907329. See also S.J. Billinge et al.: Phys. Rev. Lett. **77**, 715 (1996)

11.51 G. Biotteau, M. Hennion, F. Moussa, J. Rodriguez-Carvajal, L. Pinsard, A. Revcolevschi, Y.M. Mukovskii, and D. Shulyatev: Phys. Rev. B **64**, 104421 (2001)
11.52 C.H. Booth et al.: Phys. Rev. Lett. **80**, 853 (1998). See also C.H. Booth et al.: Phys. Rev. B **57**, 10440 (1998)
11.53 M. Hennion, F. Moussa, G. Biotteau, J. Rodriguez-Carvajal, L. Pinsard, and A. Revcolevschi: Phys. Rev. Lett. **81**, 1957 (1998), and references therein. See also F. Moussa et al.: Phys. Rev. B **60**, 12299 (1999); and [11.51]
11.54 M. Hennion et al.: cond-mat/991036
11.55 G. Allodi, R. De Renzi, and G. Guidi: Phys. Rev. B **56**, 6036 (1997). See also G. Allodi et al.: Phys. Rev. B **57**, 1024 (1998)
11.56 G. Papavassiliou, M. Fardis, M. Belesi, T.G. Maris, G. Kallias, M. Pissas, D. Niarchos, C. Dimitropoulos, and J. Dolinsek: Phys. Rev. Lett. **84**, 761 (2000)
11.57 C. Kapusta, P.C. Riedi, M. Sikora, and M.R. Ibarra: Phys. Rev. Lett. **84**, 4216 (2000). See also C. Kapusta et al.: J. Appl. Phys. **87**, 7121 (2000)
11.58 J. Dho, I. Kim, and S. Lee: Phys. Rev. B **60**, 14545 (1999); and references therein
11.59 S. Mori, C.H. Chen, and S.-W. Cheong: Phys. Rev. Lett. **81**, 3972 (1998), and references therein
11.60 A. Simopoulos, G. Kallias, E. Devlin, and M. Pissas: Phys. Rev. B **63**, 054403 (2001). See also G. Kallias, M. Pissas, E. Devlin, A. Simopoulos, and D. Niarchos: Phys. Rev. B **59**, 1272 (1999)
11.61 F. Rivadulla, M. Freita-Alvite, M.A. Lopez-Quintela, L.E. Hueso, D.R. Miguens, P. Sande, and J. Rivas: cond-mat/0105088
11.62 G. Papavassiliou et al.: Phys. Rev. B **59**, 6390 (1999). See also M. Belesi et al.: cond-mat/0004332; G. Allodi, R. De Renzi, F. Licci, and M.W. Pieper: Phys. Rev. Lett. **81**, 4736 (1998)
11.63 A. Yakubovskii, K. Kumagai, Y. Furukawa, N. Babushkina, A. Taldenkov, A. Kaul, and O. Gorbenko: Phys. Rev. B **62**, 5337 (2000)
11.64 F.C. Chou, F. Borsa, J.H. Cho, D.C. Johnston, A. Lascialfari, D.R. Torgeson, and J. Ziolo: Phys. Rev. Lett. **71**, 2323 (1993)
11.65 M.-H. Julien, F. Borsa, P. Carretta, M. Horvatic, C. Berthier, and C.T. Lin: Phys. Rev. Lett. **83**, 604 (1999)
11.66 A. Moreo, S. Yunoki, E. Dagotto: Science **283**, 2034 (1999)
11.67 M. Belesi, G. Papavassiliou, M. Fardis, M. Pissas, J.E. Wegrowe, and C. Dimitropoulos: Phys. Rev. B **63**, 180406R (2001)
11.68 G. Papavassiliou, M. Belesi, M. Fardis, and C. Dimitropoulos: Phys. Rev. Lett. **87**, 177204 (2001)
11.69 A.W. Hunt, P.M. Singer, K.R. Thurber, and T. Imai: Phys. Rev. Lett. **82**, 4300 (1999)
11.70 N.J. Curro, B.J. Suh, P.C. Hammel, M. Huecker, B. Buechner, U. Ammerahl, and A. Revcolevschi: Phys. Rev. Lett. **85**, 642 (2000)
11.71 M.-H. Julien, A. Campana, A. Rigammonti, P. Carretta, F. Borsa, P. Kuhns, A.P. Reyes, W.G. Moulton, M. Horvatic, C. Berthier: Phys. Rev. B **63**, 144508 (2001)
11.72 G. Allodi, M. Cestelli Guidi, R. de Renzi, A. Caneiro, L. Pinsard: Phys. Rev. Lett. **87**, 127206 (2001)
11.73 N. Bloembergen, T.J. Rowaland: Acta Met. **1**, 731 (1953)
11.74 C.P. Slichter: *Principles of Magnetic Resonance* (Springer, Berlin Heidelberg New York 1992). See also L.C. Hebel and C.P. Slichter: Phys. Rev. **113**, 1504 (1959)

11.75 C.H. Pennington, D.J. Durand, C.P. Slichter, J.P. Rice, E.D. Bukowski, and D.M. Ginsberg: Phys. Rev. B **39**, 274 (1989)
11.76 P.M. Singer, A.W. Hunt, A.F. Cederstrom, and T. Imai: Phys. Rev. B **60**, 15345 (1999)
11.77 G.B. Teitel'baum, I.M. Abu-Shiekah, O. Bakharev, H.B. Brom, J. Zaanen: Phys. Rev. B **63**, 020507(R) (2000)
11.78 J. Burgy, M. Mayr, V. Martin-Mayor, A. Moreo, E. Dagotto: Phys. Rev. Lett. **87**, 277202 (2001)
11.79 For a review of NMR applied to high-T_c superconductors, see A. Rigamonti, F. Borsa, and P. Carretta: Rep. Prog. Phys. **61**, 1367 (1998)
11.80 C.H. Pennington and C.P. Slichter: *Nuclear Resonance Studies of* $YBa_2Cu_3O_{7-\delta}$, Chap. 5 of *Physical Properties of High Temperature Superconductors II*, (Ed.) D.M. Ginsberg.
11.81 C.P. Poole, Jr., H.A. Farach, and R.J. Creswick: *Superconductivity* (Academic Press, 1995)
11.82 A. Barnabe et al.: Appl. Phys. Lett. **71**, 3907 (1997); B. Raveau et al.: J. Solid State Chem. **130**, 162 (1997)
11.83 G. Martins, M. Laukamp, J. Riera and E. Dagotto: Phys. Rev. Lett. **78**, 3563 (1997); and references therein
11.84 T. Kimura et al.: Phys. Rev. Lett. **83**, 3940 (1999)
11.85 For a review see L.E. Cross: Ferroelectrics **76**, 241 (1987)
11.86 G. Burns and B.A. Scott: Solid State Commun. **13**, 423 (1973)
11.87 E.V. Colla et al.: Phys. Rev. Lett. **85**, 3033 (2000), and references therein. For recent related work on nanometer-size domains in relaxor ferroelectrics, see P.M. Gehring, S.E. Park and G. Shirane: cond-mat/0011090. Diffuse scattering around Bragg points has been studied in these compounds, similarly as in manganites. See for example: D. La-Orauttapong et al.: cond-mat/0106051; K. Hirota et al.: cond-mat/0109386; S.B. Vakhrushev and S.M. Shapiro: cond-mat/0203103
11.88 H. Oshima, M. Nakamura, and K. Miyano: Phys. Rev. B **63**, 075111 (2001)
11.89 H. Oshima et al.: Phys. Rev. B **63**, 094420 (2001)
11.90 Y. Moritomo, A. Machida, S. Mori, N. Yamamoto, and A. Nakamura: Phys. Rev. B **60**, 9220 (1999). See also Y. Moritomo: Phys. Rev. B **60**, 10374 (1999)
11.91 K. Takenaka, S. Okuyama, R. Shiozaki, T. Fujita, and S. Sugai: J. Appl. Phys. **91**, 2994 (2002)
11.92 T. Katsufuji et al.: preprint
11.93 S.B. Ogale et al.: Phys. Rev. B **57**, 7841 (1998)
11.94 M. Hervieu et al.: cond-mat/0010427. See also B. Raveau, A. Maignan, C. Martin, and M. Hervieu: *The important role of crystal chemistry upon the CMR properties of manganites*, contribution to [1.21], p. 43
11.95 R.D. Merithew, M.B. Weissman, F.M. Hess, P. Spradling, E.R. Nowak, J. O'Donnell, J.N. Eckstein, Y. Tokura and Y. Tomioka: Phys. Rev. Lett. **84**, 3442 (2000); and references therein
11.96 B. Raquet, A. Anane, S. Wirth, P. Xiong, and S. von Molnar: Phys. Rev. Lett. **84**, 4485 (2000); and references therein. For a comment and reply on this paper see M. Weissman: Phys. Rev. Lett. **86**, 1390 (2001) and A. Anane and S. von Molnar: Phys. Rev. Lett. **86**, 1391 (2001)
11.97 A. Anane, B. Raquet, S. von Molnar, L. Pinsard-Godart, and A. Revcolevschi: J. Appl. Phys. **87**, 5025 (2000)
11.98 V. Podzorov, M. Uehara, M.E. Gershenson, T.Y. Koo, and S.-W. Cheong: Phys. Rev. B **61**, R3784 (2000). See also V. Podzorov, M.E. Gershenson, M. Uehara, and S.-W. Cheong: Phys. Rev. B **64**, 115113 (2001); and V.

Podzorov, C.H. Chen, M.E. Gershenson, and S.-W. Cheong: Europhys. Lett. **55**, 411 (2001)
11.99 E. Colla, L. Chao, and M. Weissman: cond-mat/0109485, and references therein
11.100 A. Rakhmanov, K.I. Kugel, Y.M. Blanter, and M.Y. Kagan: Phys. Rev. B **63**, 174424 (2001)
11.101 N. Kida et al.: Phys. Rev. B **62**, R11965 (2000). See also N. Kida et al.: cond-mat/0106129; N. Kida and M. Tonouchi: Appl. Phys. Lett. **78**, 4115 (2001)
11.102 J.-S. Zhou and J.B. Goodenough: Phys. Rev. Lett. **80**, 2665 (1998), and references therein
11.103 R.H. Heffner et al.: cond-mat/9910064, and references therein. See also R.H. Heffner et al.: Phys. Rev. B **63**, 094408 (2001)
11.104 S.H. Chun et al.: cond-mat/9906198. See also M. Jaime et al.: Phys. Rev. B **54**, 11914 (1996); M. Jaime et al.: Appl. Phys. Lett. **68**, 1576 (1996); M. Jaime et al.: Phys. Rev. Lett. **78**, 951 (1997); M. Jaime, P. Lin, S.H. Chun, and M.B. Salamon: Phys. Rev. B **60**, 1028 (1999), and references therein
11.105 Y. Hirai et al.: Univ. of Wisconsin, Madison, preprint, May 9, 2000
11.106 T. Wu, S.B. Ogale, J.E. Garrison, B. Nagaraj, A. Biswas, Z. Chen, R.L. Greene, R. Ramesh, and T. Venkatesan: Phys. Rev. Lett. **86**, 5998 (2001)
11.107 N.E. Massa, H. Tolentino, H. Salva, J.A. Alonso, M.J. Martinez-Lopez, and M.T. Casais: preprint
11.108 R. Przenioslo, I. Sosnowska, E. Suard, A. Hewat, and A.N. Fitch, J. Phys. Condens. Matt. **14**, 5747 (2002)
11.109 A. Maignan, C. Martin, F. Damay, B. Raveau, and J. Hejtmanek: Phys. Rev. B **58**, 2758 (1998)
11.110 J.J. Neumeier and J.L. Cohn: Phys. Rev. B **61**, 14319 (2000), and references therein. See also J.L. Cohn and J.J. Neumeier, cond-mat/0208050.
11.111 R. Mahendiran et al.: cond-mat/0010384 and cond-mat/0106164. In this reference, the random field Ising model was used to explain the results, similarly as previously done by A. Moreo et al.: Phys. Rev. Lett. **84**, 5568 (2000), based on theoretical considerations
11.112 Recent results by J. Burgy, E. Dagotto, and M. Mayr, cond-mat/0207560, preprint (2002), suggest that some percolative transitions can be of first order. Correlated disorder and other forms of percolation are studied in the reference mentioned above.
11.113 C. Martin, A. Maignan, M. Herview, and B. Raveau: Phys. Rev. B **60**, 12191 (1999)
11.114 P.A. Algarabel et al.: Phys. Rev. B **65**, 104437 (2002)
11.115 H. Aliaga, M.T. Causa, H. Salva, M. Tovar, A. Butera, B. Alascio, D. Vega, G. Polla, G. Leyva, and P. Konig: cond-mat/0010295
11.116 Y.-R. Chen and P.B. Allen: cond-mat/0101354
11.117 W. Bao, J.D. Axe, C.H. Chen, and S.-W. Cheong: Phys. Rev. Lett. **78**, 543 (1997)
11.118 V. Podzorov, B.G. Kim, V. Kiryukhin, M.E. Gershenson, and S.-W. Cheong: Phys. Rev. B **64**, 140406 (2001), and references therein
11.119 H.L. Liu et al.: Phys. Rev. Lett. **81**, 4684 (1998)
11.120 M.M. Savosta et al.: Phys. Rev. B **62**, 9532 (2000). See also C. Martin et al.: Phys. Rev. B **62**, 6442 (2000)
11.121 S. Yoon, M. Rubhausen, S.L. Cooper, K.H. Kim and S.-W. Cheong: Phys. Rev. Lett. **85**, 3297 (2000)
11.122 B.F. Woodfield, M.L. Wilson, and J.M. Byers: Phys. Rev. Lett. **78**, 3201 (1997)

11.123 A.L. Cornelius, B. Light, and J.J. Neumeier: submitted to Phys. Rev. B (cond-mat/0108239)
11.124 W.A. Philips: J. Low Temp. Phys. **7**, 351 (1972)
11.125 V.N. Smolyaninova, A. Biswas, X. Zhang, K.H. Kim, B.-G. Kim, S.-W. Cheong, and R.L. Greene: Phys. Rev. B **62**, R6093 (2000)
11.126 M.F. Hundley and J.J. Neumeier: Phys. Rev. B **55**, 11511 (1997); M.F. Hundley: private communication
11.127 L. Ghivelder, I.A. Castillo, M.A. Gusmão, J.A. Alonso, and L.F. Cohen: Phys. Rev. B **60**, 12184 (1999)
11.128 H. Kopferman: *Nuclear Moments* (Academic Press, New York 1958)
11.129 C. Kittel and H. Kroemer: *Thermal Physics* (Freeman, New York 1980) p. 63
11.130 M.R. Lees, O.A. Petrenko, G. Balakrishnan, and D.M. Paul: Phys. Rev. B **59**, 1298 (1999)
11.131 C. Kittel: *Quantum Theory of Solids* (Wiley, New York 1987) p. 55; E.S.R. Gopal: *Specific Heat at Low Temperatures* (Plenum, New York 1966)
11.132 C.D. Ling, J.J. Neumeier, E. Granado, J.W. Lynn, D.N. Argyriou, and P.L. Lee: submitted to Phys. Rev. B

Chapter 12

12.1 J.H. Jung, K.H. Kim, T.W. Noh, E.J. Choi, and J. Yu: Phys. Rev. B **57**, R11043 (1998)
12.2 Y. Okimoto et al.: Phys. Rev. Lett. **75**, 109 (1995). See also Y. Okimoto et al.: Phys. Rev. B **55**, 4206 (1997). Some of these results have been criticized by K. Takenaka et al.: J. Phys. Soc. Jpn. **68**, 1828 (1999). See also K. Takenaka, Y. Sawaki and S. Sugai: Phys. Rev. B **60**, 13011 (1999); K. Takenaka, Y. Sawaki, R. Shiozaki, and S. Sugai: Phys. Rev. B **62**, 13864 (2000)
12.3 H.J. Lee, J.H. Jung, Y.S. Lee, J.S. Ahn, T.W. Noh, K.H. Kim, and S.-W. Cheong, Phys. Rev. B **60**, 5251 (1999)
12.4 K.H. Kim, J.H. Jung, and T.W. Noh: Phys. Rev. Lett. **81**, 1517 (1998)
12.5 J.H. Hung, J.S. Ahn, J. Yu, T.W. Noh, J. Lee, Y. Moritomo, I. Solovyev and K. Terakura: Phys. Rev. B **61**, 6902 (2000)
12.6 J.H. Hung, H.J. Lee, T.W. Noh, E.J. Choi, Y. Moritomo, Y.J. Wang, and X. Wei: Phys. Rev. B **62**, 481 (2000)
12.7 H.J. Lee, K.H. Kim, J.H. Hung, T.W. Noh, R. Suryanarayanan, G. Dhalenne, and A. Revcolevschi: Phys. Rev. B **62**, 11320 (2000)
12.8 J.H. Hung, K.H. Kim, H.J. Lee, J.S. Ahn, N.J. Hur, T.W. Noh, M.S. Kim, and J.-G. Park: Phys. Rev. B **59**, 3793 (1999)
12.9 K.H. Kim, S. Lee, T.W. Noh, and S.-W.: Cheong, preprint
12.10 H.J. Lee, K.H. Kim, M.W. Kim, T.W. Noh, B.G. Kim, T.Y. Koo, S.-W. Cheong, Y.J. Wang, and X. Wei: Phys. Rev. B **65**, 115118 (2002)
12.11 M.W. Kim, J.H. Jung, K.H. Kim, H.J. Lee, J. Yu, T.W. Noh, and Y. Moritomo: cond-7mat/0203201. For related investigations, see T. Ishikawa, K. Tobe, T. Kimura, T. Katsufuji, and Y. Tokura: Phys. Rev. B **62**, 12354 (2000)
12.12 S.L. Cooper: *Structure and Bonding*, Vol. 98 (Springer, Berlin Heidelberg New York 2001) p. 161, and references therein
12.13 T. Katsufuji et al.: Phys. Rev. B **54**, 14230 (1996)
12.14 S. Uchida et al.: Phys. Rev. B **43**, 7942 (1991)

12.15 Results for the optical conductivity in the $D = \infty$ approximation can also be found in A.J. Millis et al.: Phys. Rev. B **54**, 5405 (1996)
12.16 F. Mack and P. Horsch: Phys. Rev. Lett. **82**, 3160 (1999). See also P. Horsch, J. Jaklic, and F. Mack: Phys. Rev. B **59**, 6217 (1999)
12.17 N. Furukawa: J. Phys. Soc. Jpn. **64**, 2734 (1995); ibid. **64**, 3164 (1995). See also N. Furukawa: J. Phys. Soc. Jpn. **63**, 3214 (1994)
12.18 H. Shiba, R. Shiina, and A. Takahashi: J. Phys. Soc. Jpn. **66**, 941 (1997). See also P.E. de Brito and H. Shiba: Phys. Rev. B **57**, 1539 (1998); H. Nakano, Y. Motome, and M. Imada: cond-mat/9905271
12.19 R. Kilian and G. Khaliullin: Phys. Rev. B **58**, R11841 (1998)
12.20 H. Nakano, Y. Motome, and M. Imada: J. Phys. Soc. Jpn. **69**, 1282 (2000)
12.21 A. Chattopadhyay, A.J. Millis, and S. Das Sarma: cond-mat/9908305
12.22 L. Ward: *The Optical Constants of Bulk Materials and Films* (Adam Hilger, 1988)
12.23 C. Kittel: *Introduction to Solid State Physics* (John Wiley & Sons, 1986)
12.24 K. Takenaka et al.: Phys. Rev. B **60**, 13011 (1999)
12.25 M. Quijada et al.: Phys. Rev. B **58**, 16093 (1998)
12.26 F. Wooten: *Optical Properties of Solids* (Academic Press, 1972)
12.27 J.R. Reitz, F.J. Milford, and R.W. Christy: *Foundations of Electromagnetic Theory* (Addison-Wesley, 1979)
12.28 H.G. Tompkins: *A User's Guide to Ellipsometry* (Academic Press, 1993)
12.29 Y. Tokura et al.: Science **288**, 462 (2000)
12.30 J. Orenstein and D.H. Rapkine: Phys. Rev. Lett. **60**, 968 (1988)
12.31 M.W. Kim et al.: (to be published)

Chapter 13

13.1 J.A. Mydosh: *Spin Glasses: an Experimental Introduction* (Taylor & Francis, London 1993)
13.2 A.P. Young (Ed.) *Spin Glasses and Random Fields* (World Scientific, Singapore 1997)
13.3 R.S. Freitas et al.: cond-mat/0010110, and references therein
13.4 I.F. Voloshin, A.V. Kalinov, S.E. Savelev, L.M. Fisher, N.A. Babushkina, L.M. Belova, D.I. Khomskii, K.I. Kugel: JETP Lett. **71**, 106 (2000)
13.5 A. K. Raychaudhuri et al., cond-mat/0007339.
13.6 S. Seiro et al.: J. Magn. Magn. Mater. **226–230**, 988 (2001). See also S. Seiro, H.R. Salva, and A.A. Ghilarducci: Jornadas SAM 2000 preprint; A. Ghilarducci et al.: preprint. Results for $Pr_{1-x}Ca_xMnO_3$ at 400 K show that above the ordering transitions the system probably remains inhomogeneous, as extensively discussed in Chap. 17 and others
13.7 R. Mathieu, P. Nordblad, A.R. Raju, and C.N.R. Rao: preprint
13.8 M. Roy, J.F. Mitchell, A.P. Ramirez, and P. Schiffer: Phys. Rev. B **58**, 5185 (1998)
13.9 M. Roy et al.: cond-mat/0001064
13.10 P. Levy, F. Parisi, G. Polla, D. Vega, G. Leyva, H. Lanza, R.S. Freitas and L. Ghivelder: Phys. Rev. B **62**, 6437 (2000)
13.11 R.S. Freitas, L. Ghivelder, P. Levy, and F. Parisi: Phys. Rev. B **65**, 104403 (2002)
13.12 F. Parisi, P. Levy, L. Ghivelder, G. Polla, and D. Vega: Phys. Rev. B **63**, 144419 (2001). See also J. Sacanell, P. Levy, L. Ghivelder, G. Polla, and F. Parisi: cond-mat/0108026

13.13 P. Levy, L. Granja, E. Indelicato, G. Polla, and F. Parisi: submitted to Phys. Rev. Lett. See also L. Granja, E. Indelicato, P. Levy, G. Polla, D. Vega, and F. Parisi: Proceedings of the LAW3M 2001 conference, Physica B, to be published
13.14 P. Levy, F. Parisi, L. Granja, E. Indelicato, and G. Polla: Phys. Rev. Lett. **89**, 137001 (2002). See also P. Levy, F. Parisi, M. Quintero, L. Granja, J. Curiale, J. Sacanell, G. Leyva, G. Polla, R.S. Freitas, and L. Ghivelder: Phys. Rev. B **65**, 140401 (2002)
13.15 M. Mézard: cond-mat/0005173 and cond-mat/0110363
13.16 A.P. Young: cond-mat/0010371
13.17 E. Dagotto, T. Hotta, and A. Moreo: Phys. Rep. **344**, 1 (2001); A. Moreo, S. Yunoki, and E. Dagotto: Science **283**, 2034 (1999) and references therein
13.18 M. Roy, J.F. Mitchell, and P. Schiffer: J. Appl. Phys. **87**, 5831 (2000)
13.19 R. von Helmolt, J. Wecker, T. Lorenz, and K. Samwer: Appl. Phys. Lett. **67**, 2093 (1995)
13.20 D. N. H. Nam, R. Mathieu, P. Nordblad, N. V. Khiem, and N. X. Phuc, Phys. Rev. B **62**, 1027 (2000).
13.21 M. Sirena et al.: Phys. Rev. B **64**, 104409 (2001)
13.22 J. López, P.N. Lisboa-Filho, W.A.C. Passos, W.A. Ortiz, and F.M. Araujo-Moreira: Phys. Rev. B **63**, 224422 (2001)
13.23 S. Balevicius, B. Vengalis, F. Anisimovas, J. Novickij, R. Tolutis, O. Kiprianovic, V. Pyragas, E.E. Tornau: J. Magn. Magn. Mater. **211**, 243 (2000)
13.24 L.M. Fisher et al.: J. Phys.: Condens. Matter **10**, 9769 (1998)
13.25 A. Anane, J.-P. Renard, L. Reversat, C. Dupas, P. Veillet, M. Viret, L. Pinsard and A. Revocolevschi: Phys. Rev. B **59**, 77 (1999)
13.26 S.B. Ogale, V. Talyansky, C.H. Chen, R. Ramesh, R.L. Greene, and T. Venkatesan: Phys. Rev. Lett. **77**, 1159 (1996)
13.27 Y.G. Zhao, J.J. Li, R. Shreekala, H.D. Drew, C.L. Chen, W.L. Cao, C.H. Lee, M. Rajeswari, S.B. Ogale, R. Ramesh, G. Baskaran, and T. Venkatesan: Phys. Rev. Lett. **81**, 1310 (1998)
13.28 R.H. Heffner, L.P. Le, M.F. Hundley, J.J. Neumeier, G.M. Luke, K. Kojima, B. Nachumi, Y.J. Uemura, D.E. MacLaughlin, and S.-W. Cheong: Phys. Rev. Lett. **77**, 1869 (1996)
13.29 M. Ziese, C. Srinitiwarawong, and C. Shearwood: J. Phys.: Condens. Matter **10**, L659 (1998)
13.30 E. Dagotto: unpublished results
13.31 M. Roy, J.F. Mitchell, S.J. Potashnik, and P. Schiffer: J. Magn. and Magn. Mater. **218**, 191 (2000)
13.32 Y.G. Joh, R. Orbach, G.G. Wood, J. Hammann, E. Vincent: Phys. Rev. Lett. **82**, 438 (1999); E. Vincent, V. Dupuis, M. Alba, J. Hammann, J.P. Bouchaud: Europhys. Lett. **50**, 674 (2000); Y.G. Joh, R. Orbach, J. Hammann: Phys. Rev. Lett. **77**, 4648 (1996)
13.33 J.M. De Teresa et al.: Phys. Rev. Lett. **76**, 3392 (1996)
13.34 A. Maignan et al.: Phys. Rev. B **60**, 15214 (1999)
13.35 I.G. Deac, S.V. Diaz, B.G. Kim, S.-W. Cheong, and P. Schiffer: cond-mat/0203102
13.36 R. Mathieu et al.: Phys. Rev. B **63**, 174405 (2001). See also R. Mathieu, P. Svedlindh, and P. Nordblad: cond-mat/0007154, submitted to Europhys. Lett.
13.37 A. Niazi, P.L. Paulose and E.V. Sampathkumaran: Phys. Rev. Lett. **88**, 107202 (2002). See also A. Niazi, E.V. Sampathkumaran, P.L. Paulose, D. Eckert, A. Handstein, and K.-H. Müller: Solid State Commun. **120**, 11 (2001); and references therein

13.38 R.S. Freitas et al.: Phys. Rev. B **64**, 144404 (2001)
13.39 I. Deac et al.: Phys. Rev. B **63**, 172408 (2001)

Chapter 14

14.1 S. Billinge and T. Egami: Phys. Rev. B **47**, 14386 (1993), and references therein. See also B.H. Toby et al.: Phys. Rev. Lett. **64**, 2414 (1990)
14.2 D. Louca, T. Egami, E.L. Brosha, H. Röder, and A.R. Bishop: Phys. Rev. B **56**, R8475 (1997)
14.3 D. Louca and T. Egami: Phys. Rev. B **59**, 6193 (1999)
14.4 T. Shibata, B. Bunker, J. Mitchell, and P. Schiffer: cond-mat/0107636, to appear in PRL
14.5 Y. Yamada, O. Hino, S. Nohdo, R. Kanao, T. Inami, and S. Katano: Phys. Rev. Lett. **77**, 904 (1996)
14.6 V.A. Ivanshin, J. Deisenhofer, H.-A. Krug von Nidda, A. Loidl, A.A. Mukhin, A.M. Balbashov, and M.V. Eremin: Phys. Rev. B **61**, 6213 (2000). See also M. Paraskevopoulos, F. Mayr, C. Hartinger, A. Pimenov, J. Hemberger, P. Lunkenheimer, A. Loidl, A.A. Mukhin, V.Y. Ivanov, and A.M. Balbashov: J. Magn. Magn. Mater. **211**, 118 (2000); A.A. Mukhin, V.Y. Ivanov, V.D. Travkin, A. Pimenov, A. Loidl, and A.M. Balbashov: Europhys. Lett. **49**, 514 (2000)
14.7 J. Matsuno, A. Fujimori, Y. Takeda, and M. Takano, Europhys. Lett. **59**, 252 (2002).
14.8 A. Pimenov, M. Biberacher, D. Ivannikov, A. Loidl, V.Y. Ivanov, A.A. Mukhin, and A.M. Balbashov: Phys. Rev. B **62**, 5685 (2000)
14.9 D. Ivannikov, M. Biberacher, H.-A. Krug von Nidda, A. Pimenov, A. Loidl, A.A. Mukhin, and A.M. Balbashov: cond-mat/0204112
14.10 S.E. Foss, M. Mayr, and E. Dagotto: preprint
14.11 P. Wagner, I. Gordon, S. Mangin, V.V. Moshchalkov, and Y. Bruynseraede: 1999 preprint, cond-mat/9908374
14.12 D.E. Cox, T. Iglesias, E. Moshopoulou, K. Hirota, K. Takahashi, and Y. Endoh: cond-mat/0010339
14.13 V. Kiryukhin, Y.J. Wang, F.C. Chou, M.A. Kastner, and R.J. Birgeneau: Phys. Rev. B **59**, R6581 (1999)
14.14 R. Cauro et al.: Phys. Rev. B **63**, 174423 (2001)
14.15 Y. Endoh, K. Hirota, S. Ishihara, S. Okamoto, Y. Murakami, A. Nishizawa, T. Fukuda, H. Kimura, H. Nojiri, K. Kaneko, and S. Maekawa: Phys. Rev. Lett. **82**, 4328 (1999). See also H. Nojiri, K. Kaneko, M. Motokawa, K. Hirota, Y. Endoh, and K. Takahashi: Phys. Rev. B **60**, 4142 (1999), and Y. Endoh, H. Nojiri, K. Kaneko, K. Hirota, T. Fukuda, H. Kimura, Y. Murakami, S. Ishihara, S. Maekawa, S. Okamoto, and M. Motokawa: J. Mater. Sci. Eng. B **56**, 1 (1999) (see also cond-mat/9812404)
14.16 T.W. Darling, A. Migliori, E.G. Moshopoulos, S.A. Trugman, J.J. Neumeier, J.L. Sarrao, A.R. Bishop, and J.D. Thompson: Phys. Rev. B **57**, 5093 (1998)
14.17 K. Takenaka, Y. Sawaki, and S. Sugai: Phys. Rev. B **60**, 13011 (1999)
14.18 R.V. Demin, L.I. Koroleva, and A.M. Balbashov: JETP Lett. **70**, 314 (1999), and references therein
14.19 J.H. Jung, K.H. Kim, H.J. Lee, J.S. Ahn, N.J. Hur, T.W. Noh, M.S. Kim, and J.-G. Park: Phys. Rev. B **59**, 3793 (1999)

14.20 Of course, the discussion here is based upon the assumption that the material is chemically homogeneous, namely, that the distribution of Sr dopants is totally random. This may not be the case in LSMO (P. Schiffer, private communication) and more work should be devoted to the verification of chemical homogeneity in manganites
14.21 L. Vasiliu-Doloc, J.W. Lynn, Y.M. Mukovskii, A.A. Arsenov, D.A. Shulyatev: J. Appl. Phys. **83**, 7342 (1998)
14.22 G.C. Xiong, S.M. Bhagat, Q. Li, M. Dominguez, H.L. Ju, R.L. Greene, T. Venkatesan, J.M. Byers, and M. Rubinstein: Solid State Commun. **97**, 599 (1996). See also M. Dominguez et al.: Solid State Commun. **97**, 193 (1996); S.I. Patil et al.: Phys. Rev. B **62**, 9548 (2000)
14.23 N. Furukawa, Y. Moritomo, K. Hirota, and Y. Endoh: 1998 preprint, cond-mat/9808076
14.24 H.S. Shin, J.E. Lee, Y.S. Nam, H.L. Ju, and C.W. Park: Solid State Commun. **118**, 377 (2001)
14.25 H.R. Salva, C.A. Ramos, A.A. Ghilarducci, R.D. Sanchez, and C. Vazquez-Vazquez: J. Magn. Magn. Mater. **226–230**, 601 (2001)
14.26 G. Allodi et al.: cond-mat/9911164
14.27 R. Mahendiran, C. Marquina, M.R. Ibarra, A. Arulraj, C.N.R. Rao, A. Maignan, and B. Raveau: cond-mat/0108078
14.28 N. Fukumoto, S. Mori, N. Yamamoto, Y. Moritomo, T. Katsufuji, C.H. Chen, and S.-W. Cheong: Phys. Rev. B **60**, 12963 (1999)
14.29 J. Joshi, A.K. Sood, S.V. Bhat, A.R. Raju, and C.N.R. Rao: cond-mat/0201336
14.30 D.E. Cox et al.: Phys. Rev. B **57**, 3305 (1998)
14.31 F. Rivadulla et al.: Solid State Commun. **110**, 179 (1999)
14.32 A.K. Raychaudhuri et al.: Solid State Commun. **120**, 303 (2001)
14.33 P.G. Radaelli et al.: cond-mat/0006190
14.34 D. Casa et al.: Europhys. Lett. **47**, 90 (1999)
14.35 V. Kiryukhin et al.: Nature **386**, 813 (1997)
14.36 A. Guha et al.: cond-mat/0005011
14.37 M.M. Savosta, A.S. Karnachev, S. Krupicka, J. Hejtmánek, Z. Jirák, M. Marysko, and P. Novák: Phys. Rev. B **62**, 545 (2000). See also S. Krupicka et al.: J. Phys.: Condens. Matter **13**, 6813 (2001)
14.38 D. Niebieskikwiat, R.D. Sánchez, and A. Caneiro, preprint, to appear in J. Magn. Magn. Mater. See also D. Niebieskikwiat, R.D. Sánchez, A. Caneiro, and B. Alascio: Phys. Rev. B**63**, 212402 (2001); R. De Renzi et al.: Physica B **289–290**, 85 (2000); D. Niebieskikwiat, R.D. Sánchez, A. Caneiro, and B. Maiorov: preprint, to be published in Physica B, proceedings of the LAW3M workshop, Bariloche, Argentina (Sept. 2001)

Chapter 15

15.1 M. Kubota, H. Yoshizawa, Y. Moritomo, H. Fujioka, K. Hirota, and Y. Endoh: J. Phys. Soc. Jpn. **68**, 2202 (1999). See also M. Kubota, H. Yoshizawa, K. Shimizu, K. Hirota, Y. Moritomo, and Y. Endoh, in *Physics in Local Lattice Distortions*, edited by H. Oyanagi and A. Bianconi (2001) p. 422
15.2 D.N. Argyriou, H.N. Bordallo, B.J. Campbell, A.K. Cheetham, D.E. Cox, J.S. Gardner, K. Hanif, A. dos Santos, and G.F. Strouse: Phys. Rev. B **61**, 15269 (2000)
15.3 R.I. Bewley et al.: Phys. Rev. B **60**, 12286 (1999)

15.4 R.H. Heffner, D.E. MacLaughlin, G.J. Nieuwenhuys, T. Kimura, G.M. Luke, Y. Tokura, and Y.J. Uemura: Phys. Rev. Lett. **81**, 1706 (1998)
15.5 O. Chauvet, G. Goglio, P. Molinic, B. Corraze, and L. Brohan: Phys. Rev. Lett. **81**, 1102 (1998)
15.6 N.H. Hur, J.-T. Kim, K.H. Yoo, Y.K. Park, J.-C. Park, E.O. Chi, Y.U. Kwon: Phys. Rev. B **57**, 10740 (1998)
15.7 T. Ishikawa, T. Kimura, T. Katsufuji, and Y. Tokura: Phys. Rev. B **57**, R8079 (1998)
15.8 H.J. Lee, K.H. Kim, J.H. Jung, T.W. Noh, R. Suryanarayanan, G. Dhalenne and A. Revcolevschi: cond-mat/0004475
15.9 A.J. Millis: Phys. Rev. Lett. **80**, 4358 (1998). See reply to this comment, T. Perring, G. Aeppli and Y. Tokura: Phys. Rev. Lett. **80**, 4359 (1998)
15.10 J.-S. Zhou, J.B. Goodenough, and J.F. Mitchell: Phys. Rev. B **58**, R579 (1998)
15.11 L. Vasiliu-Doloc, S. Rosenkranz, R. Osborn, S.K. Sinha, J.W. Lynn, J. Mesot, O.H. Seeck, G. Preosti, A.J. Fedro, and J.F. Mitchell: Phys. Rev. Lett. **83**, 4393 (1999)
15.12 D.N. Argyriou, H.N. Bordallo, J.F. Mitchell, J.D. Jorgensen, and G.F. Strouse: Phys. Rev. B **60**, 6200 (1999)
15.13 B.J. Campbell et al.: Phys. Rev. B **65**, 014427 (2001)
15.14 S. Shimomura, N. Wakabayashi, H. Kuwahara, and Y. Tokura: Phys. Rev. Lett. **83**, 4389 (1999)
15.15 M. Kubota et al.: J. Phys. Soc. Jpn. **69**, 1986 (2000)
15.16 Y. Moritomo, Y. Maruyama, T. Akimoto, and A. Nakamura: Phys. Rev. B **56**, R7075 (1997)
15.17 P.D. Battle et al.: J. Phys.: Condens. Matter **8**, L427 (1996)
15.18 Y. Moritomo, Y. Maruyama, T. Akimoto, and A. Nakamura: J. Phys. Soc. Jpn. **67**, 405 (1998)
15.19 S. Larochelle, A. Mehta, N. Kaneko, P.K. Mang, A.F. Panchula, L. Zhou, J. Arthur, and M. Greven: Phys. Rev. Lett. **87**, 095502 (2001)
15.20 T. Kimura et al.: preprint

Chapter 16

16.1 D. Stauffer and A. Aharony: *Introduction to Percolation Theory* (Taylor & Francis, London 1994)
16.2 S.R. Broadbent and J.M. Hammersley: Proc. Camb. Philos. Soc. **53**, 629 (1957)
16.3 S. Kirkpatrick: Rev. Mod. Phys. **45**, 574 (1973)
16.4 P.W. Anderson: Phys. Rev. **109**, 1492 (1958)
16.5 A.A. Abrikosov: *Fundamentals of the Theory of Metals* (North-Holland, Amsterdam 1988)
16.6 E. Abrahams, S.V. Kravchenko, and M.P. Sarachik: Rev. Mod. Phys. **73**, 251 (2001). Note that the widely held view that Anderson localization prevents a 2D system from being metallic has been shown to be wrong in recent experiments (see, for instance, S.V. Kravchenko et al.: Phys. Rev. B **50**, 8039 (1994))
16.7 Y. Avishai and J.M. Luck: Phys. Rev. B **45**, 1074 (1992)

Chapter 17

17.1 Y. Imry and S.K. Ma: Phys. Rev. Lett. **35**, 1399 (1975)

17.2 Some recent papers on the random field Ising model include E.T. Seppälä and M.J. Alava: Phys. Rev. E **63**, 066109 (2001). See also W.C. Barber and D.P. Belanger: cond-mat/0009228 and 0009049

17.3 Recently, the RFIM in its anisotropic version was also used to study the influence of quenched disorder on the charge-ordered states of manganites O. Zachar and I. Zaliznyak: cond-mat/0112240

17.4 The reader can find information about the random field Heisenberg model in A.P. Malozemoff: Phys. Rev. B **37**, 7673 (1988), including the argument leading to the critical dimension four, and other references. Another more recent paper related to the RFHM is R. Fisch: Phys. Rev. B **57**, 269 (1998)

17.5 M. Mayr, A. Moreo, J.A. Verges, J. Arispe, A. Feiguin, and E. Dagotto: Phys. Rev. Lett. **86**, 135 (2001)

17.6 S.L. Yuan et al.: J. Phys.: Condens. Matter **13**, L509 (2001); S.L. Yuan et al.: App. Phys. Lett. **79**, 90 (2001); S.L. Yuan et al.: 2002, preprint. See also P. Bastiaansen and H. Knops: J. Phys. Chem. Solids **59**, 297 (1998); L.P. Gorḱov and V.Z. Kresin: JETP Lett. **67**, 985 (1998); L.P. Gorḱov and V.Z. Kresin: J. Supercond. **12**, 243 (1999); N. Vandewalle, M. Ausloos, and R. Cloots: Phys. Rev. B **59**, 11909 (1999); W.-G. Yin and R. Tao: Phys. Rev. B **62**, 550 (2000); Y. Xiong et al.: cond-mat/0102083

17.7 M.J. Calderón, J.A. Vergés, and L. Brey: Phys. Rev. B **59**, 4170 (1999); and references therein. See also J.A. Vergés: cond-mat/9905235 (for details of the computational implementation of the Kubo formalism to calculate the conductance) and *Magnetic and electric properties of systems with colossal magnetoresistance*, M.J. Calderón's PhD thesis, Univ. Autonoma de Madrid, 2001

17.8 J.A. Vergés, V. Martín-Mayor, and L. Brey: Phys. Rev. Lett. **88**, 136401 (2002)

17.9 P.B. Allen: Commun. Condens. Matter. Phys. **15**, 327 (1992). See also P.B. Allen: *Boltzmann Theory and Resistivity of Metals*, in *Quantum Theory of Real Materials*, ed. by J.R. Chelikowsky and S.G. Louie (Kluwer, Boston 1996) Chap. 17, p. 219

17.10 H.J. Schulz, G. Cuniberti, and P. Pieri: cond-mat/9807366; and references therein

17.11 P.B. Allen: Nature **405**, 1007 (2000)

17.12 V.J. Emery and S.A. Kivelson: Phys. Rev. Lett. **74**, 3253 (1995)

17.13 P.B. Allen: cond-mat/0108343, and references therein

17.14 A. Biswas, M. Rajeswari, R.C. Srivastava, Y.H. Li, T. Venkatesan, R.L. Greene, and A.J. Millis: Phys. Rev. B **61**, 9665 (2000)

17.15 A. Biswas, M. Rajeswari, R.C. Srivastava, T. Venkatesan, R.L. Greene, Q. Lu, A.L. de Lozanne, and A.J. Millis: Phys. Rev. B **63**, 184424 (2001)

17.16 H.-T. Kim, K.-Y. Kang, and E.-H. Lee: cond-mat/0203245

17.17 N.D. Mathur and P.B. Littlewood: Solid State Commun. **119**, 271 (2001). See also L. Proville: cond-mat/0109237, and references therein

17.18 T. Egami and D. Louca: Phys. Rev. B **65**, 094422 (2002), and references therein

17.19 J. Schmalian and P. Wolynes: Phys. Rev. Lett. **85**, 836 (2000). See also H. Westfahl Jr., J. Schmalian, and P. Wolynes: cond-mat/0102285

17.20 See *Quantum Phase Transitions*, edited by S. Sachdev (Cambridge University Press, Cambridge 1999). For a brief review on quantum phase transition see also T. Vojta: *Quantum Phase Transitions*, cond-mat/0010285, and references therein. See also D. Belitz and T.R. Kirkpatrick: *Quantum Phase Transitions*, cond-mat/9811058. A discussion about quantum phase transitions and the influence of quenched disorder can be found in D. Belitz and T.R. Kirkpatrick, cond-mat/0106279 and references therein
17.21 A. Palanisami, R.D. Merithew, M.B. Weissman, and J.N. Eckstein: Phys. Rev. B **64**, 132406 (2001)
17.22 A. Machida et al.: Phys. Rev. B **62**, 3883 (2000)
17.23 J.M. Kosterlitz, D.R. Nelson, and M.E. Fisher. Phys. Rev. B **13**, 412 (1976); and references therein. See also A. Aharony: cond-mat/0201576
17.24 J. Burgy et al.: preprint, 2002
17.25 R.B. Griffiths: Phys. Rev. Lett. **23**, 17 (1969)
17.26 M.B. Salamon, P. Lin, and S.H. Chun: Phys. Rev. Lett. **88**, 197203 (2002)
17.27 Y. Tomioka and Y. Tokura: preprint, 2002, submitted to Phys. Rev. B
17.28 A. Sundaresan, A. Maignan, and B. Raveau: Phys. Rev. B **56**, 5092 (19797)
17.29 B. Raveau, A. Maignan, C. Martin, and M. Hervieu: *The important role of crystal chemistry upon the CMR properties of manganites*, contribution to [1.21], p. 43
17.30 H. Kuwahara and Y. Tokura: *First-order insulator-metal transitions in perovskite manganites with charge-ordering instability*, contribution to [1.21], p. 217
17.31 D. Niebieskikwiat, R.D. Sánchez, and A. Caneiro: to appear in J. Magn. Magn. Mater. See also D. Niebieskikwiat, R.D. Sánchez, L. Morales, and B. Maiorov: cond-mat/0205115
17.32 M. Auslender and E. Kogan: cond-mat/0105550 and cond-mat/0106032, preprints. See also B.M. Letfulov and J.K. Freericks: cond-mat/0103471
17.33 N. Nagaosa: Physica C **357-360**, 53 (2001). See also S. Murakami and N. Nagaosa, cond-mat/0209002.
17.34 G.R. Blake, L. Chapon, P.G. Radaelli, D.N. Argyriou, M.J. Gutmann, and J.F. Mitchell: cond-mat/0206082.
17.35 Details of these investigations can be found in D. Akahoshi, M. Uchida, Y. Tomioka, T. Arima, Y. Matsui, and Y. Tokura, preprint 2002.
17.36 L.M. Rodriguez-Martinez and J.P. Attfield: Phys. Rev. B **54**, R15622 (1996). See also J.P. Attfield et al.: Nature **394**, 157 (1998)
17.37 Q. Yuan and T. Kopp: cond-mat/0201059, and references therein
17.38 N.A. Babushkina et al.: Nature **391**, 159 (1998)
17.39 N.A. Babushkina et al.: J. Appl. Phys. **83**, 7369 (1988)
17.40 N.A. Babushkina et al.: Phys. Rev. B **59**, 6994 (1999)
17.41 B. Garcia-Landa et al.: Solid State Commun., **105**, 567 (1998)
17.42 A.M. Balagurov et al.: Phys. Rev. B **60**, 383 (1999)
17.43 D.I. Khomskii: Preprint Lebedev Physical Inst. No. 105, Moskow (1969)

Chapter 18

18.1 For a review see T. Timusk and B. Statt: Rep. Prog. Phys. **62**, 61 (1999) (see cond-mat/9905219). See also V.M. Loktev, R.M. Quick, and S. Sharapov: cond-mat/0012082
18.2 Z.-X. Shen and D. Dessau: Phys. Rep. **253**, 1 (1995)
18.3 A. Damascelli, Z.-X. Shen, and Z. Hussain: Rev. Mod. Phys., to be published

18.4 D.S. Dessau et al.: Phys. Rev. Lett. **81**, 192 (1998)
18.5 T. Saitoh et al.: preprint, cond-mat/9911189
18.6 A. Fujimori, T. Mizokawa, and T. Saitoh: *Electronic Structure of Perovskite-type Manganese Oxides*, contribution to [1.21], p. 279
18.7 D.B. Romero et al.: Phys. Rev. B **63**, 132404 (2001)
18.8 Y.-D. Chuang, A.D. Gromko, D.S. Dessau, T. Kimura, and Y. Tokura: Science **292**, 1509 (2001)
18.9 B. Keimer: Science **292**, 1498 (2001)
18.10 A. Moreo, A. Yunoki, and E. Dagotto: Phys. Rev. Lett. **83**, 2773 (1999)
18.11 L. Craco, M. Laad and E. Müller-Hartmann: cond-mat/0110229 and references therein
18.12 V. Perebeinos and P.B. Allen: Phys. Rev. Lett. **85**, 5178 (2000). See also P.B. Allen and V. Perebeinos: Phys. Rev. Lett. **83**, 4828 (1999); Phys. Rev. B **60**, 10747 (1999); V. Perebeinos and P.B. Allen: cond-mat/0007301
18.13 J. van den Brink, P. Horsch, and A.M. Oleś: Phys. Rev. Lett. **85**, 5174 (2000). For more recent results see J. Bała, G.A. Sawatzky, A.M. Oleś, and A. Macridin: Phys. Rev. Lett. **87**, 067204 (2001); J. Bała, A.M. Oleś, and P. Horsch: Phys. Rev. B **65**, 134420 (2002); J. Bała, A.M. Oleś and G.A. Sawatzky: Phys. Rev. B **65**, 184414 (2002)
18.14 W.-G. Yin, H.-Q. Lin, and C.-D. Gong: Phys. Rev. Lett. **87**, 047204 (2001)

Chapter 19

19.1 S. Shimomura, T. Tonegawa, K. Tajima, N. Wakabayashi, N. Ikeda, T. Shobu, Y. Noda, Y. Tomioka, and Y. Tokura: Phys. Rev. B **62**, 3875 (2000)
19.2 P. Dai, J.A. Fernandez-Baca, N. Wakabayashi, E.W. Plummer, Y. Tomioka, and Y. Tokura: Phys. Rev. Lett. **85**, 2553 (2000)
19.3 C.P. Adams, J.W. Lynn, Y.M. Mukovskii, A.A. Arsenov, and D.A. Shulyatev: Phys. Rev. Lett. **85**, 3954 (2000). See also J.W. Lynn et al.: J. Appl. Phys. **89**, 6846 (2001)
19.4 V. Kiryukhin, T.Y. Koo, A. Borissov, Y.J. Kim, C.S. Nelson, J.P. Hill, D. Gibbs, and S.-W. Cheong: Phys. Rev. B **65**, 094421 (2002)
19.5 C.S. Nelson, M. v. Zimmermann, Y.J. Kim, J.P. Hill, D. Gibbs, V. Kiryukhin, T.Y. Koo, S.-W. Cheong, D. Casa, B. Keimer, Y. Tomioka, Y. Tokura, T. Gog, and C.T. Venkataraman: Phys. Rev. B **64**, 174405 (2001)
19.6 J.P. Hill, C.S. Nelson, M. v. Zimmermann, Y.-J. Kim, D. Gibbs, D. Casa, B. Keimer, Y. Murakami, C. Venkataraman, T. Gog, Y. Tomioka, Y. Tokura, V. Kiryukhin, T.Y. Koo, and S.-W. Cheong: Appl. Phys. A **73**, 723 (2001)
19.7 T.Y. Koo, V. Kiryukhin, P.A. Sharma, J.P. Hill, and S.-W. Cheong: Phys. Rev. B **64**, 220405(R) (2001)
19.8 J.M. Zuo and J. Tao: Phys. Rev. B **63**, 060407 (2001)
19.9 M. García-Hernández, A. Mellergärd, F.J. Mompeán, D. Sánchez, A. de Andrés, R.L. McGreevy, and J.L. Martínez: preprint, cond-mat/0201436
19.10 E. Saitoh, Y. Tomioka, T. Kimura, and Y. Tokura: J. Phys. Soc. Jpn. **69**, 2403 (2000)
19.11 E. Saitoh, Y. Okimoto, Y. Tomioka, T. Katsufuji, and Y. Tokura: Phys. Rev. B **60**, 10362 (1999)
19.12 J.M. De Teresa, M.R. Ibarra, P. Algarabel, L. Morellon, B. Garcia-Landa, C. Marquina, C. Ritter, A. Maignan, C. Martin, B. Raveau, A. Kurbakov, and V. Trounov: Phys. Rev. B **65**, 100403 (R) (2002)
19.13 M.R. Ibarra et al.: Phys. Rev. B **56**, 8252 (1997)

19.14 J.F. Mitchell, D.N. Argyriou, J.D. Jorgensen, D.G. Hinks, C.D. Potter, and S.D. Bader: Phys. Rev. B **55**, 63 (1997)
19.15 J.E. Gordon, S.D. Bader, J.F. Mitchell, R. Osborn, and S. Rosenkranz: Phys. Rev. B **60**, 6258 (1999)
19.16 J. Mira, J. Rivas, L.E. Hueso, F. Rivadulla, M.A. Lopez Quintela, M.A. Senaris Rodriguez, and C.A. Ramos: Phys. Rev. B **65**, 024418 (2002)
19.17 J.L. Martinez, A. de Andres, M. Garcia-Hernandez, C. Prieto, J.M. Alonso, E. Herrero, J. Gonzalez-Calbet, and M. Vallet-Regi: J. Magn. Magn. Mater. **196–197**, 520 (1999)
19.18 K.H. Kim, S. Lee, T.W. Noh, and S.-W. Cheong: Phys. Rev. Lett. **88**, 167204 (2002)
19.19 K.H. Kim, M. Uehara, and S.-W. Cheong: Phys. Rev. B **62**, R11945 (2000); D. Bhattacharya et al.: preprint
19.20 D.N. Argyriou, J.W. Lynn, R. Osborn, B. Campbell, J.F. Mitchell, U. Ruett, H.N. Bordallo, A. Wildes, and C.D. Ling: Phys. Rev. Lett. **89**, 036401 (2002)
19.21 J. Kunze, S. Naler, J. Bäckström, M. Rübhausen, and J.F. Mitchell: preprint, 2002

Chapter 20

20.1 J.L. Gavilano et al.: Phys. Rev. Lett. **81**, 5648 (1998), and references therein
20.2 S. Süllow et al.: cond-mat/9912390. See also S. Süllow, I. Prasad, M.C. Aronson, S. Bogdanovich, J.L. Sarrao, and Z. Fisk: Phys. Rev. B **62**, 11626 (2000). For more details on the comparison manganites – EuB_6 see S. Yoon et al.: Phys. Rev. B **58**, 2795 (1998)
20.3 J.B. Torrance et al.: Phys. Rev. Lett. **29**, 1168 (1972)
20.4 S. von Molnar and S. Methfessel: J. Appl. Phys. **38**, 959 (1967); and references therein
20.5 Y. Shapira et al.: Phys. Rev. B **10**, 4765 (1974), and references therein
20.6 J.M. Tarascon et al.: Solid State Commun. **37**, 133 (1981)
20.7 H. Ohno et al.: Phys. Rev. Lett. **68**, 2664 (1992), and references therein
20.8 C.S. Snow, S.L. Cooper, D.P. Young, Z. Fisk, A. Comment, and J.-P. Ansermet: Phys. Rev. B **64**, 174412 (2001)
20.9 S. Paschen et al.: Phys. Rev. B **61**, 4174 (2000)
20.10 T. Kasuya and A. Yanase: Rev. Mod. Phys. **40**, 684 (1968)
20.11 H. Rho, C.S. Snow, S.L. Cooper, Z. Fisk, A. Comment, and J.-P. Ansermet: Phys. Rev. Lett. **88**, 127401 (2002)
20.12 Y. Maeno et al.: Nature **372**, 532 (1994)
20.13 S. Nakatsuji and Y. Maeno: Phys. Rev. B **62**, 6458 (2000)
20.14 G. Cao, S.C. McCall, J.E. Crow and R.P. Guertin: Phys. Rev. B **56**, 5387 (1997)
20.15 G. Cao, S. McCall, V. Dobrosavljevic, C.S. Alexander, J.E. Crow, and R.P. Guertin: Phys. Rev. B **61**, R5053 (2000)
20.16 T. Hotta and E. Dagotto: Phys. Rev. Lett. **88**, 017201 (2002)
20.17 S. Nakatsuji, K. Sugahara, M. Yoshioka, D. Hall and Y. Maeno: preprint 2002
20.18 G. Cao, L. Balicas, Y. Xin, E. Dagotto, J.E. Crow, C.S. Nelson, and J.P. Hill: cond-mat/0205151
20.19 M. García-Hernández et al.: Phys. Rev. Lett. **86**, 2443 (2001)
20.20 J. Mira, F. Rivadulla, J. Rivas, A. Fondado, R. Caciuffo, F. Carsughi, T. Guidi, and J.B. Goodenough: cond-mat/0201478

20.21 A.P. Ramirez and M.A. Subramanian: Science **277**, 546 (1997)
20.22 J.A. Alonso, J.L. Martinez, M.J. Martinez-Lopez, M.T. Casais and M.T. Fernandez-Diaz: Phys. Rev. Lett. **82**, 189 (1999)
20.23 G. Briceno et al.: Science **270**, 273 (1995)
20.24 R. Caciuffo et al.: Phys. Rev. B **59**, 1068 (1999). See also D. Nam et al.: Phys. Rev. B **59**, 4190 (1999) and references therein; A.V. Samoilov et al.: Phys. Rev. B **57**, R14032 (1998)
20.25 R.D. Sanchez et al.: J. Solid State Chem. **151**, 1 (2000)
20.26 F. Hellman et al.: Phys. Rev. Lett. **77**, 4652 (1996). See also W. Teizer et al.: Phys. Rev. Lett. **85**, 848 (2000), and references therein
20.27 S. Kawasaki et al.: J. Phys. Soc. Jpn. **67**, 1529 (1998), and references therein
20.28 T. Takeda et al.: Solid State Sci., in press
20.29 M.A. Subramanian et al.: Phys. Rev. Lett. **82**, 1558 (1999)
20.30 Y. Okimoto et al.: Phys. Rev. B **51**, 9581 (1995) See also Shin-ichi Shamoto et al., preprint 2002.
20.31 C.C. Hays et al.: Phys. Rev. B **60**, 10367 (1999)
20.32 For a recent reference see M. Matsuura et al.: cond-mat/0006185, and references therein
20.33 K. J. Singh et al.: cond-mat/0102199
20.34 M.A. Manekar et al.: cond-mat/0103291. See also cond-mat/0010341
20.35 J. García Soldevilla et al.: Phys. Rev. B **61**, 6821 (2000), and references therein
20.36 J.L. Gavilano et al.: Phys. Rev. B **52**, R13106 (1995). See also S. Barth et al.: Phys. Rev. B **39**, 11695 (1989)
20.37 S.D. Ganichev et al.: cond-mat/0011405; and references therein
20.38 D.P. Young, J.F. DiTusa, R.G. Goodrich, J. Anderson, S. Guo, P.W. Adams, J.Y. Chan, and D. Hall: cond-mat/0202220
20.39 S. Ilani et al.: Science **292**, 1354 (2001)
20.40 S.V. Kravchenko et al.: Phys. Rev. B **50**, 8039 (1994)
20.41 E.L. Nagaev: *Physics of Magnetic Semiconductors* (MIR Publishers, Moscow, 1983)
20.42 P. Nyhus, S. Yoon, M. Kauffman, S.L. Cooper, Z. Fisk, and J. Sarrao: Phys. Rev. B **56**, 2717 (1997)
20.43 W. Hayes and R. Loudon: *Scattering of Light by Crystals* (John Wiley and Sons, New York 1978)
20.44 P. Nyhus, S.L. Cooper, and Z. Fisk: Phys. Rev. B **51**, 15626 (1995); P. Nyhus, S.L. Cooper, Z. Fisk, and J. Sarrao: Phys. Rev. B **55**, 12488 (1997)
20.45 J.K. Freericks and T.P. Devereaux: Phys. Rev. B **64**, 125110 (2001)
20.46 H.L. Liu, S. Yoon, S.L. Cooper, S.-W. Cheong, P.D. Han, and D.A. Payne: Phys. Rev. B **58**, R10115 (1998)
20.47 P. Majumdar and P.B. Littlewood: Nature (London) **395**, 479 (1998)
20.48 S. von Molnar and M.W. Shafer: J. Appl. Phys. **41**, 1093 (1970)
20.49 C.N. Guy, S. von Molnar, J. Etourneau, and Z. Fisk: Solid State Commun. **33**, 1055 (1980)
20.50 A.A. Abrikosov and V.B. Genkin: Sov. Phys. JETP **38**, 417 (1974)
20.51 A. Zawadowski and M. Cardona: Phys. Rev. B **42**, 10732 (1990)
20.52 I.P. Ipatova, A.V. Subashiev, and V.A. Voitenko: Solid State Commun. **37**, 893 (1981)
20.53 The observation of a strong light scattering response in this symmetry also implies that other large magneto-optical signatures, e.g., large Kerr rotations, should be observed
20.54 T. Dietl, M. Sawicki, M. Dahl, D. Heiman, E.D. Isaacs, M.J. Graf, S.I. Gubarev, and D.L. Alov: Phys. Rev. B **43**, 3154 (1991)

20.55 D.L. Peterson, D.U. Bartholomew, U. Debska, A.K. Ramdas, and S. Rodriguez: Phys. Rev. B **32**, 323 (1985)
20.56 E.D. Isaacs, D. Heiman, M.J. Graf, B.B. Goldberg, R. Kershaw, D. Ridgley, K. Dwight, A. Wold, J. Furdyna, and J.S. Brooks: Phys. Rev. B **37**, 7108 (1988)

Chapter 21

21.1 S.S.P. Parkin: Annu. Rev. Mater. Sci. **25**, 357 (1995)
21.2 G.A. Prinz: Science **282**, 1660 (1998). See also G.A. Prinz: Phys. Today, April 1995, p. 58
21.3 J.L. Simonds: Phys. Today, April 1995, p. 26. For a recent review, see P. Grünberg: Phys. Today, May 2001, p. 31
21.4 T. Thurn-Albrecht et al.: Science **290**, 2126 (2000)
21.5 D. Weller and M.R. Doerner: Annu. Rev. Mater. Sci. **30**, 611 (2000)
21.6 T. Kimura and Y. Tokura: Annu. Rev. Mater. Sci. **30**, 451 (2000), and references therein
21.7 J.Z. Sun et al., IBM J. Res. Dev., **42**, No. 1; J.Z. Sun et al.: Appl. Phys. Lett. **69**, 3266 (1996); J.Z. Sun, J. Magn. Magn. Mater. **202**, 157 (1999); See also S. Sarkar, P. Raychaudhuri, A.K. Nigam, and R. Pinto: Solid State Commun. **117**, 609 (2001)
21.8 W. Prellier, P. Lecoeur, and B. Mercey: cond-mat/0111363, preprint, and references therein. See also R. Ramesh et al.: *Colossal Magnetoresistive Manganites: The Push Towards Low Field Magnetoresistance*, contribution to [1.21], p. 155, and references therein.
21.9 I.N. Krivorotov, K.R. Nikolaev, A.Y. Dobin, A.M. Goldman, and E.D. Dahlberg: Phys. Rev. Lett. **86**, 5779 (2001)
21.10 Q. Zhan, R. Yu, L.L. He, D.X. Li, J. Li, S.Y. Xu, and C.K. Ong: preprint
21.11 C.-C. Fu, Z. Huang, and N.-C. Yeh: Phys. Rev. B **65**, 2245XX (2002), to appear
21.12 H. Li, J.R. Sun, and H.K. Wong: submitted to Appl. Phys. Lett., preprint 2001
21.13 J. Nogués and I.K. Schuller: J. Magn. Magn. Mater. **192**, 203 (1999)

Chapter 22

22.1 There are other theories for manganites that are not discussed here. For example, A.S. Alexandrov and A.M. Bratkovsky: Phys. Rev. Lett. **82**, 141 (1999), proposed bipolarons as a possibility to explain CMR
22.2 R. Allub and B. Alascio: Solid State Commun. **99**, 613 (1996); R. Allub and B. Alascio: Phys. Rev. B **55**, 14113 (1997); R. Aliaga, R. Allub, and B. Alascio, 1998, preprint, cond-mat/9804248. See also J.M.D. Coey, M. Viret, L. Ranno, and K. Ounadjela: Phys. Rev. Lett. **75**, 3910 (1995); C. Varma: Phys. Rev. B**54**, 7328 (1996); and [5.1]
22.3 K. Slevin, T. Ohtsuki, and T. Kawarabayashi: Phys. Rev. Lett. **84**, 3915 (2000)
22.4 J.H. Schön, M. Dorget, F.C. Beuran, X.Z. Xu, E. Arushanov, M. Laguës, and C.D. Cavellin: Science **293**, 2430 (2001); J.H. Schön, C. Kloc, and B. Batlogg: Science **293**, 2432 (2001)

22.5 X. Hong, A. Posadas, A. Lin, and C.H. Ahn: preprint, 2002. See also C.H. Ahn, S. Gariglio, P. Paruch, T. Tybell, L. Antognazza, and J.-M. Triscone: Science **284**, 1152 (1999)

22.6 Y. Tokura, H. Kuwahara, Y. Moritomo, Y. Tomioka, and A. Asamitsu: Phys. Rev. Lett. **76**, 3184 (1996)

22.7 J.A. Fernandez-Baca, P. Dai, H. Kawano-Furukawa, H. Yoshizawa, E.W. Plummer, S. Katano, Y. Tomioka, and Y. Tokura: cond-mat/0110659

22.8 S.H. Pan, J.P. O'Neal, R.L. Badzey, C. Chamon, H. Ding, J.R. Engelbrecht, Z. Wang, H. Eisaki, S. Uchida, A.K. Gupta, K.-W. Ng, E.W. Hudson, K.M. Lang, and J.C. Davis: Nature **413**, 282 (2001)

22.9 M.-H. Julien et al.: Phys. Rev. B **63**, 144508 (2001); A.W. Hunt et al.: cond-mat/0011380; Y. Sidis et al.: cond-mat/0101095; P.M. Singer et al.: cond-mat/0108291

22.10 I. Iguchi, T. Yamaguchi, and A. Sugimoto: Nature **412**, 420 (2001)

22.11 J. Bobroff, H. Alloul, S. Ouazi, P. Mendels, A. Mahajan, N. Blanchard, G. Collin, V. Guillen, and J.-F. Marucco: cond-mat/0203225

22.12 N.-C. Yeh et al.: Physica C **364–365**, 450 (2001). See also N.-C. Yeh et al.: Phys. Rev. Lett. **87**, 087003 (2001)

22.13 R.S. Decca, D. Drew, E. Osquiguil, B. Maiorov, and J. Guimpel: Phys. Rev. Lett. **85**, 3708 (2000)

22.14 M. Strongin, R. Thompson, O. Kammerer and J. Crow: Phys. Rev. B **1**, 1078 (1970); Y. Imry and M. Strongin: Phys. Rev. B **24**, 6353 (1981); and references therein. For related more recent references see: A.F. Hebard and M.A. Paalanen: Phys. Rev. Lett. **65**, 927 (1990); A. Yazdani and A. Kapitulnik: Phys. Rev. Lett. **74**, 3037 (1995); M. Fisher, G. Grinstein, and S. Girvin: Phys. Rev. Lett. **64**, 587 (1990); A. Goldman and N. Markovic: Phys. Today **51** (11), 39 (1998)

22.15 H. Ohno: Science, **281**, (1998); T. Dietl, cond-mat/0201282. See also S.J. Potashnik et al., Appl. Phys. Lett. **79** 1495 (2001); S.J. Potashnik et al., cond-mat/0204250.

22.16 J. Schliemann and A.H. MacDonald: Phys. Rev. Lett. **88** 137201 (2002); M. Berciu and R.N. Bhatt: Phys. Rev. Lett. **87** 107203 (2001); A. Kaminski and S. Das Sarma: cond-mat/0201229; A. Chattopadhyay et al.: Phys. Rev. Lett. **87**, 227202 (2001); M.J. Calderón et al.: cond-mat/0203404

22.17 M. Mayr, G. Alvarez, and E. Dagotto: Phys. Rev. B **65** 241202 (RC) (2002). For independent work arriving to similar conclusions see A. Kaminski and S. Das Sarma, Phys. Rev. Lett. **88**, 247202 (2002).

22.18 G. Alvarez, M. Mayr, and E. Dagotto: cond-mat/0205197

22.19 T. Ishiguro, K. Yamaji, and G. Saito: *Organic Superconductors* (Springer, Berlin Heidelberg New York 1998). Phase diagrams of relevance for our discussion can be found, e.g., in Figs. 3.14, 3.15, 3.19, 3.23, 4.14, 5.67, and 7.6

22.20 T. Vuletić, P. Auban-Senzier, C. Pasquier, S. Tomić, D. Jérome, M. Héritier, and K. Bechgaard: cond-mat/0109031

22.21 S. Lefebvre et al.: Phys. Rev. Lett. **85**, 5420 (2000)

22.22 T. Sasaki, N. Yoneyama, A. Matsuyama, and N. Kobayashi: Phys. Rev. B **65** 060505 (2002)

22.23 V.A. Sidorov, M. Nicklas, P.G. Pagliuso, J.L. Sarrao, Y. Bang, A.V. Balatsky, and J.D. Thompson: cond-mat/0202251; and references therein

22.24 A.H. Castro Neto and B.A. Jones: Phys. Rev. B **62**, 14975 (2000)

22.25 J. García Soldevilla et al.: Phys. Rev. B **61**, 6821 (2000)

22.26 T. Takimoto, T. Hotta, T. Maehira, and K. Ueda: cond-mat/0204023

Index

Springer Series in Solid-State Sciences

Editors: M. Cardona P. Fulde K. von Klitzing H.-J. Queisser

1 **Principles of Magnetic Resonance**
3rd Edition By C. P. Slichter

2 **Introduction to Solid-State Theory**
By O. Madelung

3 **Dynamical Scattering of X-Rays in Crystals** By Z. G. Pinsker

4 **Inelastic Electron Tunneling Spectroscopy**
Editor: T. Wolfram

5 **Fundamentals of Crystal Growth I**
Macroscopic Equilibrium and Transport Concepts
By F. E. Rosenberger

6 **Magnetic Flux Structures in Superconductors**
2nd Edition By R. P. Huebener

7 **Green's Functions in Quantum Physics**
2nd Edition By E. N. Economou

8 **Solitons and Condensed Matter Physics**
Editors: A. R. Bishop and T. Schneider

9 **Photoferroelectrics** By V. M. Fridkin

10 **Phonon Dispersion Relations in Insulators** By H. Bilz and W. Kress

11 **Electron Transport in Compound Semiconductors** By B. R. Nag

12 **The Physics of Elementary Excitations**
By S. Nakajima, Y. Toyozawa, and R. Abe

13 **The Physics of Selenium and Tellurium**
Editors: E. Gerlach and P. Grosse

14 **Magnetic Bubble Technology** 2nd Edition
By A. H. Eschenfelder

15 **Modern Crystallography I**
Fundamentals of Crystals
Symmetry, and Methods of Structural Crystallography
2nd Edition
By B. K. Vainshtein

16 **Organic Molecular Crystals**
Their Electronic States By E. A. Silinsh

17 **The Theory of Magnetism I**
Statics and Dynamics
By D. C. Mattis

18 **Relaxation of Elementary Excitations**
Editors: R. Kubo and E. Hanamura

19 **Solitons** Mathematical Methods for Physicists
By. G. Eilenberger

20 **Theory of Nonlinear Lattices**
2nd Edition By M. Toda

21 **Modern Crystallography II**
Structure of Crystals 2nd Edition
By B. K. Vainshtein, V. L. Indenbom, and V. M. Fridkin

22 **Point Defects in Semiconductors I**
Theoretical Aspects
By M. Lannoo and J. Bourgoin

23 **Physics in One Dimension**
Editors: J. Bernasconi and T. Schneider

24 **Physics in High Magnetics Fields**
Editors: S. Chikazumi and N. Miura

25 **Fundamental Physics of Amorphous Semiconductors** Editor: F. Yonezawa

26 **Elastic Media with Microstructure I**
One-Dimensional Models By I. A. Kunin

27 **Superconductivity of Transition Metals**
Their Alloys and Compounds
By S. V. Vonsovsky, Yu. A. Izyumov, and E. Z. Kurmaev

28 **The Structure and Properties of Matter**
Editor: T. Matsubara

29 **Electron Correlation and Magnetism in Narrow-Band Systems** Editor: T. Moriya

30 **Statistical Physics I** Equilibrium Statistical Mechanics 2nd Edition
By M. Toda, R. Kubo, N. Saito

31 **Statistical Physics II** Nonequilibrium Statistical Mechanics 2nd Edition
By R. Kubo, M. Toda, N. Hashitsume

32 **Quantum Theory of Magnetism**
2nd Edition By R. M. White

33 **Mixed Crystals** By A. I. Kitaigorodsky

34 **Phonons: Theory and Experiments I**
Lattice Dynamics and Models of Interatomic Forces By P. Brüesch

35 **Point Defects in Semiconductors II**
Experimental Aspects
By J. Bourgoin and M. Lannoo

36 **Modern Crystallography III**
Crystal Growth
By A. A. Chernov

37 **Modern Chrystallography IV**
Physical Properties of Crystals
Editor: L. A. Shuvalov

38 **Physics of Intercalation Compounds**
Editors: L. Pietronero and E. Tosatti

39 **Anderson Localization**
Editors: Y. Nagaoka and H. Fukuyama

40 **Semiconductor Physics** An Introduction
6th Edition By K. Seeger

41 **The LMTO Method**
Muffin-Tin Orbitals and Electronic Structure
By H. L. Skriver

42 **Crystal Optics with Spatial Dispersion, and Excitons** 2nd Edition
By V. M. Agranovich and V. L. Ginzburg

43 **Structure Analysis of Point Defects in Solids**
An Introduction to Multiple Magnetic Resonance Spectroscopy
By J.-M. Spaeth, J. R. Niklas, and R. H. Bartram

44 **Elastic Media with Microstructure II**
Three-Dimensional Models By I. A. Kunin

45 **Electronic Properties of Doped Semiconductors**
By B. I. Shklovskii and A. L. Efros

46 **Topological Disorder in Condensed Matter**
Editors: F. Yonezawa and T. Ninomiya

Springer Series in Solid-State Sciences

Editors: M. Cardona P. Fulde K. von Klitzing H.-J. Queisser

Springer Series in Solid-State Sciences

Editors: M. Cardona P. Fulde K. von Klitzing H.-J. Queisser

Managing Editor: H. K.V. Lotsch

90 **Earlier and Recent Aspects of Superconductivity**
Editors: J. G. Bednorz and K. A. Müller

91 **Electronic Properties of Conjugated Polymers III** Basic Models and Applications
Editors: H. Kuzmany, M. Mehring, and S. Roth

92 **Physics and Engineering Applications of Magnetism** Editors: Y. Ishikawa and N. Miura

93 **Quasicrystals** Editors: T. Fujiwara and T. Ogawa

94 **Electronic Conduction in Oxides** 2nd Edition
By N. Tsuda, K. Nasu, F. Atsushi, and K.Siratori

95 **Electronic Materials**
A New Era in Materials Science
Editors: J. R. Chelikowsky and A. Franciosi

96 **Electron Liquids** 2nd Edition By A. Isihara

97 **Localization and Confinement of Electrons in Semiconductors**
Editors: F. Kuchar, H. Heinrich, and G. Bauer

98 **Magnetism and the Electronic Structure of Crystals** By V.A. Gubanov, A.I. Liechtenstein, and A.V. Postnikov

99 **Electronic Properties of High-T_c Superconductors and Related Compounds**
Editors: H. Kuzmany, M. Mehring, and J. Fink

100 **Electron Correlations in Molecules and Solids** 3rd Edition By P. Fulde

101 **High Magnetic Fields in Semiconductor Physics III** Quantum Hall Effect, Transport and Optics By G. Landwehr

102 **Conjugated Conducting Polymers**
Editor: H. Kiess

103 **Molecular Dynamics Simulations**
Editor: F. Yonezawa

104 **Products of Random Matrices** in Statistical Physics By A. Crisanti, G. Paladin, and A.Vulpiani

105 **Self-Trapped Excitons**
2nd Edition By K. S. Song and R. T. Williams

106 **Physics of High-Temperature Superconductors**
Editors: S. Maekawa and M. Sato

107 **Electronic Properties of Polymers**
Orientation and Dimensionality of Conjugated Systems Editors: H. Kuzmany, M. Mehring, and S. Roth

108 **Site Symmetry in Crystals**
Theory and Applications 2nd Edition
By R. A. Evarestov and V. P. Smirnov

109 **Transport Phenomena in Mesoscopic Systems** Editors: H. Fukuyama and T. Ando

110 **Superlattices and Other Heterostructures**
Symmetry and Optical Phenomena 2nd Edition
By E. L. Ivchenko and G. E. Pikus

111 **Low-Dimensional Electronic Systems**
New Concepts
Editors: G. Bauer, F. Kuchar, and H. Heinrich

112 **Phonon Scattering in Condensed Matter VII**
Editors: M. Meissner and R. O. Pohl

113 **Electronic Properties of High-T_c Superconductors**
Editors: H. Kuzmany, M. Mehring, and J. Fink

114 **Interatomic Potential and Structural Stability**
Editors: K. Terakura and H. Akai

115 **Ultrafast Spectroscopy of Semiconductors and Semiconductor Nanostructures**
2nd Edition By J. Shah

116 **Electron Spectrum of Gapless Semiconductors**
By J. M. Tsidilkovski

117 **Electronic Properties of Fullerenes**
Editors: H. Kuzmany, J. Fink, M. Mehring, and S. Roth

118 **Correlation Effects in Low-Dimensional Electron Systems**
Editors: A. Okiji and N. Kawakami

119 **Spectroscopy of Mott Insulators and Correlated Metals**
Editors: A. Fujimori and Y. Tokura

120 **Optical Properties of III–V Semiconductors**
The Influence of Multi-Valley Band Structures
By H. Kalt

121 **Elementary Processes in Excitations and Reactions on Solid Surfaces**
Editors: A. Okiji, H. Kasai, and K. Makoshi

122 **Theory of Magnetism**
By K. Yosida

123 **Quantum Kinetics in Transport and Optics of Semiconductors**
By H. Haug and A.-P. Jauho

124 **Relaxations of Excited States and Photo-Induced Structural Phase Transitions**
Editor: K. Nasu

125 **Physics and Chemistry of Transition-Metal Oxides**
Editors: H. Fukuyama and N. Nagaosa

Druck: Strauss Offsetdruck, Mörlenbach
Verarbeitung: Schäffer, Grünstadt

1-MONTH